EXPERIMENTS IN POLYMER SCIENCE

Experiments in Polymer Science

EDWARD A. COLLINS

Adjunct Professor of Chemistry
Rensselaer Polytechnic Institute

and

Development Consultant
B. F. Goodrich Chemical Company
Development Center
Avon Lake, Ohio

JAN BAREŠ

former Postdoctoral Fellow
Rensselaer Polytechnic Institute

and

Scientist
Xerox Corporation
Webster, New York

FRED W. BILLMEYER, JR.
Professor of Analytical Chemistry
Rensselaer Polytechnic Institute
Troy, New York

A WILEY-INTERSCIENCE PUBLICATION

JOHN WILEY & SONS, New York • London • Sydney • Toronto

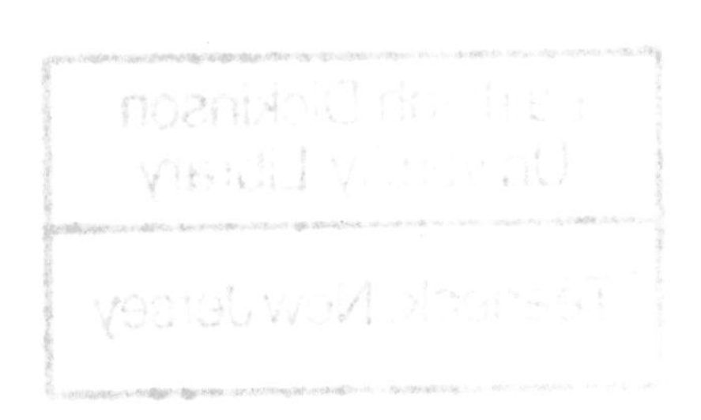

Library of Congress Cataloging in Publication Data:

Collins, Edward A.
Experiments in polymer science.

"A Wiley-Interscience publication."
1. Polymers and polymerization—Laboratory manuals.
I. Bareš, Jan, joint author. II. Billmeyer, Fred W., joint author. III. Title.

QD385.C64 547'.84'028 73-650
ISBN 0-471-16584-0
ISBN 0-471-16585-9 (pbk)

Printed in the United States of America

10 9 8 7 6 5 4 3 2 1

Preface

Experiments in Polymer Science was written with two groups of readers in mind. Although they are different and will use the book in different ways, their needs are basically similar; thus we feel that this book can do double duty easily and efficiently.

The first and obvious user is the student, graduate or advanced undergraduate, who will use this book as a laboratory manual in a course in polymer science or engineering. His interest will center on the experiments in Part III, which he must perform in the laboratory. But he is hereby warned (if he chances to read this far in such an unlikely place as a Preface) that he will have a much better chance of understanding the experiment, completing it successfully in the allotted time, and preparing a satisfactory report, if he studies both the experimental procedure and the background material in Part I or Part II in advance of coming to the laboratory. The instructor should see that assignments to this end are made and adhered to.

The second reader to whom *Experiments in Polymer Science* is addressed is the industrial polymer scientist or engineer, which indeed the student may become in a few years. His approach to the book should, we feel, reflect our philosophy that the industrial career of the technically trained scientist or engineer--whether in research, production, sales, or another branch of the organization--is a continuing educational experience in which he is sometimes the student and sometimes the instructor. In either role, he will find (if we have been successful) much that will make his job easier and more meaningful, and much that will contribute to more effective communication with his co-workers and a clearer understanding of the problems they face. He will no doubt be interested first in Parts I and II, but he will find study of the experiments in Part III rewarding as well.

Experiments in Polymer Science is intended as a companion book to F. W. Billmeyer's *Textbook of Polymer Science*, 2nd edition, 1971, and frequent reference is made to topics discussed more fully in the *Textbook*, as well as to books dealing more specifically with polymer laboratory practice, and to the general literature.

The present book takes an integrated approach to the polymer science laboratory, by providing experiments leading to the synthesis of several polymers which are then used in the molecular, physical, and thermal characterization experiments that follow. Major emphasis, however, is on characterization and property measurement, for two reasons: First, the organic polymer chemist will find ample supplementary material allowing him to modify experiments and to broaden this phase of the course as desired; examples of such supplementary material are Sorensen and Campbell's *Preparative Methods of Polymer Chemistry*, Braun, Cherdon, and Kern's *Techniques of Polymer Syntheses and Characterization*, and the series *Macromolecular Syntheses*. Second, the experiments on synthesis can be omitted--for example, for students with more interest in the polymer engineering disciplines--with readily available commercial polymer samples substituted for those prepared in the laboratory where required in the experiments on characterization and property measurement.

All experiments presented in this book have been tested in the laboratory. They are based on a schedule of one 3-hour laboratory period per week. Enough material is presented to provide a two-semester sequence, although the instructor may wish to select experiments for completion in one semester or an even shorter time period, based on interest and the availability of equipment. Selected experiments can also be included in other undergraduate and graduate laboratory courses, including those in organic and physical chemistry, instrumental analysis, and materials science and engineering.

The preparation of *Experiments in Polymer Science* has been in many respects a team effort. It grew from notes for a Polymer Science Laboratory course offered in Rensselaer's Department of Chemistry. Concepts and much background material were supplied by E. A. Collins, who taught the course with the assistance of J. Bareš and A. Mehta as laboratory assistant. F. W. Billmeyer coordinated and drafted the final manuscript. Several generations of students, at Rensselaer and elsewhere, contributed invaluable experience. Special thanks must go to Shirley Papp and Diane Rym at B. F. Goodrich Chemical Co. and Edis Hite for typing many drafts, and Helen Hayes at Rensselaer who typed the manuscript you are reading. Acknowledgement is also due to L. A. Chandler, C. A. Daniels, J. A. Davidson, H. V. Flint, R. A. Jones, R. Marsh, A. P. Metzger, R. G. Raike, and C. Wilkes at B. F. Goodrich Chemical Co., to P. H. Geil and C. E. Rogers at Case Western Reserve University, and to B. Wunderlich at Rensselaer Polytechnic Institute. The encouragement and assistance of Walter H. Bauer, Dean Emeritus of Rensselaer's School of Science, is gratefully acknowledged. One of us (E.A.C.) thanks B. F. Goodrich Chemical Co. for permission to co-author this book.

Edward A. Collins
Jan Bareš
Fred W. Billmeyer, Jr.

Troy, New York
March 1973

Contents

Part II Introduction to Polymer Characterization Techniques

Part III Experiments

EXPERIMENTS IN POLYMER SCIENCE

Part I

INTRODUCTION TO POLYMER SYNTHESIS TECHNIQUES

1

Polymerization Mechanisms

On close examination, the history of polymers goes back farther than that of any other group of substances known to man. From the fig leaf in the Garden of Eden, man has been dependent upon naturally-occurring polymers for food, clothing, shelter, and communication. Today, his dependence on synthetic polymers, the primary subject of this book, is even more fundamental.

The history of synthetic polymers is relatively short, however, because of the complexity of their structure. Even their basic long-chain structure was still questioned only 50 years ago; it was only through the pioneering work of Staudinger (1920) and the quantitative studies of Carothers (1929, 1931) that the macromolecular concept was finally accepted. Long before this time, synthetic polymers were being produced, and natural polymers were being altered chemically, without a clear understanding of the processes involved. Styrene was first polymerized in 1839, and in that same year Goodyear discovered the vulcanization of rubber. Cellulose nitrate was first made in 1838 and commercialized in 1840. The first synthetic polymers to be produced commercially were phenol-formaldehyde resins, in 1907.

Today, with polymer science and engineering well developed, we can undertake the study of these important materials in the laboratory with a background of knowledge which allows, in most cases, a full understanding of the experiments we perform and, what is more important, the ability to define the areas which are not yet fully understood as the basis for future research. The objective of this book is to provide information on the experimental techniques and their scientific background for polymer synthesis and characterization.

As general reference to these introductory paragraphs we list a few general books on polymer science. Chief among these is *Textbook of Polymer Science* (Billmeyer 1971), hereafter referred to as the *Textbook*, which is cited extensively as a companion book to this laboratory manual.

GENERAL REFERENCES

Textbook, Chap. 1; Flory 1953, Chap. II; Mark 1966*a*; Margerison 1967; Rodriguez 1970; Seymour 1971; Billmeyer 1972.

A. *Classification of Polymers*

Before entering a discussion of the ways in which polymers are synthesized, it is convenient to classify them, for simplification, in one or another way. There are many schemes of classification possible. First is that already mentioned: synthetic versus natural polymers. In this book, which is based on synthesis and the characterization of the resulting polymers, such a classification is of little practical value. Synthetic polymers could be further classified by monomer type, preparative techniques, polymer structure, physical properties, processing techniques, or end use, among others. For example, a classification according to processing techniques might first divide polymers into thermoplastics and thermosets, and carry on from there. A classification by end use might attempt to associate polymer classes to specific industries. For example, diene polymers are associated with the rubber industry, the olefin polymers with the sheet, film, and fiber industry, and the acrylics with coatings and decorative materials. Similar identifications could be made for the vinyls, polyesters, polyamides, alkyds, cellulosics, phenolics, urethanes, and siloxane polymers. These classifications are useful in a practical sense, but offer little or no insight into the scientific principles distinguishing one polymer type from another.

In accord with our present concern with polymerization mechanisms, we shall adopt the widely used classification of Carothers (1929; also Mark 1940), in which polymers are divided into *addition* and *condensation* types on the basis of a difference in elemental composition between the polymer and its monomer. In the original meaning, any polymer formed from a polyfunctional monomer by a condensation reaction with the elimination of a small molecule such as water was classified as a condensation polymer.

The discovery of many more polymer types and many more ways of synthesizing polymers has required that this classification be modified, and the grounds for it clarified. Perhaps the most useful scheme for our purposes is based on polymerization mechanisms, as formulated by Mark (1950) and discussed in the *Textbook*, Chap. 8A. This classification divides polymers into those formed by *step-reaction* and by *chain* mechanisms. Polymers made by step-reaction mechanisms are formed by a series of reactions, often of the condensation type, in which any two species (monomers, dimers, trimers, etc.) can react at any time, leading to a larger molecule. Elimination of a small molecule is a possible but not an essential feature. Usually, but not essentially, such polymers contain functional groups (ester, amide, urethane, etc.) as part of the main chain.

Chain-reaction polymers, which can often be identified with Carothers' addition polymers, are produced by reactions in which monomers are added, one after another, to a rapidly growing chain. Usually they do not contain functional groups in the chain, and no small molecule is lost at each step in the polymerization. The differences between chain and step polymerization mechanisms are shown in Table 1-1, taken from the *Textbook*.

It is clear that no single classification can be all-inclusive, and we do not attempt to use any one rigorously; more important is the development of an understanding of the need for different classifications and the interrelations among them. In the course of this

TABLE 1-1. Distinguishing features of chain- and step-polymerization mechanisms

Chain Polymerization	Step Polymerization
Only growth reaction adds repeating units one at a time to the chain.	Any two molecular species present can react.
Monomer concentration decreases steadily throughout reaction.	Monomer disappears early in reaction: at DP[a] 10, less than 1% monomer remains.
High polymer is formed at once; polymer molecular weight changes little throughout reaction.	Polymer molecular weight rises steadily throughout reaction.
Long reaction times give high yields but affect molecular weights little.	Long reaction times are essential to obtain high molecular weights.
Reaction mixture contains only monomer, high polymer, and about 10^{-8} part of growing chains.	At any stage all molecular species are present in a calculable distribution.

[a]Degree of polymerization.

book we shall further subdivide and reclassify as the occasion demands. Step reactions will be classified by reaction type as condensation, addition, ring-opening, amidation, ester interchange, or interfacial; and chain reactions by initiation mechanism as free-radical, anionic, cationic, or coordination. In Chap. 2, we reclassify polymers by preparative methods as bulk, suspension, solution, emulsion, or interfacial. The polymers synthesized or characterized in the experiments in this book, plus a few others of commercial importance, are classified as step-reaction (condensation) and chain-reaction (addition) types in Table 1-2.

B. Step-Reaction (Condensation) Polymerization

Step-reaction or condensation polymerizations take place by a variety of reactions, as described in the *Textbook*, Chap. 8B and the references cited therein. All of them proceed by the stepwise reaction between functional groups on the reacting molecules. In general, these reactions fall into two classes: *polycondensation*, in which a small molecule is eliminated at each step, and *polyaddition*, in which this does not occur. If A and B are functional groups which can react, attached to a nonreactive portion of the molecule R or R', these

TABLE 1-2. Classification of common polymers as resulting from step or chain reactions

Step Reaction	Chain
Polyesters	Polystyrene
Polyamides[a]	Poly(methyl methacrylate)
Polyurethanes	Poly(vinyl chloride)
Phenol-formaldehyde resins	Polyethylene
Urea-formaldehyde resins	Poly(vinyl acetate)
Melamine-formaldehyde resins	Polyisoprene
Polyacetals	Polybutadiene
Polysulfides	Polypropylene

[a] Can also be formed by chain mechanisms, such as ionic ring-opening reactions.

can be represented as follows: polycondensation

$$A\text{-}R\text{-}A + B\text{-}R'\text{-}B \rightarrow A\text{-}R\text{-}R'\text{-}B + AB$$

and polyaddition

$$A\text{-}R\text{-}A + B\text{-}R'\text{-}B \rightarrow A\text{-}R\text{-}AB\text{-}R'\text{-}B$$

An example of polycondensation is the formation of a polyamide from a diamine and a dibasic acid with the elimination of water, the first step of which is

$$NH_2(CH_2)_xNH_2 + HOOC(CH_2)_{x'}COOH \rightarrow$$

$$H[\text{-}NH(CH_2)_xNHCO(CH_2)_{x'}CO\text{-}]OH + H_2O$$

The formation of a polyurethane by the reaction of a glycol and a diisocyanate is an example of polyaddition. Again, the first step is

$$HO(CH_2)_xOH + OCN(CH_2)_{x'}NCO \rightarrow$$

$$[\text{-}O(CH_2)_xOCONH(CH_2)_{x'}NHCO\text{-}]$$

Other examples are given in Table 8-2 of the *Textbook*. These reactions are well known in small molecules; their application to polymerization requires only that the reactants be difunctional so that the product of each reaction step can participate in further reactions.

The *degree of polymerization* $\bar{x}_n$, that is, the number-average* number of monomers or repeat units in the chain, can easily be predicted from kinetic or statistical considerations (*Textbook*, Chap. 8) in terms of the *extent of reaction* p. The latter is easily measured in most cases as the fraction of the functional groups used up. The two quantities are related by the Carothers equation,

$$\bar{x}_n = 1/(1-p) \tag{1-1}$$

from which it follows that the *number-average molecular weight* is given by

$$\bar{M}_n = M_0/(1-p) \tag{1-2}$$

where M_0 is the molecular weight of the repeat unit, usually that of the monomer less the weight of any small molecule lost at each reaction step.

Examination of this equation will show that the preparation of high-molecular-weight polymers by step-reaction polymerization places unusually demanding and rigorous requirements on the purity of the reactants. Stoichiometry must be exactly balanced; that is, the number of each type of functional group must be the same. Conversion must be high, and there must be no monofunctional reactants present and little or no tendency to form cyclic compounds. Practically, high reaction rates are desirable. Some of the characteristics of step-reaction polymerization made understandable by the Carothers equation are listed in Table 1-1.

It is obvious that an excess of one type of functional group limits the maximum chain length obtainable, and this technique is occasionally used industrially to provide molecular-weight control. The resulting value of $\bar{x}_n$ can be predicted if the initial ratio of functional-group concentrations is known, as is amplified in the *Textbook*. If the stoichiometry were exact, infinite chain length would theoretically be possible. Practically, the highest molecular weight attainable is limited by two factors: First, the reaction is one of equilibrium, and to shift it in the direction of high polymer requires essentially complete removal of the water or other small molecule formed. In practice, step-reaction polymerizations in bulk systems are often finished with a long high-temperature high-vacuum step to accomplish this. Second, the reaction slows down at higher conversions as the concentrations of the reactants are reduced; thus, the final equilibrium is approached more and more slowly.

Kinetic analysis shows that the time dependence of the extent of reaction is given by

$$p = kc_0t/(1 + kc_0t) \tag{1-3}$$

*The concepts of the number- and weight-averages are defined and discussed with respect to molecular weight in Chap. 7.

where k is the rate constant, which can be demonstrated to be independent of p over most of the range of interest, and c_0 is the initial functional-group concentration. It follows that

$$\bar{x}_n = kc_0t + 1 \tag{1-4}$$

Since the constant term is small, the degree of polymerization increases linearly with time. These considerations apply to catalyzed polymerizations, as discussed in the *Textbook*, Chap. 8C.

If, as is usual, all molecules have equal reactivity, step-reaction polymerization is a completely random process in that any species present can react with any other. Statistical considerations (*Textbook*, Chap. 8C, or for a simple treatment, Billmeyer 1972) show that the distribution of chain lengths or molecular weights achieved at equilibrium is a function of p given by

$$w_x = x(1-p)^2\, p^{x-1} \tag{1-5}$$

where w_x is the weight fraction of the polymer consisting of chains with degree of polymerization x. From this distribution function one can calculate the *weight-average degree of polymerization* $\bar{x}_w$:

$$\bar{x}_w = (1 + p)/(1 - p) \tag{1-6}$$

and of course $\bar{M}_w = M_0\,\bar{x}_w$. The ratio of weight-average to number-average molecular weight (or degree of polymerization) is a convenient measure of the breadth of the molecular-weight distribution,

$$\bar{x}_w/\bar{x}_n = \bar{M}_w/\bar{M}_n = 1 + p \tag{1-7}$$

and can be seen to approach 2 at high conversion as p approaches unity.

GENERAL REFERENCES

Textbook, Chap. 8C; Flory 1953; Sokolov 1968; Carmichael 1970; Odian 1970; Rodriguez 1970, Chap. 4; Braun 1971, Chap. 4; Williams 1971, Chap. 3.

C. *Radical Chain (Addition) Polymerization*

In this section we discuss chain polymerization in which the growing polymer molecule is a free radical, that is, a substance containing one unpaired electron, but no net electric charge. Other types of growing chains are possible, including both cations and anions; these are discussed in Sec. D. Regardless of the nature of the chain itself, there are three fundamental steps in any chain reaction, and chain polymerization is no exception. They are *initiation*, in which the chain is formed; *propagation*, in which (in polymerization) monomer is added to the growing chain; and *termination*, in which the growth activity of the chain is destroyed, leaving the polymer molecule. Other reactions may be involved, and some are discussed below, but they are not essential. The steps of initiation, propagation, and termination refer to molecular reactions of a single chain and not to the system as a whole; all occur concurrently in the polymerizing mass. The time between initiation and termination of a given chain is typically from a few tenths second to a few seconds; during this time, thousands or tens of thousands of monomers add to the growing chain.

Initiation

Initiation in radical-chain polymerization is the process by which a monomer molecule acquires an active site for the propagation of the chain reaction, that is, becomes a free radical. This can take place as the result of a number of processes, including oxidation-reduction reactions, reaction of another free radical with the monomer, interaction of high-energy radiation with the monomer, or in some cases thermal decomposition of the monomer without the need for any other agent. Some of these agents and mechanisms are discussed further in Chap. 3B; here we take as an example the most common form of initiation: the production of a free radical by the thermal decomposition of a relatively unstable *initiator* molecule, and the subsequent addition of a monomer to this primary radical to form the first chain radical. This is a two-step process, but we shall assume, as is usually the case, that the second step is rapid compared to the first.

Among the common thermal initiators for free-radical polymerization are peroxides, such as benzoyl peroxide, and azo compounds, such as azobisisobutyronitrile, whose structures are given in Chap. 3B. If we designate such a molecule R:R, where the two dots represent the electrons of the unstable bond, e.g., the O-O bond in benzoyl peroxide (and note that the two fragments R do not have to be identical), the primary initiation step is the thermal rupture of this bond, one electron remaining with each radical fragment:

$$\mathrm{R{:}R} \rightarrow 2\,\mathrm{R}\cdot$$

where the single dot represents the unpaired electron characteristic of a free radical. The second step of initiation, in which R· adds to a monomer M, may be represented schematically as

$$\mathrm{R}\cdot + \mathrm{M} \rightarrow \mathrm{RM}\cdot$$

In terms of molecular structure, in the typical case of a *vinyl* monomer of the general structure CH_2=CHX, where X is hydrogen in ethylene, CH_3 in propylene, Cl in vinyl chloride, phenyl in styrene, etc., the radical R· adds to the double bond, regenerating the unpaired electron at the substituted end (usually) of the monomer:

$$\mathrm{R}\cdot + \mathrm{CH_2 = CHX} \rightarrow \mathrm{RCH_2 - \dot{C}HX}$$

As a result of this step, the initiator fragment R is chemically bonded to the growing chain and thus incorporated into the final polymer molecule.

If the first step is rate determining, the rate of initiation v_i is proportional to the initiator concentration [I]. In the second step, however, all the radicals produced by the decomposition of the initiator may not be effective in reacting with monomer to produce active chains; an *initiator efficiency* factor f is introduced into the expression for v_i to account for this, f being the fraction of initiator fragments that produce active chains. We can then write

$$v_i = 2 f k_d [I] \tag{1-8}$$

The factor two arises from the fact that the decomposition of one initiator molecule yields two radicals, and k_d is the rate constant for thermal decomposition of the initiator. Note that k_d is independent of the presence of monomer, but for thermal initiators depends strongly on temperature. If the second step of addition to monomer is slow, v_i is also proportional to the monomer concentration [M].

Propagation

The free-radical chain formed in the initiation step is capable of adding other monomers in succession:

$$RM\cdot + M \rightarrow RM_2\cdot$$

or, in general,

$$RM_x\cdot + M \rightarrow RM\cdot_{x+1}$$

The rate constant for propagation, k_p, is assumed to be independent of the number, x, of monomers in the growing chain. This is a good assumption except for the first few additions, which have little effect on the final result. We write for the rate of propagation

$$v_p = k_p [M] [M\cdot] \tag{1-9}$$

where [M•] represents the sum of the concentrations of all chain radicals of the type $M\cdot_x$.

Termination

The growing-chain radicals are extremely reactive toward one another; only their low concentration (typically about 10^{-8} M) prevents their early reaction, pairing the two odd electrons and destroying their activity. Eventually, this bimolecular reaction occurs in one of two possible ways: *combination*, in which the two chains yield a single polymer molecule,

$$-CH_2\dot{C}HX + \cdot CHXCH_2- \rightarrow -CH_2CHXCHXCH_2-$$

or *disproportionation*, in which hydrogen transfer occurs with the formation of two polymer molecules,

$$-CH_2\dot{C}HX + \cdot CHXCH_2- \quad \rightarrow \quad -CH_2CH_2X + HXC{=}CH-$$

It can be seen that if termination is by combination, each end of the resulting polymer chain is terminated by an initiator fragment R, whereas if it is by combination, each of the two polymer molecules formed contains one R group, and one of them is terminated on the other end by a vinyl group, -CH=CHX.

Both types of termination are known. For example, polystyrene is predominantly terminated by combination, whereas poly(methyl methacrylate) is terminated by disproportion if polymerized at temperatures above 60°C, and partly by each mechanism at lower temperatures.

Separate equations for the rate of termination can be written for the cases of combination and disproportionation,

$$v_{t,c} = 2\,k_{t,c}\,[M\cdot]^2 \;;\quad v_{t,d} = 2\,k_{t,d}\,[M\cdot]^2 \qquad (1\text{-}10)$$

but where there is no need to distinguish between the two it is customary to write simply

$$v_t = 2\,k_t\,[M\cdot]^2 \qquad (1\text{-}11)$$

where $k_t = k_{t,c} + k_{t,d}$. Although there is no way of distinguishing between the two mechanisms of termination by measurement of v_t or, as shown below, the over-all rate of polymerization, the two do lead to significant differences in the molecular-weight distribution of the resulting polymer.

Over-all polymerization rate

As there is no simple way to determine the concentration [M·] of growing-chain radicals, it is desirable to eliminate this quantity from the expression for the rate of propagation, Eq. 1-9. This is done by means of the *steady-state assumption* that, during most of the polymerization, radicals must be generated and used up at the same rate if the reaction is to avoid the fates of dying out or explosively increasing in rate. It follows that $v_i = v_t$, and from this relation one can solve for [M·] and substitute for it in Eq. 1-9, yielding

$$v_p = k_p\,(f\,k_d\,[I]/k_t)^{1/2}\,[M] \qquad (1\text{-}12)$$

Since monomers are used up thousands of times faster by propagation than by initiation, this is essentially the over-all rate of disappearance of monomer, often written $-d[M]/dt$. The bimolecular nature of the termination step leads directly to the conclusion that the over-all polymerization rate is proportional to the square root of the initiator concentration, a characteristic feature of radical-chain polymerization. A number of other characteristics of this type of reaction are listed in Table 1-1. A few values of k_p and k_t for common monomers are given in Table 9-3 of the *Textbook*.

Chain transfer

Several reactions can occur in chain polymerization in addition to the three essential ones described above. The most important of these additional possibilities is *chain transfer*, in which the growing chain reacts with some other molecule to produce a polymer molecule and a new species capable of further propagation. No radicals are created or destroyed, hence there is no effect on the rate of polymerization, but the reaction leads to a limitation on the chain length obtainable.

Chain transfer usually involves cleavage of the weakest bond in the molecule attacked, with transfer of a hydrogen or other atom, plus one electron, to the original radical and production of an unpaired electron on the donor molecule. The latter can be solvent (if present), monomer, initiator, or polymer. In the latter case, which may be either intermolecular or intramolecular (as described in the *Textbook*, Chap. 13A), no new polymer molecule is formed but a branch grows from the point of transfer.

The mechanism of action of *inhibitors* and *retarders*, described in the *Textbook*, Chaps. 9A and 9C, is often one of chain transfer. The effects of chain-transfer reactions on chain polymerization is described in Table 1-3.

In general terms, a chain-transfer reaction can be written

$$M_x\cdot + XP \rightarrow M_xX + P\cdot$$

$$P\cdot + M \rightarrow PM\cdot, \quad \text{etc.}$$

where P may be monomer, initiator, solvent, or other added chain-transfer agent. Transfer to polymer is omitted here since no new molecule is formed. Since chain-transfer reactions do not use up monomer or change the number of free radicals, they have no effect on the over-all polymerization rate.

Kinetic chain length and degree of polymerization

Having now defined all the reactions leading to polymer molecules, we may consider the calculation of the degree of polymerization. We first define the *kinetic chain length* ν as the number of monomer units used up per active chain. Thus, $\nu = v_p/v_i = v_p/v_t$. Use of Eqs. 1-8 and 1-12 leads to

$$\nu = k_p\,[M]/2(f\,k_d\,k_t)^{1/2}\,[I]^{1/2} \tag{1-13}$$

In the absence of chain transfer, the kinetic chain length is directly related to the degree of polymerization $\bar{x}_n$, the relation depending upon the type of termination. For disproportionation, $\bar{x}_n = \nu$ and for combination, $\bar{x}_n = 2\nu$.

TABLE 1-3. Effect of chain-transfer reactions on chain length and structure

Reaction	Effect on			
	v_p	$\overline{M}_n$	$\overline{M}_w$	*Architecture*
To small molecule, giving active radical	None	Decreases	Decreases	None
To small molecule, retardation or inhibition	Decreases	May decrease or increase	May decrease or increase	None
To polymer, intermolecular	None	None	Increases	Produces long branches
To polymer, intramolecular	None	None	Increases	Produces short branches

If chain transfer is present, however, the kinetic chain is regenerated in the transfer step and can be many times larger than the degree of polymerization. The latter is best considered as defined by the ratio of the rate of propagation to the sum of the rates of all reactions leading to polymer molecules. As is indicated in the *Textbook*, Eq. 9-15, this leads to a relation which, when inverted, shows that $1/\overline{x}_n$ is given by a series of terms for normal termination and transfer to any molecular species which may be present (again, transfer to polymer is omitted since no new polymer molecules are formed):

$$1/\overline{x}_n = k_t\, v_p/k_p^2\, [\mathrm{M}]^2 + C_M + C_S\, [\mathrm{S}]/[\mathrm{M}] \qquad (1\text{-}14)$$

where the first term represents termination by combination (and is multiplied by 2 if termination is by disproportionation), the second transfer to monomer, and the third transfer to solvent or any added chain-transfer agent. (Transfer to initiator is possible, as described in the *Textbook*, but is omitted here for simplicity.) The *transfer constants* C are ratios of the rate constant for the transfer reaction involved to k_p. A table of chain-transfer constants for common monomers and solvents is given in the *Textbook*, Table 9-1.

Molecular-weight distribution

The distribution of molecular weights arising in radical-chain polymerization restricted to low conversion (so that [M] and consequently ν and $\bar{x}_n$ do not change during the reaction) is given by Eq. 1-5 for the case of termination by disproportionation; formally, this is the same equation as for step-reaction polymerization, but its interpretation requires that p be defined as the probability that a growing chain will propagate rather than terminate at the next reaction step. Thus, $p = v_p/(v_p + v_t)$. If high polymer is formed as the result of a value of p near 1, then $\bar{x}_w/\bar{x}_n = 2$. For termination by combination, a somewhat different distribution function is found, and $\bar{x}_w/\bar{x}_n = 1.5$; the distribution is narrowed because of the higher probability of one long and one short chain combining in termination, compared to the cases of two short or two long chains.

GENERAL REFERENCES

Textbook, Chap. 9; Flory 1953; Eastmond 1967; Lenz 1967; Bagdasar'yan 1968; Odian 1970; Rodriguez 1970, Chap. 4; Braun 1971, Chap. 3; Williams 1971, Chap. 4.

D. *Ionic Chain Polymerization*

Addition polymerization involving anionic or cationic active chains (*Textbook*, Chap. 10A-C) is not as well understood as free-radical polymerization, for a number of reasons. The mechanism and kinetics are very much influenced by the nature of the reaction medium, particularly as it affects the ion-pair structure of the active chain end and its counterion of opposite sign (sometimes referred to as the gegenion). Heterogeneous catalysts are often used, making the analysis more complicated; the reactions are often very rapid; and the results may depend strongly on temperature and the nature of the solvent, counterion, and presence of trace amounts of small molecules classed as cocatalysts.

Cationic and anionic polymerization have many similar characteristics. Both depend, of course, on the formation and propagation of an ionic species, and both often depend similarly on the influences mentioned above. Cationic polymerization usually requires vinyl monomers having electron-releasing substituents, such as phenyl, alkoxy, or 1,1-dialkyl groups. Typical examples of commercial importance include polyisobutylene, polyoxymethylene, polyepichlorohydrin, poly(vinyl ethers), and furane polymers. Anionic polymerization can be carried out with monomers having electron-withdrawing groups such as carboxyl, phenyl, or nitrile. (Note that some monomers such as styrene polymerize well by either mechanism; see the *Textbook*, Table 10-1.) Some specific examples of commercial importance are poly(ethylene oxide), styrene-butadiene block copolymers, poly(*cis*-1,4-isoprene), and polybutadiene.

Cationic polymerization

Cationic polymerization proceeds by the attack on monomer of acidic or electrophilic propagating species such as carbonium ions, oxonium ions, sulfonium ions, or ammonium ions. Carbonium ions are by far the most important chain carriers in commercial cases. The monomers must be basic relative to the acidic chain carriers.

Cationic polymerization can be initiated by many different chemicals, including Brönsted acids such as hydrogen halides or mineral acids and Lewis-acid complexes such as $BF_3 \cdot H_2O$ or $TiCl_4 \cdot H_2O$. Most cationic polymerizations appear to require the presence of a small molecule in trace amounts known as a cocatalyst, for example H_2O, HCl, or RCl.

The mechanism of cationic polymerization is illustrated by the polymerization of isobutylene by boron trifluoride catalyst in the presence of water as the cocatalyst. The steps of importance are:

Catalyst activation

$$BF_3 + H_2O \rightleftarrows H^+(BF_3OH)^-$$

Initiation

$$H^+(BF_3OH)^- + (CH_3)_2C{=}CH_2 \rightarrow (CH_3)_3C^+(BF_3OH)^-$$

Propagation

$$(CH_3)_3C^+ (BF_3OH)^- + CH_2{=}C(CH_3)_2 \rightarrow$$

$$CH_3{-}C(CH_3)_2{-}C(CH_3)_2^+ (BF_3OH)^-$$

This step may best be considered a "push-pull" attack of the carbonium ion-counterion pair on the double bond of the monomer.

Termination

$$CH_3[(CH_3)_2CCH_2]_x\ C(CH_3)_2^+\ (BF_3OH)^- \rightarrow$$

$$CH_3[(CH_3)_2CCH_2]_x\ C({=}CH_2)(CH_3) + H^+(BF_3OH)^-$$

which is a spontaneous decomposition of the carbonium ion chain, regenerating the catalyst-cocatalyst complex.

Chain transfer to monomer can also occur (*Textbook*, p. 314).

The kinetics of the system may be described as follows:

$$v_i = k_i\ [C][M]$$

$$v_p = k_p\ [M^+][M] \qquad (1\text{-}15)$$

$$v_t = k_t\ [M^+]$$

where [C] is written for the concentration of the catalyst-cocatalyst complex. At the steady state, it follows that the over-all reaction rate is given by

$$v_p = (k_i\, k_p/k_t)\ [C][M]^2 \tag{1-16}$$

Note that the effect of the monomolecular termination step is to make the rate dependent on the first power, rather than the square root, of the catalyst concentration. Expressions for $\bar{x}_n$ vary depending on whether transfer or termination predominates, and are given in the *Textbook*.

Summary Some of the important characteristics of cationic polymerization are:

The rate of polymerization is, as stated above, proportional to the first power of the catalyst concentration.

Reaction rates are typically quite fast, even compared to radical-chain polymerization, due to the high reactivity of the carbonium ion.

Low polymerization temperature favors propagation over competing reactions resulting from the high carbonium-ion reactivity.

The degree of polymerization is shown by kinetic analysis to be independent of the catalyst concentration as a result of the monomolecular termination step.

Termination by a bimolecular step involving two carbonium ions is ruled out by the electrostatic repulsion between ions of like sign.

Because of the ion-counterion nature of the growing chain, the dielectric constant of the medium has a profound effect on the polymerization. Solvents with higher dielectric constant favor initiation and propagation, and lead to polymer of high molecular weight.

Anionic polymerization

Anionic polymerization proceeds by the attack on monomer of basic or nucleophilic propagating species such as carbanions. A variety of basic catalysts can be used to initiate anionic polymerization, including hydroxides, alkoxides, covalent or ionic metal amides, aliphatic organic anions, Grignard reagents, alkali metals, and metal ketyls.

The mechanism of anionic polymerization is much the same as that of cationic polymerization except that the initiating species is a base rather than an acid, the propagating species is a carbanion rather than a carbonium ion, and the charge-transferring species is a hydride ion rather than a proton. The simplest anionic systems are those based on organo-alkali compounds in homogeneous systems. Initiation can be by electron transfer from alkali metals (heterogeneous), alkali-metal complexes (homogeneous), or organo-alkali catalysts (homogeneous or heterogeneous).

Initiation by alkali metals involves transfer of an electron from the metal to the monomer, forming a radical-ion intermediate which quickly dimerizes to form a di-anion capable of propagating at both ends. Termination can occur only by chain transfer, because of the extreme unlikelihood of recombination or transfer of the charge carrier when the latter is H^- rather than H^+ as in cationic polymerization.

This leads to the possibility, realized in practice with enough care, of excluding all molecules to which transfer can occur and achieving nonterminating or "living" polymer chains. This type of polymerization is illustrated by the polymerization of styrene by metallic sodium:

Initiation

$$Na + CH_2{=}CH(C_6H_5) \rightarrow \cdot CH_2{-}CH(C_6H_5)^- \; Na^+$$

$$2 \cdot CH_2{-}CH(C_6H_5)^- \; Na^+ \rightarrow Na^+ \; {}^-HC(C_6H_5)CH_2CH_2CH(C_6H_5)^- \; Na^+$$

Propagation

$$Na^+ \; {}^-HC(C_6H_5)CH_2CH_2CH(C_6H_5)^- \; Na^+ + (x + y) \; CH_2{=}CH(C_6H_5) \rightarrow$$

$$Na^+ \; {}^-HC(C_6H_5)CH_2[CH(C_6H_5)CH_2]_x - [CH_2CH(C_6H_5)]_yCH_2CH(C_6H_5)^- \; Na^+$$

If there are no impurities present to which transfer can occur or which will react with excess sodium (such as water, alcohols, oxygen, or acids), the polystyryl disodium chains continue to grow at both ends as long as monomer is present. When it is used up, a different monomer may be added to produce a block copolymer (*Textbook*, Chap. 11D and p. 406), or polymers with reactive end groups such as carboxyl or hydroxyl can be synthesized by adding CO_2 or ethylene oxide, respectively.

Initiation by electron transfer in a heterogeneous system occurs at the metal surface and is relatively slow. In contrast, quite rapid electron-transfer initiation in a homogeneous system can be obtained by use of a soluble aromatic-metal complex such as sodium naphthalide. This has the unusual (for chain polymerization) feature that all the chains are initiated at about the same time. This leads to the production of polymer having an extremely narrow molecular-weight distribution; values of $\overline{M}_w/\overline{M}_n$ around 1.03 can be obtained (Exp. 9).

Initiation by an organoalkali compound, such as a lithium alkyl, proceeds by direct anionic attack on the double bond:

Initiation

$$RLi + CH_2{=}CH(C_6H_5) \rightarrow RCH_2CH(C_6H_5)^- \; Li^+$$

Propagation

$$RCH_2CH(C_6H_5)^- \; Li^+ + x \; CH_2{=}CH(C_6H_5) \rightarrow$$

$$R{-}[CH_2CH(C_6H_5)]_xCH_2CH(C_6H_5)^- \; Li^+$$

Again, there is no chain termination in the absence of materials having an active hydrogen. The distinguishing feature here is that the chain grows from only one end.

The kinetics of anionic polymerization is more complex than that of other types, primarily because the nature of the solvent and counterion can have a profound effect. Use of solvents that solvate the cation, and of more electronegative alkali metals, tend to increase the polymerization rate.

Summary The following features are characteristic of anionic polymerization:

The rate of polymerization shows a variable dependence on the catalyst concentration, ranging from fractional to first order, and is very dependent on the nature of the metal counterion and the solvent.

There is no termination in the absence of impurities, leading to the concept of "living" polymers.

Molecular-weight distributions can be quite broad in the case of heterogeneous initiation, or very narrow if the initiation step is homogeneous.

GENERAL REFERENCES

Textbook, Chaps. 10B, C; Lenz 1967; Kennedy 1968-1969; Odian 1970; Braun 1971, Chap. 3; Williams 1971, Chap. 4; and specifically:
Cationic polymerization: Plesch 1963; Pepper 1964; Eastham 1965.
Anionic polymerization: Overberger 1965; Szwarc 1968; Morton 1969; Smid 1969.

E. *Coordination and Ring-Opening Polymerization*

Coordination polymerization

Coordination polymerization is a chain-addition reaction that usually involves a propagation step in which monomer is held in a coordinated way on the surface of a solid catalyst. There are, however, many variations of the process and the presence of a heterogeneous system is not essential. Most prominent among the large number of catalyst systems are those discovered in the 1950's by Ziegler (1964), who found that mixtures of transition-metal halides and aluminum alkyls polymerize ethylene to high molecular weight under mild conditions of temperature and pressure. The many Ziegler catalyst systems now known are obtained by the interaction of an organometallic compound or halide of a Group I-III metal with a halide or other derivative of a Group IV-VI transition metal.

In 1964, Ziegler shared the Nobel Prize with Natta for work on coordination polymerization. Natta's contribution (Natta 1965) was the discovery that stereoregular polymers (*Textbook*, Chap. 5A) can be obtained by this technique; in fact, the coordination step between monomer and catalyst is the essential event in the mechanism of stereospecific polymerization. Coordination catalysts both supply the initiating species and provide the counterion surroundings which account for the regularity of addition to the growing chain. Coordination of the monomer on the catalyst activates the growth reaction and brings the monomer into the proper steric arrangement for stereospecific addition.

Numerous mechanisms have been postulated for stereoregular coordination polymerization; some of them are described in the *Textbook*,

Chap. 10D. The chain is usually anionic or cationic, but free-radical carriers are also possible. Among the monomers of industrial importance polymerized in this way are ethylene, propylene, 1-butene, butadiene, and isoprene.

Ring-opening polymerization

Polymerization of cyclic compounds involving opening of the ring is an example of the importance of considering carefully the classification of polymers described in Sec. A. Structurally, these are condensation polymers because of the presence of functional groups, such as amide or ether linkages, in the chain; but kinetically and mechanistically most of them are ionic chain-reaction polymers, and their polymerizations show all the characteristics of ionic polymerization, such as the strong influences of solvent and counterion type. Cyclic ethers, esters, amides, acetals, siloxanes, and epoxides are polymerized commercially in this way.

Initiation of ring-opening polymerization may be written in general form as

$$\widehat{R - Z} + C \rightarrow M^*$$

where Z is the functional group in the monomer and C is the catalyst. The initiated species M* may be either an ion or a neutral molecule depending on the nature of the catalyst, which itself can be either an ion, such as OH^-, OR^-, H^+, or a neutral molecule such as Na, BF_3, or H_2O.

The initiated species grows by successive ring-opening addition of monomer. Only monomer adds to the growing chain in the propagation step, as is characteristic of chain reactions. This step may be atypically slow, however, and the superficial behavior of the system may resemble that of a condensation polymerization; for example, molecular weight may increase slowly throughout the course of the reaction.

Ring-opening polymerization of the cyclic acetal trioxane is carried out in Exp. 6. Another typical example is the anionic polymerization of ethylene oxide by a variety of catalysts which can be represented by M^+A^-.

Initiation

$$\overset{\overset{O}{/\;\;\backslash}}{H_2C - CH_2} + M^+\ A^- \rightarrow ACH_2CH_2O^-\ M^+$$

Propagation

$$ACH_2CH_2O^-\ M^+ + \overset{\overset{O}{/\;\;\backslash}}{H_2C - CH_2} \rightarrow ACH_2CH_2OCH_2CH_2O^-\ M^+$$

Many ring-opening reactions of this type have the characteristics of "living" polymerizations in that there is often no termination unless a terminator is deliberately added.

GENERAL REFERENCES

Textbook, Chaps. 10D, E; Lenz 1967; Odian 1970; Braun 1971, Chap. 8; and specifically:
Coordination polymerization: Natta 1966; Reich 1966; Goodman 1967; Ketley 1967.
Ring-opening polymerization: Furukawa 1963; Frisch 1969.

F. *Copolymerization*

In the polymerization of a mixture of two or more monomers, the rates at which different monomers add to the growing chain determine the composition and hence the properties of the resulting copolymer. The order as well as the ratio of amounts in which monomers add are determined by their relative reactivities in the chain-growth step, which in turn are influenced by the nature of the end of the growing chain, depending on which monomer added previously. Among the possibilities are random and regular alternating addition as well as block formation.

In condensation copolymerization, the reactivities of functional groups are often independent of other considerations and the results are trivial: Random addition occurs in the same ratio as the concentrations of the monomers in the mixture. The situations for free-radical and other types of chain polymerization are similar, and the following discussion is limited, as in the *Textbook* (Chap. 11), to the free-radical case.

With two monomers present, there are four possible propagation reactions, assuming that growth is influenced only by the nature of the end of the growing chain and of the monomer:

$$M_1\cdot + M_1 \xrightarrow{k_{11}} M_1\cdot$$

$$M_1\cdot + M_2 \xrightarrow{k_{12}} M_2\cdot$$

$$M_2\cdot + M_1 \xrightarrow{k_{21}} M_1\cdot$$

$$M_2\cdot + M_2 \xrightarrow{k_{22}} M_2\cdot$$

With the definition of *reactivity ratios* $r_1 = k_{11}/k_{12}$ and $r_2 = k_{22}/k_{21}$ and the application of the steady-state assumption, one can derive a *copolymer equation* relating the instantaneous composition of copolymer being formed, $d[M_1]/d[M_2]$, to the monomer concentrations:

$$d[M_1]/d[M_2] = [M_1](r_1[M_1] + [M_2])/[M_2]([M_1] + r_2[M_2]) \qquad (1\text{-}17)$$

A useful approximation to this equation for low conversion, where $[M_1]/[M_2]$ is essentially constant, is

$$r_2 = r_1 H^2/h + H(1 - h)/h \qquad (1\text{-}18)$$

where $H = [M_1]/[M_2]$ and $h = d[M_1]/d[M_2]$. This equation is used in Exp. 5 as an aid in evaluating r_1 and r_2.

The reactivity ratios are ratios of the rate constants for a given radical adding its own monomer to that for its adding the other monomer. A value of $r > 1$ means that the radical prefers to add its own monomer, and vice versa. In the system styrene-methyl methacrylate, Exp. 5, $r_1 = 0.52$ and $r_2 = 0.46$; each radical adds the other monomer about twice as fast as its own. A few reactivity ratios are given in the *Textbook*, Table 11-1, and an extensive tabulation in Mark 1966b.

Copolymerizations are classified according to values of the product r_1r_2. When $r_1r_2 = 0$, neither monomer radical will add its own monomer, and propagation can continue only by adding first one, then the other monomer to produce an *alternating* copolymer. The usual case is $0 < r_1r_2 < 1$, with both reactivity ratios less than one, as in the example given above. When $r_1r_2 = 1$, the copolymerization is said to be *ideal*, since $r_1 = 1/r_2$ and each radical shows the same preference for one of the monomers. The sequence of monomers in the copolymer is completely random, determined only by the composition of the comonomer feed. When one reactivity ratio is greater than unity, the copolymer contains a larger proportion of the more reactive monomer, and as the difference in reactivity of the two monomers increases, it becomes more and more difficult to produce copolymers containing appreciable amounts of both monomers. In those rare cases when both reactivity ratios are greater than one, there is a tendency to produce block copolymers, but these are better prepared by the anionic "living polymer" techniques described in Sec. D.

Polymerizations involving three or more monomers can be considered similarly, but the kinetics is beyond the scope of this book (Ham 1964). Such cases are important commercially, however. One example is the tough, rigid ABS resins (*Textbook*, Chap. 14A), terpolymers of acrylonitrile, butadiene, and styrene made as composite (two-phase) materials reinforced by graft copolymerization. Another is the EPDM (ethylene-propylene-diene monomer) rubbers, where the diene is added to provide sites for vulcanization (*Textbook*, Chap. 13G). In the coatings field, as many as five or more monomers are often copolymerized to obtain the desired end-use properties of, say, a paint film.

GENERAL REFERENCES

Textbook, Chap. 11; Ham 1964, 1966; Lenz 1967; Odian 1970; Braun 1971, Chap. 3; Williams 1971, Chap. 5.

G. *Chain Branching and Crosslinking*

Step-reaction polymerization

Branching or crosslinking (formation of network polymers) occurs in stepwise polymerization when one or more of the monomers has functionality greater than two, with the possibility of chain growth at three or more sites. This complicates and broadens the molecular-weight distribution but has little effect on polymerization kinetics. Common examples of network step-reaction polymers are the phenol-formaldehyde resins and the urea-formaldehyde and melamine-formaldehyde resins, which have similar structural architecture.

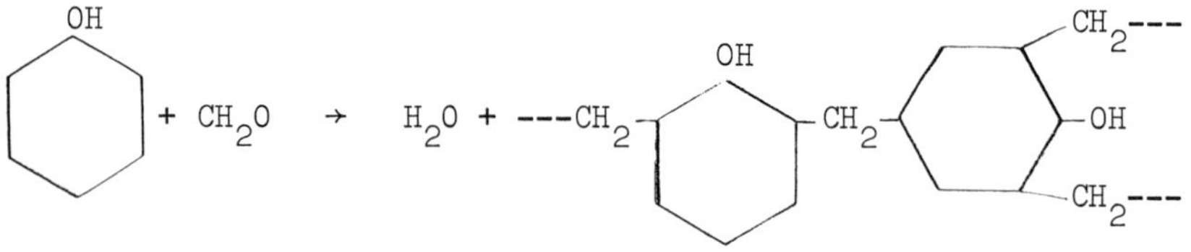

$$C_6H_5OH + CH_2O \rightarrow H_2O + \text{---}CH_2\text{-}C_6H_3(OH)\text{-}CH_2\text{-}C_6H_2(OH)(CH_2\text{---})_2$$

An important characteristic of the polymerizations leading to crosslinked networks is the *gel point*, the extent of reaction at which the polymer becomes insoluble. This point can be calculated from stoichiometric data (*Textbook*, Chap. 8D) with fair accuracy.

A second mode of branch or network formation in step-reaction polymerization is through reaction of an active site formed during the polycondensation. An example is the formation of a polyurea from a diisocyanate and a diamine,

$$OCNRNCO + H_2NR'NH_2 \rightarrow OCNRNHCONHR'NH_2$$

followed by reaction of the secondary amine group with another isocyanate to form a branch:

$$OCNRN(CONHRNCO)CONHR'NH_2$$

Chain polymerization

Branching can occur in chain polymerization by chain transfer to polymer (Table 1-3), or chain transfer to monomer followed by polymerization of that monomer. Usually these reactions do not lead to crosslinking. This can be accomplished as desired, either during polymerization by the use of polyfunctional monomers such as divinyl benzene (Exp. 8), allyl dimethacrylate, or triallyl cyanurate; or in a post-polymerization step as described below.

An example of chain transfer to polymer leading to branch growth is hydrogen abstraction from, say, polystyrene by a second chain radical:

$$M_n\cdot + \text{---}\ CH_2CH\phi CH_2CH\phi CH_2CH\phi\ \text{---} \rightarrow$$

$$M_nH + \text{---}\ CH_2CH\phi CH_2\dot{C}\phi CH_2CH\phi\ \text{---}$$

where ϕ is phenyl. Growth at the new radical site yields the branch.

In the case of polyethylene, the transfer to polymer can be intramolecular rather than intermolecular, leading to production of a short (2- or 4-carbon) branch, as described in the *Textbook*, Chap. 13A. Transfer to monomer followed by polymerization is the mechanism for branch formation in poly(vinyl chloride).

In both step and chain polymerization, the probability of branch or network formation increases with conversion, that is, with increased concentration of polymer chains. To avoid these features, polymerization can be stopped at low conversion or carried out in the presence of diluents.

Post-treatment

Crosslinking in a post-polymerization treatment is standard practice in rubber vulcanization. The mechanisms are complex (*Textbook*, Chap. 19B), but in the presence of sulfur, crosslinks of the following type are formed in natural rubber:

$$\begin{array}{l} \text{---}\ CHC(CH_3){=}CHCH_2\ \text{---} \\ \quad\quad\ | \\ \quad\quad\ S \\ \quad\quad\ | \\ \quad\quad\ S \\ \quad\quad\ | \\ \text{---}\ CHC(CH_3){=}CHCH_2\ \text{---} \end{array}$$

Other mechanisms active at the same time involve radical reactions at the double bonds giving further crosslinking.

The formation of active sites on many polymer chains can be effected by exposure to high-energy radiation in the form of gamma rays, x-rays, or ultraviolet light; or to beams of electrons or neutrons; or by adding to the polymer free-radical initiators which are stable until the mass is subsequently heated. For example, polyethylene can be crosslinked either by radiation or by heat treatment after the addition of dicumyl peroxide as an initiator. Not all polymers crosslink on exposure to radiation, however, as discussed in the *Textbook*, Chap. 12F.

In many cases, branching or network formation is undesirable from the standpoint of processability and end-use properties of the polymer. There are some notable exceptions, of course, a few of which have been mentioned.

GENERAL REFERENCES

Textbook, Chaps. 8D, 11D; Schultz 1966; Temin 1966; Lenz 1967.

BIBLIOGRAPHY

Bagdasar'yan 1968. Kh. S. Bagdasar'yan, *Theory of Free Radical Polymerization,* Israel Science Translations, Jerusalem, Israel, 1968.

Billmeyer 1971. Fred W. Billmeyer, Jr., *Textbook of Polymer Science,* 2nd ed., Interscience Div., John Wiley and Sons, New York, 1971.

Billmeyer 1972. Fred W. Billmeyer, Jr., *Synthetic Polymers,* Doubleday and Co., New York, 1972.

Braun 1971. Dietrich Braun, Harald Cherdon, and Werner Kern, *Techniques of Polymer Syntheses and Characterization,* Wiley-Interscience, New York, 1971.

Carmichael 1970. Jack B. Carmichael, "Step-Reaction Polymerization," pp. 1-13 in Herman F. Mark, Norman G. Gaylord, and Norbert M. Bikales, eds., *Encyclopedia of Polymer Science and Technology,* Vol. 13, Interscience Div., John Wiley and Sons, New York, 1970.

Carothers 1929. W. H. Carothers, "An Introduction to the General Theory of Condensation Polymers," *J. Am. Chem. Soc. 51,* 2548-2559 (1929).

Carothers 1931. Wallace H. Carothers, "Polymerization," *Chem. Revs. 8,* 353-426 (1931).

Eastham 1965. A. M. Eastham, "Cationic Polymerization," pp. 35-59 in Herman F. Mark, Norman G. Gaylord, and Norbert M. Bikales, eds., *Encyclopedia of Polymer Science and Technology,* Vol. 3, Interscience Div., John Wiley and Sons, New York, 1965.

Eastmond 1967. G. C. Eastmond, "Free-Radical Polymerization," pp. 361-431 in Herman F. Mark, Norman G. Gaylord, and Norbert M. Bikales, eds., *Encyclopedia of Polymer Science and Technology,* Vol. 7, Interscience Div., John Wiley and Sons, New York, 1967.

Flory 1953. Paul J. Flory, *Principles of Polymer Chemistry,* Cornell University Press, Ithaca, New York, 1953.

Frisch 1969. Kurt C. Frisch and Sidney L. Reegan, eds., *Ring-Opening Polymerization,* Marcel Dekker, New York, 1969.

Furukawa 1963. Junji Furukawa and Takeo Saegusa, *Polymerization of Aldehydes and Oxides,* Interscience Div., John Wiley and Sons, New York, 1963.

Goodman 1967. Murray Goodman, "Concepts of Polymer Stereochemistry," pp. 73-156 in Norman L. Allinger and Ernest L. Eliel, eds., *Topics in Stereochemistry,* Vol. 2, Interscience Div., John Wiley and Sons, New York, 1967.

Ham 1964. George E. Ham, "Theory of Copolymerization," Chap. I in George E. Ham, ed., *Copolymerization,* Interscience Div., John Wiley and Sons, New York, 1964.

Ham 1966. George E. Ham, "Copolymerization," pp. 165-244 in Herman F. Mark, Norman G. Gaylord, and Norbert M. Bikales, eds., *Encyclopedia of Polymer Science and Technology,* Vol. 4, Interscience Div., John Wiley and Sons, New York, 1966.

Kennedy 1968-1969. Joseph P. Kennedy and Erik G. M. Tornqvist, eds., *Polymer Chemistry of Synthetic Elastomers,* Interscience Div., John Wiley and Sons, New York; Part I, 1968; Part II, 1969.

Ketley 1967. A. D. Ketley, ed., *The Stereochemistry of Macromolecules,* Vol. 1, Marcel Dekker, New York, 1967.

Lenz 1967. Robert W. Lenz, *Organic Chemistry of Synthetic High Polymers,* Interscience Div., John Wiley and Sons, New York, 1967.

Margerison 1967. D. Margerison and G. C. East, *An Introduction to Polymer Chemistry*, Pergamon Press, New York, 1967.

Mark 1940. H. Mark and G. Stafford Whitby, eds., *Collected Papers of Wallace Hume Carothers on High Polymeric Substances*, Interscience Publishers, New York, 1940.

Mark 1950. H. Mark and A. V. Tobolsky, *Physical Chemistry of High Polymeric Systems*, Interscience Publishers, New York, 1950

Mark 1966a. Herman F. Mark and the editors of *Life*, *Giant Molecules*, Time, Inc., New York, 1966.

Mark 1966b. Herman Mark, B. Immergut, E. H. Immergut, L. J. Young, and K. I. Benyon, "Copolymerization Reactivity Ratios," pp. II-142 to II-289, with an Appendix by L. J. Young, pp. II-291 to II-340, in J. Brandrup and E. H. Immergut, eds., with the collaboration of H.-G. Elias, *Polymer Handbook*, Interscience Div., John Wiley and Sons, New York, 1966.

Morton 1969. Maurice Morton, "Anionic Polymerization," Chap. 5 in George E. Ham., ed., *Vinyl Polymerization*, Part I, Vol. 2, Marcel Dekker, New York, 1969.

Natta 1965. Giulio Natta, "Macromolecular Chemistry: From the Stereospecific Polymerization to the Symmetric Autocatalytic Synthesis of Molecules," *Science 147*, 261-272 (1965).

Natta 1966. Giulio Natta and Umberto Giannini, "Coordinate Polymerization," pp. 137-150 in Herman F. Mark, Norman G. Gaylord, and Norbert M. Bikales, eds., *Encyclopedia of Polymer Science and Technology*, Vol. 4, Interscience Div., John Wiley and Sons, New York, 1966.

Odian 1970. George Odian, *Principles of Polymerization*, McGraw-Hill Book Co., New York, 1970.

Overberger 1965. C. G. Overberger, J. E. Mulvaney, and Arthur M. Schiller, "Anionic Polymerization," pp. 95-137 in Herman F. Mark, Norman G. Gaylord, and Norbert M. Bikales, eds., *Encyclopedia of Polymer Science and Technology*, Vol. 2, Interscience Div., John Wiley and Sons, New York, 1965.

Pepper 1964. D. C. Pepper, "Polymerization," pp. 1293-1348 in George A. Olah, ed., *Friedel-Crafts and Related Reactions*, Vol. II, Part 2, Interscience Div., John Wiley and Sons, New York, 1964.

Plesch 1963. P. H. Plesch, ed., *The Chemistry of Cationic Polymerization*, Macmillan, New York, 1963.

Reich 1966. Leo Reich and A. Schindler, *Polymerization by Organometallic Compounds*, Interscience Div., John Wiley and Sons, New York, 1966.

Rodriguez 1970. Ferdinand Rodriguez, *Principles of Polymer Systems*, McGraw-Hill Book Co., New York, 1970.

Seymour 1971. Raymond B. Seymour, *Introduction to Polymer Chemistry*, McGraw-Hill Book Co., New York, 1971.

Shultz 1966. Allan R. Shultz, "Crosslinking with Radiation," pp. 398-414 in Herman F. Mark, Norman G. Gaylord, and Norbert M. Bikales, eds., *Encyclopedia of Polymer Science and Technology*, Vol. 4, Interscience Div., John Wiley and Sons, New York, 1966.

Smid 1969. J. Smid, "Elementary Steps in Anionic Vinyl Polymerization," pp. 345-488 in Teiji Tsuruta and Kenneth F. O'Driscoll, eds., *Structure and Mechanisms of Vinyl Polymerization*, Marcel Dekker, New York, 1969.

Sokolov 1968. L. B. Sokolov, *Synthesis of Polymers by Polycondensation,* Israel Scientific Translations, Jerusalem, Israel, 1968.

Staudinger 1920. H. Staudinger, "Polymerization" (in German), *Ber. 53B,* 1073-1085 (1920).

Szwarc 1968. Michael Szwarc, *Carbanions, Living Polymers and Electron-Transfer Processes,* John Wiley and Sons, New York, 1968.

Temin 1966. Samuel C. Temin, "Crosslinking," pp. 331-398 in Herman F. Mark, Norman G. Gaylord, and Norbert M. Bikales, eds., *Encyclopedia of Polymer Science and Technology,* Vol. 4, Interscience Div., John Wiley and Sons, New York, 1966.

Williams 1971. David J. Williams, *Polymer Science and Engineering,* Prentice-Hall, Englewood Cliffs, New Jersey, 1971.

Ziegler 1964. Karl Ziegler, "Consequences and Development of an Invention" (in German), *Angew. Chem. 76,* 545-553 (1964).

2

Polymerization Conditions

The purpose of this chapter is to introduce a somewhat different classification of polymerization reactions, which has value for consideration of the synthesis experiments provided later in this book. This is a classification on the basis of homogeneity or heterogeneity of the polymerizing system. For the most part, however, we exclude discussion of systems in which the heterogeneous character results only from the presence of a solid catalyst; these were discussed in Chaps. 1D and 1E, to which reference should be made. The present discussion will center mainly on free-radical chain polymerization, in Secs. A and B, and to step-reaction polymerization in Sec. C.

A typical example of a homogeneous polymerization is the polymerization of styrene in benzene solution, with benzoyl peroxide as the initiator. Monomer, polymer and initiator are all soluble in the solvent, and only one phase is present. The mass polymerization of methyl methacrylate (Exp. 3) and the copolymerization of styrene and methyl methacrylate (Exp. 5) are further examples of homogeneous polymerization. The emulsion polymerization of styrene with potassium persulfate as the initiator (Exp. 4) is, however, an example of a heterogeneous polymerization, in which both monomer and initiator radicals diffuse to the reaction site through a second phase. The polymerization of vinyl chloride in bulk, although initially homogeneous, becomes heterogeneous as soon as polymer is formed, since the polymer is insoluble in the monomer. The polymerization proceeds in both the monomer and the swollen polymer phase.

The statements made in Chap. 1C regarding the kinetics of radical-chain polymerization generally hold if the field of observation is small enough to include only a single phase, even in a heterogeneous system. However, the local conditions may be quite different from those in the system as a whole, and this must be taken into account. For example, the concentration of monomer in a polymerizing emulsion particle, or a precipitated polymer particle in vinyl-chloride polymerization, may be quite different from that calculated for the system as a whole. Thus the local kinetics, and the nature of the polymer formed, are determined by the environment of the reactants and thus the choice of the preparative technique. The properties of a polymer made in emulsion may be quite different from the product from the same monomer polymerized in solution.

When the over-all system is considered rather than a microscopic portion of it, the kinetics of the polymerization change somewhat depending upon heterogeneity or homogeneity. It is convenient to consider three cases: *homogeneous kinetics*, *emulsion kinetics*, and *precipitation kinetics*, the latter being the case of polymer insoluble in the remainder of the system. Table 2-1 classifies the common types of polymerizing systems according to heterogeneity and to kinetics, using this scheme.

TABLE 2-1. *Classification of polymerization systems by heterogeneity and kinetics, with examples of suitable monomers and solvents*

Monomer-polymer phase relation	Monomer phase location	
	Continuous	Dispersed
In same phase	*Solution polymerization* Solution kinetics Methyl methacrylate in toluene; acrylic acid in water; styrene in benzene	*Suspension polymerization* Solution kinetics Styrene, methyl methacrylate with water
	Bulk polymerization Solution kinetics Styrene, methyl methacrylate	
	Solid-state polymerization Solution kinetics Trioxane	
In different phases	*Bulk polymerization with polymer precipitating* Precipitation kinetics Vinyl chloride, acrylonitrile	*Suspension polymerization with polymer precipitating* Precipitation kinetics Vinyl chloride, acrylonitrile
	Solution polymerization with polymer precipitating Precipitation kinetics Acrylic acid in hexane	*Emulsion polymerization* Emulsion kinetics Vinyl acetate, vinyl chloride, styrene, butadiene, acrylonitrile

In addition, the various types of polymerization systems offer practical advantages and disadvantages, different for each, from the standpoints of control of the reaction and the isolation and purity of the polymer. These considerations are summarized in Table 2-2.

A. *Homogeneous Polymerization*

Solution polymerization

Because the system remains homogeneous and of low viscosity throughout the reaction, solution polymerization provides the most straightforward conformation to the polymerization kinetics developed in Chap. 1C. The three steps of initiation, propagation, and termination interact with the effective concentration of each species essentially equal to its concentration in the bulk; that is, concentrations are maintained at the sites of reaction unimpeded by diffusion control. The steady-state approximation is readily fulfilled, and the over-all polymerization rate is in good agreement with that predicted by Eq. 1-12:

$$-d[M]/dt = k_p(f\, k_d\, [I]/k_t)^{1/2}\, [M] \qquad (2\text{-}1)$$

The concentrations of monomer and initiator at the start of the polymerization are known, and their subsequent concentrations at any later time can be calculated by integrating this equation. Since [I] varies with conversion, it must be eliminated, and Eq. 1-13 for the kinetic chain length allows this to be done conveniently. The result is

$$-d[M]/dt = k_p^{\,2}\, [M]^2/2\, k_t\, \nu \qquad (2\text{-}2)$$

whence

$$1/[M] - 1/[M]_0 = (k_p^{\,2}/2\, k_t\, \nu)(t - t_0) \qquad (2\text{-}3)$$

Of the three rate constants and the efficiency factor f appearing in these equations, the rate constant for the decomposition of initiator, k_d, is the only one which is readily measured, since it can be determined in a separate experiment with the initiator dissolved in a solvent but without the complicating presence of polymerization. The rate of initiator decomposition is dependent upon its concentration and the choice of solvent, however, because of self- and solvent-induced decomposition and viscosity effects. Decomposition rates of a given initiator in an inert solvent may be two to ten times faster, typically, than in the presence of monomer. Generally, the decomposition-rate data in an inert solvent can be used to compare the reactivity of one initiator to that of another, but many exceptions occur.

The determination of the other constants is even more difficult. The efficiency factor f is usually determined from knowledge of k_d, [I], and v_i, the latter being determined from knowledge of v_p and the kinetic chain length ν. There is no way to separate k_p from k_t,

TABLE 2-2. Comparison of polymerization systems

Type	Advantages	Disadvantages
Homogeneous		
Bulk (batch type)	Low impurity level	Difficult to control thermally
	Low residual initiator	Difficult to remove traces of monomer and initiator
	Simple equipment for making castings	Autoacceleration can broaden MWD[a] and lead to violent or explosive reactions
		Product may require reduction to useful particle size
Bulk (continuous)	Better thermal control than in bulk	Difficult to isolate polymer
	Narrower MWD than in bulk	Requires agitation, material transfer, monomer recycling
Solution	Easy thermal control	Difficult to remove solvent and other ingredients
	Easy mixing due to low viscosity	Cost of solvent recovery
	Improved initiator efficiency	
	Solution may be useful directly	
Heterogeneous		
Suspension	Low viscosity throughout	Reaction highly sensitive to agitation
	High purity product compared to emulsion systems	Particle size and surface characteristics difficult to control

Type	Advantages	Disadvantages
	Simple polymer isolation	Possible contamination by stabilizers
	Easy thermal control	Polymer may require washing, drying, and compaction
	Polymer may be of directly usable particle size	
Emulsion	Low viscosity throughout	Emulsifiers, surfactants and coagulants hard to remove
	Easy thermal control	High residual impurity level may degrade polymer properties
	Latex may be directly usable	
	Virtually 100% conversion may be achieved	Higher cost than suspension systems
	High MW[a] and narrow MWD possible, at high rates	Polymer may require washing, drying and compacting
	Small particle size product	
	Operable with soft or tacky polymers	
Polymer precipitating	Viscosity controllable by polymer precipitation	Removal of solvent or monomer difficult
	Polymer recovery easy	Difficult to obtain uniform polymer
	MW generally increases with conversion	
	Rate generally increases with conversion	

[a]MW = molecular weight; MWD = molecular-weight distribution.

however, using steady-state solution kinetics. The separate determination of these quantities is described in the *Textbook*, Chap. 9D.

The choice of solvent for a solution polymerization is, of course, conditioned by the possibility of chain transfer to that solvent. Usually, care must be taken to avoid this reaction, and the kinetic considerations outlined above assume that it is not significant.

Bulk polymerization

In bulk polymerization, unlike other methods, the charge to the reactor is essentially pure monomer or monomers, with only small amounts of initiator and other modifiers. We consider here the case in which the polymer is soluble in its monomer (and vice versa), so that the polymerizing system remains homogeneous throughout the reaction. This means, among other things, that the polymer is most likely noncrystalline. Bulk polymerization is usually used when some special property, such as high purity, optical clarity, or very high molecular weight, is required in the polymer.

Bulk polymerization in step-reaction systems is relatively simple and is often the method of choice. These reactions are only mildly exothermic, so that heat removal is not a problem; in fact, heat must be applied to raise the mass to reaction temperature and keep it there. The viscosity of the mixture is relatively low until high conversion is reached (see Table 1-1) so that mixing, heat transfer, and release of volatile by-products is easy.

In contrast, bulk radical-chain polymerization is much more difficult to control. The process is highly exothermic, with about 15-20 kcal/mole being released in the conversion of the double bond of the monomer to a single bond. For a monomer of about 100 in molecular weight (styrene, methyl methacrylate, vinyl acetate) and a reasonable specific heat, one can calculate that with no heat transfer to the surroundings, a temperature rise well in excess of 100°C could be expected during the polymerization. If the reaction rate were reasonable, heat transfer would be no problem, but with the usual thermally decomposed initiators, the rate of initiation and therefore the overall polymerization rate is highly temperature dependent. Additionally, since high polymer is formed immediately, the reaction mass can become quite viscous early in the reaction, increasing the difficulty of mixing and heat transfer. "Hot spots" can form, and unless precautions are taken, runaway or explosive reactions can result.

Autoacceleration When certain monomers such as methyl acrylate and methyl methacrylate are polymerized in bulk, a significant rate acceleration is noted early in the reaction. The polymer formed during the period of high rate has substantially higher molecular weight than that formed previous to the onset of the effect. The explanation lies in the viscous nature of the polymerizing mixture. As the viscosity increases, the mobility of the polymer radicals decreases, and with it the probability of their mutual termination. The average lifetime of the growing chain increases, and with it both the rate of polymerization and the molecular weight of the polymer. The increased rate produces a larger heat evolution, and the resulting temperature rise results in a still higher rate, leading to the term autoacceleration. The phenomenon is often referred to as the Trommsdorff effect (Trommsdorff 1948).

GENERAL REFERENCES

Textbook, Chaps. 10D, 12A; Schildknecht 1956; Ringsdorf 1965; Lenz 1967; Odian 1970; Braun 1971, Chap. 2; Williams 1971, Chap. 4.

B. *Heterogeneous Polymerization*

Suspension polymerization

Suspension polymerization, sometimes called pearl or bead polymerization, can be considered to be bulk polymerization carried out with the monomer dispersed in small droplets, typically 10-1000 micrometers in diameter. The droplets are suspended in a medium in which the monomer is essentially insoluble (usually water) by means of carefully controlled agitation and with the aid of a dispersant or protective colloid. This is usually a water-soluble polymer such as poly(vinyl alcohol) or methyl cellulose in the aqueous phase, but an oil-soluble dispersant may also be added to the monomer phase. The dispersant prevents the aggregation of partially polymerized monomer particles at the stage when they are quite sticky. The size of the particles is determined by the rate of agitation as well as the amount and nature of the dispersant used.

Monomer-soluble initiators are used in suspension polymerization. The basic polymerization mixture consists of monomer, initiator, dispersant, and suspending medium. The low viscosity of the mixture and the diluting effect of the suspending medium allow good heat removal and consequent temperature control. Often, the product can be isolated merely by filtration, but washing and drying are usually required.

Provided the particles are not too small and initiation is restricted to the monomer phase, suspension polymerization is well described by the solution kinetics outlined previously. If the particle size becomes too small, so that there are only a few free radicals in each polymerizing particle, the kinetics is modified to become more like that of emulsion polymerization.

Emulsion polymerization

The critical difference in the polymerizing mixture of an emulsion system compared to all others is the presence of soap. When a soap or detergent is added to water in increasing amounts, a limit of solubility for single soap molecules is soon reached. Above this limit, additional soap forms loose aggregates called *micelles* containing 50-100 soap molecules, in some arrangement in which their hydrocarbon "tails" are directed inward from the surface. This forms a hydrocarbon region whose surface consists of the hydrophilic ends of the soap molecules.

When a monomer of limited (typically 0.01-1%) water solubility is added to such a soap solution, and the system is agitated, some of the monomer dissolves in the soap micelles, swelling them to perhaps twice their original size. The rest of the monomer (beside the small amount soluble in the water) is emulsified, that is, dispersed into relatively stable droplets a few micrometers in diameter, stabilized by the soap. The system then contains three phases: water containing small amounts of dissolved soap and monomer; monomer droplets stabilized by soap; and much smaller soap micelles saturated with monomer.

If a free-radical source is added to the aqueous phase, for example the water-soluble initiator potassium persulfate, there are three possible sites for polymerization. The radicals formed can react with the dissolved monomer in the aqueous phase, or can diffuse into the monomer droplets or the micelles. Although there is much more monomer in the droplets, there are many more micelles, typically around 10^{18} per cm^3, with a surface area of 50-100 m^2. This is 10-1000 times greater than the surface area of the monomer droplets. The diffusion rate of the radicals, which is directly proportional to the surface area, is far greater to the micelles, and thus there is virtually no initiation in the monomer droplets. Calculation shows that there is very little initiation in the aqueous phase, as a result of the low monomer concentration there; if an initiator radical should add monomer in the aqueous phase, the short chain resulting is likely to diffuse quickly into the more favorable hydrophobic surrounding of a micelle.

Thus propagation proceeds in micelles as radicals diffuse in. The supply of monomer in these micelles is replenished by diffusion, through the aqueous phase, from the monomer droplets which act as reservoirs. Only about one micelle in a thousand becomes initiated in this way; hereafter we refer to these as monomer-polymer particles. As these particles grow, they increase in surface area and require more soap to saturate their surface. This demand is satisfied at the expense of the uninitiated micelles, which disappear at 2-20% conversion, depending on the details of the system.

Propagation continues in a monomer-polymer particle until a second radical diffuses into it, or (less likely) the radical already there diffuses out. One can calculate that with two radicals present, the probability is high that termination will take place immediately. Later, of course, another radical can enter the same particle and initiate a new chain; as most of the micelles disappear, this is highly probable.

Effectively, the monomer-polymer particles isolate the sites of chain growth from one another. This prevents termination at normal rates, and allows many more sites of polymerization to exist simultaneously than would otherwise be found. Both rate of polymerization and molecular weight are high, although these are usually inversely related, as can be seen from Eq. 2-2. A new kinetics is needed to describe emulsion polymerization.

In this kinetics (Smith 1948*a*, *b*), the rate of polymerization is expressed in terms of the number of monomer-polymer particles per liter, N. Only half of these are active at any moment, on the average, the rest being terminated. Thus, the *effective* concentration of radicals at the polymerization site is $N/2N_0$, where N_0 is Avogadro's number. From Eq. 1-9,

$$v_p = k_p\,[\mathrm{M}]\,N/2N_0 \tag{2-4}$$

It can also be shown that the kinetic chain length is

$$\nu = k_p \, [M] \, N/2v_i N_0 \qquad (2\text{-}5)$$

and that N is proportional to the 0.6 power of the soap or emulsifier concentration [E] and the 0.4 power of [I]. Thus, both rate and molecular weight increase with increasing amounts of soap.

Polymerization with polymer precipitating

When the polymer is insoluble in its own monomer, a polymerization which begins with a homogeneous solution of monomer and initiator becomes heterogeneous as the polymer precipitates as a second phase. This is the case in the bulk or suspension polymerization of vinyl chloride, vinylidene chloride, and acrylonitrile, among others. The kinetics of these polymerizations is different from that of ordinary bulk polymerization. In the polymer phase, diffusion rates are so low that termination is greatly slowed down. The number of polymerization sites increases, and with it both rate and molecular weight, until the concentration of monomer drops off at high conversion.

The kinetics is best considered as the sum of independent reactions in the two phases. In the monomer phase, normal solution kinetics is obeyed. In the polymer phase, which is swollen by monomer to the limit of solubility of monomer in the polymer, the exact kinetics depends upon such factors as surface area and the location of the trapped radicals. In vinyl-chloride polymerization, these appear to be limited to the surface regions of the polymer particles, at least at higher conversions. Termination is postulated to occur primarily in the monomer phase, as a result of transfer to monomer within the polymer particle and diffusion of the small monomer radical to the liquid phase. In acrylonitrile polymerization, however, chain transfer to monomer is slow, and many radicals remain permanently trapped in the polymer until at some later time the diffusion rates are increased, for example by heating the polymer.

GENERAL REFERENCES

Textbook, Chap. 12B; Schildknecht 1956; Lenz 1967; Odian 1970; Braun 1971, Chap. 2; Williams 1971, Chap 4; and specifically:
Suspension polymerization: Trommsdorff 1956; Farber 1970.
Emulsion polymerization: Duck 1966; Vanderhoff 1969.
Polymerization with polymer precipitating: Mickley 1962; Talamini 1966; Jenkins 1967; Crosato-Arnaldi 1968; Farber 1968; Meeks 1969; Ugelstad 1971.

C. *Interfacial Polymerization*

Interfacial polymerization is the term used to describe step-reaction polymerization carried out at the interface between two immiscible liquid phases. Each phase contains one monomer; for example, an organic phase may contain an acid chloride and an aqueous phase a diamine or glycol. Since HCl is given off, the aqueous phase usually also contains an acid acceptor such as NaOH or Na_2CO_3. In Exp. 2, poly(hexamethylene sebacamide), 610 nylon, is prepared in this way.

Interfacial polymerization can be carried out in several ways. In one (the "nylon rope trick," Morgan 1959), the two phases are carefully layered over one another, and a film or filament of polymer can be withdrawn continuously from the interface. The polymerization rate, controlled by the rates of diffusion of the reactants to the interface, is quite rapid, and in contrast to the usual step polymerization, high-molecular-weight polymer is produced at room temperature. If the two phases are interdispersed by violent agitation, the interfacial area is greatly increased, and with proper selection of reagents and concentrations, the entire charge can be polymerized in a few minutes.

Interfacial polymerization is generally less sensitive to the requirements of monomer purity and stoichiometry than is convenional stepwise polymerization. The high rates, high molecular weights, and low polymerization temperatures characteristic of this method are additional advantages. For example, it is especially useful for preparing condensation polymers that cannot be made conventionally because they are unstable at the elevated temperatures required. The increased cost of solvent removal and recovery, and of purification of the polymer from acid reaction products, have deterred the commercial application of the technique.

GENERAL REFERENCES

Textbook, Chap. 8B; Morgan 1965.

BIBLIOGRAPHY

Braun 1971. Dietrich Braun, Harald Cherdon, and Werner Kern, *Techniques of Polymer Syntheses and Characterization*, Wiley-Interscience, New York, 1971.

Crosato-Arnaldi 1968. A. Crosato-Arnaldi, P. Gasparini, and G. Talamini, "The Bulk and Suspension Polymerization of Vinyl Chloride," *Makromol. Chem. 117*, 140-152 (1968).

Duck 1966. Edward W. Duck, "Emulsion Polymerization," pp. 801-859 in Herman F. Mark, Norman G. Gaylord, and Norbert M. Bikales, eds., *Encyclopedia of Polymer Science and Technology*, Vol. 5, Interscience Div., John Wiley and Sons, New York, 1966.

Farber 1968. Elliott Farber and Marvin Koral, "Suspension Polymerization Kinetics of Vinyl Chloride," *Polymer Eng. Sci. 8*, 11-18 (1968).

Farber 1970. Elliott Farber, "Suspension Polymerization," pp. 552-571 in Herman F. Mark, Norman G. Gaylord, and Norbert M. Bikales, eds., *Encyclopedia of Polymer Science and Technology*, Vol. 13, Interscience Div., John Wiley and Sons, New York, 1970.

Jenkins 1967. A. D. Jenkins, "Occlusion Phenomena in the Polymerization of Acrylonitrile and Other Monomers," Chap. 6 in George E. Ham, ed., *Vinyl Polymerization*, Part I, Marcel Dekker, New York, 1967.

Lenz 1967. Robert W. Lenz, *Organic Chemistry of Synthetic High Polymers*, Interscience Div., John Wiley and Sons, New York, 1967.

Meeks 1969. M. R. Meeks, "An Analog Computer Study of Polymerization Rates in Vinyl Chloride Suspensions," *Polymer Eng. Sci. 9,* 141-151 (1969).

Mickley 1961. Harold S. Mickley, Alan S. Michaels, and Albert L. Moore, "Kinetics of Precipitation Polymerization of Vinyl Chloride," *J. Polymer Sci. 60,* 121-140 (1962).

Morgan 1959. Paul W. Morgan and Stephanie L. Kwolek, "The Nylon Rope Trick," *J. Chem. Educ. 36,* 182-184 (1959).

Morgan 1965. Paul W. Morgan, *Condensation Polymers: by Interfacial and Solution Methods,* Interscience Div., John Wiley and Sons, New York, 1965.

Odian 1970. George Odian, *Principles of Polymerization,* McGraw-Hill Book Co., New York, 1970.

Ringsdorf 1965. H. Ringsdorf, "Bulk Polymerization," pp. 642-666 in Herman F. Mark, Norman G. Gaylord, and Norbert M. Bikales, eds., *Encyclopedia of Polymer Science and Technology,* Vol. 2, Interscience Div., John Wiley and Sons, New York, 1965.

Schildknecht 1956. Calvin E. Schildknecht, ed., *Polymer Processes,* Interscience Publishers, New York, 1956.

Smith 1948a. Wendell V. Smith, "The Kinetics of Styrene Emulsion Polymerization," *J. Am. Chem. Soc. 70,* 3695-3702 (1948).

Smith 1948b. Wendell V. Smith and Roswell H. Ewart, "Kinetics of Emulsion Polymerization," *J. Chem. Phys. 16,* 592-599 (1948).

Talamini 1966. Gianpietro Talamini, "The Heterogeneous Bulk Polymerization of Vinylchloride," *J. Polymer Sci. A-2 4,* 535-537 (1966).

Trommsdorff 1948. Ernst Trommsdorff, Herbert Köhle, and Paul Lagally, "On the Polymerization of Methyl Methacrylate" (in German), *Makromol. Chem. 1,* 169-198 (1948).

Trommsdorff 1956. Ernst Trommsdorff and C. E. Schildknecht, "Polymerization in Suspension," Chap III in Calvin E. Schildknecht, ed., *Polymer Processes,* Interscience Publishers, New York, 1956.

Ugelstad 1971. J. Ugelstad, H. Lervik, B. Gardinovacki and E. Sud, "Radical Polymerization of Vinyl Chloride: Kinetics and Mechanism of Bulk and Emulsion Polymerization," *Pure and Applied Chem. 25,* 121-152 (1971).

Vanderhoff 1969. John W. Vanderhoff, "Mechanism of Emulsion Polymerization," Chap. 1 in George E. Ham, ed., *Vinyl Polymerization,* Vol. 1, Part II, Marcel Dekker, New York, 1969.

Williams 1971. David J. Williams, *Polymer Science and Engineering,* Prentice-Hall, Englewood Cliffs, New Jersey, 1971.

3

Materials and Their Purification

The purpose of this chapter is to describe, by class, the various materials used in typical polymer syntheses and their roles in the reactions. Particular emphasis is placed on the requirements for their purity and the means by which these requirements are met, since chemical purity and freedom from interfering contaminants is especially important if one is to produce polymer of high molecular weight in good yield.

A. *Monomers, Diluents, and Solvents*

Monomers

The purity of commercially available monomers is usually not high enough for their use, without treatment, in well-controlled polymerization reactions. Even 10^{-2} - 10^{-4}% impurity can have a significant effect on polymerization kinetics. Anionic polymerizations are especially susceptible to impurities at even lower concentrations than these. Consequently, monomers must be further purified and their purity checked before they are used. Purified monomers must be stored with care, usually in sealed tubes under an inert atmosphere, and kept at low temperatures in the dark. Even the closures for flasks or containers should be selected for their inertness, aluminum foil and polytetrafluoroethylene being good examples of inert materials.

Common impurities in monomers may originate during their production (for example, ethyl benzene or divinyl benzene in styrene), as the result of oxidation (peroxides in dienes, benzaldehyde in styrene), from handling and storage (material from which the containers are made), or from partial polymerization to oligomers.

In stepwise polymerization, the presence of monofunctional reactants, which can use up functional groups and destroy stoichiometric equivalence, must be avoided. In addition polymerization, it is especially important to avoid water and compounds containing hydroxyl or carboxyl groups; these can be tolerated only in amounts less than about

0.05% by weight in most cases.

Purification Although each monomer and polymerization method has its own special requirements, some useful purification procedures include simple distillation or distillation under inert gas (Perry 1965; Tipson 1965; Diamond 1966; Krell 1966), recrystallization (Tipson 1957), and chromatography. A typical distillation setup (Fig. 3-1) should have a column of 20-30 plates, a cold trap to protect the vacuum pump, a manostat to maintain the pressure desired (usually 100 torr or less), and a means for introducing inert gas through a thin capillary into the boiler; this prevents bumping and sweeps out oxygen.

Some specific purification routines include the following: Water is removed from hydrocarbon monomers by treating them with activated alumina, metal hydrides or oxides, or alkali metals such as sodium. Tables 3-1 to 3-4 give useful properties of these and other commonly used drying agents. Olefin impurities are eliminated by washing the monomer with 36*N* sulfuric acid. Aromatic impurities are removed by use of a nitrating mixture. To eliminate impurities containing carbonyl or hydroxyl groups, the monomer can be passed through columns containing

Fig. 3-1. A typical distillation apparatus (courtesy Kontes/Martin).

TABLE 3-1. Suitable drying agents for some common classes of organic liquids (Broughton 1966)

Drying Agent	Suitable for Drying	Unsuitable for Drying
Phosphorus pentoxide	Alkyl halides, hydrocarbons, halogenated hydrocarbons, CS_2	Bases, ketones, aldehydes or other materials where polymerization may be caused
Sulfuric acid	Alkyl halides, saturated hydrocarbons, halogenated hydrocarbons	Bases, ketones, alcohols, aldehydes, phenols, etc.
Calcium chloride	Ethers, esters, alkyl halides, aryl halides, etc.	Alcohols, amines, phenols, aldehydes, amides, fatty acids
Potassium hydroxide	Bases	Ketones, aldehydes, esters, acids
Potassium carbonate	Bases, some halides, ketones	Fatty acids, esters
Sodium sulfate	Most materials	
Magnesium sulfate	Most materials	
Anhydrous copper sulfate	Ethers, alcohols, etc.	Amines
Sodium	Ethers, saturated hydrocarbons	Alcohols, amines, esters
Calcium sulfate	Most materials	

activated alumina, molecular sieves, or silica gel. For olefin monomers, the recommended purification method is distillation from reducing agents such as metal hydrides or aluminum alkyls (Borchert 1969). Monomers for radical polymerization usually contain inhibitors when received. These can be removed by washing the monomer with dilute acid or base, depending on the type of inhibitor. For example, phenolic inhibitors such as hydroquinone are usually washed out with dilute NaOH. Drying prior to distillation is done with the use of anhydrous salts (Table 3-2).

TABLE 3-2. Vapor pressures of hydrates of some common drying agents at 25°C (Broughton 1966)

Hydrate	Vapor Pressure, torr	Hydrate	Vapor Pressure, torr
$BaO \cdot H_2O$	10^{-16}	$KOH \cdot H_2O$	1.5
$CaSO_4 \cdot 1/4H_2O$	0.004	$ZnCl_2 \cdot 1^1/2H_2O$	2.3
$CaCl_2 \cdot H_2O$	0.04	$CuSO_4 \cdot H_2O$	0.8
$NaOH \cdot H_2O$	0.7	$MgSO_4 \cdot H_2O$	1
$CaO \cdot H_2O$	0.8	H_2SO_4, 95%	0.001
$K_2CO_3 \cdot 1^1/2H_2O$	1.1	$Na_2SO_4 \cdot 10H_2O$	22.3

TABLE 3-3. Suitable drying agents for gases (Broughton 1966)

Drying Agent	Gases
CaO	NH_3, amines
$CaCl_2$	H_2, HCl, CO_2, CO, SO_2, N_2, CH_4, O_2, paraffins, ethers, olefins, alkyl chlorides
P_2O_5	H_2, O_2, CO_2, CO, SO_2, N_2, CH_4, C_2H_4, paraffins
H_2SO_4	H_2, N_2, CO_2, Cl_2, CO, CH_4, paraffins
Fused KOH	NH_3, amines

Finally, monomer of high purity can often be obtained by distilling monomer which has been partially polymerized, say to about 10% conversion; vinyl acetate is a typical example. The majority of the unwanted impurities is used up in reactions leading to polymer, and is not carried over in the distillation.

Diluents and Solvents

The criteria for the purity of monomers also apply to any diluents and solvents used in a polymerization. These materials must be purified by distillation, crystallization, or the other methods described for monomers. Diluents for anionic polymerization are distilled from an excess of the initiator; for example, tetrahydrofuran from sodium naphthalene or benzene from butyl lithium (see also Chap. 4A, Fig. 4-3). Peroxides are removed from these solvents by treatment with sodium stearate.

TABLE 3-4. Equilibrium water-vapor content of gases dried over common reagents at 25°C (Broughton 1966)

Reagent	Mg/liter gas	% by Volume
P_2O_5	2×10^{-5}	--
$Mg(ClO_4)_2$(anhyd.)	5×10^{-4}	--
$Mg(ClO_4)_2 \cdot 3H_2O$(30% water)	0.002	0.0002
MgO	0.008	0.0007
BaO	0.00065	--
$Ba(ClO_4)_2$(anhyd.)	0.82	0.094
H_2SO_4 (95%)	0.003	0.0003
H_2SO_4 (80%)	0.20	0.021
Alumina	0.003	0.0003
Silica gel (dry)	0.003	0.0003
CaO	0.003	0.0003
$CaCl_2$ (granulated)	0.14-0.245	0.0149-0.0264
$CaCl_2$ (fused)	0.36	0.0395
$CaSO_4$ (anhyd.)	0.005	0.0005
$CaBr_2$	0.20	0.021
$ZnBr_2$	1.1	0.124
$ZnCl_2$ (sticks)	0.8	0.092
$CuSO_4$	1.4	0.165
KOH (fused)	0.014	0.0015
NaOH (fused)	0.16	0.0170

Further examples of purification of monomers, solvents, and diluents, in which the final stages are carried out in the polymerization apparatus prior to polymerization, are given in Chap. 4A.

GENERAL REFERENCES

Broughton 1956; Marsden 1963; Leonard 1970; Riddick 1970; Braun 1971.

B. *Initiators and Catalysts*

In this book as well as the *Textbook* (for example, on p. 311), we distinguish between initiators and catalysts as agents helping to start, maintain, and regulate polymerization reactions. The word *initiator* is used to describe a substance which, usually by its decomposition, starts a polymerizing chain. A fragment of the initiator is usually chemically bonded to the chain. Free-radical initiators furnish the common example. We speak of a *catalyst* when the chemical substance exerts its influence throughout the growth of the chain, as is common in ionic polymerization. Only the solid surface-active catalysts used in coordination polymerization, however, fit the classic definition of a catalyst as a substance which does not take part in the reaction.

Free-radical initiators

Free-radical initiators are low-molecular-weight organic or inorganic compounds which readily decompose to fragments containing unpaired electrons. The fragments react with the monomer's double bond as described in Chap. 1C. Among the common types of free-radical initiators are the following:

Peroxide initiators The best known representatives of this group are benzoyl peroxide (I), dicumyl peroxide (II), and the inorganic potassium persulfate $K_2S_2O_8$ (III).

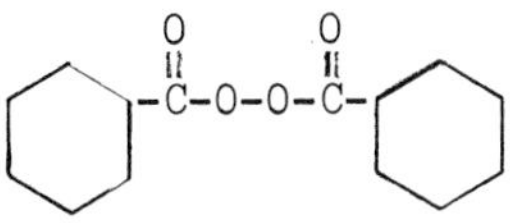

```
      O     O                 CH3  CH3             O     O
      ‖     ‖                  |    |              ‖     ‖
 ⬡ -C-O-O-C- ⬡           ⬡ -C-O-O-C- ⬡        KO-S-O-O-S-OK
                               |    |              ‖     ‖
                              CH3  CH3             O     O

        I                        II                   III
```

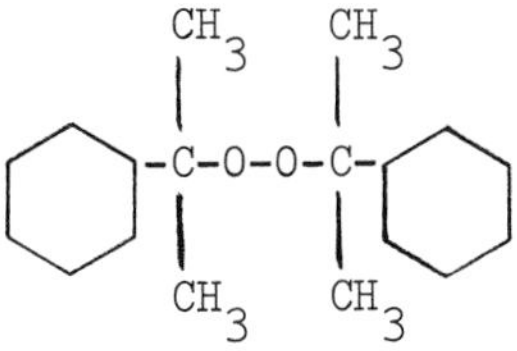

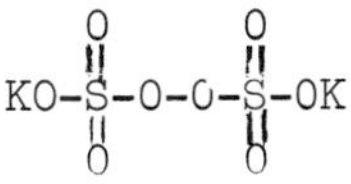

These materials are inexpensive and relatively easy to handle (BUT OBSERVE CAUTION: ALL INITIATORS ARE DANGEROUS, POTENTIALLY EXPLOSIVE COMPOUNDS), but their strong oxidizing nature may lead to unwanted side reactions.

Azo initiators In this group, 2,2'-azobisisobutyronitrile (AIBN) (IV) is the most widely used, with 2,2'-azobis-2-methylbutyronitrile (V) next.

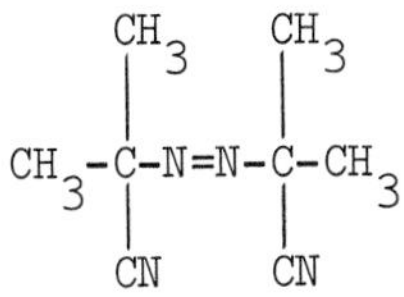

```
      CH3   CH3                   C2H5  C2H5
       |     |                     |     |
  CH3-C-N=N-C-CH3             CH3-C-N=N-C-CH3
       |     |                     |     |
       CN    CN                    CN    CN

         IV                          V
```

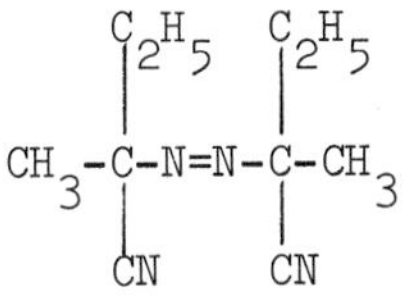

These initiators decompose cleanly into fragments with the elimination of nitrogen:

$$R\text{-}N{=}N\text{-}R' \rightarrow R\cdot + R'\cdot + N_2$$

Redox initiators The initiators mentioned so far decompose into free-radical fragments by reactions in which chemical bonds are broken, usually by the application of heat. Free radicals can also appear as transient intermediates in an oxidation-reduction reaction. Systems of two substances, a reducing and an oxidizing agent, which react in this way are called redox initiators (*Textbook*, p. 363). A common example is the reaction

$$2\ Fe^{++} + H_2O_2 \rightarrow 2\ Fe^{+++} + 2\ OH^-$$

which proceeds in part as follows:

$$Fe^{++} + H_2O_2 \rightarrow Fe^{+++} + OH^- + HO\cdot$$

In the presence of monomer, the initiation reaction uses the hydroxyl radicals formed in this way.

Thermal and radiation initiation Thermal decomposition of pure monomers to radicals is known, styrene being the most common example. The exact mechanism is not understood, and the susceptibility of the system to the presence of trace impurities, similarly decomposed, makes this technique unsuitable for controlled polymerization.

The use of high-energy radiation is, in contrast, one of the most readily controlled forms of radical initiation (*Textbook*, pp. 287-8), since the generation of radicals can be started or stopped instantaneously by controlling the radiation, in contrast to the finite time necessary to heat or cool a thermally controlled system. Although a wide variety of energetic sources can be used, including x-rays and many energetic particles (*Textbook*, pp. 372-3), a convenient laboratory source is ultraviolet radiation. "Black-light" fluorescent lamps are available emitting radiation at either of two wavelength regions of the mercury arc, approximately 2540 A and 3660 A. The 3660 A photons are energetic enough to decompose some thermal initiators such as AIBN, while the 2540 A radiation can decompose monomers by attack at the double bond.

Ionic Catalysts

The addition of an ionic group to a vinyl monomer can initiate polymerization, as described in Chap. 1D. The processes of ionic initiation and growth are selective and often involve the solvent or another substance (cocatalyst) present in trace amounts. The components of such a polymerization must be selected properly and have high purity.

Anionic catalysts Substances such as alkali metal amides, lithium alkyls, Grignard reagents, and alkali metal ketyls form carbanions with monomers as a result of group addition, or radical ions as a result of electron transfer. Typical examples involving group addition are potassium amide, $K^+ NH_2^-$, in liquid ammonia (*Textbook*, p. 318); butyl lithium, $C_4H_9^- Li^+$, in benzene; and phenyl magnesium bromide in tetrahydrofuran.

Alkali metals and their complexes with aromatic hydrocarbons act as anionic catalysts through electron transfer. Examples are the polymerization of styrene by sodium metal or by sodium naphthalide in tetrahydrofuran. In either case, the same ion pair is formed; the reaction with sodium naphthalide is

$$C_{10}H_8^- Na^+ + C_6H_5CH{=}CH_2 \rightarrow C_6H_5CH{=}CH_2^- Na^+ + C_{10}H_8$$

Cationic catalysts The formation of a carbonium ion by addition of a cation fragment to a monomer is in analogy to the anionic case, except that in the simplest case, a proton rather than an electron is transferred. Cationic catalysts include common strong acids and Friedel-Crafts catalysts, the latter usually requiring the presence of a cocatalyst. An example of initiation by a strong acid is the system perchloric acid, styrene, and ethylene dichloride as solvent. Complexes of a Friedel-Crafts catalyst and a cocatalyst which can initiate styrene are stannic chloride plus *t*-butyl chloride, $SnCl_5^-$ $(CH_3)_3C^+$, and boron trifluoride plus water (or any proton donor), BF_3OH^- H^+. Boron trifluoride is commonly used as the etherate, which is thought to react with water to give the above complex plus diethyl ether. Other Friedel-Crafts catalysts used in polymerization are $AlCl_3$, $TiCl_4$, and PF_5.

Purity of ionic catalysts The anionic catalysts, often containing metal-carbon bonds, are readily destroyed by water, CO_2, oxygen, alcohols, amines, and acids. The Friedel-Crafts cationic catalysts, however, require trace amounts of a cocatalyst such as water, HCl, a chlorinated hydrocarbon, or a proton donor. Larger amounts of these same substances inhibit the reactions.

Ring-opening catalysts

Ring opening (Chap. 1E) can be initiated by such groups as OH, NH_2, COOH, or sometimes water. An ionic complex which favors the ring opening may be involved. There are many common examples, including the use of water, hexamethylene diamine, ω-aminocaproic acid, or sebacic acid to polymerize caprolactam. Monomers of this type can also be initiated by anionic or cationic catalysts, such as sodium metal, sodium methoxide, and many others.

Catalysts for condensation polymerization

Although condensation polymerization can proceed independently through self-catalysis (*Textbook*, p. 265), strong acids, organometallic catalysts, or metal oxides are often used. Examples are the use of Sb_2O_3 in the polymerization of glycolic acid and PbO in the formation of linear aliphatic polyesters.

Purification and storage of initiators

Initiators for radical polymerization, such as benzoyl peroxide and AIBN, are usually purified by recrystallization from anhydrous methanol or some other convenient solvent (CAUTION IS REQUIRED). Organometallic catalysts and such species as $TiCl_4$ can be distilled under low pressure. Catalysts and initiators should be stored at low temperatures and in the dark.

GENERAL REFERENCES

Mark 1965; Zand 1965; O'Driscoll 1969.

C. Other Ingredients

In this section we describe all the variety of minor ingredients in a polymerization recipe, some essential and others nonessential. For convenience, we divide them into active ingredients, which take part in the chemical reactions of polymerization, and additives, which do not. Many additives are introduced only in a compounding step after the completion of the polymerization; these we describe only by reference.

Active Ingredients

Chain-transfer agents or modifiers A chain-transfer agent is a substance which terminates a growing chain by hydrogen transfer and simultaneously produces a new radical capable of chain growth. (Ionic chain transfer is also known, but for simplicity we refer only to the case of radical polymerization.) Chain-transfer agents are primarily used to control molecular weight. Mercaptans are among the transfer agents commonly used. All the other components of the polymerizing mixture can undergo transfer, including solvent, monomer, initiator, and even polymer, but their contributions are usually small.

The term *modifier* is sometimes used interchangeably with chain-transfer agent, especially when the purpose of the agent is to prevent chain-branching and crosslinking reactions, as in the production of styrene-butadiene rubber by emulsion polymerization.

Inhibitors, retarders, and terminators This group of substances has the ability to destroy free radicals (or ions) and consequently to slow down or stop the polymerization process (*Textbook*, pp. 285-6, 296). They differ in the degree, rather than the kind, of their action. *Inhibitors* prevent the polymerization from taking place at all by destroying radicals so quickly that polymerization cannot start until all the inhibitor is used up. Typical inhibitors include quinone, hydroquinone, and chloranil (tetrachloroquinone), and such inorganic compounds as cuprous chloride, which is often used in the distillation of vinyl monomers.

In contrast to inhibitors, *retarders* merely slow down the polymerization reaction, since the rate of their reactions with radicals is too slow to use up all that are generated under the polymerization conditions.

A strong inhibitor, added late in the polymerization to stop the reaction sharply at a desired conversion, is called a *terminator* or, colloquially, a shortstop. Many of the substances cited in Sec. B as inhibiting ionic polymerizations are effective terminators.

Additives

We include here, as additives, substances which are added to polymerization mixtures to facilitate the reaction or the processing of the resulting polymers.

The course of the polymerization reaction often depends on the acidity of the medium; this is especially true in emulsion polymerization. *Buffers* are therefore used to control pH. For example, styrene polymerizes best in a weakly alkaline medium. In emulsion polymerization, adjustment of pH allows control of the size of the emulsion particles and the stability and viscosity of the emulsion. Phosphates, carbonates, or acetates, added at about the 2% level, are often used.

To obtain and preserve the suspension or emulsion of monomer during these types of polymerization, *surface-active agents* (soaps, emulsifiers) are added in concentrations of 0.1-0.2%. In addition to soaps, protective colloids may be added to stabilize the final latex product. Here, and also in suspension polymerization, natural or synthetic water-soluble polymers are used, such as gelatin, poly(vinyl alcohol), or methyl cellulose.

To aid in the coagulation of an emulsion in order to isolate the polymer from it, for example in Exp. 4, use is made of *flocculants* or *coagulants*. These substances cause the small particles of polymer to clump together, forming a *floc* which can be removed by filtration. The process is one of simultaneous adsorption of the flocculant particles or molecules onto many emulsion particles. The most widely used flocculant is alum, a common name for aluminum sulfate or potassium aluminum sulfate, $K_2SO_4 \cdot Al_2(SO_4)_3 \cdot 24H_2O$. Ferric salts and some polyelectrolytes, such as polyacrylamide, can be used as flocculants.

Other additives added late in the polymerization or during the isolation of the polymer, for reasons obvious from their names, are foaming agents, antioxidants, and thermal stabilizers. For a description of still other additives such as fillers, plasticizers, and colorants, see the *Textbook*, pp. 499-503.

GENERAL REFERENCES

Palit 1965; Dixon 1967; Goldfinger 1967; Gellner 1970.

D. *Influence of Impurities*

Air and water

Although the influence of air or water depends upon the type of polymerization, in general they both should be removed from the system to insure good control. Oxygen has an adverse effect on almost all types of polymerization, although its adventitious behavior as an initiator is thought to have played a role in some early uncontrolled commercial polymerizations. At high reaction temperatures, it can cause oxidation of monomer and polymer and lead to degradation, branching, and crosslinking. In radical polymerization, it can inhibit or accelerate the reaction, depending on the amount present and the types of system. In pure monomers, it leads to the formation of peroxides which can cause inhibition or retardation. In ionic polymerizations, oxygen (and water and CO_2 also) readily destroy the carbon-metal bond of the catalyst and act as inhibitors.

While traces of water can act as a catalyst or cocatalyst, as described in Sec. B, larger amounts have an inhibiting influence. Excess water, for example, decomposes boron trifluoride etherate to boric and hydrofluoric acids plus ether, in contrast to the action of trace amounts as a cocatalyst.

Inert gases

Inert gases are used to provide an inert atmosphere during polymerization, or to remove volatile by-products of the reaction (for example, water) produced in condensation reactions.

Nitrogen is usually satisfactory for these uses if it contains less than 10^{-2}% residual oxygen; some grades containing about 10^{-4}% oxygen are available. Argon is also satisfactory; it usually contains less oxygen than does nitrogen, and its higher density makes it suitable for blanketing reactions. It is, however, more expensive than nitrogen.

If necessary, inert gases can be purified. Oxygen can be removed by passing the gas over copper heated to 900°C, or by using other contact catalysts or metal-organic compounds, such as aluminum alkyls in a boiling solvent. The gases can be dried by passing them over molecular sieves, by the use of conventional drying agents (Table 3-3), or by passing them through a cold trap (Fig. 3-2) kept below -80°C.

Determination of purity

Almost all the methods of analytical and physical chemistry can in principle be used to determine the purity of the materials discussed in this chapter. Among the tests which are the simplest and easiest to perform, and are therefore used most frequently, are infrared, ultraviolet, atomic absorption, and nuclear magnetic resonance spectroscopy, mass spectrometry, polarography, gas and liquid chromatography, and differential refractometry. Because impurity concentrations are usually quite small, the most powerful methods are those with high sensitivity in the detection of a specific impurity. The spectroscopic and chromatographic methods are examples. Other techniques for determining impurities in low concentrations are freezing-curve methods (to 10^{-3} mole %), melting points (to 10^{-4}), and differential boiling points (to 2×10^{-3}). The most universally applicable method for water determination is the Karl Fischer technique, though cloud-point determinations are sometimes applicable. Solvent purity is readily checked by gas chromatography or infrared spectroscopy. For some specific determinations, titration can be used, as in functional-group determination (to about ± 2%) in the reactants for condensation polymerization.

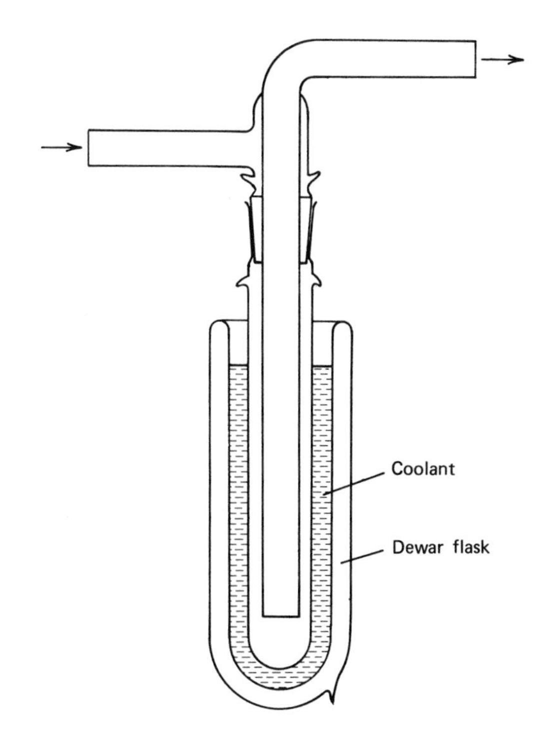

Fig. 3-2. A cold trap.

GENERAL REFERENCES

Kolthoff 1959-1972 and continuing.

BIBLIOGRAPHY

Borchert 1969. A. E. Borchert, "Polymerization Procedures, Laboratory," pp. 304-311 in Herman F. Mark, Norman G. Gaylord, and Norbert M. Bikales, eds., *Encyclopedia of Polymer Science and Technology*, Vol. 11, Interscience Div., John Wiley and Sons, New York, 1969.

Braun 1971. Dietrich Braun, Harald Cherdon, and Werner Kern, *Techniques of Polymer Syntheses and Characterization*, Interscience-Wiley, New York, 1971.

Broughton 1956. Geoffrey Broughton, "Solvent Removal, Evaporation and Drying," pp. 787-839 in Arnold Weissberger, ed., *Separation and Purification*, Vol. III of A. Weissberger, ed., *Technique of Organic Chemistry*, 2nd ed., Part I, Interscience Div., John Wiley and Sons, New York, 1956.

Diamond 1966. P. S. Diamond and R. F. Denman, *Laboratory Techniques in Chemistry and Biochemistry*, D. Van Nostrand Co., Princeton, New Jersey, 1966.

Dixon 1967. J. K. Dixon, "Flocculation," pp. 64-78 in Herman F. Mark, Norman G. Gaylord, and Norbert M. Bikales, eds., *Encyclopedia of Polymer Science and Technology*, Vol. 7, Interscience Div., John Wiley and Sons, New York, 1967.

Gellner 1970. Otto Gellner, "Surface-Active Agents," pp. 477-486 in Herman F. Mark, Norman G. Gaylord, and Norbert M. Bikales, eds., *Encyclopedia of Polymer Science and Technology*, Vol. 13, Interscience Div., John Wiley and Sons, New York, 1970.

Kolthoff 1959-1972. I. M. Kolthoff and Philip J. Elving, eds., with the assistance of Ernest B. Sandell, *Treatise on Analytical Chemistry*, Interscience Div., John Wiley and Sons, New York, 1959-1972 and continuing.

Kontes/Martin. Kontes/Martin Co., Evanston, Illinois 60204.

Krell 1966. Erich Krell and E. C. Lumb, *Handbook of Laboratory Distillation*, Elsevier Publishing Co., New York, 1966.

Leonard 1970. Edward Charles Leonard, ed., *Vinyl and Diene Monomers*, Interscience Div., John Wiley and Sons, New York, 1970.

Mark 1965. H. F. Mark, "Catalysis," pp. 26-34 in Herman F. Mark, Norman G. Gaylord and Norbert M. Bikales, eds., *Encyclopedia of Polymer Science and Technology*, Vol. 3, Interscience Div., John Wiley and Sons, New York, 1965.

Marsden 1963. C. Marsden and Seymour Mann, *Solvents Guide*, 2nd ed., Interscience Div., John Wiley and Sons, New York, 1963.

O'Driscoll 1969. Kenneth F. O'Driscoll and Premamoy Ghosh, "Initiation in Free Radical Polymerization," pp. 59-97 in Teiji Tsuruta and Kenneth F. O'Driscoll, eds., *Structure and Mechanisms in Vinyl Polymerization*, Marcel Dekker, New York, 1969.

Palit 1965. Santi R. Palit, Satya R. Chatterjee, and Asish R. Mukherjee, "Chain Transfer," pp. 575-610 in Herman F. Mark, Norman G. Gaylord, and Norbert M. Bikales, eds., *Encyclopedia of Polymer Science and Technology*, Vol. 3, Interscience Div., John Wiley and Sons, New York, 1965.

Perry 1965. E. S. Perry and A. Weissberger, eds., *Distillation*, Vol. IV of A. Weissberger, ed., *Technique of Organic Chemistry*, 2nd ed., Interscience Div., John Wiley and Sons, New York, 1965.

Riddick 1970. John A. Riddick and William B. Bunger, *Organic Solvents, Physical Properties and Methods of Purification*, Vol. II of Arnold Weissberger, ed., *Techniques of Chemistry*, 3rd ed., Interscience Div., John Wiley and Sons, New York, 1970.

Tipson 1957. R. Stuart Tipson, "Crystallization and Recrystallization," pp. 395-562 in Arnold Weissberger, ed., *Separation and Purification*, Vol. III of A. Weissberger, ed., *Technique of Organic Chemistry*, 2nd ed., Part I, Interscience Div., John Wiley and Sons, New York, 1957.

Tipson 1965. R. Stuart Tipson, "Distillation under Moderate Vacuum," pp. 511-534 in E. S. Perry and A. Weissberger, eds., *Distillation*, Vol. IV of A. Weissberger, ed., *Technique of Organic Chemistry*, 2nd ed., Interscience Div., John Wiley and Sons, New York, 1965.

Zand 1965. Robert Zand, "Azo Catalysts," pp. 278-295 in Herman F. Mark, Norman G. Gaylord, and Norbert M. Bikales, eds., *Encyclopedia of Polymer Science and Technology*, Vol. 2, Interscience Div., John Wiley and Sons, New York, 1965.

4

Experimental Methods and Apparatus

A. *Polymerization*

The methods and apparatus used for polymerization vary widely depending upon a number of factors: the type of polymerization--for example, bulk, solution, or emulsion; the mechanism--radical, ionic, or step-reaction; the nature of the monomer--gas, liquid, or solid; the size of the batch or number of samples required; and the objectives of the experiment--preparation of a small sample with high purity for kinetic or structural studies, or polymerization of a large batch for fabrication into samples with which end-use properties can be measured. For these reasons, no universal description of the apparatus and procedure for polymerization can be written. Instead, we describe several typical methods of rather different sorts in this section.

Some general statements can be made. Oxygen, CO_2, and water must almost always be carefully excluded to obtain well-controlled polymerizations. This can be accomplished by purging the polymerization apparatus with an inert gas (Chap. 3D) and blanketing the apparatus with inert gas during the reaction. A typical procedure is therefore to evacuate the polymerization vessel, fill it with inert gas, and transfer the reagents into the cooled vessel under a blanket of inert gas, usually in the following order: diluent (if any), initiator, additives (if any), and finally monomer. In ionic polymerizations, the catalyst is introduced last and the reaction starts immediately. Some reagents--for example, the catalyst for an ionic polymerization--must be handled in a glove box (Fig. 4-1) or in disposable glove bags (I^2R) under an inert atmosphere.

Bulk radical polymerization

Free-radical polymerization of liquid monomers in bulk or with only small amounts of diluents is a simple technique exemplifying the general principles outlined above. An apparatus for the bulk radical polymerization of methyl methacrylate (Exp. 3) is shown in Fig. 4-2. Inhibitor-free and predried monomer is distilled under an inert atmosphere directly into a graduated separatory funnel (1) as shown.

Fig. 4-1. A glove box (courtesy Fisher).

Provision is made at (2) to evacuate the remainder of the apparatus and fill it with inert gas. The ampoules (3) (previously prepared with weighed amounts of initiator and any other additives) are filled through a multi-port distillation receiver (4). They are then cooled by raising a Dewar flask with acetone-dry ice mixture or liquid nitrogen to freeze the contents, and sealed off at the constriction with a torch. Polymerization is carried out by immersing the sealed tubes in a constant-temperature bath.

Ionic polymerization

Typical apparatus and procedure The rigorous requirements for high purity and complete protection of reagents needed in ionic polymerization, particularly when control of molecular-weight distribution is desired, necessitate the use of procedures and methods that are much

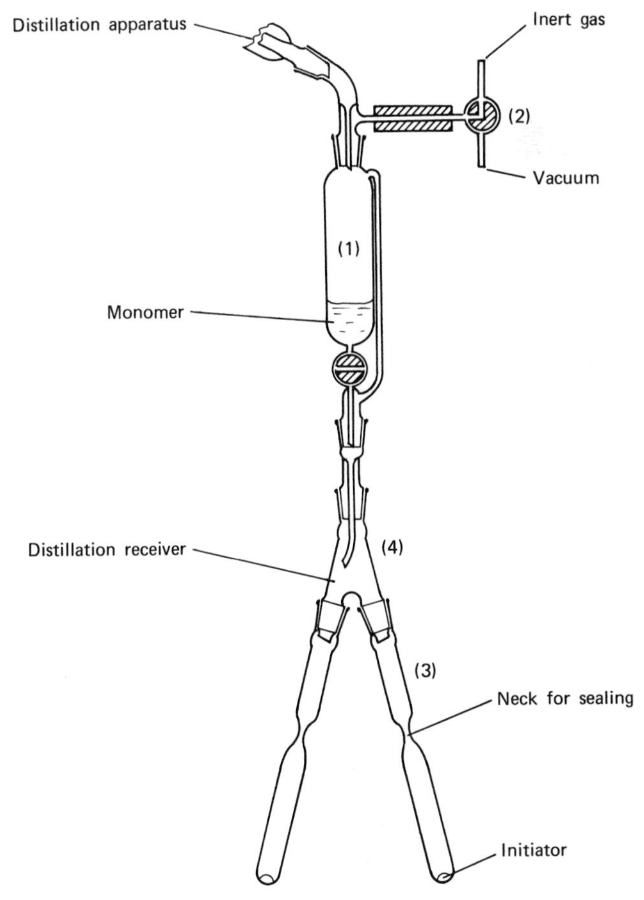

Fig. 4-2. Apparatus for the bulk radical polymerization of such monomers as methyl methacrylate. For this and the following figures, see the text for details of the use of the equipment.

more complicated and time consuming than is commensurate with this course. Therefore we describe these techniques in somewhat more detail.

Equipment for the anionic polymerization of styrene with butyl lithium catalyst is shown in Fig. 4-3. The all-glass apparatus includes a section for the distillation of the benzene solvent from an excess of the catalyst. Dry benzene is placed with butyl lithium in flasks A and B. Inert gas is introduced through A, which is a scrubber to purify the gas, and then blankets the solvent reservoir B and reaction vessel C. Back diffusion is prevented by a mineral-oil check valve or bubbler at D. Benzene is distilled from B to C, being condensed by the condenser between points 1 and 2. Distilled benzene is also transferred to the graduated separatory funnel by means of a hypodermic syringe, which pierces rubber serum stoppers at 6 (to load) and 5 (to deliver).

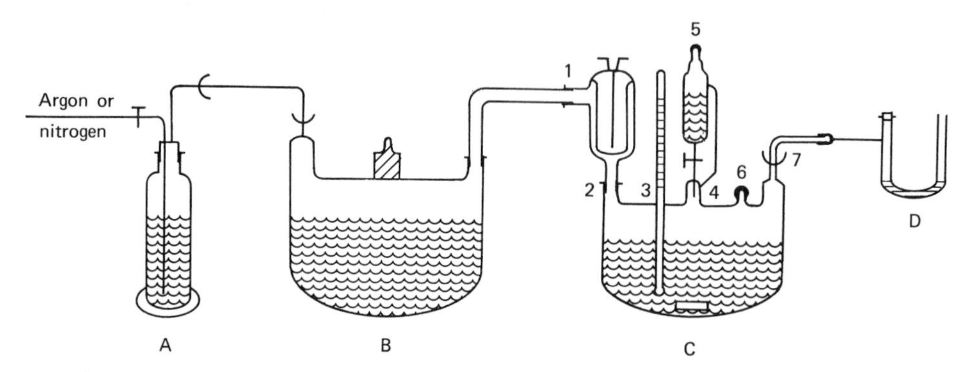

Fig. 4-3. Equipment for solvent purification followed by anionic polymerization (Sorenson 1968).

Flask C is now isolated and dry styrene (previously distilled) is transferred into C. Catalyst is added to the separatory funnel, again using a syringe. The polymerization is initiated by adding the catalyst solution to the monomer solution in C.

Styrene purification Preparation of rigorously pure dry styrene can be carried out in the apparatus shown in Fig. 4-4. Styrene is washed 3 times with 10% aqueous NaOH to remove inhibitor, then washed 3 times with water, predried over anhydrous $CaCl_2$, and dried over CaH_2 for a week. It is then vacuum distilled and transferred to flask A. Liquid sodium-potassium alloy is placed in B, and the filling tube sealed. The alloy is subsequently freed of organic impurities by keeping it

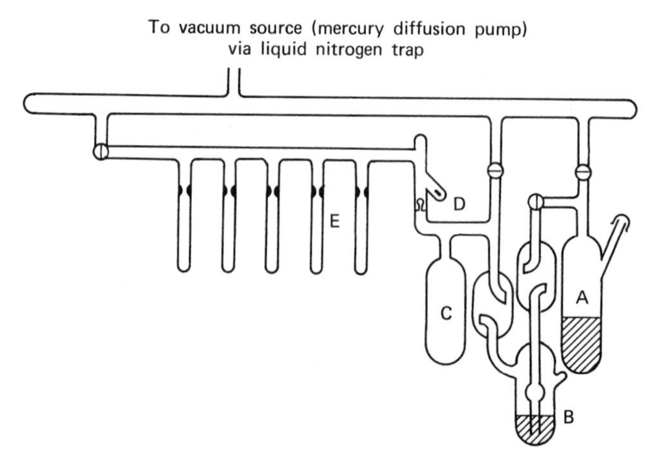

Fig. 4-4. Apparatus for the preparation of pure dry styrene (Ueno 1966).

under high vacuum for 20 hours. Then the middle fraction of the styrene is distilled through the Na-K alloy into cooled flask C and there degassed. Finally, it is distilled from C through a breakable joint D into polymerization ampoules E, which are sealed off at a pressure below 10^{-5} torr. For copolymerization experiments, another set of flasks A, B, C, and joint D can be added at the far end of the manifold holding the ampoules E. Graduated ampoules allow the monomers to be mixed in known amounts.

Precautions for highest purity When the highest purity of the reagents is required, all greased ground joints and stopcocks are eliminated and replaced by proper sequences of all-glass seals and breakseals. An example of this type of apparatus, used in a study of the mechanism of the cationic polymerization of *p*-methyl styrene in toluene (Kennedy 1971) is as follows: Degassed, predried, and partially polymerized monomer is placed in flask A (Fig. 4-5) and degassed once more. The line is then sealed at B. Flask D, containing BaO, is baked for several hours at 300°C under high vacuum. Seal C is broken and the monomer is distilled into D, where it is dried with occasional stirring for three days. The monomer is then distilled into a measuring buret F, and the line between D and F is sealed. The monomer is finally distilled into the polymerization vessel Q through line L and the line sealed at G. Toluene is purified by successive washing with H_2SO_4, $NaHCO_3$, and distilled water, dried over sodium wire, fractionated, and similarly distilled into Q. To start the polymerization, the system is sealed off at R, the vial S with catalyst (in this case $AlBr_3$) is broken, and the catalyst is rapidly mixed into the monomer-toluene system. To stop the polymerization, an excess of NH_3-C_2H_5OH mixture is added through break-seal M.

A similar "break-seal" technique was used (Inoue 1969) for the anionic preparation of styrene-isoprene block copolymers with narrow distributions of block lengths. As Fig. 4-6 shows, tetrahydrofuran is released from flask S through a break-seal into an evacuated flask P, and thence to N for purification with sodium. After this step, N is sealed off, and the purified monomer distilled from P through the manifold L into R. The reaction flask R is then sealed off from the manifold. Styrene monomer is added from A, followed by catalyst solution (butyl lithium in hexane) from I. The reaction starts at once and continues until the styrene monomer is used up. The flask now contains polystyrene with active ends capable of further growth—"living polymer." Isoprene is now added from B to form polyisoprene blocks. The sequence polystyrene-polyisoprene can be continued several times if additional monomer reservoirs are used. Finally, the reaction is terminated by adding *n*-butyl alcohol in hexane from T.

Simplifications for routine work Where less emphasis need be placed on purity, several simplifications can be made to these techniques. The monomer can be distilled separately before polymerization, and polymerization can be carried out in a glass tube sealed with a rubber serum cap. A carefully cleaned and dried tube is capped while hot. Air is flushed out by purging (through syringe needles) with inert gas, or by alternate evacuation and filling with inert gas. The tube is cooled—for example, in an acetone-dry ice mixture—and reagents introduced with syringes. Then the tube is sealed under low pressure (ca.

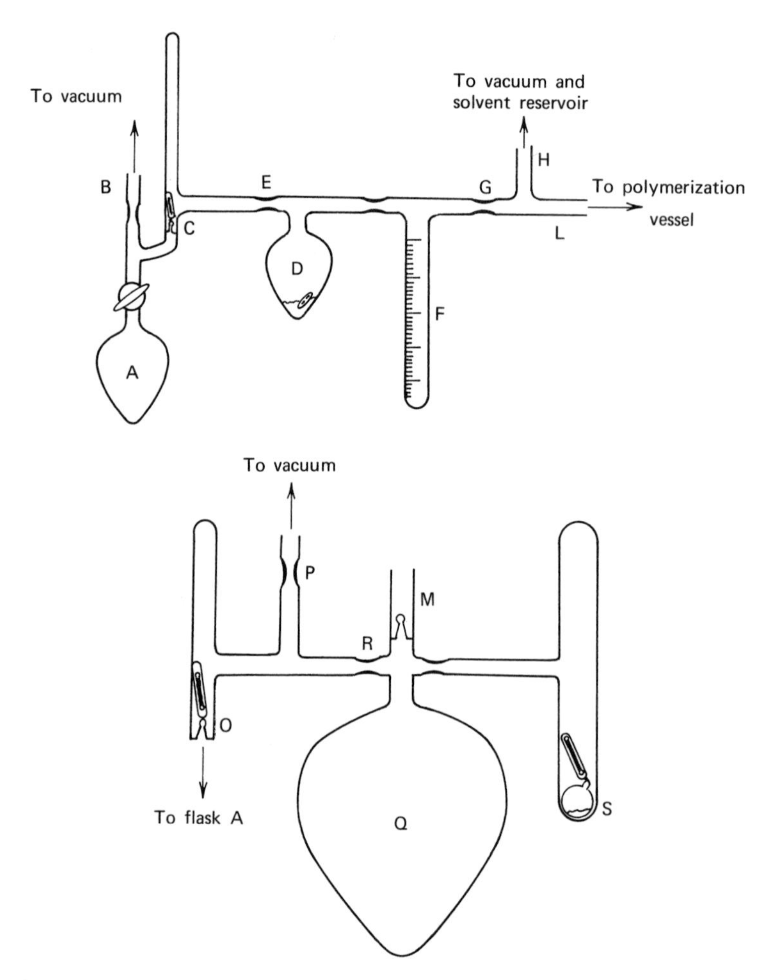

Fig. 4-5. Equipment for cationic polymerization with rigorously-controlled purity (Kennedy 1971).

10 torr) of inert gas (Fig. 4-7). Needles providing for the flow of inert gas can either be removed for complete sealing or left in place during polymerization.

While ordinary syringes (Fig. 4-8*a*, *b*) often suffice, they can be replaced by microliter syringes (Fig. 4-8*c*) to handle small amounts of catalysts, transfer agents, or other additives, or by special syringes (Fig. 4-8*d*) for gases. For ease of handling, a "Luer-Lok" connection between needle and syringe (Fig. 4-8*b*) is recommended. Since the pressure in the tubes and flasks may be above or below atmospheric, care is required to prevent unexpected movement of the plunger, as shown in Fig. 4-9. For extensive or precise use, syringes with valve and "stirrup" (Fig. 4-10) are available.

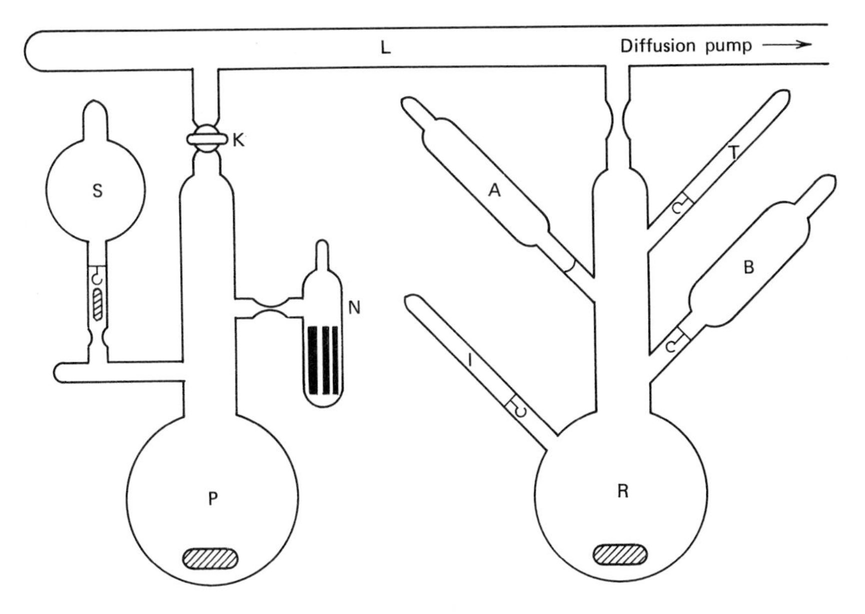

Fig. 4-6. Apparatus for the anionic preparation of block copolymers (Inoue 1969).

If the polymerization is to be carried out in a sealed rather than a capped tube, a constriction for sealing is prepared ahead of time (Fig. 4-7*a*). Reagents, particularly solid, must be added carefully to prevent contamination of the neck (Fig. 4-11), since subsequent heating here can decompose fragments of the reagents and lead to contamination. In this procedure, care must be taken to remove all air after the solid reagent is added, by flushing with inert gas or repeated cycles of freezing, evacuation, and melting under vacuum.

Polymerization of gaseous monomers

Somewhat different techniques are used for the polymerization of gaseous monomers. Usually, the monomer is bubbled through the stirred mixture of catalyst and solvent. Ethylene is polymerized with a Ziegler-Natta catalyst (Chap. 1D) in this way (Fig. 4-12). Ethylene and dry nitrogen are washed by hexane in scrubbers before entering the reactor. The gas outlet is provided with a hexane bubbler to prevent back-diffusion of air. The flask is filled with purified hexane, the neck is closed with a rubber serum stopper, and the assembly is purged with nitrogen. A solution of aluminum triethyl in cyclohexane is injected through the serum cap, using a syringe flushed with nitrogen. Subsequently, a $TiCl_4$ solution is added slowly to the stirred mixture to form the insoluble catalyst. The reaction proceeds as the ethylene flow is started. To stop the reaction, absolute ethanol is injected.

Fig. 4-7. Polymerization in test tubes capped with serum stoppers and purged with inert gas.

Emulsion polymerization

Emulsion polymerization can be carried out in several types of apparatus depending on the size of the batch and the vapor pressure of the monomer. Amounts of 300-400 ml of emulsion, using monomers boiling above or not too far below room temperature, for example styrene (Exp. 4) or vinyl chloride, are conveniently polymerized in soda bottles (Fig. 4-13). A drilled bottle cap with a rubber gasket permits sampling in systems where the internal pressure is not too high. Care must be taken to select an inert gasket material: Acrylate or other rubbers ("Geon" 102, "Hycar" OR, "Viton" A) are satisfactory for styrene and butadiene. For monomers with vapor pressure lower than one atmosphere in the emulsion, larger amounts can be polymerized in a 3-neck flask equipped with stirrer, nitrogen inlet, and condenser (Fig. 4-14).

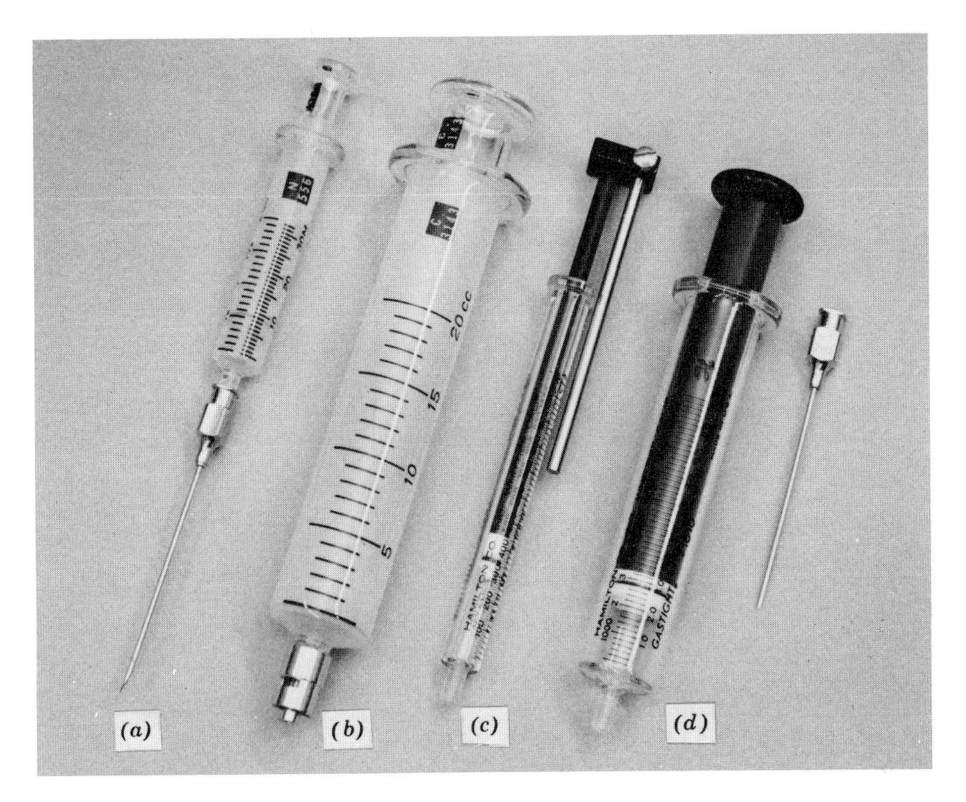

Fig. 4-8. Syringes useful in test-tube polymerization manipulations.

Polycondensation

Step-reaction polymerization of small batches is easily carried out in test tubes (Exp. 1). Agitation and removal of water is effected by bubbling inert gas through the mixture, as in Fig. 4-7*b*. Larger batches are polymerized in a kettle fitted with stirrer, thermometer, nitrogen inlet, separatory funnel, and reflux condenser with trap so that the amount of water eliminated can be measured to follow the course of the polymerization (Fig. 4-15).

High-pressure polymerization

Polymerizations conducted at pressures significantly above atmospheric require special apparatus the description of which is beyond the scope of this book.

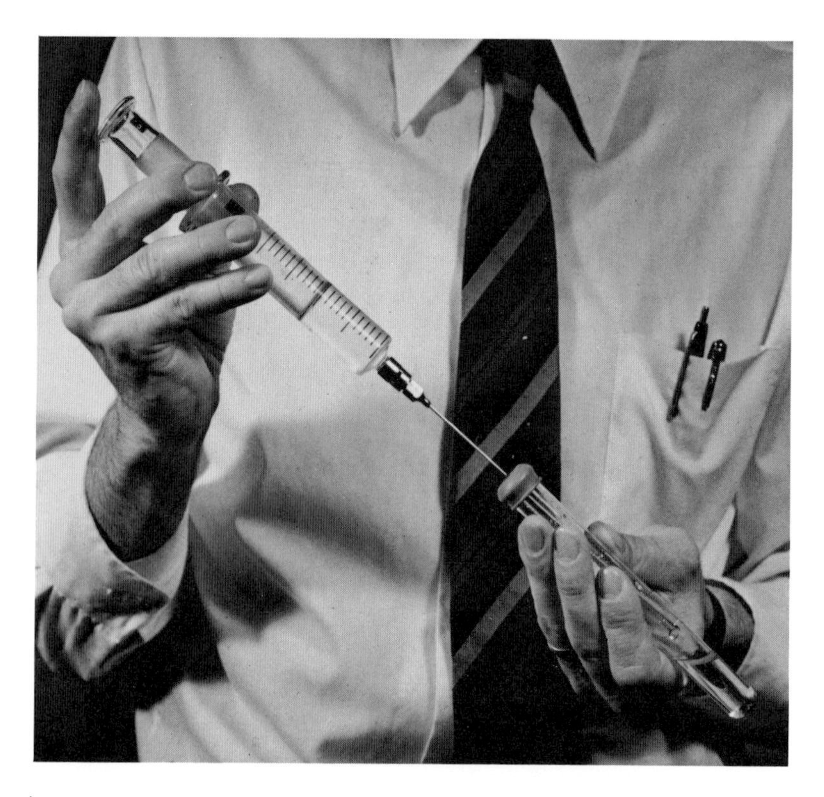

Fig. 4-9. Holding the syringe to control flow against internal pressure.

Agitation

A few polymerizations proceed smoothly without the need for any stirring. These, however, are the exception, and are usually limited to small quantities of material. Some examples are the solid-state polymerization of trioxane to polyoxymethylene (Exp. 6); ring-opening polymerization in bulk, such as the polymerization of ε-caprolactam to 6 nylon; and the bulk or solution polymerization of small amounts (up to about 50 ml) of monomers such as styrene or methyl methacrylate in test tubes (Exps. 3 and 5).

For most polymerizations, continuous or interrupted stirring is needed for thermal control and to maintain uniform conditions throughout the mixture. An example is the polycondensation of large batches. Note, for example, the type of stirrer used in Fig. 4-15 for handling a highly viscous polycondensation mixture. For smaller amounts of these materials, polymerized in test tubes, agitation by inert-gas bubbles as in Fig. 4-7*b* is adequate.

In still other polymerizations, such as emulsion and suspension systems, stirring is absolutely essential and critical. For small

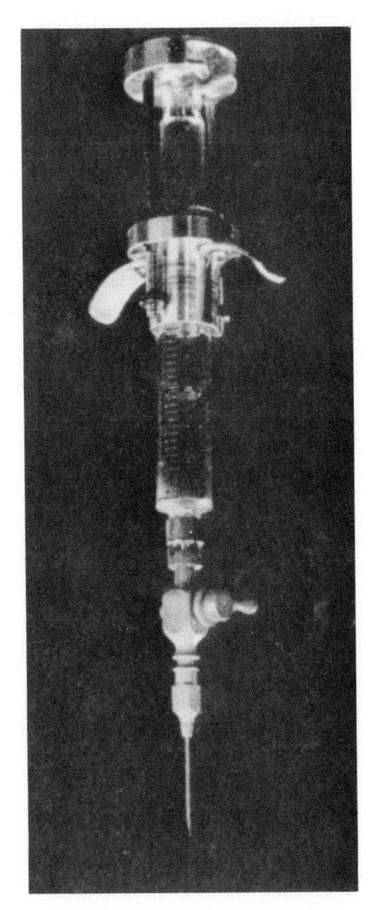

Fig. 4-10. Syringe with valve and "stirrup" (Houston 1948).

amounts (less than about 500 ml) any of several methods of agitation can be used, including a rocking bath (Fig. 4-16), tumbling or rotation, or a shaker (Fig. 4-17). For larger amounts, a stirred reactor like that of Fig. 4-14 is used. Ionic or coordination polymerizations in which the catalyst is a slurry of solid particles also require adequate stirring. Interfacial polymerization (Exp. 2) requires violent agitation to renew continuously the surface between the two phases. A "Waring" blender is effective.

GENERAL REFERENCES

Pinner 1961; Overberger 1963; Elliott 1966; Sorenson 1968; Borchert 1969; Gaylord 1969; McCaffery 1970.

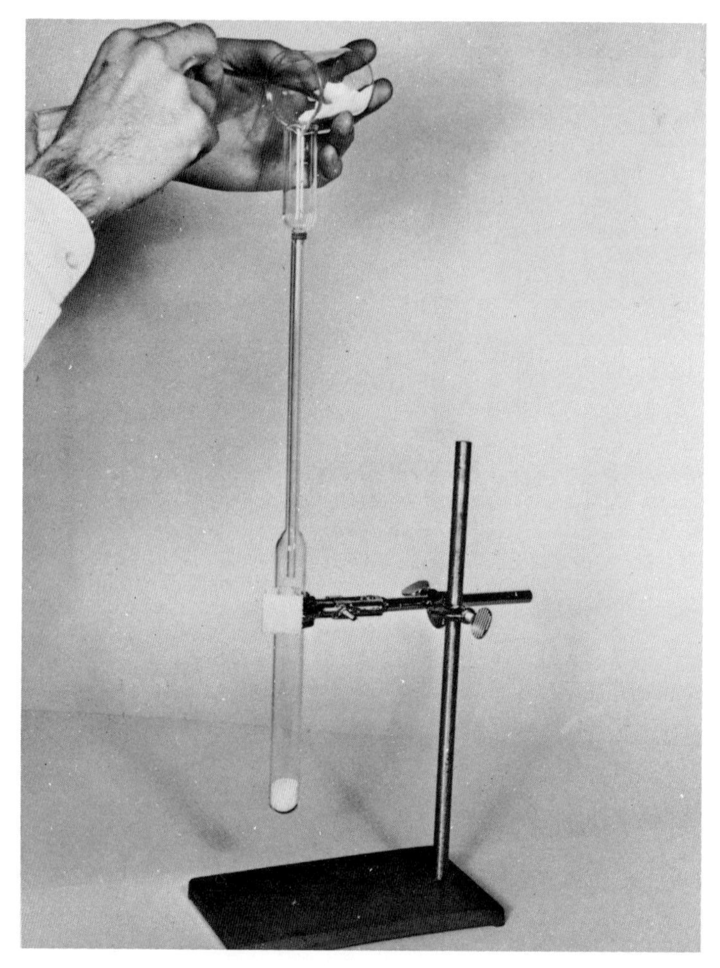

Fig. 4-11. Filling a polymerization tube (Sorenson 1968).

B. *Isolation and Purification of Polymer*

Isolation

Various techniques are used to isolate from the reaction mixture polymer prepared in solution, suspension, or emulsion systems. If the polymer is already precipitated, it can be collected by filtration, using ordinary filter paper for aqueous systems or fritted-glass filters if organic solvents were used. Suspension polymers can be filtered or centrifuged directly.

Polymers made in emulsion systems usually have a very small particle size (Chap. 2B), and coagulation is necessary before they can be removed by filtration. For less stable emulsions, high-shear

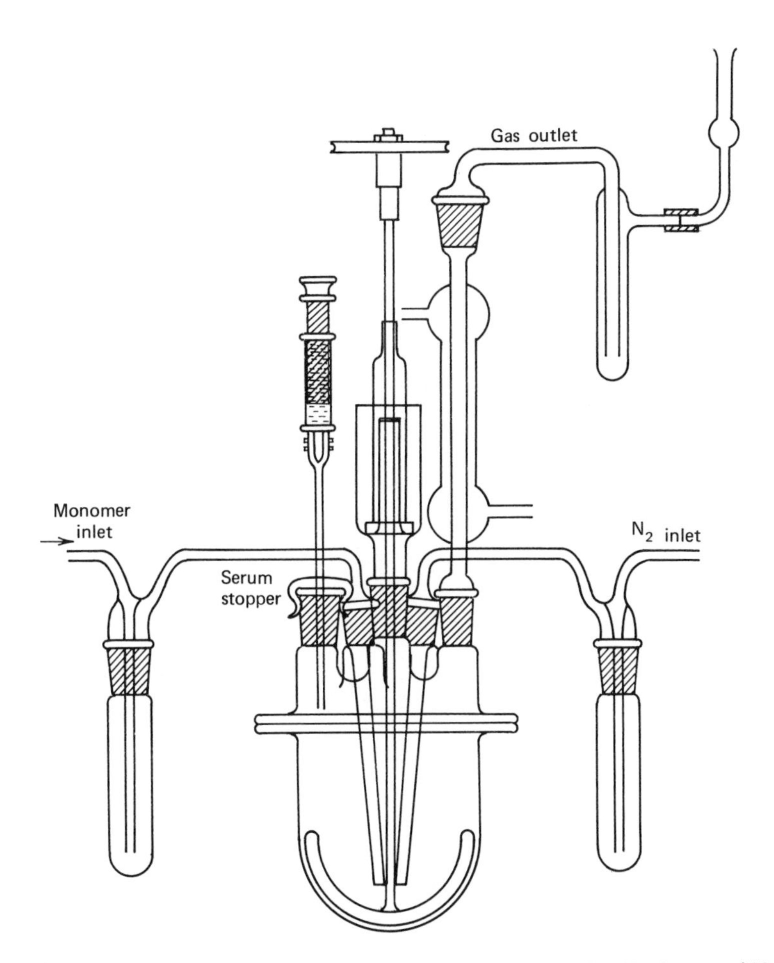

Fig 4-12. Apparatus for the polymerization of ethylene with a Ziegler-Natta catalyst (Pinner 1961).

agitation with alternate heating and cooling may cause coagulation, but usually a coagulant or flocculant such as NaCl or alum is used.

Polymers made in solution can be precipitated by the addition of 5-10 volumes of a nonsolvent, with agitation. The precipitant must be miscible with the polymerization solvent, a nonsolvent for the polymer, and ideally a solvent for the initiator or catalyst, additives, and oligomers. If the polymer precipitates in a very finely divided form, it may behave like an emulsion and require coagulation as described above. After collection by filtration or centrifugation, the polymer should be washed or extracted by various solvents to remove traces of other reactants.

Polymers may also be recovered by evaporation of the solvent or solvents by high temperature, vacuum, or both. The polymer then retains all the nonvolatile polymerization ingredients and may well be

Fig. 4-13. "Soda-bottle" polymerization equipment.

too impure for further use. This method is sometimes used, however, to determine conversion or yield.

Purification

Polymers are usually purified by repeated cycles of dissolving, precipitation, and washing, or by extraction with suitable solvents. Table 4-1 lists some solvents and nonsolvents useful for these treatments of the polymers prepared in the experiments in Part III.

Depending on the nature and amount of impurities, the solution-precipitation-washing cycle can be repeated one or more times with the most suitable solvent-nonsolvent pair, or for greater effectiveness with several different pairs. Usually, a solution of about 5-10%

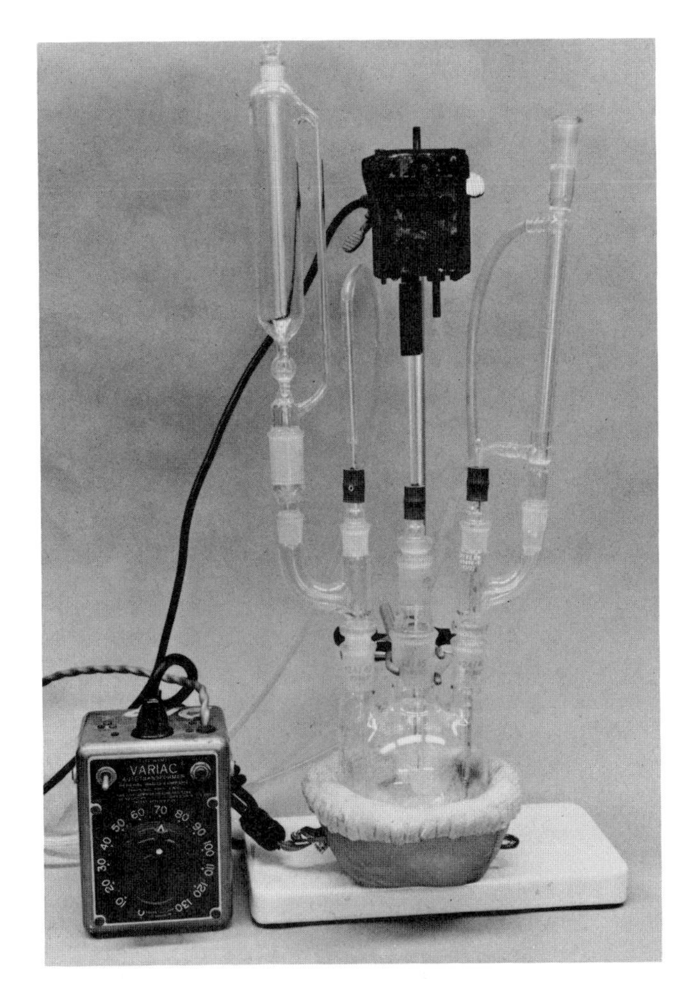

Fig. 4-14. Three-neck flask apparatus for emulsion polymerization.

concentration is prepared and added to 5-10 volumes of violently agitated nonsolvent; a "Waring" blender is effective. Ideally all traces of catalyst or initiator, residual monomer, additives, and oligomers can be removed.

Extraction is usually resorted to only if the above procedure cannot be used successfully. The extraction of bulk polymer, even if crosslinked and highly swollen, is time consuming, and it is advisable to use finely ground material to speed up the procedure by the increase in surface area. The usual extractor is the Soxhlet (Fig. 4-18), which extracts continuously with pure solvent which is simultaneously distilled into the extraction reservoir. By evaporating this solvent, nonvolatile extracted materials such as oligomers can be recovered.

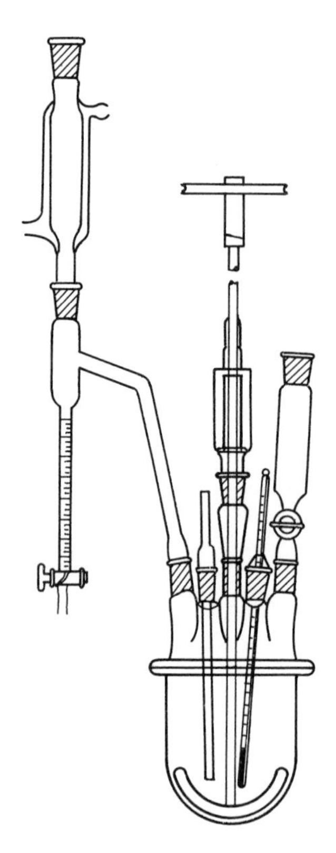

Fig. 4-15. Equipment for polycondensation (Pinner 1961).

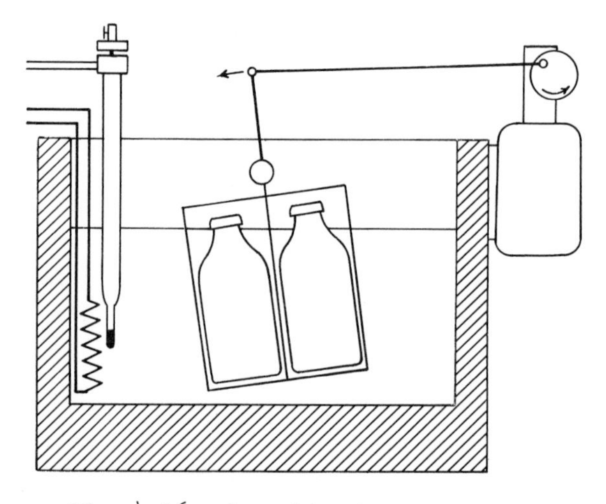

Fig. 4-16. A rocking bath agitator.

Fig. 4-17. A shaker bath for providing agitation.

Solvent removal

The removal of solvents from polymers is controlled by two factors, the solubility of the solvent in the polymer (or the polymer-solvent equilibrium) and the diffusion constant of the solvent molecules in the polymer. The product of these two is the permeability (*Textbook*, p. 133), a measure of the rate of movement of solvent into or out of the polymer.

To dry a polymer means to lower its solvent concentration to a prescribed level, set by the method of detection or the requirements of the use to be made of the sample. To make this concentration as low as possible, the concentration of solvent vapor in the drying space must be made correspondingly low. For example, the polymer-water equilibrium for 6 or 66 nylon at room temperature and 60% relative humidity corresponds to about 1% water in the polymer. The water content of these nylons cannot be brought below this level except at lower relative humidities and will, in fact, increase to this level if a drier sample is exposed to these conditions.

These are equilibrium considerations. The rate at which equilibrium is reached depends on the solvent diffusion rate, which is typically extremely low, especially in glassy polymers, where the mobility of solvent molecules in the polymer is severely restricted. The diffusion rates, however, increase greatly with increasing temperature.

TABLE 4-1. Solvent and nonsolvent pairs for various polymers (Braun 1971; see also Guzmán 1961, Giesekus 1967).

Polymer	Solvent	Nonsolvent
Polymeric carbohydrates[a]	Decahydronaphthalene, aromatic hydrocarbons	Acetone, ether, lower alcohols
Polyisobutylene	Hexane, carbon tetrachloride, chloroform, ether	Lower alcohols, ketones, some esters
Polystyrene	Benzene, toluene, chloroform, cyclohexanone, butyl acetate, carbon disulfide	Lower alcohols, ether, acetone
Vinyl polymers	Aromatic hydrocarbons, chlorinated hydrocarbons, ketones, ethers	Lower alcohols, esters
Poly(vinyl chloride)	Tetrahydrofuran, dioxane, cyclohexanone, dimethyl formamide	Methanol, water, acetone
Poly(vinyl acetate)	Benzene, toluene, chloroform, acetone, butyl acetate	Ether, methanol, petroleum ether
Acrylic polymers	Ketones, esters, cyclic ethers	Lower alcohols, petroleum ether
Polyacrylates and polymethacrylates	Chloroform, acetone, ethyl acetate, tetrahydrofuran, toluene	Methanol, ether, hexane, petroleum ether
Polyacrylonitrile	Dimethyl formamide, dimethyl sulfoxide	Alcohols, ether, water
Polyacrylamide	Water	Methanol, acetone
Polymeric acids	Water, dilute alkalis	Organic liquids
Polymeric alcohols	Water, dimethyl formamide	Methanol, ether
Polyesters	Chlorinated hydrocarbons, formic acid	Methanol, ether, petroleum ether
Polyurethanes	Chlorinated hydrocarbons	Methanol, ether, petroleum ether

Polymer	Solvent	Nonsolvent
Polyamides	Formic acid, *m*-cresol, perfluorinated alcohols	Methanol, ether

[a]Italics denotes families of polymers.

These considerations lead to the following conclusions regarding the efficient drying of polymers. The temperature of drying should be as high as possible without risk of degrading the polymer, and if at all possible, above the glass transition temperature. The selection of final solvent to be removed, if a choice is available, should favor small, compact molecules with high diffusion rates. The solvent vapor pressure in the space around the polymer should be minimized. For example, the removal of water is best accomplished in the presence of a drying agent (Table 3-1) or in vacuum. Even so, the "complete" removal

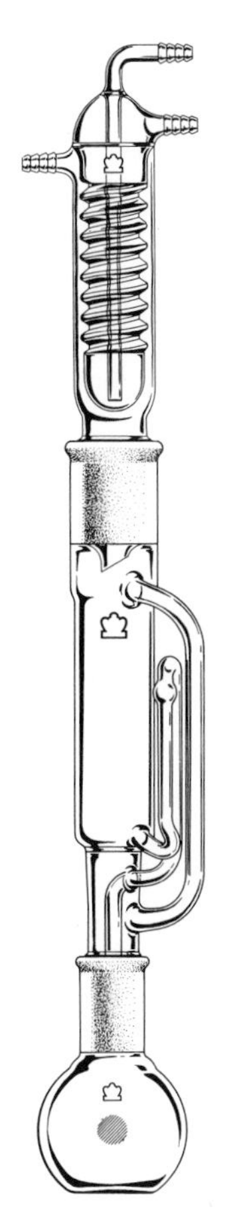

Fig. 4-18. The Soxhlet extractor (courtesy Kontes).

of solvent, including water, is almost impossible, and adequate removal can take days. For 66 or 6 nylon, the removal of water to a concentration lower than 0.05% from a film only 0.1 mm thick takes 12-24 hours at 100°C under a vacuum of 10^{-2} torr. The removal of traces of *m*-cresol from nylon is even more difficult because of its larger size and correspondingly lower diffusion constant.

GENERAL REFERENCES

Broughton 1956; Craig 1956; Diamond 1966; Sorenson 1968; Braun 1971.

C. *Auxiliary Apparatus*

Vacuum Apparatus

The purification of materials and the need for protection from air and moisture often dictates the use of vacuum equipment. This consists of a rotary and possibly also a diffusion pump, together with traps, vacuum gauges, check valves, manifolds, and associated components, referred to collectively as a *vacuum bench*. A typical setup is shown in Figs. 4-19 and 4-20. The rotary pump alone can evacuate equipment to a pressure of 10^{-2} to 10^{-4} torr, and the diffusion pump, to at least 10^{-6} torr. A cold trap, using acetone-dry ice or liquid nitrogen, protects the system from mercury or oil vapors from the diffusion pump, or oil vapors from the rotary pump, and conversely protects both pumps from condensing organic vapors. To exploit the pumping capacity of the diffusion pump adequately, the tubing in the vacuum system must be short and large (at least 3 cm) diameter.

As the diffusion pump works efficiently only at pressures below 1 torr, the apparatus must first be evacuated with the rotary pump, the diffusion pump being bypassed. Then stopcocks 2 and 3 are turned to place the diffusion pump in the line. No air can be let into the system while the diffusion pump is hot, for the air flow would carry oil vapors from the pump throughout the apparatus, where they would condense. Since this oil has a very low vapor pressure, it cannot be removed except by disassembling the system completely and cleaning it.

The high vacuum part of the bench should be all glass or metal, with ground joints lubricated with vacuum grease, or metal-flange joints with O-ring vacuum gaskets. Details of the manifold will vary depending on the experiments to be carried out. Pressures are usually measured by a thermistor gauge down to about 10^{-3} torr and by an ionization gauge at lower pressures. For absolute pressure readings, a MacLeod gauge is sometimes used.

The need for a diffusion pump should be carefully considered, because it always makes the operation of the bench more complicated. Several newer types of mechanical or sorption pumps are capable of replacing it.

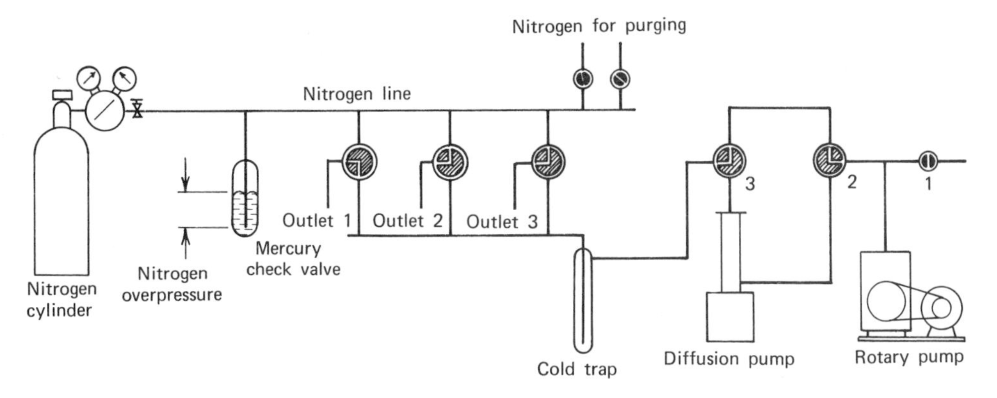

Fig. 4-19 Sketch of the components of a typical vacuum bench.

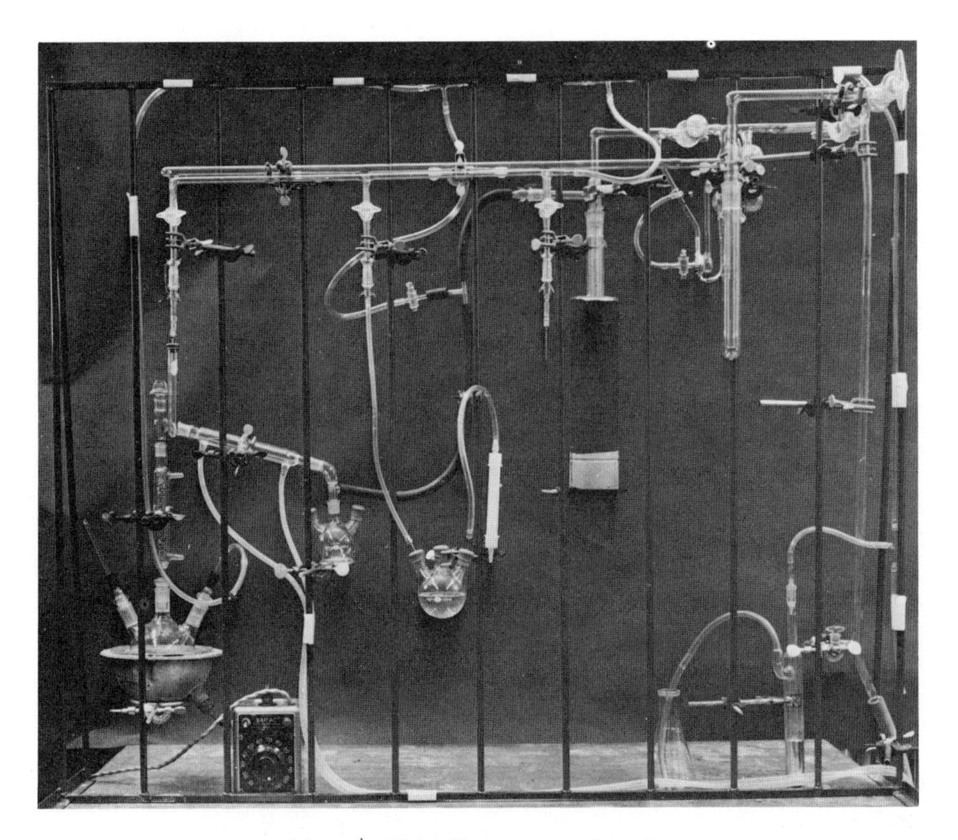

Fig. 4-20. The vacuum bench.

Thermal-control equipment

Heating A wide variety of heating methods can be used in polymerization. In addition to water baths for temperatures up to 100°C, vapor baths (Fig. 4-21) can be used for small-scale polymerizations. Table 4-2 lists some of the high-boiling liquids that can be used. Safety precautions must be taken when flammable liquids are used.

Immersion baths, with built-in or separate temperature controllers and thermometers, are used for almost all laboratory-scale polymerizations. The liquids used include water, hydrocarbon or silicone oils, molten salts, or Wood's metal. A high-temperature bath for use up to 275°C is shown in Fig. 4-22. A silicone oil (General Electric SF 1153) commonly used in it can be heated to 230°C without decomposition. A high-temperature alternative to liquid baths is the hot sand bath, where heating is effected by percolating hot air through the sand. Heating mantles are not often used, since they can develop overheated areas if the level of the liquid in the vessel is lowered during the reaction.

Cooling For polymerization at low temperatures, ice, ice-salt mixtures (Table 4-3), and the acetone-dry ice mixture are frequently used to achieve temperatures between 0°C and -78°C. For small-scale polymerizations not too far below room temperature, cold plates are very

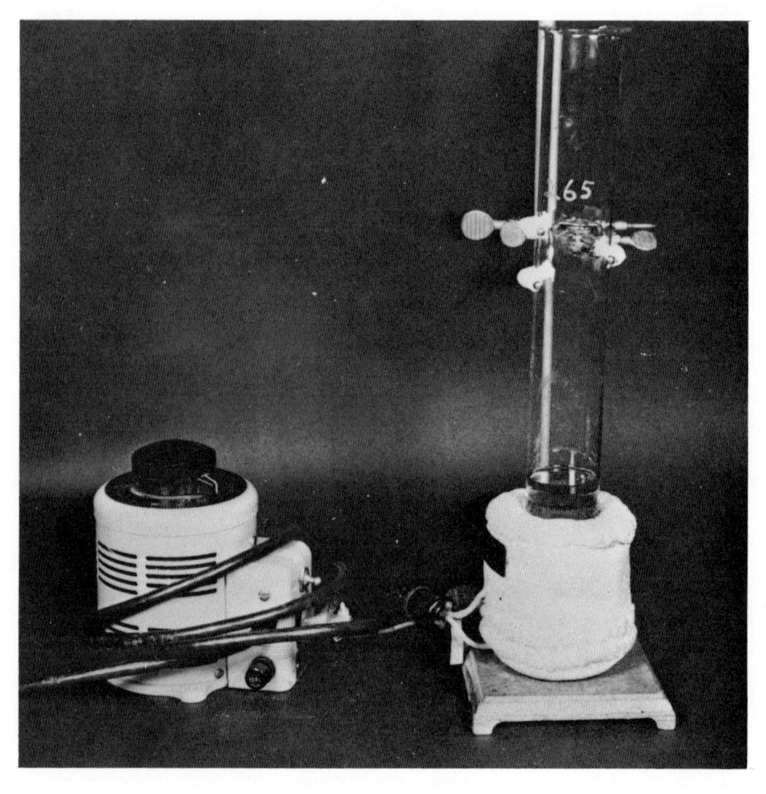

Fig. 4-21. A vapor bath (Sorenson 1968).

TABLE 4-2. Liquids suitable for use in vapor baths (Sorenson 1968)

Compound	Boiling Point, °C	Compound	Boiling Point, °C
Water	100	Diethylene glycol	245
Toluene	111	Diphenyl	255
n-Butanol	117	Diphenyl ether	259
Perchloroethylene	121	Diphenylmethane	265
Methyl Cellosolve[a]	125	*o*-Chlorodiphenyl	268
Chlorobenzene	133	Diphenylmethane-	
Ethyl Cellosolve	135	*o*-hydroxydiphenyl (60/40)	270
m-Xylene	139	Methyl naphthyl ether	275
Anisole	152	Biphenyl/diphenylene	
Cyclohexanone	156	oxide (25/75)	275
Cyclohexanol	160	Acenaphthene	277
Phenetole	166	Triethylene glycol	282
Butyl Cellosolve	171	Dimethyl phthalate	283
p-Cymene	176	Diphenylethane	284
o-Dichlorobenzene	179	*o*-Hydroxydiphenyl	285
Phenol	181	Diphenylene oxide	288
Decahydronaphthalene	190	Fluorene	295
Ethylene glycol	197	Benzophenone	305
m-Cresol	202	*p*-Hydroxydiphenyl	308
Tetrahydronaphthalene	206	Hexachlorobenzene	310
Naphthalene	218	Arochlor 1242[b]	325
Methyl salicylate	222	*o*-Terphenyl	330
Butyl Carbitol[a]	231	Arochlor 1248	340
n-Decyl alcohol	231	Anthracene	340
Methylnaphthalene	242	Arochlor 1254	365
		Anthraquinone	380

[a]Trademarks for Union Carbide's ether solvents.

[b]Trademark for Monsanto's chlorinated aromatic hydrocarbon.

convenient. Cryostats with suitable liquids (Table 4-4) covering a wide range of temperatures can be used in special experiments.

Drying apparatus

The removal of solvent can be carried out in a simple air convection oven or a vacuum oven. If the latter is used, a trap must be utilized to protect the vacuum pump, as described above. It is often a simplification to remove most of the solvent while using a water aspirator to provide the vacuum, and use the rotary pump only to remove the last traces. Depending on the amount of sample, test tubes, flasks, or other containers may replace the vacuum oven. For small quantities, a freeze drier is eminently convenient and successful.

Fig. 4-22. A high-temperature immersion bath.

GENERAL REFERENCES

Egly 1957; Miller 1957; Guthrie 1963; Tipson 1965; Diels 1966; Espe 1966, 1968.

D. Laboratory Safety

ALMOST WITHOUT EXCEPTION, POLYMERIZATION REACTIONS ARE INHERENTLY DANGEROUS IN SOME RESPECT. SAFETY PRECAUTIONS BEYOND THOSE NORMALLY OBSERVED IN EVERY CHEMICAL LABORATORY ARE OFTEN REQUIRED, AND VIGILANT ATTENTION TO POSSIBLE DANGERS MUST NEVER BE RELAXED. THE CONTENTS OF THIS SECTION MUST BE STUDIED WITH CARE, AND THE WARNINGS STATED AND IMPLIED IN IT ADHERED TO WITHOUT FAILURE.

TABLE 4-3. Temperatures obtainable by adding the indicated substance to crushed ice (Egly 1957)

Substance Added to Ice	Parts (by wt.) Added per 100 Parts Ice	Resultant Temp., °C
Na_2CO_3	20	-2
KCl	30	-11
NH_4Cl	25	-15
NH_4NO_3	50	-17
$NaNO_3$	50	-18
38% HCl	50	-18
Conc. H_2SO_4	25	-20
NaCl	33 to 100	-20 to -22
$NaNO_3 + NH_4NO_3$	55 + 52	-26
$KNO_3 + NH_4CNS$	9 + 67	-28
$CaCl_2 \cdot 6H_2O$	100	-29
Tech. KCl	100	-30
$NH_4Cl + KNO_3$	13 + 38	-31
KNO_3 + KCNS	2 + 112	-34
$NaNO_3 + NH_4CNS$	55 + 40	-37
66% H_2SO_4	100	-37
Dil. HNO_3	100	-40
$CaCl_2 \cdot 6H_2O$	150	-49
$CaCl_2 \cdot 6H_2O$	500	-54

Toxicity

The chemical reagents used in polymer synthesis and processing are sometimes highly toxic. The toxicity of acrylonitrile is comparable to that of cyanide, for example, while vinyl chloride and *m*-cresol among others, are also highly toxic. In other cases, the chemicals used for polymerization can cause skin irritation through prolonged or repeated direct contact or exposure to vapors. Styrene, methyl methacrylate, aromatic compounds, and even some common solvents fall in this class.

Particularly insidious are the common solvents which cause chronic poisoning on prolonged exposure, including benzene in particular, and also nitrobenzene and carbon tetrachloride. The concentration at which the odor of CCl_4 is detected is already above the permissible safety level. The risk of poisoning by nitrobenzene is increased by its rapid diffusion through undamaged skin. Phenol also quickly penetrates skin and mucous membranes, causing burns and possible chronic poisoning. Initiators such as quinone, hydroquinone, benzoyl peroxide, and the azo compounds are both toxic and irritant.

Generally, almost all reagents should be considered harmful, not only if ingested but especially if they come in contact with the eyes. In some cases, the risk of damage to the eyes is enhanced by the fact

TABLE 4-4. Liquids for low-temperature baths (Egly 1957)

Substance	Invariant Point (m.p. or eutectic), °C.	Boiling Point, °C.	Flammability[a]
Benzyl bromide	-4.0	198	+
Methyl salicylate	-8.6	223	+
Diethylene glycol	-10.5	244.5	+
tert-Amyl alcohol	-11.9	101.8	+
Glycerol	-15.6	290	+
Benzal chloride	-16.1	214	+
Ethylene glycol	-17.4	197.4	+
Butyl benzoate	-22.4	250	+
Carbon tetrachloride	-22.8	76-77	-
Ethyl sulfate	-24.5	208	+
Pentachloroethane	-29	162	-
Dipropyl ketone	-32.6	144	+
Ethyl chloride	-35.3	83.5	+
Benzyl chloride	-39	179.4	+
Diethyl ketone	-42	102.7	+
Acetylene tetrachloride	-43.8	146.3	-
Chlorobenzene	-45.2	132	+
Ethyl malonate	-49.8	198.9	+
Diethylcarbitol	-52	115.6	+
Diacetone	-55	165	+
Isopropyl ether	-60	67.5	+
Chloroform	-63.5	61.2	-
Trichloroethylene	-73	87	-
Butyl acetate	-76.8	126.5	+
49% carbon tetrachloride + 51% chloroform	-81	--	-
Ethyl acetate	-83.6	77.1	+
Isopropyl alcohol	-88.5	82.4	+
sec-Butyl alcohol	-89	99.5	+
Acetone	-94.6	56.5	+
Toluene	-95.1	110.8	+
Methanol	-97.8	64.6	+
31% chloroform + 69% trichloroethylene	-100	--	-
27% chloroform + 60% methylene chloride + 13% carbon tetrachloride	-111	--	-
Ethyl ether	-116.3	34.5	+
Ethyl alcohol	-117.3	78.4	+
Ethyl bromide	-119	38.0	-
n-Butyl chloride	-123.1	78	+
Methylcyclohexane	-126.4	100.3	+
sec-Butyl chloride	-131.3	68	+
Ethyl chloride	-138.7	12.2	+

Substance	Invariant Point (m.p. or eutectic), °C.	Boiling Point	Flammability[a]
20% chloroform + 14% *trans*-dichloroethylene + 21% trichloroethylene + 45% ethyl bromide	-139	--	-
Isobutane	-145	-10.2	+
14.5% chloroform + 25.3% methylene chloride + 33.4% ethyl bromide + 10.4% *trans*-dichloroethylene + 16.4% trichloroethylene	-150	--	-
18% chloroform + 13% *trans*-dichloroethylene + 20% trichloroethylene + 41% ethyl bromide + 8% ethyl chloride	<-150	--	-
25% methyl chloride + 75% methyl ether	-154	-20	+
Ethylene	-169.4	<-103.9	+
64.5% vol. *n*-pentane + 24.4% methylcyclohexane + 11.1% *n*-propyl alcohol	<-180	--	+
Ethane	-183.2	-89.0	+
Propane	-187.6	-42.1	+

[a] + means flammable; - means nonflammable.

that the substances do not cause immediate or strong pain. The hazards of breathing vapors should always be kept to a minimum by working in well-ventilated areas and limiting handling of reagents to short intervals, as is usual in the experiments. SAFETY GOGGLES MUST BE WORN DURING ALL CHEMICAL EXPERIMENTS.

Fire and explosives

Most polymerization reagents are highly flammable, including the majority of monomers and organic solvents. Many reagents, moreover, either ignite spontaneously in contact with air or moisture, or are actually explosives. Many metal-organic catalysts decompose explosively slightly above room temperature or in the presence of traces of oxygen or moisture. Generally, these compounds must be handled under inert liquids or in a glove box blanketed with inert gas. Specific precautions vary with the nature of the catalyst and are usually stated in each recipe. Peroxide initiators, and to a lesser extent azo initiators, must invariably be handled in minute quantities and protected from mechanical shock or flame to avoid the possibility of serious explosive decomposition.

The hazards in dealing with flammable organic vapors and monomers are more common but no less severe. These liquids or their vapors can be ignited not only by open flames but by sparks in electrical contacts such as motors, switches, or poorly protected wiring. Stirring motors and blenders, unless of the approved explosion-proof type, are typically hazardous in this respect. Unprotected motors of this type can be used, after careful deliberation, only if it can be demonstrated that the ventilation of the working space is adequate to guarantee sufficiently low concentrations of flammable vapors. Such unprotected motors can, for example, be purged by an inert gas flow when their use is essential.

Heat of reaction

Addition polymerization is invariably highly exothermic, since roughly the difference in enthalpy between a single and a double carbon-carbon bond (about 15-20 kcal) is released per mole of monomer polymerized (*Textbook*, pp. 16, 307). Unless this release of thermal energy is controlled, by heat transfer to the surroundings or by the boiling and refluxing of a volatile solvent, the reaction may go out of control with the distinct possibility of an explosion. For this reason, only moderate amounts are polymerized under laboratory conditions, and then only with the reaction vessels properly protected.

Test tubes must be wrapped in metal screen or wire mesh, or in some cases placed inside pipe "bombs" (Fig. 4-23), as for example in the bulk polymerization of styrene or methyl methacrylate. Bottles and flasks containing the mixes for emulsion or suspension polymerization are also placed in metal containers or screened by heavy wire mesh (Fig. 4-13). Serum stoppers, if used, may prevent damage by

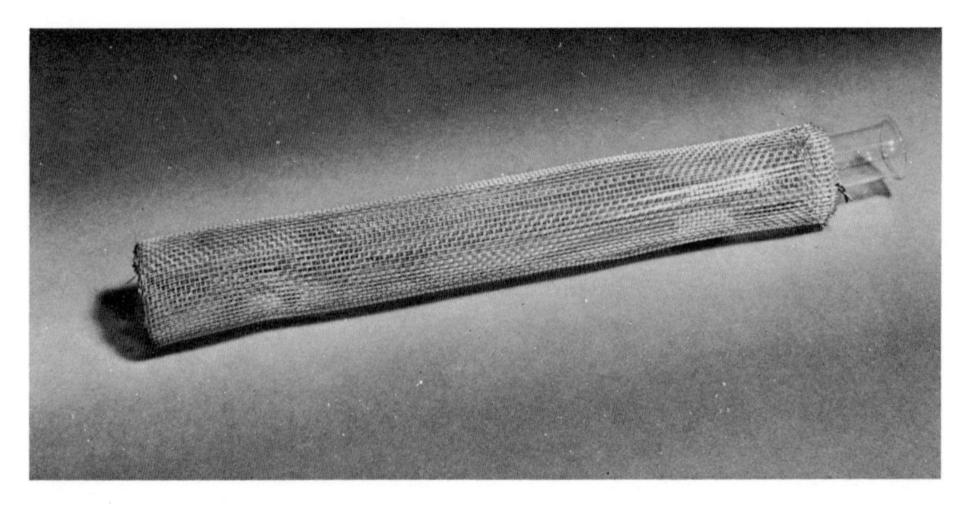

Fig. 4-23. A protective screen for bulk polymerization in test tubes.

blowing off. If the vessel must be handled during polymerization (for example, to withdraw a sample), it must always be wrapped in a cloth or towel. The entire polymerization system, including bath and reactor, must be placed behind a sufficiently large safety screen of tempered glass or plastic. SAFETY GOGGLES MUST BE WORN DURING ALL CHEMICAL EXPERIMENTS.

There is also danger of implosion of large evacuated glass vessels. Flasks or desiccators of over 300-500 ml capacity should be protected by wire mesh or wrapped in cloth before being evacuated. It is also advisable to inspect them for scratches or other stress concentrations before evacuating them.

Electrical and other hazards

All equipment must have proper electrical wiring and insulation, according to standard electrical-code regulations. No improvised wiring should be permitted at any time.

Some characterization equipment exhibits special hazards in addition to those described above--for example, the danger of exposure to x-radiation during the use of x-ray apparatus for structure analysis. All necessary precautions, as outlined in the experiments, must be observed.

GENERAL REFERENCES

Bohme 1961; Steere 1967*a*, *b*; Sax 1969; NFPA 1971*a*, *b*; MCA 1972.

BIBLIOGRAPHY

Bohme 1961. Charles W. Bohme, *Fire Protection for Chemicals*, National Fire Protection Association Publication No. SPP-3, Boston, Massachusetts, 1961.

Borchert 1969. A. E. Borchert, "Polymerization Procedures, Laboratory," pp. 304-311 in Herman F. Mark, Norman G. Gaylord, and Norbert M. Bikales, eds., *Encyclopedia of Polymer Science and Technology*, Vol. 11, Interscience Div., John Wiley and Sons, New York, 1969.

Braun 1971. Dietrich Braun, Harald Cherdron, and Werner Kern, *Techniques of Polymer Syntheses and Characterization*, Wiley-Interscience, New York, 1971.

Broughton 1956. Geoffrey Broughton, "Solvent Removal, Evaporation and Drying," pp. 787-839 in Arnold Weissberger, ed., *Separation and Purification*, Vol. III of A. Weissberger, ed., *Technique of Organic Chemistry*, 2nd ed., Part I, Interscience Div., John Wiley and Sons, New York, 1956.

Craig 1956. Lyman C. Craig and David Craig, "Laboratory Extraction and Countercurrent Distribution," pp. 149-331 in Arnold Weissberger, ed., *Separation and Purification*, Vol. III of A. Weissberger, ed., *Technique of Organic Chemistry*, 2nd ed., Part I, Interscience Div., John Wiley and Sons, New York, 1956.

Diamond 1966. P. S. Diamond and R. F. Denman, *Laboratory Techniques in Chemistry and Biochemistry*, D. Van Nostrand Co., Princeton, New Jersey, 1966.
Diels 1966. K. Diels and R. Jaeckel, *Leybold Vacuum Handbook*, Pergamon Press, New York, 1966.
Egly 1957. Richard S. Egly, "Heating and Cooling," pp. 51-182 in Arnold Weissberger, ed., *Laboratory Engineering*, Vol. III of A. Weissberger, ed., *Technique of Organic Chemistry*, 2nd ed., Part III, Interscience Div., John Wiley and Sons, New York, 1957.
Elliott 1966. J. R. Elliott, ed., *Macromolecular Syntheses*, Vol. 2, John Wiley and Sons, New York, 1966.
Espe 1966, 1968. Werner Espe, *Materials of High Vacuum Technology*, Pergamon Press, New York, Vol. 1, 1966; Vols. 2 and 3, 1968.
Fisher. Fisher Scientific Co., Pittsburgh, Pennsylvania 15219.
Gaylord 1969. Norman G. Gaylord, ed., *Macromolecular Syntheses*, Vol. 3, John Wiley and Sons, New York, 1969.
Giesekus 1967. Hanswalter Giesekus, "Turbidimetric Titration," Chap. C.1 in Manfred J. R. Cantow, ed., *Polymer Fractionation*, Academic Press, New York, 1967.
Guthrie 1963. Andrew Guthrie, *Vacuum Technology*, John Wiley and Sons, New York, 1963.
Guzmán 1961. G. M. Guzmán, "Fractionation of Polymers," pp. 113-183 in J. C. Robb and F. W. Peaker, eds., *Progress in High Polymers*, Vol. 1, Academic Press, New York, 1961.
Houston 1948. Robert J. Houston, "Improved Polymerization Techniques," *Anal. Chem. 20*, 49-51 (1948).
Inoue 1969. Takahashi Inoue, Toshiichi Soen, Takeji Hashimoto, and Hiromichi Kawai, "Thermodynamic Interpretation of Domain Structure in Solvent-Cast Films of A-B Type Block Copolymers of Styrene and Isoprene," *J. Polymer Sci. A-2 7*, 1283-1302 (1969).
I^2R. Instruments for Research and Industry, Cheltenham, Pennsylvania 19102.
Kennedy 1971. Joseph P. Kennedy, P. L. Magagnini, and Peter H. Plesch, "Criticism of Claims in the Field of Isomerization Polymerization. I. Polymerization of *p*-Methylstyrene," *J. Polymer Sci. A-1 9*, 1635-1646 (1971).
Kontes. Kontes Glass Co., Vineland, New Jersey 08360.
MCA 1972. General Safety Committee of the Manufacturing Chemists Association, *Guide for Safety in the Chemical Laboratory*, 2nd ed., D. Van Nostrand Co., Princeton, New Jersey, 1972.
McCaffery 1970. Edward M. McCaffery, *Laboratory Preparation for Macromolecular Chemistry*, McGraw-Hill Book Co., New York, 1970.
Miller 1957. Glenn H. Miller, "Operations with Gases," pp. 283-374 in Arnold Weissberger, ed., *Laboratory Engineering*, Vol. III of A. Weissberger, ed., *Technique of Organic Chemistry*, 2nd ed., Part II, Interscience Div., John Wiley and Sons, New York, 1957.
NFPA 1971a. *Manual of Hazardous Chemical Reactions 1971*, National Fire Protection Association Publication No. 491 M, Boston, Massachusetts, 1971.
NFPA 1971b. *Hazardous Chemicals Data 1971*, National Fire Protection Association Publication No. 49, Boston, Massachusetts, 1971.
Overberger 1963. C. G. Overberger, ed., *Macromolecular Syntheses*, Vol. 1, John Wiley and Sons, New York, 1963.

Pinner 1961. S. H. Pinner, *A Practical Course in Polymer Chemistry*, Pergamon Press, New York, 1961.

Sax 1969. N. Irving Sax, ed., *Dangerous Properties of Industrial Materials*, 3rd ed., Reinhold Publishing Co., New York, 1969.

Sorenson 1968. Wayne R. Sorenson and Tod W. Campbell, *Preparative Methods of Polymer Chemistry*, 2nd ed., Interscience Div., John Wiley and Sons, New York, 1968.

Steere 1967a. Norman V. Steere, ed., *Safety in the Chemical Laboratory*, American Chemical Society, Division of Chemical Education, Easton, Pennsylvania, 1967.

Steere 1967b. Norman V. Steere, ed., *CRC Handbook of Laboratory Safety*, Chemical Rubber Company, Cleveland, Ohio, 1967.

Ueno 1966. K. Ueno, K. Hayashi and S. Okamura, "Studies on Radiation-Induced Ionic Polymerization III-Polymerization of Styrene at Room Temperature," *Polymer 7*, 431-458 (1966).

5

Following the Course of Polymerization

As indicated in Chap. 1 and the *Textbook* (for example, p. 287), the experimentally available quantities for following the course of a polymerization are only the polymerization rate and the molecular weight of the resulting polymer. From these two measurements one can derive many facts about the nature of the polymerization kinetics and mechanism, as outlined in part in Chaps. 1 and 2 and in part in the *Textbook*. In this chapter we describe the most important ways by which the polymerization rate is determined, through measurement of the extent of polymerization or conversion as a function of time.

Following the usual practice (Chap. 1B and the *Textbook*, p. 265), the *extent of reaction, p,* is defined as the fraction of the functional groups (including the double bonds in vinyl polymerization) which has reacted. The *conversion* is essentially the same quantity, though often expressed as per cent, and the two terms can be used interchangeably. In this book, however, we reserve the word conversion to describe the result of chain or addition polymerization, in which monomer is converted directly to high polymer in a single rapid step without the isolation of products of intermediate degree of polymerization.

The measurement of molecular weight is described in Chapter 7.

GENERAL REFERENCES

Kline 1959; Fettes 1964.

A. *Direct Weighing*

Probably the most direct method of determining the conversion of monomer to polymer is to stop the polymerization, and isolate and weigh the resulting polymer. This is practiced, for example, in Exps. 3 and 5, but it has some disadvantages, notably in that further polymerization of the same sample is not possible.

In some cases, the polymerization must be stopped rapidly by the addition of a fast-acting shortstop (Chap. 3C). For small quantities of material, however, as in the experiments cited, it may be adequate to dilute the system rapidly in the presence of air and moisture or other reagents which will react rapidly with any remaining initiator or catalyst.

The isolation of the polymer can be done in either of two ways. One is distillation of remaining monomer and other volatile materials from the polymer (Chap. 4B). This procedure usually requires the use of a shortstop, and must be done under conditions where no further polymerization, depolymerization, or thermal degradation of the polymer takes place. A correction must be made for the weight of any non-volatile components, such as initiator or shortstop, that may remain behind. Distillation can be used for recovering polymer from solution, emulsion, suspension, or low-conversion bulk polymerizations of such monomers as styrene, methyl methacrylate, or vinyl acetate. It cannot be used, however, in some bulk polycondensations, for example that of ω-aminoundecanoic acid (Exp. 1) even for low extent of polymerization (because too much of the polymer is still present as volatile oligomers), or for bulk polymerizations at high extent of reaction (because of the difficulty of removing monomer from the viscous high-polymer mass; see Chap. 4B). The determination is also time-consuming.

Alternatively, as in Exps. 3 and 5, the polymer can be isolated by precipitation and subsequent drying, or by the extraction of residual monomer and other volatiles from finely divided polymer. These techniques are also discussed in Chap. 4B. Again there are disadvantages, for here a correction should be made for the solubility of oligomers under the conditions used, but this is not usually known. In addition, the extra handling of the polymer in the steps of precipitation or extraction, filtration, washing, and drying, may well lead to unavoidable losses reducing the precision attainable.

B. *Chemical Methods*

The chemical determination of the remaining monomer functional groups at some stage of the polymerization is often practiced, particularly for step reactions. In fact, as in Exp. 9, this method leads not only to the extent of reaction, but through the Carothers equation relating p and degree of polymerization, to the number-average molecular weight of the polymer in the case of linear polycondensation. Double bonds can also be determined chemically, so that this class of methods is applicable to chain polymerization as well.

The chemical determinations are usually carried out by titration; although any suitable analytical technique can be used. The titration can be direct; examples are the titration of remaining carboxyl groups in polyesterification by means of standard base (Patton 1962), and the titration of amine groups in the polycondensation of ω-aminoundecanoic acid (Exp. 1) by standard acid.

Back titration is often useful, and may be preceded by the reaction of the monomer or functional group. One example from polyesterification is the reaction of hydroxyl groups with acetic anhydride in pyridine solution, followed by dilution and back titration of the resulting acetic acid and carboxyl end groups on the polymer with standard base. This analysis provides the sum of the hydroxyl and carboxyl ends (Sorenson 1968, pp. 154-5).

A second example, from addition polymerization, is the determination of such monomers as styrene or methyl methacrylate by bromination of the double bonds (Kolthoff 1947). In one case, the styrene remaining in a solution polymerization in CCl_4 was reacted with a standard solution of bromine in acetic acid. The excess bromine was reacted with potassium iodide, and the free iodine titrated with thiosulfate. In another study, small amounts of styrene were extracted from the polymer by vacuum drying and trapped in a cold trap, then subsequently determined by bromination (Lewis 1945).

GENERAL REFERENCES

Price 1959; Hellman 1962; Brauer 1965; Palit 1968.

C. *Dilatometry*

It can readily be shown that there is a linear relationship between the degree of conversion in a polymerizing system and its volume, provided only that the monomer and polymer mix without change in volume. The volume change on polymerization can be assumed to be a constant increment for each monomer used up. For chain polymerization, this increment may be thought of as being the difference in volume occupied by two "ends," one of a monomer and the other that of the growing chain, and the volume of the carbon-carbon bond formed when one monomer is added to the chain. It is clear that when a small molecule is formed as in step-reaction polymerization, some of this volume increment is recovered. In addition, the small molecule may go into the vapor phase. Hence, dilatometry is of little value, and we restrict this discussion to addition polymerization.

Using specific volumes $\overline{V}$* (the inverse of densities) expressed in ml/g, and adopting the subscripts 1 for monomer, 2 for polymer and p for the mixture at extent of reaction p, the rule of additivity of volumes during mixing leads to

$$\overline{V}_p = c_1 \overline{V}_1 + c_2 \overline{V}_2 \tag{5-1}$$

where c is concentration, and in a system containing monomer and polymer only $c_1 + c_2 = 1$. Solving,

$$c_2 = p = (\overline{V}_p - \overline{V}_1) \,/\, (\overline{V}_2 - \overline{V}_1) \tag{5-2}$$

Since the denominator is a constant, the polymer concentration, which is the conversion, is proportional to the change in volume as the reaction proceeds. This is amply confirmed experimentally. If there is a diluent in the system, it can be shown that the proportionality is

*The symbol differs slightly from that used in the *Textbook*.

retained, but the value of the constant denominator is changed.

The measurement of the volume change is complicated if there are any gases dissolved in the monomer or polymer, which may separate as bubbles at the temperature of polymerization, or if a gas is generated in the reaction, as, for example, in the decomposition of AIBN (Chap. 3B). Dissolved gases should be removed by degassing the monomer and diluents before polymerization. Another complication can arise if the polymer is insoluble in the polymerizing mixture, since the density of the precipitated form may not be well known.

Dilatometers

For following the volume change during polymerization, the reaction is carried out in a *dilatometer*, a simple form of which is shown in Fig. 5-1. The design and construction of these small glass instruments depend on a number of considerations, such as the expected volume

Fig. 5-1. A simple dilatometer.

change and the accuracy with which it must be measured, the rate of polymerization, the corresponding rate of heat evolution and its effect on temperature uniformity, the viscosity of the polymerizing mixture, and the need for stirring. If, as is usual, a capillary is used to display the volume change, the dilatometer becomes a good thermometer, and it is essential both to control the temperature to about ± 0.01°C and to maintain it uniform throughout the vessel. The diameter and length of the capillary needed can be estimated, for a given dilatometer volume, from the anticipated degree of conversion and the specific volumes or densities involved. The latter are given in Table 5-1.

In practice the volume change is detected by measuring, with a vertically aligned cathetometer, the height of the liquid meniscus in the dilatometer capillary. Keeping the same subscripts, it is easy to show that Eq. 5-2 transforms into

$$p = (h_p - h_1) / (h_2 - h_1) \qquad (5\text{-}3)$$

Since h_2 is not usually available, it is more convenient to retain the denominator of Eq. 5-2, writing

$$p = (h_p - h_1)\, s / (\overline{V}_2 - \overline{V}_1)\, m \qquad (5\text{-}4)$$

where s is the cross-sectional area of the capillary and m the mass of the sample.

If the viscosity of the polymerizing mixture is low (usually implying a restriction to low conversion), the rate of polymerization is not too fast to avoid problems of heat transfer, and no stirring is required, a simple dilatometer like that of Fig. 5-1 suffices. Where better control of purity of the system is required, the instrument need be only slightly more complex. For example, a study of the rate of vinyl-acetate polymerization was made in the dilatometer shown in Fig. 5-2 (Starkweather 1930). Monomer was introduced into bulb B and frozen. The dilatometer was evacuated and sealed off at E. Then bulb C was cooled and B allowed to warm to room temperature, so that volatile impurities collected in C. The dilatometer was then sealed off at D, and the monomer distilled into A, leaving nonvolatile impurities behind. Finally, the instrument was sealed off or disconnected at F.

More complicated dilatometers have been developed featuring elaborate filling procedures assuring higher purity (Goldfinger 1948), low thermal inertia so that rapid reactions can be studied (Schulz 1947), stirring which is essential in emulsion polymerization (Corrin 1947, Paoletti 1964), and automatic recording of the volume change (Bell 1961, Niezette 1971).

GENERAL REFERENCES

Rubens 1965, 1966; Suh 1968; Bauer 1971.

TABLE 5-1. Densities of monomers and polymers and volume changes on polymerization

Monomer	Density, g/ml, 25°C		Volume Change, %
	Monomer	Polymer	
Vinyl chloride	0.919	1.406	34.4
Acrylonitrile	0.800	1.17	31.0
Vinylidene bromide	2.178	3.053	28.7
Vinylidene chloride[a]	1.213	1.71	28.6
Vinyl bromide	1.512	2.075	27.3
Methacrylonitrile	0.800	1.10	27.0
Methyl acrylate	0.952	1.223	22.1
Vinyl acetate[a]	0.934	1.191	21.6
Methyl methacrylate	0.940	1.179	20.6
Diallyl succinate	1.056	1.30	18.8
Ethyl methacrylate	0.911	1.11	17.8
Diallyl maleate	1.077	1.30	17.2
Ethyl acrylate	0.919	1.095	16.1
n-Butyl acrylate	0.894	1.055	15.2
n-Propyl methacrylate	0.902	1.06	15.0
Styrene	0.905	1.062	14.5
n-Butyl methacrylate	0.889	1.055	14.3

[a] 20°C.

D. *Refractometry*

The determination of extent of reaction by means of refractive-index measurement is based upon the dependence of the refractive index of liquids upon their chemical composition and molecular structure. Since the structure of a polymer differs from that of the monomer because of the rearrangement of chemical bonds, the refractive index of

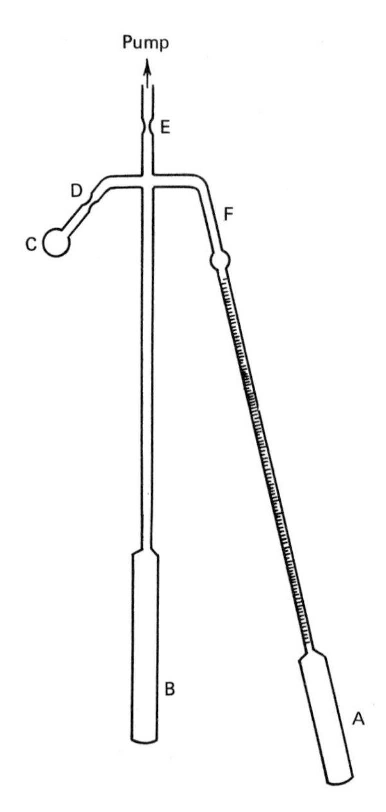

Fig. 5-2. Dilatometer used in vinyl-acetate polymerization (Starkweather 1930).

a polymer differs from that of the monomer. Refractive index also depends significantly on the temperature, and slightly (negligibly for liquids) on the pressure. To obtain precision in the fourth decimal of refractive index requires temperature control to ± 0.02°C.

Generally, the dependence of the refractive index on conversion in a monomer-polymer or monomer-diluent-polymer system is best determined by calibration using systems of known concentration. In some cases, however, calculations based on the additivity of the specific refraction of the components give sufficiently accurate answers; the fact that monomer and polymer are chemically similar, except for functional groups, is favorable.

The specific refraction r is given most accurately by the Lorentz-Lorenz equation

$$r = \overline{V}\,(n^2 - 1)\,/\,(n^2 + 2) \tag{5-5}$$

where n is refractive index. The additivity of r is expressed by the equation

$$r_p = c_1 r_1 + c_2 r_2 \qquad (5\text{-}6)$$

where the subscripts have the same meaning as in Sec. C. Since monomer, polymer and the polymerizing mixture all have similar values of n, the expression for r can be simplified to the Gladstone-Dale rule, $r = \overline{V}\ (n-1)$, for use in Eq. 5-6, leading to

$$\overline{V}_p\ (n_p - 1) = c_1\ \overline{V}_1\ (n_1 - 1) + c_2\ \overline{V}_2\ (n_2 - 1) \qquad (5\text{-}7)$$

Remembering that $c_1 + c_2 = 1$, and using the additivity of specific volumes expressed in Eq. 5-1 allows solution for the conversion as

$$p = c_2 = \frac{\overline{V}_1\ (n_p - n_1)}{\overline{V}_2\ n_2 - \overline{V}_1\ n_1 - (\overline{V}_2 - \overline{V}_1)n_p} \qquad (5\text{-}8)$$

Values of the density $d = 1/\overline{V}$ and the refractive index n for some common monomers and polymers are given in Tables 5-1 and 5-2. For the system methyl methacrylate-poly(methyl methacrylate), inserting the numerical values leads to the relation

$$p = 1.0684\ (n_p - 1.4140)\ /\ (0.2137\ n_p - 0.2359) \qquad (5\text{-}9)$$

which is used to determine conversion from refractive index in Exp. 3.

Although it requires removal of a drop of sample per determination, refractometry is a simple and rapid means of following polymerization rate. A variation of the technique, allowing the use of sealed cylindrical tubes as reaction vessels and the measurement of refractive index in situ, was described by Naylor (1953) and Pulley (1968).

GENERAL REFERENCES

Bauer 1960; Brauer 1962.

E. *Other Methods*

In addition to the four methods described above, each of which requires only simple equipment and a minimum of calibration, several other techniques for following the course of polymerization are useful from time to time.

Viscometry

The use of viscosity measurements (Chap. 7C) for the determination of the extent of polymerization is complicated by the fact that the viscosity depends upon both polymer concentration and molecular weight.

TABLE 5-2. Refractive indices for some polymers and monomers

Material	n_d^{20}	
	Monomer	Polymer
Vinyl chloride[a]	1.380	1.545
Acrylonitrile[b]	1.3888	1.518
Vinyl acetate	1.3956	1.4667
Methacrylonitrile[b]	1.401	1.520
Methyl acrylate	1.4021	1.4725
Ethyl acrylate	1.4068	1.4685
Ethyl methacrylate[b]	1.4143	1.485
Methyl methacrylate	1.4147	1.492
n-Butyl acrylate	1.4190	1.4634
n-Propyl methacrylate	1.4191	1.484
n-Butyl methacrylate	1.4239	1.4831
Vinylidene chloride	1.4249	1.654
Styrene	1.5458	1.5935

a n_d^{15}.

b n_d^{25}.

An increase in viscosity of the polymerization mixture reveals only that conversion is increasing, unless reference is made to a calibration curve obtained by using another method.

Infrared spectroscopy

Absorption of infrared radiation (*Textbook*, Chap. 4B) is usually utilized to identify and determine the concentrations of specific functional groups through characteristic absorption frequencies associated with vibrational and rotational modes. Because of the change in structure between monomer and polymer, their infrared spectra often show characteristic differences which can be used to follow the course of polymerization with moderate accuracy.

An example is the polymerization of methyl methacrylate. The infrared spectra of the monomer and polymer are shown in Fig. 5-3. Bands characteristic of the monomer which disappear on polymerization include that due to C=C stretching vibrations in the $>C=CH_2$ group at about 1630 cm^{-1}, to CH_2 wagging in the same group at 1300 and 1330 cm^{-1}, and to CH bending modes in the same group (shifted from their normal frequencies by the nearby double bond and oxygens) at 930 cm^{-1} and 800 cm^{-1}. Frequency shifts also occur in the C-O-C stretching absorptions in the 1100-1300 cm^{-1} region as the double bonds disappear. At the same time, a band characteristic of the polymer resulting from rocking modes in isolated $-CH_2-$ groups appears at 750 cm^{-1}, and the methyl rock band, shifted to 1020 cm^{-1} by the presence of the double bond in the monomer, returns to its normal frequency of 980 cm^{-1} (Day 1972).

One advantage to the use of the infrared absorption method is that, using a special polymerization cell mounted in the spectrophotometer, the measurements can be made without interfering with the polymerization. Otherwise, the reaction mixture can be sampled, with the spectroscopic measurements following. In some cases, useful information on the course of polymerization can be obtained by measurement in the ultraviolet region, but this is less common.

NMR and EPR spectroscopy

These absorption spectroscopic methods (*Textbook*, Chap. 4D) are useful for identifying the chemical surroundings of protons (or some other nuclei less widely studied), in *nuclear magnetic resonance* spectroscopy, or unpaired electrons, in *electron paramagnetic* or *electron spin resonance* spectroscopy. The absorption bands are at radio frequencies, for example 60 MHz, and result from a transition in the spin of the nucleus or electron between orientations parallel and antiparallel to an external magnetic field. The exact absorption frequencies depend upon the immediate chemical surroundings of the nuclei or electrons in the material studied, and as these change during polymerization, the spectrum changes as well.

The equipment for NMR and EPR is relatively complicated and expensive, and its use is justified only in special cases. NMR studies

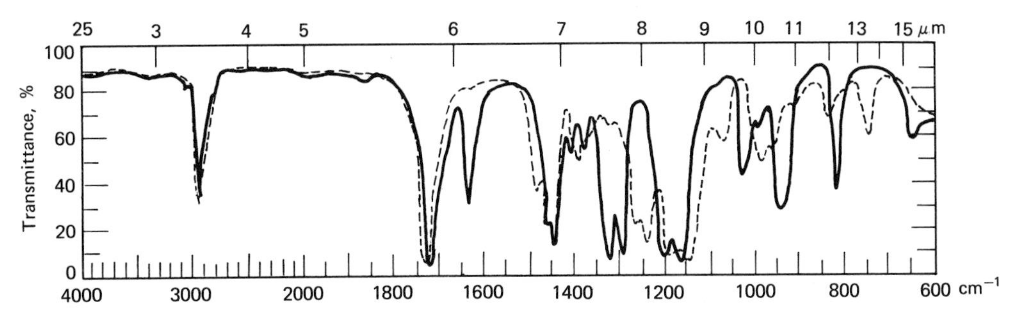

Fig. 5-3. Infrared spectra of methyl methacrylate monomer and polymer (Stanley 1959).

are of particular interest because this is one of the very few methods which gives information on stereoregularity and tacticity (*Textbook*, Chaps. 5A, 10D). EPR gives information about the concentration of free radicals, which is otherwise unknown in steady-state polymerizations (*Textbook*, Chap. 9D).

Dielectric constant and conductance

During the course of a polymerization both the dielectric constant (and the related loss factor, both defined in the *Textbook*, p. 132) and the conductance of the mixture may change. Conductivity measurement is well suited for polymerizations in which the monomer participates in charge-transfer reactions, such as the polycondensation of alkyd resins (Kienle 1934). The determination of dielectric constant or loss requires a nonconductive monomer. These quantities can be expected to decrease as polymerization proceeds, as the mobility of the dipole of the monomer becomes severely restricted when it is incorporated into the chain.

The methods require that the polymerization be carried out in a conductometric cell or a measuring condenser. Because the quantitative interpretation of changes in the electrical properties in terms of extent of polymerization is difficult, it is best to calibrate these methods against another technique. Both determinations are rather sensitive to small concentrations of impurities, but they may be useful when it is not possible to remove samples during the polymerization.

Gas chromatography

Gas chromatography is perhaps better suited for following the course of polymerization than any other method, yet to date it has not been extensively used. This is surprising, since it is simple and rapid, has more than adequate resolution, and utilizes widely available equipment. No difficulty is experienced in analyzing multicomponent systems, such as copolymers with several monomers, for which other methods may not be applicable or require extensive calibration.

This method is of course ideally suited for gaseous monomers, though few if any monomers have such low volatility that they cannot be analyzed. Minor but not insurmountable problems can occur when the polymer is soluble in its monomer, or when the analysis requires a sample inlet temperature high enough to degrade or depolymerize the polymer.

GENERAL REFERENCES

Viscometry: Embrie 1951; Bamford 1958.
Infrared spectroscopy: Tryon 1962; Zbinden 1964; Hummel 1966; Kössler 1967; Elliott 1969.
NMR and EPR: Wall 1962; Bresler 1966; Bovey 1968.

Dielectric constant and conductance: Böttcher 1952; Smyth 1955; Stock 1965; Aélion 1968; Vaughan 1971.

Gas chromatography: Guillot 1968; Stevens 1969; Johnston 1970; Mano 1970.

BIBLIOGRAPHY

Aélion 1968. M. R. Aélion, "Preparation and Structure of Some New Types of Polyamides" (in French), *Ann. Chim.* Ser. 14 *3*, 5-6 (1968).

Bamford 1958. C. H. Bamford, W. G. Barb, A. D. Jenkins, and P. F. Onyon, *The Kinetics of Vinyl Polymerization by Radical Mechanisms,* Butterworths, London, 1958, pp. 32-38.

Bauer 1960. N. Bauer, K. Fajans and S. Z. Lewin, "Refractometry," pp. 1139-1282 in Arnold Weissberger, ed., *Physical Methods of Organic Chemistry,* Vol. I of A. Weissberger, ed., *Technique of Organic Chemistry,* 2nd ed., Part II, Interscience Div., John Wiley and Sons, New York, 1960.

Bauer 1971. Norman Bauer and Seymour Z. Lewin, "Determination of Density," pp. 57-124 in Arnold Weissberger and Bryant W. Rossiter, eds., *Determination of Mass Transport, and Electrical-Magnetic Properties,* Vol. I of A. Weissberger, ed., *Techniques of Chemistry,* 3rd ed., Part 4, Interscience Div., John Wiley and Sons, New York, 1971.

Bell 1961. C. L. Bell, "Automatic Recording Dilatometer," *J. Sci. Instr. 38,* 27-28 (1961).

Bovey 1968. F. A. Bovey, "Nuclear Magnetic Resonance," pp. 356-396 in Herman F. Mark, Norman G. Gaylord, and Norbert M. Bikales, eds., *Encyclopedia of Polymer Science and Technology,* Vol. 9, Interscience Div., John Wiley and Sons, New York, 1968.

Böttcher 1952. Carl Johan Friedrich Böttcher, *Theory of Electric Polarization,* Elsevier Publishing Co., New York, 1952.

Brauer 1962. G. M. Brauer and E. Horowitz, "Index of Refraction," pp. 45-53 in Gordon M. Kline, ed., *Analytical Chemistry of Polymers,* Part III, Interscience Div., John Wiley and Sons, New York, 1962.

Brauer 1965. G. M. Brauer and G. M. Kline, "Chemical Analysis," pp. 632-655 in Herman F. Mark, Norman G. Gaylord and Norbert M. Bikales, eds., *Encyclopedia of Polymer Science and Technology,* Vol 3, Interscience Div., John Wiley and Sons, New York, 1965.

Bresler 1966. S. E. Bresler and E. N. Kazbekow, "Electron-Spin Resonance," pp. 669-692 in Herman F. Mark, Norman G. Gaylord, and Norbert M. Bikales, eds., *Encyclopedia of Polymer Science and Technology,* Vol. 5, Interscience Div., John Wiley and Sons, New York, 1966.

Corrin 1947. M. L. Corrin, "Kinetic Treatment of Emulsion Polymerization," *J. Polymer Sci. 2*, 257-262 (1947).

Day 1972. Band assignments kindly furnished by Charles Day, Du Pont Company, January 1972.

Elliott 1969. Arthur Elliott, *Infrared Spectra and Structure of Organic Long-Chain Polymers,* St. Martin's Press, New York, 1969.

Embrie 1951. W. H. Embrie, J. M. Mitchell and H. Laverne Williams, "Compositional Heterogeneity of Butadiene-Acrylonitrile Copolymers Prepared in Emulsion at 5°C," *Can. J. Chem. 29,* 253-260 (1951).

Fettes 1964. E. M. Fettes, ed., *Chemical Reactions of Polymers,* Interscience Div., John Wiley and Sons, New York, 1964.

Goldfinger 1948. G. Goldfinger and K. E. Lauterback, "Initial Rate of Polymerization of Styrene," *J. Polymer Sci. 3,* 145-156 (1948).

Guillot 1968. Jean Guillot, "Kinetic Study of Polymerization Reactions by Gas Chromatography" (in French), *Ann. Chim.* Ser. 14 *3,* 441-456 (1968).

Hellman 1962. Max Hellman and Leo A. Wall, "End Group Analysis," pp. 397-419 in Gordon M. Kline, ed., *Analytical Chemistry of Polymers,* Part III, *Identification Procedures and Chemical Analysis,* Interscience Div., John Wiley and Sons, New York, 1962.

Hummel *1966.* Dieter O. Hummel, *Infrared Spectra of Polymers: in the Medium and Long Wavelength Region,* John Wiley and Sons, New York, 1966.

Johnston 1970. H. Kirk Johnston and Alfred Rudin, "Reactivity Ratios from Gas-Liquid Chromatography," *J. Paint Technol. 42,* 429-434 (1970).

Kienle 1934. R. H. Kienle and H. H. Race, "The Electrical, Chemical, and Physical Properties of Alkyd Resins," *Trans. Electrochem. Soc. 65,* 87-107 (1934).

Kline 1959. Gordon M. Kline ed., *Analytical Chemistry of Polymers,* Part I, Interscience Publishers, New York, 1959.

Kolthoff 1947. I. M. Kolthoff and F. A. Bovey, "Amperometric Titration of Styrene with Potassium Bromate," *Ind. Eng. Chem., Anal. Ed. 19,* 498-500 (1947).

Kössler 1967. Ivo Kössler, "Infrared-Absorption Spectroscopy," pp. 620-642 in Herman F. Mark, Norman G. Gaylord, and Norbert M. Bikales, eds., *Encyclopedia of Polymer Science and Technology,* Vol. 7, Interscience Div., John Wiley and Sons, New York, 1967.

Lewis 1945. Frederick M. Lewis and Frank R. Mayo, "Precise Method for Isolation of High Polymers," *Ind. Eng. Chem., Anal. Ed. 17,* 134-136 (1945).

Mano 1970. Eloisa Mano and Roberto Riva De Almeida, "A Convenient Technique for Determination of Reactivity Ratios," *J. Polymer Sci. A-1 8,* 2713-2716 (1970).

Naylor 1953. M. A. Naylor and F. W. Billmeyer, Jr., "A New Apparatus for Rate Studies Applied to the Photopolymerization of Methyl Methacrylate," *J. Am. Chem. Soc. 75,* 2181-2185 (1953).

Nichols 1950. Frank E. Nichols and Ralph G. Flowers, "Prediction of Shrinkage in Addition Polymerizations," *Ind. Eng. Chem. 42,* 292-295 (1950).

Niezette 1971. J. Niezette and V. Desreux, "An Automatic Dilatometer for Radiation Polymerizations," *J. Applied Polymer Sci. 15,* 1981-1983 (1971).

Palit 1968. Santi R. Palit and Broja Mohan Mandal, "End Group Studies Using Dye Techniques," *J. Macromol Sci.--Revs. Macromol. Chem. C2,* 225-277 (1968).

Paoletti 1964. K. P. Paoletti and F. W. Billmeyer, Jr., "Absolute Propagation Rate Constants for the Radical Polymerization of Substituted Styrenes," *J. Polymer Sci. A2,* 2049-2062 (1964).

Patton 1962. Temple C. Patton, *Alkyd Resin Technology: Formulating Techniques and Allied Calculations,* Interscience Div., John Wiley and Sons, New York, 1962.

Price 1959. G. F. Price, "Techniques of End-Group Analysis," Chap. 7 in P. W. Allen, ed., *Techniques of Polymer Characterization*, Butterworths, London, 1959.

Pulley 1968. Jack Pulley, "Polymerization Kinetics of Methyl Methacrylate at High Conversions," M. Sc. Thesis, Rensselaer Polytechnic Institute, Department of Materials, 1968.

Rubens 1965. L. C. Rubens and R. E. Skochdopole, "Continuous Measurement of Polymerization Rates Over the Entire Conversion Range with a Recording Dilatometer," *J. Applied Polymer Sci. 9*, 1487-1497 (1965).

Rubens 1966. L. C. Rubens and R. E. Skochdopole, "Dilatometry," pp. 83-98 in Herman F. Mark, Norman G. Gaylord, and Norbert M. Bikales, eds., *Encyclopedia of Polymer Science and Technology*, Vol. 6, Interscience Div., John Wiley and Sons, New York, 1966.

Schulz 1947. G. V. Schulz and G. Harborth, "On a Dilatometric Method of Following the Progress of Polymerization" (in German), *Angew. Chem. 59A*, 90-92 (1947).

Smyth 1955. Charles Phelps Smyth, *Dielectric Behavior and Structure; Dielectric Constant and Loss, Dipole Moment and Molecular Structure*, McGraw-Hill Book Co., New York, 1955.

Sorenson 1968. Wayne R. Sorenson and Tod W. Campbell, *Preparative Methods of Polymer Chemistry*, 2nd ed., Interscience Div., John Wiley and Sons, New York, 1968.

Stanley 1959. Edward L. Stanley, "Acrylic Plastics," pp. 1-16 in Gordon M. Kline, ed., *Analytical Chemistry of Polymers*, Part I, Interscience Publishers, New York, 1959.

Starkweather 1930. Howard W. Starkweather and Guy B. Taylor, "The Kinetics of the Polymerization of Vinyl Acetate," *J. Am. Chem. Soc. 52*, 4708-4714 (1930).

Stevens 1969. Malcolm P. Stevens, *Techniques and Methods of Polymer Evaluation*, Vol. 3, *Characterization and Analysis of Polymers by Gas Chromatography*, Marcel Dekker, New York, 1969.

Stock 1965. John Thomas Stock, *Amperometric Titrations*, Interscience Div., John Wiley and Sons, New York, 1965.

Suh 1968. K. W. Suh, V. E. Meyer, L. C. Rubens and R. E. Skochdopole, "Note on the Continuous Measurement of Polymerization Rates with a Recording Dilatometer," *J. Applied Polymer Sci. 12*, 1803-1805 (1968).

Tryon 1962. M. Tryon and E. Horowitz, "Infrared Spectrophotometry," Chap. VIII in Gordon M. Kline, ed., *Analytical Chemistry of Polymers*, Part II, Interscience Div., John Wiley and Sons, New York, 1962.

Vaughan 1971. Worth E. Vaughan, Charles P. Smyth, and Jack Gordon Powles, "Determination of Dielectric Constant and Loss," pp. 351-396 in Arnold Weissberger and Bryant W. Rossiter, eds., *Determination of Mass Transport and Electrical-Magnetic Properties*, Vol. I of A. Weissberger, ed., *Techniques of Chemistry*, 3rd ed., Part 4, Interscience Div., John Wiley and Sons, New York, 1971.

Wall 1962. Leo A. Wall and Roland E. Florin, "Magnetic Resonance Spectroscopy," Chap. XII in Gordon M. Kline, ed., *Analytical Chemistry of Polymers*, Part II, Interscience Div., John Wiley and Sons, New York, 1962.

Zbinden 1964. Rudolf Zbinden, *Infrared Spectroscopy of High Polymers*, Academic Press, New York, 1964.

II

INTRODUCTION TO POLYMER CHARACTERIZATION TECHNIQUES

6

Preliminary Evaluation of Polymer Properties

All scientists coming into contact with a new polymeric material for the first time are faced with the same problem: What can be found out about the useful chemical and physical properties of this substance with a minimum expenditure of time and effort, using little in the way of expensive or unusual equipment, and requiring only a small amount of the material? Although ultimate goals may be different, this question supplies immediate objectives equally well for a student in a polymer science laboratory course, an industrial polymer chemist who has synthesized a new material, or an engineer or designer concerned with the proper selection of a material for a specific end use.

The purpose of this chapter is to provide a partial answer to the question in as systematic a way as possible, through the development of a minimal scheme for the initial evaluation of a polymeric material. This scheme (which is not unique; see Sherr 1965, Saunders 1966) attempts to provide the maximum amount of meaningful information commensurate with the restrictions outlined in the question above. The interpretation of this information will permit the user to make an intelligent estimate of the potential usefulness of an experimental material, provide a basis for decision regarding the desirability of further testing, and aid in the selection of tests to provide more quantitative information about the polymer.

The importance of carrying out the minimal-scheme evaluation firsthand cannot be emphasized strongly enough. There is no substitute for "getting the feel" of a polymer by direct handling and keen observation using rather simple techniques. Again, this is important both for the student meeting polymers for the first time and for the practicing polymer scientist evaluating a new (to him) material. No scheme, of course, can take the place of originality and discretion in the evaluation of a polymer. The scientist must use his judgment in eliminating tests that are not applicable to his material, and in devising others as needed. He should be hesitant to discard a new polymer (if such a decision is called for) on the basis of qualitative tests alone, or if the material appears to have a single deficiency. It is often possible to find uses for polymers which have one or more apparent deficiencies. They may be overcome, in many cases, by slight modifications of structure or by incorporating additives in a compounding step. Or, an application may be found where the deficiencies are of little consequence or even provide a unique combination of properties which is highly desirable for a specific end use.

The wide variety of physical and chemical properties which might be explored in a minimal evaluation scheme fall into five groups: thermal properties, solubility, stability, mechanical behavior, and spectral evidence for composition and molecular architecture. Each of these five is discussed in the sections to follow. In keeping with the objective to provide simple tests, many standard techniques for the analysis and characterization of polymers are not discussed, but reference is made to the literature at this point.

GENERAL REFERENCES

Kline 1959-1962; Nielsen 1962; Ke 1964; Haslam 1965; Schmitz 1965-1968; Saunders 1966; Slade 1966-1970; Porter 1968.

A. Thermal Characteristics

If any single property of polymers could be singled out as most important, it would without doubt be their thermal behavior, for no other class of materials shows similar characteristic changes as a function of temperature, and most of the end-use applications of polymers take advantage of their unique thermal behavior, for example in processing. This section describes three types of evaluation of thermal behavior: a general exploratory approach; the problem of inflammability, an end-use characteristic of great importance in today's ecology; and the use of thermal degradation or pyrolysis as an aid to analysis. These properties are easy to measure or use qualitatively and at least semiquantitatively in a variety of ways. Thus, the minimal scheme is not rigidly fixed, but can be modified to reflect the availability of equipment, the precision of results needed, the properties of interest, and the time available for testing.

Qualitative thermal analysis

A surprisingly large number of thermal properties can be explored quickly and easily using a "hot block," Fisher-Johns melting-point apparatus, hot stage microscope, or Kofler hot bench. Studies with one of these simple pieces of apparatus (Figs. 6-1 and 6-2) form the heart of this minimal scheme, and yield the following types of information from a few milligrams of sample in 10-20 minutes.

Thermoplasticity It is possible to differentiate between thermoplastic and thermosetting materials. In the absence of degradation (see below), a thermoplastic can be heated and cooled repeatedly without appreciable changes in fluidity or plasticity, whereas a thermoset will quickly harden and become intractable.

Fig. 6-1. Determination of polymer melt temperature using a hot block (Sorenson 1968).

Softening temperature Probing a small sample with forceps or spatula while the hot block is being heated slowly (1-5°/min) will provide a good estimate of both the softening temperature and its range, and the nature of the melt as highly viscous, highly elastic, or intermediate. These considerations apply to both amorphous and crystalline polymers, though other evidence (next paragraph) allows the two to be distinguished, especially when crystallinity is high. The terms softening temperature and softening range are preferably applied to amorphous or nearly amorphous polymers.

Crystalline melting point The melting range of crystalline or semicrystalline polymers is usually much narrower than that for amorphous materials. However, even the crystalline materials do go through a softening range before melting (see, for example, the *Textbook*, pp. 169-170). They often show large and obvious increases in melt fluidity at temperatures slightly above the melting point, in contrast to amorphous materials for which fluidity increases slowly and continuously with temperature.

In the absence of degradation, repeated melting and cooling or annealing a semicrystalline polymer can lead to an increase in crystal perfection and a higher melting point. This should not be mistaken for a thermosetting reaction. In addition, heating a crystalline

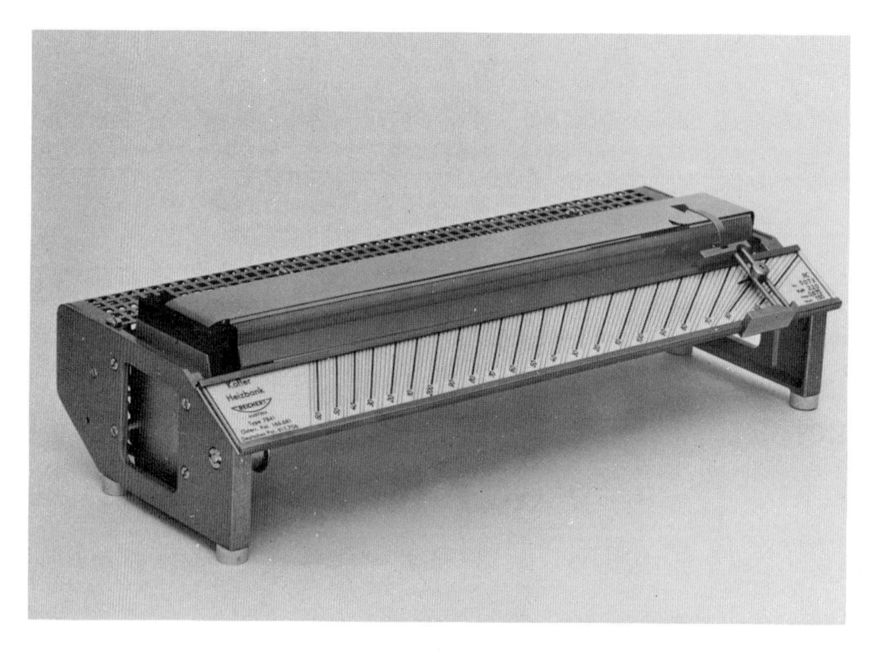

Fig. 6-2. Hot-block apparatus for determining the thermal properties of polymers (courtesy Hacker).

polymer above its melting point followed by rapid quenching may lead to the formation of an amorphous glass which, on subsequent heating above the glass-transition temperature, may crystallize (see the *Textbook*, pp. 121-2). The crystalline melting point will then be observed as the temperature is raised to the appropriate level.

Flow temperature The flow temperature or approximate temperature of processing the polymer can be observed as the temperature at which the polymer leaves a molten trail when drawn over the hot metal surface with moderate pressure. A simple flow test such as that using the melt indexer (ASTM D 1238) may be advisable to provide additional quantitative data on flow and processability.

Thermal stability Evidence for instability at elevated temperatures can be seen as darkening, gas evolution, embrittlement, or an irreversible increase or decrease in fluidity. Again, these must not be confused with thermosetting reactions. The temperature at which degradation first occurs provides a guide to an upper temperature limit for processing or use. (See also Sec. C.)

Optical properties In addition to the observation of color changes indicating degradation or the presence of impurities, assessment can be made of the transparency or opacity of the material.

Fiber-forming properties An attempt can be made to draw a fiber out of the molten mass with forceps or a spatula (Fig. 6-3). Subsequent drawing of the fibers after they have cooled to room temperature (*Textbook*, pp. 175-6) can provide an insight into the molecular weight of the sample: If the fibers can be cold drawn, the molecular weight is high enough to insure good mechanical properties.

Tack temperature The temperature at which two freshly broken pieces of the sample will adhere when pressed together is defined as the tack temperature of the material. It is asserted (McLaren 1951) to be the temperature at which the viscosity of the sample is 10^8 poise.

Adhesive properties An indication of the adhesive properties of the sample can be obtained by probing the polymer with a clean glass rod or metal spatula and observing how difficult it is to remove the sample from the probe after cooling.

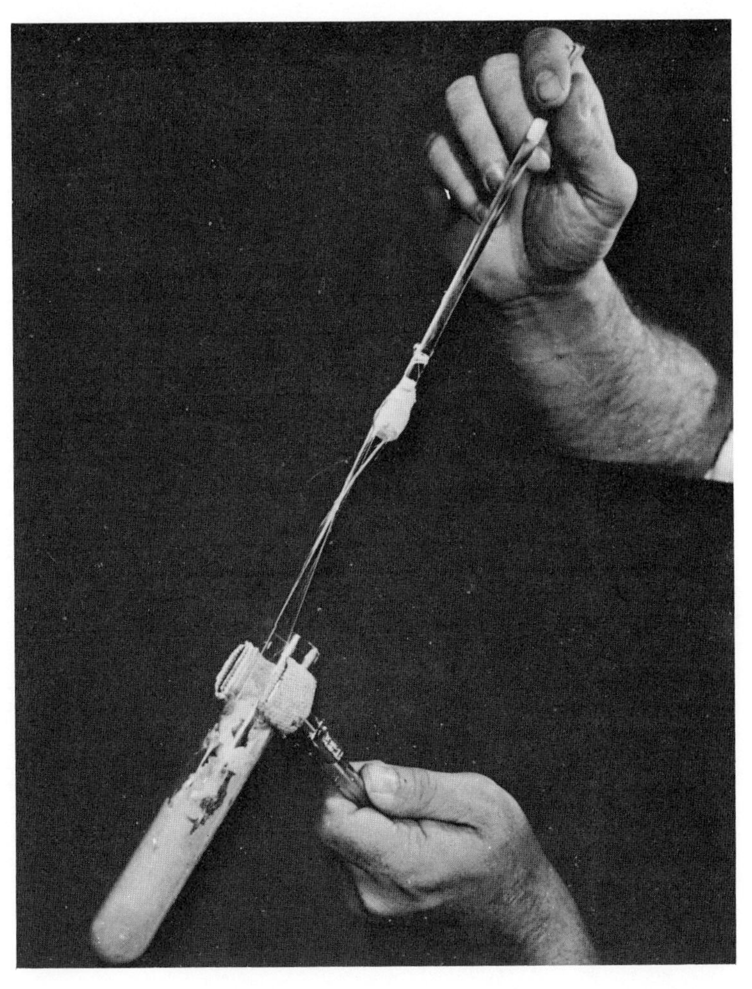

Fig. 6-3. Drawing out fibers from a polymer melt (Sorenson 1968).

Homogeneity The presence of more than one phase can often be detected by observing how opacity changes with temperature. This should not be confused with the obvious example of a sudden decrease in opacity at the crystalline melting point.

Relation to DTA Some of the above data can be obtained quantitatively by differential thermal analysis (Chap. 9, Exp. 20), for example softening temperature, glass-transition temperature, crystalline melting point, and decomposition temperature. The sample size and time of analysis are similar. Nevertheless, the "hot block" tests are still important for the additional information obtainable by firsthand observation of the material.

Flammability

Solid polymers do not burn in the absence of degradation, for a necessary prerequisite to sustained combustion is the formation of volatile fuels from the polymer. When a high enough heat flux is directed to any polymer, degradation to volatile products will occur. If the main components of this volatile pyrolizate are nonflammable, the polymer will not sustain combustion; examples are poly(vinyl chloride) and other chlorine-containing materials which release HCl, and poly(acrylic acid), which releases CO_2 and water. Most polymers, however, do burn upon degradation and must be specifically compounded with certain additives to obtain flame retardancy, if not nonflammability.

In the initial characterization of a new polymer, a number of observations can lead to an estimate of the inherent flame retardancy of the material. Generally, a polymer will be flammable if it melts below 200°C, for the main cause of flame spread is through the melt; if it releases highly volatile degradation products such as aliphatic, olefinic, or alcoholic compounds; or if it produces little char (carbonaceous residue) when heated above 400°C in an oxidative atmosphere. On the other hand, a polymer will be inherently flame retardant if it chars heavily at low temperatures; if it is thermally stable above 400°C; if it does not have a low melting temperature; and if the degradation products at temperatures below 400°C contain significant amounts of nonflammable gases such as CO_2, HCl, H_2O, etc.

Burning a polymer may often provide clues to its identity: Many polymers produce significant amounts of their monomers on burning (*Textbook*, p. 372). Some of these, for example styrene and methyl methacrylate, have characteristic odors which rapidly become familiar.

Pyrolysis

One of the major analysis techniques of modern times is gas chromatography, whose use is of course limited to the separation and identification of volatile compounds. It might seem that its applicability to polymers would be limited to analysis of residual monomer, solvents, diluents, plasticizers, and other small organic molecules such as the byproducts of condensation reactions or other volatile reaction products. When controlled pyrolysis of the polymer through the use of a heated sample inlet system is added to the gas chromatograph, however, the technique can be used both to identify polymers and to study their thermal stability and degradation (Guillet 1960, Cobler 1962, Hulot 1964, Haken 1969).

Depending on their structure, polymers can be expected to pyrolyze in one of several different ways: They can depolymerize to monomer, depolymerize to fragments or byproducts of monomer, or decompose to small fragments by random scission (*Textbook*, Chap. 12E). Examples of polymers which depolymerize predominantly to monomer include polystyrene, poly(methyl methacrylate), polyisobutylene, and other poly(methacrylic esters). In contrast, poly(vinyl chloride) and poly(vinyl acetate) also depolymerize, but the products here include quantitative yields of HCl and acetic acid, respectively. Polymers decomposing by random scission include poly(acrylic esters), poly(vinyl alcohol), polyethers, polyurethanes, polyamides, and polyethylene.

If this type of information is to be used to identify polymers, it is essential to control the rate and temperature of degradation, since variations in these quantities can result in different ratios of amounts of the decomposition fragments or even different products. Rapid pyrolysis and maximization of total fragmentation is usually achieved at temperatures of 400-800°C.

Residual monomer in such materials as polystyrene or poly(methyl methacrylate) can be determined in one of several ways: A small amount of the polymer can be dissolved in a known amount of a solvent such as methylene chloride, and the polymer precipitated by adding a fixed amount of a nonsolvent such as methanol. An aliquot of the supernatant liquid is injected into the gas chromatograph after careful filtration. Alternatively, an aliquot of the polymer solution can be injected, avoiding the possible loss of monomer by adsorption onto the polymer, but it is necessary to remove the polymer from the instrument by subsequent pyrolysis. Another approach is to perform a low-temperature pyrolysis in the chromatograph, in which the polymer is heated enough to soften it so that monomer can be transferred readily to the mobile phase, but not enough to cause chain scission. Quantitative analysis in any case requires the use of a calibration based upon known amounts of the component sought.

GENERAL REFERENCES

Grassie 1966; Harris 1966; Jellinek 1966; Slade 1966-1970; Porter 1968; Loan 1972, Warren 1972; and specifically:
Pyrolysis-gas chromatography: Barlow 1961, 1963, 1967; Nelson 1962; Stanley 1962; Jones 1967; Stevens 1968; Brauer 1970, Vukovik 1970.

B. *Solubility and the Preparation of Solutions*

Study of the property of *solubility* or the closely related *solvent resistance* is fundamental to any polymer evaluation. High solvent resistance, implying a polymer insoluble in most common solvents, is often a desirable end-use property for a polymer, but this same insolubility may mean that it is difficult or impossible to study its molecular structure, for example by the techniques discussed in Chap. 7.

By definition, a solution is a *molecular* dispersion of the solute in a solvent. It is often very difficult indeed to determine whether a polymer-solvent mixture is a true solution. Obviously, close visual examination to insure the absence of any undissolved material is a first requisite to such a decision. However, the refractive indices of the solvent and the swollen polymer may be nearly alike, so that undispersed particles are very difficult to see. If the polymer is dispersed into submicroscopic aggregates of only a few molecules, it may be even more difficult to distinguish this dispersion from a true solution. Quantitative studies of light scattering as a function of temperature may be helpful, but this is scarcely part of a minimal scheme.

The process of solution of a polymer is complex and governed as much by kinetic as by thermodynamic considerations. At least three steps are involved: First, the polymer is introduced into the solvent, but no obvious interactions take place at once. Next, the polymer becomes swollen as the solvent molecules diffuse into it. Swelling is most rapid if the polymer is finely divided, maximizing its surface area, and the solvent is a small, compact molecule. This is often the rate-determining step in the solution process. The volume of the polymer phase increases as solvent is imbibed, but few polymer molecules enter the solvent phase because of their lower diffusion rates.

Finally, the swollen polymer particles disintegrate, and the individual polymer molecules diffuse apart until a true solution is established; the system becomes a single homogeneous phase for the first time. Agitation can be very helpful in speeding up this process, though it is of little value in the second stage because of the high viscosity of the swelling polymer compared to the solvent phase. This final stage of solution can also be quite slow if the molecular weight of the polymer is unusually high.

A number of structural parameters of the polymer affect the ease and speed of polymer solution:

Chemical structure This parameter is a major determinant of polymer solubility. To a first approximation, the rule of thumb of "like dissolves like" is obeyed, and this is quantified in the solubility-parameter concept described below.

Molecular weight Solubility decreases as the molecular weight of the polymer increases, and this is the basis of most schemes for fractionating polymers, as discussed in Chap. 7D and in the *Textbook*, Chap. 2E. The effect is usually small compared to the influence of chemical composition on solubility.

Crystallinity As a class, crystalline polymers are quite insoluble because crystal forces as well as polymer-polymer interaction forces must be overcome to effect solution. Except for highly polar polymers (for example, polyamides soluble at room temperature in such solvents as *m*-cresol), crystalline polymers often dissolve only above or near their melting points; polyethylene is a common example.

Crosslinking Crosslinked polymers do not dissolve (unless degraded), but only swell.

For the minimal scheme, small amounts of the polymer (to produce a solution of 1-2% concentration) in finely divided form should be tested with a set of common solvents covering a variety of chemical structures. A typical set is given in Table 6-1. EXTREME CAUTION SHOULD BE EXERCISED IN USING THESE SOLVENTS. AS WITH MOST ORGANIC MATERIALS, MANY OF THEM ARE TOXIC, CORROSIVE, OR FLAMMABLE. Since solution is a slow process, the polymer-solvent mixtures should be allowed to stand for at least 24 hours before final observations are made. Heating the mixture to the boiling point of the solvent for 10-15 minutes will speed up solubilization. CAUTION: NEVER HEAT ORGANIC SOLVENTS OVER AN OPEN FLAME.

Solubility parameters

From the thermodynamic point of view, solubility occurs when the free energy of mixing of solute and solvent,

$$\Delta G = \Delta H - T\Delta S \tag{6-1}$$

is negative. For many years, it was thought that both ΔS and ΔH for mixing were always positive. It is now known (*Textbook*, Chap. 2C) that either of these quantities can be negative under some circumstances, but these are unusual and do not invalidate the considerations to follow.

The heat of mixing ΔH can be written several ways, depending on the choice of polymer-solvent interaction parameter used. Here we follow the treatment of Hildebrand (1950) and write, per unit volume,

$$\Delta H = v_1 \, v_2 \, (\delta_1 - \delta_2)^2 \tag{6-2}$$

TABLE 6-1. A typical solvent set for minimal-scheme testing

Acetone	Isopropanol
Carbon tetrachloride	Methylene dichloride
Cyclohexanone	Monochlorobenzene
Dimethyl acetamide	Nitrobenzene
Dimethyl formamide	Perchloroethylene
Dimethyl sulfoxide	Tetrahydrofuran
Dioxane	"Tetralin"[a]
Hexane	Toluene

[a] 1,2,3,4-tetrahydronaphthalene.

where the v's are volume fractions and the δ's are *solubility parameters*; δ^2 is known as the *cohesive energy density* of a substance and for small molecules is approximately its energy of vaporization per unit volume. Following usual practice, the subscript 1 denotes the solvent and 2 the polymer.

The value of the solubility parameter is that, in the absence of strong hydrogen bonding, whether or not a polymer is soluble in a given solvent can be predicted from values of δ_1 and δ_2, and these can be calculated or looked up in tables. A general rule is that solubility can be expected if $\delta_1 - \delta_2$ is less than 1.7-2.0, but not if it is appreciably larger. Solubility parameters are also useful in predicting polymer compatibility (Krause 1972).

Some typical values of solubility parameters are given in Table 6-2. Extensive tabulations are given in the literature (Burrell 1966, Hoy 1970). Perhaps the easiest way to obtain δ, however, is to calculate it from molar-attraction constants (Hoy 1970). A short table of these group contributions to δ is given in the *Textbook*, Table 2-2.

These considerations work best when there are no strong polymer-solvent interactions, but it is possible to extend them by considering more parameters, the most widely used being hydrogen-bonding tendency. Some common solvents are classified by hydrogen-bonding tendency and solubility parameter in Table 6-3. A third parameter has been introduced by Hansen (1967*a, b, c*) and Bagley (1971), but this refinement is seldom necessary. It was, however, used by Beerbower (1969) to predict swelling behavior.

Other ways of measuring polymer-solvent interactions are discussed in Chaps. 7A and 7B. An elaborate scheme for identification of polymers by their solubility is the basis of Exp. 1 in McCaffery 1970.

TABLE 6-2. Typical values of the solubility parameter δ for some common polymers and solvents

Solvent	δ_1, $(cal/cm^3)^{1/2}$	Polymer	δ_2, $(cal/cm^3)^{1/2}$
n-Hexane	7.24	Polytetrafluoroethylene	6.2
Carbon tetrachloride	8.58	Poly(dimethyl siloxane)	7.3
Toluene	8.93	Polyethylene	7.9
2-Butanone	9.04	Polypropylene	8.1
Benzene	9.15	Polybutadiene	8.6
Cyclohexanone	9.3	Polystyrene	8.6
Styrene	9.3	Poly(methyl methacrylate)	9.1
Chlorobenzene	9.5	Poly(vinyl chloride)	9.5
Acetone	9.71	Poly(vinyl acetate)	10.6
Tetrahydrofuran	9.9	Poly(ethylene terephthalate)	10.7
Methanol	14.5	66 Nylon	13.6
Water	23.4	Polyacrylonitrile	15.4

TABLE 6-3. Classification of solvents by solubility parameter and hydrogen-bonding tendency

Solvent	δ_1, $(cal/cm^3)^{1/2}$	Hydrogen Bonding Tendency
n-Heptane	7.4	Low
Carbon tetrachloride	8.58	Low
Benzene	9.2	Low
Acetonitrile	11.9	Low
4-Methyl-2-pentanone	8.4	Medium
2-Butanone	9.04	Medium
Ethyl acetate	9.1	Medium
Cyclohexanone	9.3	Medium
Dioxane	9.9	Medium
Tetrahydrofuran	9.9	Medium
Dimethyl sulfoxide	12.9	Medium
t-Butanol	10.6	High
Dimethyl formamide	12.1	High
Ethylene glycol	14.2	High
Methanol	14.5	High

*GENERAL REFERENCES**

Textbook, Chap. 2A; Hildebrand 1950, 1962; Allen 1959; Gardon 1965; Burrell 1970.

C. *Chemical, Environmental, and Thermal Stability*

There are many different tests available for evaluating the stability of polymers. For the purposes of minimal-scheme analysis, emphasis is placed on methods which are rapid and yield results which are easily recorded. Among the variables which can be measured for the evaluation of stability are weight, viscosity, solubility, color, melting point, brittleness, and elongation. In any one test, observation of any two of these variables is generally sufficient to draw the necessary conclusions.

The tests to be described require samples fabricated in the form of thin films. These should be prepared by compression molding or pressing in a small laboratory press (Carver or PHI) with steam- or electrically-heated platens; this procedure also yields clues to such thermal properties as flow and processability (Sec. A). This is not a difficult operation and requires only 2-3 g sample (Fig. 6-4). The alternative preparation of films by solvent casting is not recommended,

*A new ASTM *Method of Test for the Solubility Range of Resins and Polymers* lists 90 solvents or solvent mixtures in order of solubility parameter, hydrogen bonding tendency, and dipole moment.

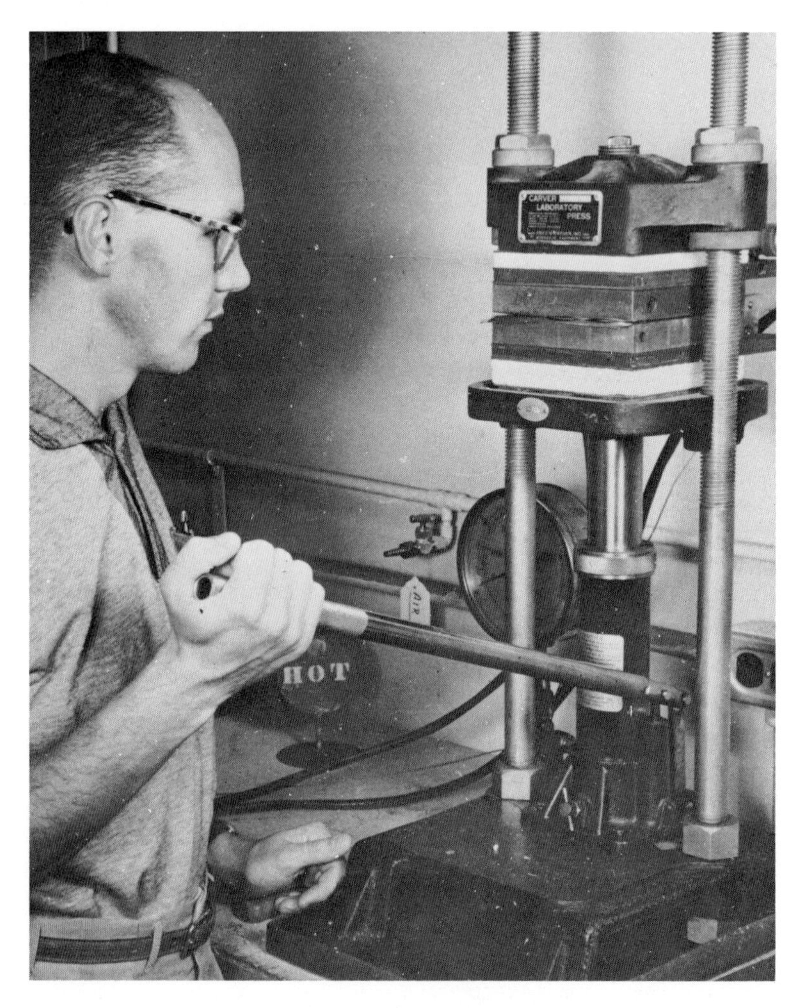

Fig. 6-4. Use of a laboratory press to prepare a polymer film specimen (Sorenson 1968).

even though solutions prepared as part of testing for solubility are at hand, because of the great difficulty in removing solvent completely (Chap. 4B).

Chemical stability

The stability of the polymer in the presence of chemical reagents is determined by immersing strips of molded film for a predetermined time (often 24 hr) in cool water, boiling water, and 10% solutions of acetic acid, NaCl, H_2SO_4, and NaOH. Stability is evaluated by observing changes in weight and flexibility of the films.

Environmental stability

The stability of polymers exposed to environmental conditions of sunlight, humidity, rain, and the like can only be determined with accuracy by actual experiment, usually long-term. Many so-called accelerated weathering techniques have been proposed (Hawkins 1972*c*), but none of them duplicates natural conditions adequately, many are not appreciably faster than outdoor exposure, and all are better described as artificial rather than accelerated tests. Devices such as the Fade-Ometer (radiation only) or the Weather-Ometer (radiation plus water spray) simulate the sunlight spectrum with a carbon or xenon arc, but the relation between exposure in these units and outdoor exposure varies quantitatively and even qualitatively from polymer to polymer.

For minimal-scheme purposes, exposure to a laboratory mercury arc or a germicidal lamp can serve to identify very unstable materials. Although examination of the ultraviolet absorption spectrum of a polymer gives positive indication of a tendency for the material to absorb damaging radiation, it does not provide a reliable prediction of what photo-induced degradation in physical properties may result.

The stability of polymers to weathering or to ultraviolet radiation is usually evaluated by changes in color, gloss, or surface texture, or such mechanical properties as tensile strength and elongation.

Thermal stability

The thermal stability of polymers can be evaluated qualitatively for minimal-scheme purposes by exposing samples in a laboratory circulating-air oven. Exposure for several hours (often 24) at 120°C will serve to screen out materials of little practical utility, while polymers withstanding exposure at 180°C are of unusual interest. The test can be repeated in a nitrogen atmosphere to differentiate the effect of heat alone from that of thermal plus oxidative degradation; the difference is particularly noticeable in polymers possessing residual unsaturation.

These tests can be made on film or powder samples. Changes in weight, color, and solubility are usually followed. Quantitative testing for thermal stability is discussed in Chap. 9E, while Sec. A describes clues to stability obtained in minimal-scheme determination of thermal properties.

GENERAL REFERENCES

Gesner 1972; Hawkins 1972*a,b,c*; Loan 1972; Shelton 1972.

D. Mechanical Properties

Polymers are most widely used because of their mechanical properties, and some measure of these is essential for inclusion in minimal-scheme testing. Tensile measurements at several strain rates provide the maximum amount of information with a minimum of time, sample preparation, and equipment needs. A tensile testing machine, such as the Instron, is required, however. Best results, including comparison among materials and with literature values, are obtained if samples are prepared and tested according to ASTM standard methods (ASTM D 638, ASTM D 882), but qualitative results obtained with films are also useful. Complete characterization beyond minimal-scheme objectives would include many other tests and evaluation at a variety of temperatures.

The stress-strain curve of a polymer (*Textbook*, Chap. 4G) provides many important clues to its mechanical behavior. As Fig. 6-5 shows, the material can be classified generally as rubbery, glassy, or (the largest class) in-between these extremes. Compounding or changes in structure can have a profound effect on the stress-strain curve and properties of a given type of polymer.

Tensile testing for the minimal scheme involves obtaining the stress-strain curve at room temperature and strain rates of 0.2, 2, and 20 in./min. From each curve are obtained the six quantities defined in Fig. 6-6: The *modulus* of elasticity obtained from the initial slope of the curve; the *yield stress*, or peak stress in the early part of the curve; the *stress at failure*; the *elongation* (or strain) *at yield*, sometimes called the yield strength; the *elongation at break*, also known as the ultimate elongation or strain at failure; and the *energy to break* or impact energy, measured by the area under the stress-strain curve up to the point of failure. Table 6-4 relates some of these characteristics to the general properties of the polymer material.

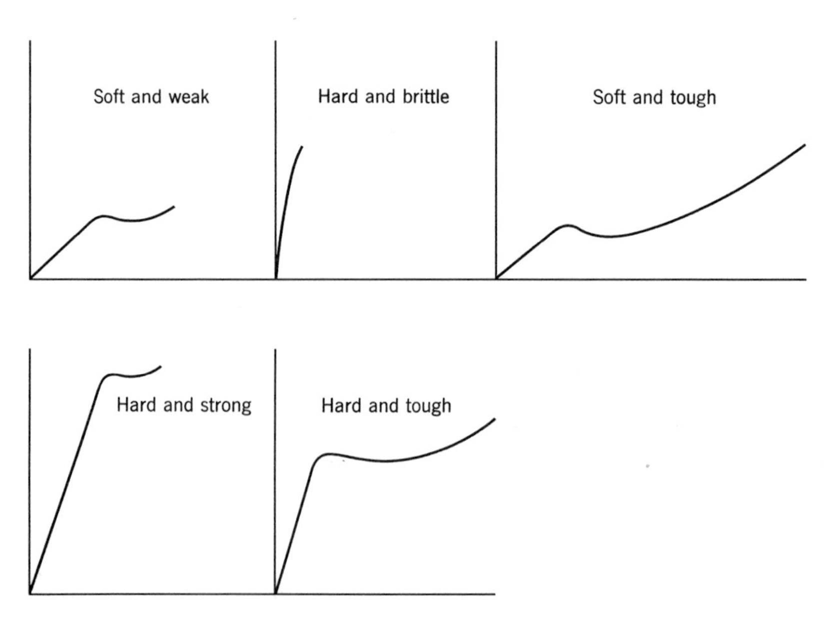

Fig. 6-5. Tensile stress-strain curves showing features typical of several types of polymers (*Textbook*, Fig. 4-11, from Winding 1961).

Because of the importance of the behavior of polymers under impact, some additional comments are in order. A major reason for performing the tensile test at 20 in./min, the highest speed possible with the usual testing equipment, is to obtain some information on this property. The pertinent feature of the stress-strain curve is the energy or work to break, which is a measure of toughness. Even though the work to break varies considerably with the strain rate, and these tests are carried out at speeds far short of impact strain rates, it has been shown (Evans 1960) that they provide a quantitative prediction of falling-weight impact resistance which is accurate to better than 20%. A more meaningful indication of impact strength or toughness can, of course, be obtained from tests specifically designed around this property (ASTM D 256, Bragaw 1956). The tensile tests at slower speeds provide knowledge of the mechanical behavior of the polymer in ways not related to impact.

GENERAL REFERENCES

Nielsen 1962; Simpson 1964; Thorkildsen 1964; Bikales 1971; Ives 1971; Ward 1971.

E. Ultraviolet and Infrared Spectroscopy

In the minimal scheme, ultraviolet and infrared spectroscopy are primarily used for the identification of samples through comparison to reference spectra (for example, Sadtler, Gevantman 1972).

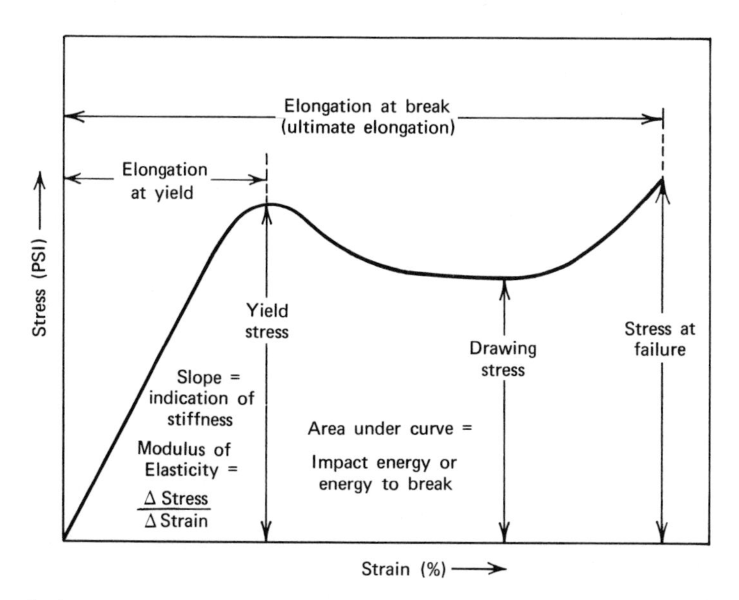

Fig. 6-6. Generalized tensile stress-strain curve for plastics.

TABLE 6-4. Characteristic features of stress-strain curves as related to polymer properties (Winding 1961)

Description of Polymer	Characteristics of Stress-Strain Curve			
	Modulus	Yield Stress	Ultimate Strength	Elongation at Break
Soft, weak	Low	Low	Low	Moderate
Soft, tough	Low	Low	Yield stress	High
Hard, brittle	High	None	Moderate	Low
Hard, strong	High	High	High	Moderate
Hard, tough	High	High	High	High

Ultraviolet spectra

Except for aromatic structures, many synthetic polymers do not absorb in the ultraviolet region. Such polymers are likely to be useful for outdoor applications. In these cases, detection of small amounts of absorbing impurities can be a useful advantage. Solvents such as water, alcohols, or saturated hydrocarbons likewise do not absorb and can be used as required for ultraviolet analysis, whereas they cannot be used in the infrared region.

Infrared spectra

The well-known advantages of infrared spectroscopy apply to polymers as well as to other systems. The method is rapid, direct, nondestructive, and requires only small amounts of sample. Thin films can be measured directly; solution of the sample is not required. Information on physical as well as chemical structure can be obtained. Infrared analysis is a good choice for the initial characterization of a polymerization product.

Some of the information obtainable by infrared analysis on polymers is: identification of the major components of the chain structure by functional-group analysis; distinguishing among configurational isomers, such as *cis* versus *trans* or 1,2 versus 1,4 addition in diene polymers; semiquantitative determination of stereoregularity; detection and measurement of crystallinity; determination of copolymer composition and sequence distribution; measurement of chain branching; end-group analysis; and detection of chemical reactions such as oxidation or degradation.

Although some of these measurements can be made more easily by other techniques--for example, crystallinity by x-ray diffraction and sequence distribution and stereoregularity by nuclear magnetic resonance spectroscopy--the general availability of infrared equipment and the relative ease of obtaining and interpreting the data make this a very widely used method.

Sample preparation for infrared analysis becomes difficult for crosslinked polymers, for they often cannot be ground up, dissolved, or otherwise reduced to thin-film form. An alternative here is to pyrolyse the sample with care and analyze the liquid pyrolizate by infrared spectroscopy as an alternative to the gas-chromatographic method discussed in Sec. A.

GENERAL REFERENCES

Textbook, Chap. 4B; Tryon 1962; Zbinden 1964; Hummel 1966, 1971; Kössler 1967; Elliott 1969; Smith 1969, Spagnolo 1969; Bovey 1972.

F. Summary

The five steps suitable for the general evaluation of polymer properties in a minimal scheme should be considered only a base upon which more specific characterization schemes can be built according to the needs of the occasion. If, for example, the polymer scientist is searching for materials suitable for coating electrical wiring, it is obvious that additional tests would be required for dielectric constant and loss factor, volume resistivity, abrasion resistance, hardness, flexural strength, stiffness, and weatherability.

Consideration of minimal-scheme properties is also of value to the student or industrial scientist or engineer who, without formal training in polymer science, needs to acquire some familiarity with the wide variety of properties found in polymers. For this purpose, minimal-scheme properties of representative polymers covering a cross section of major types have been assembled in Table 6-5. More extensive tabulations are provided in the *Textbook* and elsewhere.

Attention is also called to Appendix I, which gives information on a number of sources of supply of standard, well-defined polymers suitable for use as substitutes, in characterization by the minimal scheme or Exps. 10-32, for the polymers prepared in Exps. 1-9. Appendix II is an expansion of Table 6-5 to include additional polymers and the results of many more standard tests.

BIBLIOGRAPHY

Allen 1959. P. W. Allen, "Introductory (Solubility and the Choice of Solvent; the Preparation and Handling of Polymer Solutions)," Chap. 1 in P. W. Allen, ed., *Techniques of Polymer Characterization*, Butterworths, London, 1959.

ASTM D 256. *Standard Methods of Test for Impact Resistance of Plastics and Electrical Insulating Materials*. ASTM Designation: D 256. American Society for Testing and Materials, Philadelphia, Pennsylvania.

TABLE 6-5. Minimal-scheme properties of some important polymers

Polymer Property	Polyurethane	Poly(methyl methacrylate)	Polyoxy-methylene	Polystyrene	Polyethylene	Polycarbonate	Poly(vinyl chloride)	Nylon 610	Nylon 11
Color	Light yellow	Colorless	Slightly yellow	Colorless	Almost colorless	Colorless	Colorless	Slightly yellow	Slightly tan
Clarity	Transparent	Transparent	Opaque	Transparent	Translucent	Transparent	Transparent	Opaque	Opaque
Approx. melt temp. °C[a]	140-160	170-200	180-190	160-170	105-115	210	150-175	220	190
Flow or processing temp. °C[a]	150-180	175-210	200-210	150-200	150-210	250	170-200	200-210	180-190
Crystallinity	Low	None	High	None	Moderate to high	Low	Low	Moderate	Moderate
Burning rate	Burns	Slow	Slow	Slow	Very slow	Self-exting.	Self-exting.	Self-exting.	Self-exting.
Degradation temp. °C[a]	280	305	250	320	270	350	240	280	250
<u>Solubility</u>									
Hexane	Insol.	Insol.	Insol.	Insol.	Insol.	Insol.	Insol.	Insol.	Insol.
Carbon tetrachloride	Insol.	Sol.	Insol.	Sol.	Insol.	Insol.	Insol.	Insol.	Insol.
Toluene	Sol.	Sol.	Insol.	Sol.	Insol.	Insol.	Insol.	Insol.	Insol.
Nitrobenzene	Insol	Insol.	Insol.	Sol.	Insol.	Insol.	Sol.	Insol.	Insol.
Cyclohexanone	Sol.	Sol.	Insol.	Sol.	Insol.	Insol.	Sol.	Insol.	Insol.
Tetrahydrofuran	Sol.	Sol.	Insol.	Sol.	Insol.	Insol.	Sol.	Insol.	Insol.
Dimethyl formamide	Sol.	Sol.	Sol. hot	Sol.	Insol.	Insol.	Sol.	Sol.	Insol.
Acetone	Insol.	Sol.	Insol.	Sol.	Insol.	Insol.	Partially Sol.	Insol.	Insol.
Perchloroethylene	Swells	Sol.	Insol.	Sol.	Insol.	Insol.	Sol.	Insol.	Insol.
Decahydronaphthalene	Insol.	Insol.	Sol. hot	Sol.	Sol. hot	Insol.	Insol.	Sol.	Insol.

ASTM D 638. *Standard Method of Test for Tensile Properties of Plastics.* ASTM Designation: D 638. American Society for Testing and Materials, Philadelphia, Pennsylvania.

ASTM D 882. *Standard Methods of Test for Tensile Properties of Thin Plastic Sheeting.* ASTM Designation: D 882. American Society for Testing and Materials, Philadelphia, Pennsylvania.

ASTM D 1238. *Standard Method of Measuring Flow Rates of Thermoplastics by Extrusion Plastometer.* ASTM Designation: D 1238. American Society for Testing and Materials, Philadelphia, Pennsylvania.

Bagley 1971. E. B. Bagley, T. P. Nelson, and J. M. Scigliano, "Three-Dimensional Solubility Parameters and their Relationship to Internal Pressure Measurements in Polar and Hydrogen Bonding Solvents," *J. Paint Technol. 43* (555), 35-42 (April, 1971).

Barlow 1961. A. Barlow, R. S. Lehrle, and J. C. Robb, "Direct Examination of Polymer Degradation by Gas Chromatography: I. Application to Polymer Analysis and Characterization," *Polymer 2*, 27-40 (1961).

Barlow 1963. A. Barlow, R. S. Lehrle, and J. C. Robb, "Polymer Degradation Examined by Gas Chromatography," pp. 267-283 in *Techniques of Polymer Science* (S.C.I. Monograph No. 17), Gordon and Breach Science Publishers, New York, 1963.

Barlow 1967. A. Barlow, R. S. Lehrle, J. C. Robb, and D. Sunderland, "Direct Examination of Polymer Degradation by Gas Chromatography: II. Development of the Technique for Quantitative Kinetic Studies," *Polymer 8*, 523-535 (1967).

Beerbower 1969. A. Beerbower and J. R. Dickey, "Advanced Methods for Predicting Elastomer/Fluids Interactions," *Am. Soc. Lubr. Eng. Trans. 12*, 1-20 (1969).

Bikales 1971. Norbert M. Bikales, ed., *Mechanical Properties of Polymers (Encyclopedia Reprints)*, Interscience Div., John Wiley and Sons, New York, 1971.

Bovey 1971. Frank A. Bovey, *High Resolution NMR of Macromolecules*, Academic Press, New York, 1972.

Bragaw 1956. C. G. Bragaw, "Tensile Impact--A Simple, Meaningful Impact Test," *Modern Plastics 33* (10), 199-204 (June, 1956).

Brauer 1970. G. M. Brauer, "Pyrolysis Gas Chromatographic Techniques for Polymer Identification," Chap. 2 in Philip E. Slade, Jr., and Lloyd T. Jenkins, eds., *Thermal Characterization Techniques*, Marcel Dekker, New York, 1970.

Burrell 1966. H. Burrell and B. Immergut, "Solubility Parameter Values," pp. IV-341 to IV-368 in J. Brandrup and E. H. Immergut, eds., with the collaboration of H.-G. Elias, *Polymer Handbook*, Interscience Div., John Wiley and Sons, New York, 1966.

Burrell 1970. H. Burrell, "Solubility of Polymers," pp. 618-626 in Herman F. Mark, Norman G. Gaylord, and Norbert M. Bikales, eds., *Encyclopedia of Polymer Science and Technology*, Vol. 12, Interscience Div., John Wiley and Sons, New York, 1970.

Carver. Fred S. Carver, Inc., W142N9050 Fountain Blvd., Menomonee Falls, Wisconsin 53051.

Cobler 1962. John G. Cobler and E. P. Samsel, "Gas Chromatography. A New Tool for the Analysis of Plastics," *SPE Trans. 2*, 145-151 (1962).

Elliott 1969. Arthur Elliott, *Infrared Spectra and Structure of Organic Long-Chain Polymers,* St. Martin's Press, New York, 1969.
Evans 1960. Robert M. Evans, Harry R. Nara, and Edward G. Bobalek, "Prediction of Impact Resistance from Tensile Data," *SPE J. 16* (1), 76-83 (1960).
Gardon 1965. J. L. Gardon, "Cohesive-Energy Density," pp. 833-862 in Herman F. Mark, Norman G. Gaylord, and Norbert M. Bikales, eds., *Encyclopedia of Polymer Science and Technology*, Vol. 3, Interscience Div., John Wiley and Sons, New York, 1965.
Gesner 1972. B. D. Gesner, "Stabilization Against Chemical Agents," Chap. 8 in W. Lincoln Hawkins, ed., *Polymer Stabilization*, Wiley-Interscience, New York, 1972.
Gevantman 1972. Lewis H. Gevantman, "Survey of Analytical Spectral Data Sources and Related Data Compilation Activities," *Anal. Chem. 47* (7), 30A-48A (June, 1972).
Grassie 1966. N. Grassie, "Degradation," pp. 647-716 in Herman F. Mark, Norman G. Gaylord, and Norbert M. Bikales, eds., *Encyclopedia of Polymer Science and Technology*, Vol. 4, Interscience Div., John Wiley and Sons, New York, 1966.
Guillet 1960. J. E. Guillet, W. C. Wooten, and R. L. Combs, "Analysis of Polymethacrylates by Gas Chromatography," *J. Applied Polymer Sci. 3*, 61-64 (1960).
Hacker. W. J. Hacker & Co., Inc., West Caldwell, New Jersey 07006.
Haken 1969. J. K. Haken, "Gas Chromatography," Chap. 6 in Raymond R. Myers and J. S. Long, eds., *Treatise on Coatings,* Vol. 2, *Characterization of Coatings: Physical Techniques,* Part I, Marcel Dekker, New York, 1969.
Hansen 1967a. Charles M. Hansen, "The Three Dimensional Solubility Parameter--Key to Paint Component Affinities: I. Solvents, Plasticizers, Polymers, and Resins," *J. Paint Technol. 39,* 104-117 (1967).
Hansen 1967b. Charles M. Hansen, "The Three Dimensional Solubility Parameter--Key to Paint Component Affinities: Dyes, Emulsifiers, Mutual Solubility and Compatibility, and Pigments," *J. Paint Technol. 39*, 505-510 (1967).
Hansen 1967c. Charles M. Hansen and K. Skaarup, "The Three Dimensional Solubility Parameter--Key to Paint Component Affinities: III. Independent Calculation of the Parameter Components," *J. Paint Technol. 39,* 511-514 (1967).
Haslam 1965. John Haslam and Harry A. Willis, *Identification and Analysis of Plastics,* D. Van Nostrand Co., Princeton, New Jersey, 1965
Hawkins 1972a. W. Lincoln Hawkins, ed., *Polymer Stabilization,* Wiley-Interscience, New York, 1972.
Hawkins 1972b. W. Lincoln Hawkins, "Environmental Deterioration of Polymers," Chap. 1 in W. Lincoln Hawkins, ed., *Polymer Stabilization*, Wiley-Interscience, New York, 1972.
Hawkins 1972c. W. Lincoln Hawkins, "Methods for Measuring Stabilizer Effectiveness," Chap. 10 in W. Lincoln Hawkins, ed., *Polymer Stabilization*, Wiley-Interscience, New York, 1972.
Hildebrand 1950. Joel H. Hildebrand and Robert L. Scott, *The Solubility of Nonelectrolytes,* 3rd ed., Reinhold Publishing Corp., New York, 1950 (reprinted by Dover Publications, New York, 1964).
Hildebrand 1962. J. H. Hildebrand and R. L. Scott, eds., *Regular*

Solutions, Prentice-Hall, New York, 1962.
Hoy 1970. K. L. Hoy, "New Values of the Solubility Parameters from Vapor Pressure Data," *J. Paint Technol. 42*, 76-118 (1970).
Hulot 1964. H. Hulot and P. Lebel, "Analysis of Elastomers by Gas Chromatography," *Rubber Chem. Tech. 37*, 297-309 (1964).
Hummel 1966. Dieter O. Hummel, *Infrared Spectra of Polymers: in the Medium and Long Wavelength Region*, John Wiley and Sons, New York, 1966.
Hummel 1971. Dieter O. Hummel, *Infrared Analysis of Polymers, Resins and Additives. An Atlas*, Wiley-Interscience, New York, 1971.
Instron. Instron Corp., 2500 Washington St., Canton, Massachusetts 02021.
Ives 1971. G. C. Ives, J. A. Mead and M. M. Riley, *Handbook of Plastic Test Methods*, CRC Press, Cleveland, 1971.
Jellinek 1966. H. H. G. Jellinek, "Depolymerization," pp. 740-793 in Herman F. Mark, Norman G. Gaylord, and Norbert M. Bikales, eds., *Encyclopedia of Polymer Science and Technology*, Vol. 4, Interscience Div., John Wiley and Sons, New York, 1966.
Jones 1967. C. E. R. Jones and G. E. J. Reynolds, "Quantitative Polymer Pyrolysis," pp. 260-271 in *Advances in Polymer Science and Technology* (S.C.I. Monograph No. 26), Gordon and Breach Science Publishers, New York, 1967.
Ke 1964. Bacon Ke, ed., *Newer Methods of Polymer Characterization*, Interscience Div., John Wiley and Sons, New York, 1964.
Kline 1959-1962. Gordon M. Kline, ed., *Analytical Chemistry of Polymers*, Interscience Div., John Wiley and Sons, New York; Part I, 1959; Parts II and III, 1962.
Kössler 1967. Ivo Kössler, "Infrared Spectroscopy," pp. 620-642 in Herman F. Mark, Norman G. Gaylord, and Norbert M. Bikales, eds., *Encyclopedia of Polymer Science and Technology*, Vol. 7, Interscience Div., John Wiley and Sons, New York, 1967.
Krause 1972. Sonja Krause, "Polymer Compatibility," *J. Macromol. Sci.--Macromol. Chem. C7*, 251-314 (1972).
Loan 1972. L. D. Loan and F. H. Winslow, "Thermal Degradation and Stabilization," Chap. 3 in W. Lincoln Hawkins, ed., *Polymer Stabilization*, Wiley-Interscience, New York, 1972.
McCaffery 1970. Edward M. McCaffery, *Laboratory Preparation for Macromolecular Chemistry*, McGraw-Hill Book Co., New York, 1970.
McLaren 1951. A. D. McLaren, G. G. Li, Robert Roger, and H. Mark, "Adhesion: IV. The Meaning of Tack Temperature," *J. Polymer Sci. 7*, 463-471 (1951).
Nelson 1962. D. F. Nelson, J. L. Lee, and P. L. Kirk, "The Identification of Plastics by Pyrolysis and Gas Chromatography," *Microchem. J. 6*, 225-231 (1962).
Nielsen 1962. Lawrence E. Nielsen, *Mechanical Properties of Polymers*, Reinhold Publishing Corp., New York, 1962.
Porter 1968. Roger S. Porter and Julian F. Johnson, *Analytical Calorimetry*, Plenum Press, New York, 1968.
PHI. Pasadena Hydraulics, Inc., 200 North Berry St., Brea, California 92621.
Sadtler. Sadtler Research Laboratories, Inc., 3316 Spring Garden St., Philadelphia, Pennsylvania 19104.
Saunders 1966. K. J. Saunders, *The Identification of Plastics and Rubbers*, Chapman and Hall, London, 1966.

Schmitz 1965-1968. John V. Schmitz, ed., *Testing of Polymers,* Interscience Div., John Wiley and Sons, New York; Vol. I, 1965; Vol. II, 1966; Vol. III, 1968.

Shelton 1972. J. Reid Shelton, "Stabilization against Thermal Oxidation," Chap. 2 in W. Lincoln Hawkins, ed., *Polymer Stabilization,* Wiley-Interscience, New York, 1972.

Sherr 1965. A. E. Sherr, "Preliminary Evaluation of Polymer Properties," *SPE J. 21* (1), 67-74 (1965).

Simpson 1964. E. H. Simpson and R. E. Polleck, "Tensile Impact Properties of Thermoplastics," *SPE Trans. 4,* 25-28 (1964).

Slade 1966-1970. Philip E. Slade, Jr., and Lloyd T. Jenkins, eds., *Techniques and Methods of Polymer Evaluation,* Marcel Dekker, New York; Vol. 1, *Thermal Analysis,* 1966; Vol. 2, *Thermal Characterization Techniques,* 1970.

Smith 1969. Clara D. Smith, "Infrared Spectroscopy," Chap. 10 in Raymond R. Myers and J. S. Long, eds., *Treatise on Coatings,* Vol. 2, *Characterization of Coatings: Physical Techniques,* Part I, Marcel Dekker, New York, 1969.

Sorenson 1968. Wayne R. Sorenson and Tod W. Campbell, *Preparative Methods of Polymer Chemistry,* 2nd ed., Interscience Div., John Wiley and Sons, New York, 1968.

Spagnolo 1969. Frank Spagnolo and Edward R. Sheffer, "Ultraviolet and Visible Spectroscopy," Chap. 11 in Raymond R. Myers and J. S. Long, eds., *Treatise on Coatings,* Vol. 2, *Characterization of Coatings: Physical Techniques,* Part I, Marcel Dekker, New York, 1969.

Stanley 1962. C. W. Stanley and W. R. Peterson, "Polymer Analysis using Gas Chromatographic Separation of Pyrolysates on a Capillary Column," *SPE Trans. 2,* 298-301 (1962).

Stevens 1969. Malcolm P. Stevens, *Techniques and Methods of Polymer Evaluation,* Vol. 3, *Characterization and Analysis of Polymers by Gas Chromatography,* Marcel Dekker, New York, 1969.

Thorkildsen 1964. R. L. Thorkildsen, "Mechanical Behavior," Chap. 5 in Eric Baer, ed., *Engineering Design for Plastics,* Reinhold Publishing Corp., New York, 1964.

Tryon 1962. M. Tryon and E. Horowitz, "Infrared Spectroscopy," Chap. VIII in Gordon M. Kline, ed., *Analytical Chemistry of Polymers,* Part II, Interscience Div., John Wiley and Sons, New York, 1962.

Vukovic 1970. R. Vukovic and V. Gnjatovic, "Characterization of Styrene-Acrylonitrile Copolymer by Pyrolysis-Gas Chromatography," *J. Polymer Sci. A-1 8,* 139-146 (1970).

Ward 1971. I. M. Ward, *Mechanical Properties of Solid Polymers,* Interscience Div., John Wiley and Sons, New York, 1971.

Warren 1972. P. C. Warren, "Stabilization Against Burning," Chap. 7 in W. Lincoln Hawkins, ed., *Polymer Stabilization,* Wiley-Interscience, New York, 1972.

Winding 1961. Charles C. Winding and Gordon D. Hiatt, *Polymeric Materials,* McGraw-Hill Book Co., New York, 1961.

Zbinden 1964. Rudolf Zbinden, *Infrared Spectroscopy of High Polymers,* Academic Press, New York, 1964.

7

Molecular Weight and Its Distribution

In this chapter, the commonly used methods of measuring the molecular weight and molecular-weight distribution of high polymers are reviewed. Particular emphasis is placed on the techniques used in Exps. 10-17. These methods, and the theory behind them, are discussed further in the *Textbook*, Chaps. 2 and 3, and such reference books as Flory (1953), Tanford (1961), and Morawetz (1965). Some corresponding experiments are found in McCaffery (1970).

By way of review, it will be recalled (*Textbook*, Chap. 1A) that, as the result of random processes at some stage in polymerization, all synthetic polymers, and all naturally occurring ones except perhaps a few of biological interest, have a distribution of molecular weights. A typical situation is depicted in Fig. 7-1. This figure shows also the approximate positions of the several important average molecular weights yielded by the various experiments in this manual and others. These averages are defined in the references cited and in some instances later in this chapter. Only two points need be made here: (1) The z-average molecular weight is directly accessible only from ultracentrifuge experiments, which are not discussed in this book. Like all the other averages, however, it can be calculated from fractionation data. (2) The value of the viscosity-average molecular weight, and its relation to the others, varies with the polymer-solvent interaction forces in the experiment in which it is determined. This is discussed further in Sec. C.

It is important to note the significance of the absolute and relative values of the various average molecular weights in relation to polymer properties. First, the properties of strength, toughness, and low sensitivity to chemical attack characteristic of polymers as a class of materials are not well developed until a molecular-weight level of around 10,000 is reached. Second, it is possible to obtain information on the breadth of the molecular-weight distribution from the relative values of two different averages. Because of their accessibility, the ratio of the weight average, $\overline{M}_w$, to the number average, $\overline{M}_n$, is usually used. Table 7-1 shows how the quantity $\overline{M}_w/\overline{M}_n$ is related to the synthesis conditions of polymers, as was discussed in Chaps. 1 and 2. Finally, the various averages may be highly sensitive

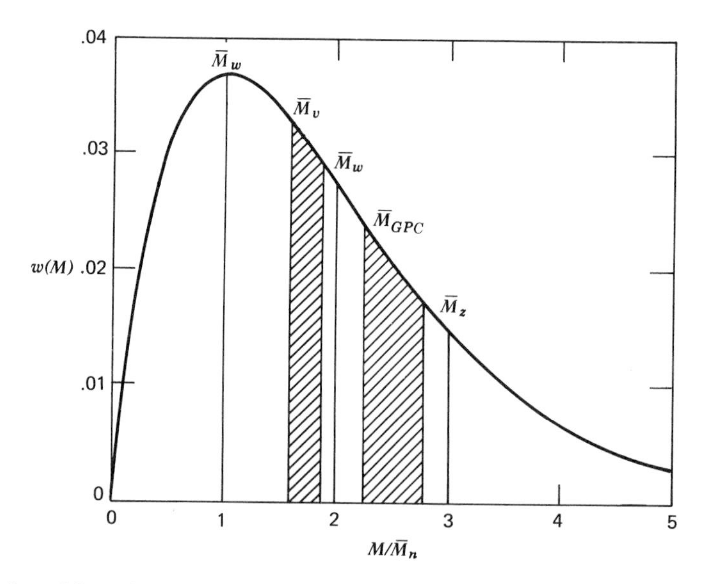

Fig. 7-1. Distribution of molecular weights in a typical polymer, showing the positions of important averages.

to the presence of a small fraction of material in an extreme range of molecular weight. Specifically, $\overline{M}_n$ is highly sensitive to the presence of a small *number* fraction of low-molecular-weight material; and $\overline{M}_w$ is similarly sensitive to small amounts by *weight* of high-molecular-weight polymer. These sensitivities, which become important limitations in some instances, are illustrated in Table 7-2. The calculations are made as described in Sec. D.

In considering random-coil polymers, it is necessary to distinguish between their size, or dimensions, and their mass. The two are not necessarily simply related, and their relative magnitudes can change with the structure of the molecule and with experimental variables such as the polymer-solvent interaction forces mentioned above. Thus it is expedient to distinguish between experiments measuring size, which may be correlated empirically with molecular weight in restricted circumstances, and those yielding directly one of the average molecular weights shown in Fig. 7-1.

To obtain information about the molecular-weight distribution, other than through comparison of two or more average molecular weights, it is necessary to separate the molecular species in a sample by some sort of fractionation process, and to determine the amounts and molecular weights of all the fractions. From these data, the distribution curve (like that of Fig. 7-1) can be plotted and the various average molecular weights calculated. The separation and determination steps may be carried out independently, or be combined, as in a chromatographic process with suitable calibration such as gel permeation chromatography (GPC, Exp. 16), or in the ultracentrifuge (*Textbook*, Chap. 3E). The separation may result from the molecular-weight

TABLE 7-1. Typical ranges of $\overline{M}_w/\overline{M}_n$ in synthetic polymers (Billmeyer 1965)

Polymer	$\overline{M}_w/\overline{M}_n$
Hypothetical monodisperse polymer	1.000
Actual "monodisperse" "living" polymers	1.01-1.05
Addition polymer, termination by coupling	1.5
Addition polymer, termination by disproportionation, or condensation polymer	2.0
High conversion vinyl polymers	2-5
Polymers made with autoacceleration	5-10
Coordination polymers	8-30
Branched polymers	20-50

TABLE 7-2. Effect of small portions of high- and low-molecular-weight polymer on calculated values of $\overline{M}_w$ and $\overline{M}_n$. The new values result from the addition of the indicated components to a monodisperse sample with M = 100,000

Component Added		Resulting Values of		
Amount	Molecular Weight	$\overline{M}_n$	$\overline{M}_w$	$\overline{M}_w/\overline{M}_n$
20% by number	10,000	85,000	98,000	1.15
20% by weight	10,000	40,000	85,000	2.1
20% by number	1,000,000	250,000	700,000	2.8
20% by weight	1,000,000	118,000	250,000	2.1
20% by number	Each *M*	216,000	695,000	3.2
20% by weight	Each *M*	46,000	216,000	4.7

dependence of solubility, as in classical fractionation (*Textbook*, Chap. 2E); of molecular size, as in GPC; or, in rare cases, of some other property. Caution is required in interpreting the results of fractionation experiments directly in terms of molecular weight unless it can be demonstrated that the separation step is independent of other variables which might alter the relationship between the driving force for separation and molecular weight.

A. *Number-Average Molecular Weight*

Methods for the determination of the number-average molecular weight fall into two categories: chemical or physical methods based on *end-group analysis;* and those based on measurement of one of the *colligative properties;* vapor-pressure lowering, freezing-point depression (*cryoscopy*), boiling-point elevation (*ebulliometry*), and the *osmotic pressure.*

Little needs to be said here about end-group analysis. The chemical reactions used to count chain ends are either identical with or closely related to those utilized to follow the course of polymerization and described in Chap. 5B and Exp. 10. These and some of the physical methods which can be applied are discussed in the *Textbook,* Chap. 3A. The major requirements for the application of the method are that it be known that the polymer is linear, that is, each molecule has only two ends; whether the groups to be determined occupy both ends or only one end of the chain; and that all of the end groups are accounted for. The calculations relating the count of end groups to the molecular weight are straightforward and are described in Exp. 10.

A review of Chap. 3B in the *Textbook* will show that, among the colligative properties, the osmotic pressure provides the largest and most precisely measured effect. Limitations of available osmotic membranes are such, however, that this technique (hereafter referred to as *membrane osmometry*) is accurate only for polymers with rather high molecular weight. To cover the range of interest, another method must be used for samples with low molecular weights, and the technique of choice, largely because of the availability of commercial equipment, is *vapor-phase osmometry*, in which vapor-pressure lowering is indirectly measured. Only these two colligative methods are discussed.

It may also be recalled that the number-average molecular weight $\overline{M}_n$ is the simple counting average in which the mass of the sample, expressed in atomic mass units or *daltons*, is divided by the number of molecules it contains: $\overline{M}_n = w/N$. Expressing N as the sum over all species of the number N_i of molecules of the ith kind, and w similarly as Σw_i where $w_i = N_i M_i$, the defining equation for the number-average molecular weight is usually written

$$\overline{M}_n = \Sigma N_i M_i / \Sigma N_i \tag{7-1}$$

Proof that the colligative methods yield $\overline{M}_n$ is given in the *Textbook.*

Membrane osmometry

The principle of the osmotic method and the nature and operation of the simple two-compartment osmometer are described in the *Textbook,* pp. 69-70, and it is assumed that the reader has reviewed this section. The apparent simplicity of the experiment and the magnitude of the osmotic pressure have resulted in widespread use of this technique, beginning many decades before the nature of polymer molecules was clarified. But the properties of available membrane materials impose severe limitations on the method, and this phase of its technology cannot be considered to have yielded to the scientific approach.

Despite this warning, the polymer scientist is often forced to use membrane osmometry, since it is the only technique with which values of $\overline{M}_n$ higher than a few tens of thousands can reliably be measured. The upper limit of the method, corresponding to the smallest osmotic height that can be measured, is in the range 500,000 to 1,000,000. The lower limit is set by the permeability of available membranes, as described below, at a rather indefinite 10,000-50,000, depending upon the sample and membrane.

Membranes A wide variety of materials has been used as the semipermeable barrier in membrane osmometry, but this discussion is limited to modern cellulosic membranes, which are virtually the only ones likely to be found in use. Perhaps the most widely used of these materials is *gel cellophane*, the modifier indicating material taken from the production line in a water-soaked condition, free from plasticizer and other additives. Depending on the speed of equilibration and retentiveness to low-molecular-weight species desired, it may never be allowed to dry, or may be dried and resoaked. Other commercially available membranes include cellulose regenerated by the deacetylation of acetylcellulose, and bacterial celluloses. Armstrong (1968) describes these materials and their sources and properties.

In either case, the membrane is usually obtained water-wet, and must be *conditioned* if it is to be used in an organic solvent. In this procedure, the water is replaced by the solvent to be used; if the latter is not miscible with water, the conditioning is carried out in two steps, using an intermediate solvent miscible with both water and the final solvent. A common choice for the intermediate solvent is a lower alcohol such as ethanol or isopropanol, or another substance such as acetone.

Most conditioning procedures recommend soaking the membrane in a series of mixtures of water and intermediate solvent, the latter in increasing amounts until in, say, the fourth mixture it is present alone. A similar series of soakings in mixtures of the intermediate and final solvents follows. Suggested times for each stage vary widely, from a few minutes up to 24 hours. The procedures are entirely empirical, and few systematic studies of the processes have been made.

Osmometers With the great rise in popularity of high-speed automatic osmometers in the last few years, much of the interest in earlier designs, described in the general references, has died out. This discussion is limited to the three versions of the automatic osmometer which are currently available commercially. The major advantages of these instruments are equilibration within 5-10 min, compared to several hours for simpler osmometers, and automatic indication and recording of the osmotic pressure. The three designs are described in order of their appearance.

The Mechrolab design (Steele 1963), available from Hewlett-Packard, is shown in Fig. 7-2. A horizontal membrane separates the solvent (below) and solution compartments. Both are open to the atmosphere, eliminating all valves except that draining the solution compartment. Below the solvent compartment, a section of glass capillary connects, through a flexible tube, to a solvent reservoir in a tower at the left.

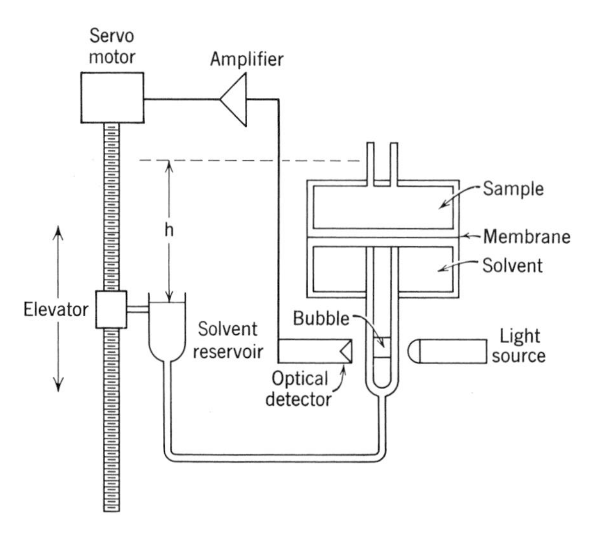

Fig. 7-2. Sketch of the essential components of the Mechrolab membrane osmometer. See the text for description (*Textbook*, Fig. 3-4, courtesy Hewlett-Packard Corp.).

A bubble of air is introduced into this capillary before the osmometer is assembled, and placed so that its meniscus is between a light source and detector. In operation, any movement of the meniscus in response to solvent flow through the membrane alters the amount of light reaching the photodetector, generating a signal which operates a servo system moving the solvent reservoir to the point which restores the meniscus to its original position. The difference in reservoir levels with solution and pure solvent in the upper (solution) compartment is the osmotic pressure of the solution. Without doubt, the greatest disadvantage of the Mechrolab design is the need for the operator to develop considerable skill in introducing and maintaining the bubble before satisfactory operation can be achieved.

The Shell Development Laboratories osmometer (Rolfson 1964) is shown in Figs. 7-3 and 7-4. It is this instrument, in the version made by J. V. Stabin, that is used in Exp. 11. The instrument is commercially available from Hallikainen. This instrument uses a horizontal membrane to separate the solvent and solution compartments, but here the solution compartment (below the membrane) is completely closed by valves. It has a flexible metal diaphragm as one wall. Transfer of solvent through the membrane deflects this diaphragm, which is made one electrode of a capacitor in a tuned circuit. A deflection of approximately 25 nm (one microinch) is enough to cause a change in frequency of the circuit, generating a signal which operates a servomotor to apply an appropriate hydrostatic pressure (negative in this case) to the solvent side of the membrane. An advantage of this instrument over the Mechrolab design is that both compartments (volume approximately 2 ml each) can be flushed completely without disassembling the cell.

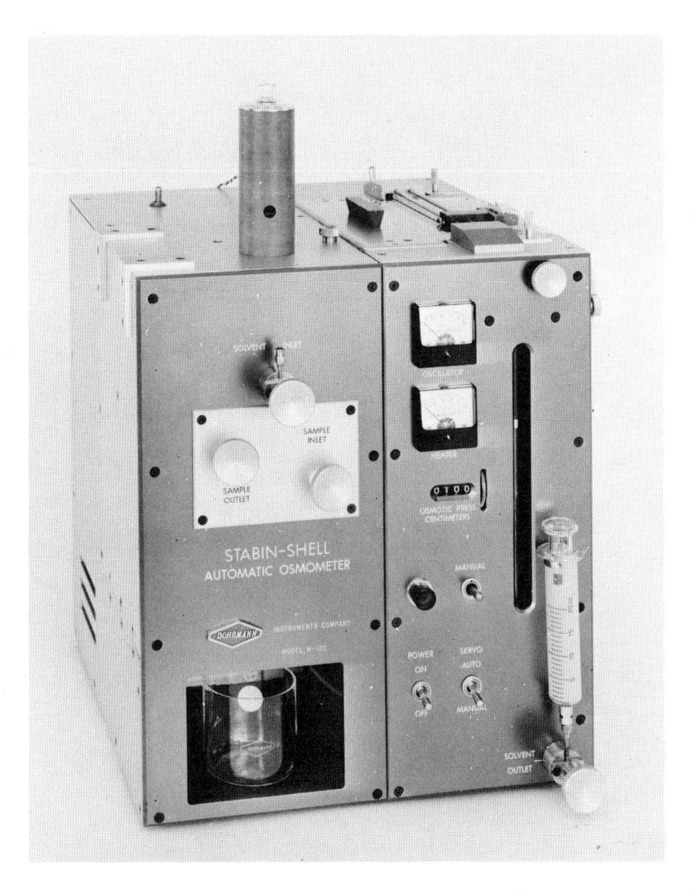

Fig. 7-3. The Shell-design membrane osmometer (Billmeyer 1965, courtesy J. V. Stabin).

The Reiff (1959) or Melabs osmometer (Fig. 7-5), now produced by Wescan, is similar to the Shell design except that no servo system is used. Instead, the osmotic pressure is measured directly by means of a strain gauge coupled to a flexible diaphragm in the solvent compartment. The solution compartment is open, as in the Mechrolab osmometer.

Armstrong (1968) describes the design and performance of these three osmometers. Krigbaum (1967) and Elias (1968) describe a number of test procedures to insure the proper operation of osmometers and membranes.

Theory In the osmotic experiment, with solvent and solution separated by a membrane permeable to the solute only, the chemical potential μ_1 of the solvent on the solution side is lower than that of the pure solvent because of the presence of the solute molecules. This chemical potential can be increased by the application of pressure to the solution, and thermodynamics identifies the rate of change of μ_1 with

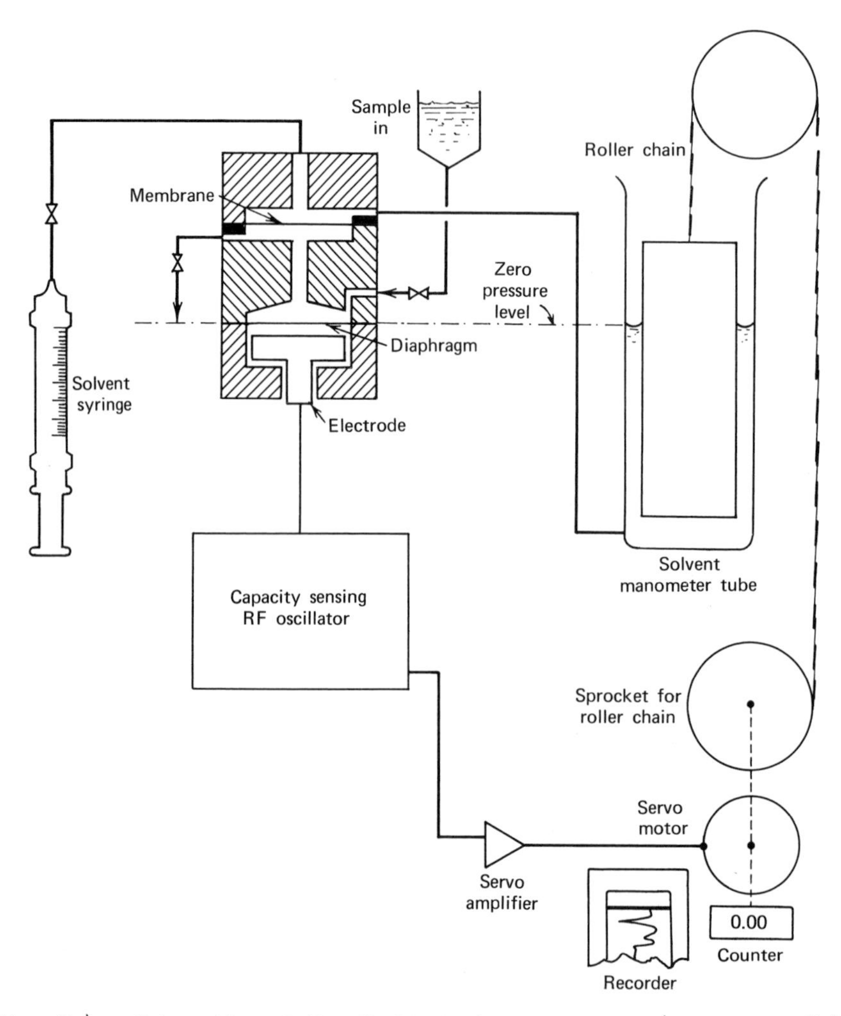

Fig. 7-4. Schematic of the Shell-design osmometer (Rolfson 1964).

pressure as the partial molar volume of the solvent. For dilute solutions, this is nearly equal to the actual molar volume of the solvent V_1 and is independent of pressure in the range of osmotic pressures. One can then write the difference in solvent chemical potential across the membrane as

$$-\Delta\mu_1 = \pi V_1 = -RT \ln a_1 \qquad (7\text{-}2)$$

where π, the osmotic pressure, is the pressure difference between solution and solvent. For dilute solutions, application of Raoult's law allows the activity to be replaced with the mole fraction of solvent,

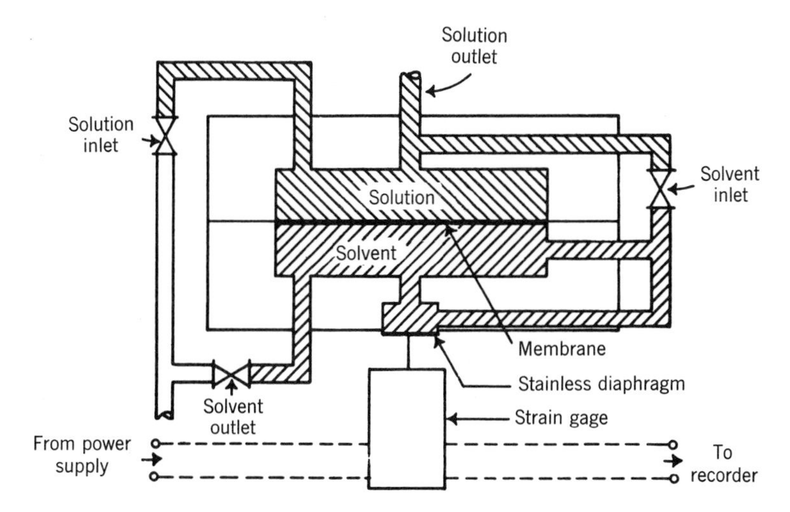

Fig. 7-5. Schematic of the Reiff osmometer (courtesy Melabs, Inc.).

n, and expansion of the logarithm yields

$$-\ln n_1 \simeq 1 - n_1 = n_2 = c\, V_1/M \tag{7-3}$$

At infinite dilution all the approximations become exact, and van't Hoff's limiting law results:

$$(\pi/c)_0 = RT/M \tag{7-4}$$

The concentration dependence of the osmotic pressure can be written by expanding Eq. 7-4 in a power series in c,

$$\pi = RT(A_1 c + A_2 c^2 + A_3 c^3 + \cdots) \tag{7-5}$$

where the A's are known as *virial coefficients*. By comparing Eqs. 7-4 and 7-5 it can be seen that $A_1 = 1/M$. The *Textbook* (pp. 65-66) shows that M is the number-average molecular weight $\overline{M}_n$. If, as is often the case, A_3 and all higher virial coefficients are negligibly small, Eq. 7-5 can be rearranged to

$$\pi/RTc = 1/\overline{M}_n + A_2 c \tag{7-6}$$

The second virial coefficient A_2 is a measure of the polymer-solvent interactions, and can be related to parameters describing them such as Flory's χ (see also the *Textbook*, Chaps. 2C and 3B):

$$A_2 = (\rho_1/M_1\ \rho_2^2)(1/2 - \chi_1) \tag{7-7}$$

In this respect, A_2 is a valuable predictor of the thermodynamic "goodness" of a solvent for a given polymer. When A_2 is high, the thermodynamic drive for solution to take place is also high, and conversely. (The *rate* of dissolution, however, may depend on other factors, as discussed in the *Textbook*, p. 26.) As A_2 goes down, the solvent becomes "poorer" until, at $A_2 = 0$, polymer of infinitely high molecular weight just precipitates from the solution. Molecules of lower molecular weight are more soluble, however, so that with real samples one can experience negative values of A_2 when precipitation is imminent. In the usual case, A_2 decreases with decreasing temperature, but the opposite may be true, as discussed in the *Textbook*, Chap. 2D. The second virial coefficient also decreases slowly with increasing polymer molecular weight. Figures 3-5 and 3-6 in the *Textbook* illustrate this behavior. (These considerations are not limited to membrane osmometry, for A_2 is determined in several other methods, as described below.)

Treatment of data The results of the osmotic experiment are the osmotic heights, measured in cm solvent, at a series of polymer concentrations. Some typical data are given in Table 7-3. To utilize these data in Eq. 7-6, it is easiest to plot π/c versus c and obtain the intercept $(\pi/c)_0$, using π and c in the units given. Then, $\overline{M}_n$ can be found from the relation

$$\overline{M}_n = RT/(\pi/c)_0 \tag{7-8}$$

It is only necessary to express R in the proper units. Starting from the handbook value $R = 0.08205$ ℓ at/mole K, recalling that 1 at = 1033 cm H_2O, and using 0.785 for the density of xylene at 105°C,

$$M_n = (0.08205 \times 1033/0.785)(105 + 273)/(\pi/c)_0 \tag{7-9}$$

There is some latitude possible in the extrapolation of the data in Table 7-3 to $c = 0$; this is also typical. Taking $(\pi/c)_0 = 1.00$ leads to $\overline{M}_n = 40{,}800$. A least-squares analysis of these data is given in App. IV. Typical plots of osmotic data are given in the *Textbook*, Figs. 3-5 and 3-6.

The second virial coefficient can be obtained from values of π/c at any two concentrations:

$$A_2 = [(\pi/c)_{c_2} - (\pi/c)_{c_1}] / RT(c_2 - c_1) \tag{7-10}$$

On plotting the data one finds that a reasonable straight line yields $\pi/c = 1.43$ at $c = 7.50$ as well as $\pi/c = 1.00$ at $c = 0$. Then $A_2 = 1.4 \times 10^{-5}$ ml mole/g^2, expressed in the usual units. This is rather small; typically, values of A_2 near 1×10^{-4} are found. Again, see App. IV for a least-squares treatment and analysis of errors.

TABLE 7-3. Typical data from membrane osmometry: Linear polyethylene in xylene at 105°C

c, g/ℓ	π, cm xylene	π/c
1.81	2.00	1.10
2.53	2.88	1.14
3.00	3.49	1.16
4.42	5.54	1.25
5.64	7.60	1.34
6.26	8.33	1.33
7.00	10.15	1.45

Limitation of the method By far the most serious limitation of membrane osmometry is the diffusion of low-molecular-weight species (which may or may not be polymer) through the membrane. If any species is present which diffuses through the membrane slowly enough that its concentration on the two sides of the membrane has not reached the same value during the time of the experiment, the apparent osmotic pressure observed will not be the true osmotic pressure of the sample. Even though the osmotic pressure changes only a few % per hour, serious errors can result. Moreover, it is not correct to extrapolate a changing osmotic pressure back to the time of the start of the experiment in an attempt to correct for diffusion. The only recourse one has is to abandon the experiment and try again with a more retentive membrane, or with a sample from which the diffusible species have been removed by fractionation or extraction.

Vapor-phase osmometry

A brief description of this technique is given in the *Textbook*, pp. 67-68. References to original papers on the theory and application of the method are cited there, but no complete discussion of vapor-phase osmometry has yet appeared; thus, no general references can be given.

"Vapor-pressure osmometers" Commercially available instruments for vapor-phase osmometry are known as "vapor pressure osmometers." Two are marketed in the U.S. The older Mechrolab design (Pasternak 1962), produced by Hewlett-Packard, is shown in Figs. 7-6 and 7-7. This is the instrument used in Exp. 12. It consists of two parts: a thermally insulated measurement chamber (Fig. 7-6, left) and an electronics unit (right). Within the measurement chamber (Fig. 7-7), two matched thermistors are suspended in a temperature-controlled cell that is saturated with solvent vapor. A drop of polymer solution is placed on one thermistor, and a drop of pure solvent on the other, by means of hypodermic syringes. The vapor pressure of the solvent in the solution drop is lowered by the presence of the solute. As a result, there is an excess condensation of solvent from the vapor in the chamber onto

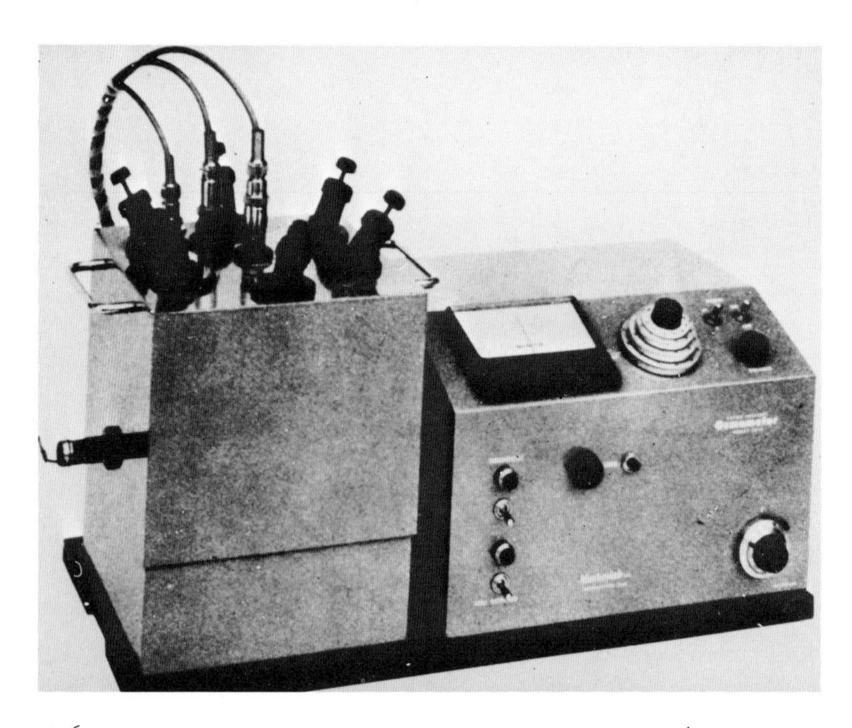

Fig. 7-6. The Mechrolab-design vapor-phase osmometer (Billmeyer 1965, courtesy Hewlett-Packard Corp.).

this drop, resulting in the release of heat of vaporization and raising the temperature of this drop very slightly. Since the thermistors are electrical elements having a high temperature coefficient of resistance, the temperature difference between the solution and solvent drops can be detected as an unbalance of a Wheatstone-bridge circuit in the electronics unit.

The Hitachi-Perkin-Elmer design instrument, marketed by Coleman, is similar except for placement of the thermistors and syringes, and details of the electronics.

Theory From Raoult's law, which states that in an ideal solution the partial vapor pressure of each component is proportional to its mole fraction, it can be shown that the vapor-pressure lowering is given by

$$p_1^0 - p_1 = p_1^0 n_2 = p_1^0 c V_1/M \qquad (7\text{-}11)$$

where the superscript zero indicates the pure solvent. Thermodynamics relates the vapor-pressure lowering to the temperature difference observed in this experiment by means of the Clapeyron equation. As was stated above, this temperature difference is proportional to the difference in electrical resistance, Δr, between the two thermistors. If

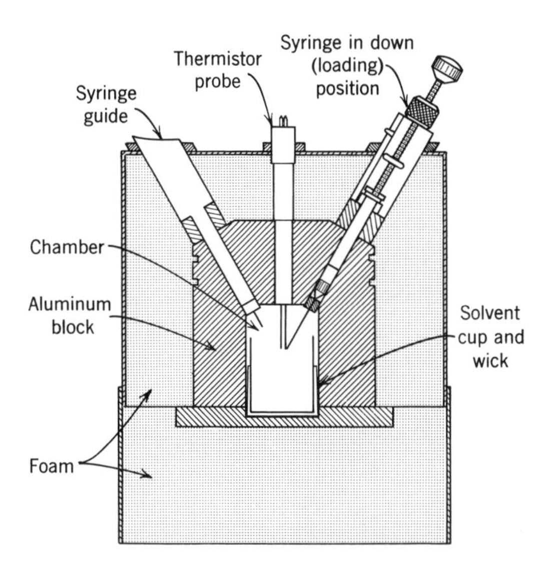

Fig. 7-7. Measurement chamber of the Mechrolab-design vapor-phase osmometer. See text for explanation (*Textbook*, Fig. 3-1, from Pasternak 1962).

all the excess heat of vaporization were utilized to produce the observed temperature difference, these relations would allow a limiting law, analogous to van't Hoff's law in membrane osmometry, to be written explicitly. This is not the case, however, since some heat is lost, for example by conduction along the thermistor leads. This can be shown to be a constant fraction of the total, and typically 70-80% of the theoretical temperature difference is achieved. In addition, the temperature coefficient of resistance of the thermistors, at the temperature of the experiment, is usually not precisely known. It is customary to combine all the necessary constants and proportionality factors, explicitly determined or not, into a single calibration constant k, and to write as the limiting law

$$(\Delta r/c)_0 = k/\overline{M}_n \tag{7-12}$$

where $\overline{M}_n$ is written at once by the argument applying to all the colligative methods. Virial expansion of Eq. 7-12 leads, in analogy to Eqs. 7-5 and 7-6, to

$$\Delta r/kc = 1/\overline{M}_n + A_2 c \tag{7-13}$$

To determine k, one measures $(\Delta r/c)_0$ for a low-molecular-weight substance of known molecular weight and uses Eq. 7-12.

Treatment of the data Table 7-4 gives typical data obtained with the vapor-phase osmometer. The calibration data, obtained with tristearin,

TABLE 7-4. Typical data from vapor-phase osmometry: calibration with tristearin, and two samples of linear polyethylene, measured in o-dichlorobenzene at 130°C

Time, min	Δr at Concentration c, g/ℓ			
	Tristearin			
	c = 4.41	c = 10.27	c = 15.29	c = 19.88
2	3.18	7.80	11.94	15.12
4	3.34	7.79	11.88	15.06
6	3.39	7.82	11.86	15.02
8	3.41	7.83	11.84	14.97
	Polyethylene Sample L			
	c = 9.45	c = 22.35	c = 33.54	c = 45.82
2	1.70	4.30	6.82	10.16
4	1.88	4.40	6.95	10.05
6	1.89	4.62	6.99	10.04
8	1.91	4.62	6.99	10.00
10	1.91	4.62	6.99	10.00
	Polyethylene Sample H			
	c = 12.40	c = 20.36	c = 30.54	c = 40.17
2	0.60	1.20	2.75	4.85
4	0.64	1.40	2.88	4.90
6	0.60	1.44	2.89	4.90
8	0.61	1.44	2.89	4.90

M = 891, show so little dependence of $\Delta r/c$ on concentration that it can be neglected. This is typical of results with low-molecular-weight substances. Calculation of $\Delta r/c$ from the steady-state values of Δr at longer times, determination of k from the tristearin data using Eq. 7-12, and the subsequent determination of $\overline{M}_n$ and A_2 using Eq. 7-13, are all routine. For these data, $k = 0.775 \times 891 = 690$. For sample L, $(\Delta r/c)_0 = 0.197$ and $\overline{M}_n = 3500$, and for sample H, $(\Delta r/c)_0 = 0.018$ and $\overline{M}_n = 38{,}400$. The calculation of A_2 is left as an exercise. Sample H is, in fact, the same polymer described in Table 7-3; its molecular weight by membrane osmometry is 40,800. The difference may reflect either experimental error or, possibly, slight diffusion through the osmotic membrane. (These examples are taken from Billmeyer 1964*b*.)

For reasons not fully understood, values of $\Delta r/c$ may scatter or exhibit curvature as a function of c instead of falling on a straight line. Although the theory does not call for it, experience attributes this to solvent effects and suggests that a plot of Δr versus c may yield a straight line which does not pass through the origin. If so, the intercept of this line at c = 0 may be used as a small correction to values of Δr. It is often found that, with this correction, values of $\Delta r/c$ follow Eq. 7-13 more satisfactorily, and values of $\overline{M}_n$ so obtained are correct for known samples.

Limitations of the method In vapor-phase osmometry, all solute molecules which are nonvolatile under the conditions of the experiment contribute to $\overline{M}_n$. In one sense this is an advantage, for the method is useful for low-molecular-weight solutes, providing a rapid method of measuring the molecular weight of rather small quantities of material. In another sense, however, it represents a limitation, since a very small weight percent of low-molecular-weight impurity, which would rapidly equilibrate in membrane osmometry and have no effect, can lead to serious errors in $\overline{M}_n$ determined by vapor-phase osmometry. In fact, experience has shown that very few polymers with $\overline{M}_n > 50{,}000$ by membrane osmometry can be obtained pure enough to give their true molecular weights in vapor-phase osmometry, even though the sensitivity of research instrumentation is adequate to measure them (Wachter 1969). Commercial instruments lack sensitivity to measure $\overline{M}_n$ much above 20,000 except in unusual circumstances.

Other limitations, resulting from the mode of operation of the instruments, require careful control of such variables as drop size and measurement time, as described in Exp. 12.

GENERAL REFERENCES

Textbook, Chap. 3B; and specifically:

Membrane osmometry: Chiang 1964; Krigbaum 1967; Coll 1968; Elias 1968; Overton 1971.

B. *Weight-Average Molecular Weight*

The weight-average molecular weight, defined as

$$\overline{M}_w = \Sigma w_i M_i = \Sigma c_i M_i / c = \Sigma N_i M_i^2 / \Sigma N_i M_i \qquad (7\text{-}14)$$

is shown in the *Textbook* (Chaps. 3C and 3E) to be accessible from light-scattering and equilibrium ultracentrifugation experiments. Since ultracentrifugation, although the method of choice for many biological materials, is seldom used for random-coil synthetic polymers, we discuss only light scattering here.

Light scattering

In the light-scattering experiment, measurement is made of the difference in scattered-light intensity between a polymer solution and its solvent. Theory (*Textbook*, Chap. 3C) shows that this scattered intensity depends upon both concentration (in the usual way for polymers) and the angle between the incident and scattered light beams. The second requirement sets the major design features of light-scattering photometers.

Preparation of solutions A task of major importance preliminary to

Fig. 7-8. Filtering equipment for clarifying solutions for light scattering and other characterization techniques: (*a*) fritted-glass filter with nitrogen supply at reduced pressure; (*b*) cellulosic filter in holder attached to syringe.

measurement of light scattering is the proper preparation of the solvent and polymer solutions to be used. Any material, particularly dust and dirt, in a liquid can be expected to scatter light. Since the scattered intensity is proportional to the square of the mass of the scattering particle, and to the square of its difference in refractive index from the liquid, large dust particles can be expected to scatter far more light than polymer molecules. It is essential that all foreign material be removed completely from the solutions and solvent to be measured. This is never easy, and for the most sophisticated work requires a variety of purification techniques and a "clean-room" environment.

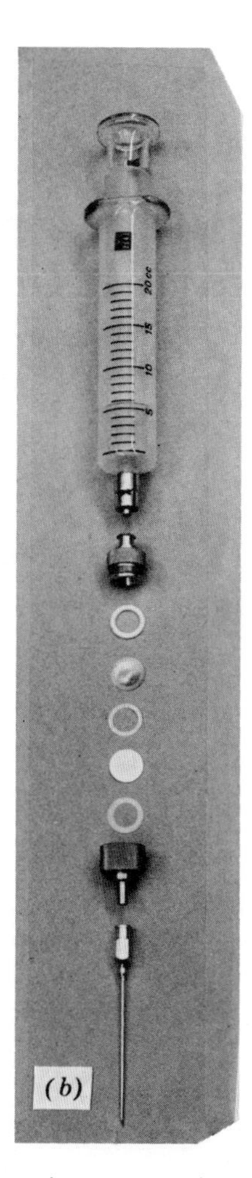

Fig. 7-8 (continued). Part *b*.

Practically, one relies upon the simpler of several available techniques to clarify solutions and solvent, followed by more than ordinary care in their subsequent handling. The filtration of solutions through ultrafine filters, with pore sizes of a few tenths micrometer, when properly carried out, usually suffices to remove superficial dust and foreign particles. The filters are conveniently supported in holders such as those shown in Fig. 7-8, and the filtered liquid is led directly into the scattering cell. It is important that solutions never be forced through filters by pressure, but allowed to flow by gravity or with *very* slight vacuum on the exit side. In more stubborn cases, the solution may be centrifuged in a laboratory centrifuge or preparative ultracentrifuge; however, problems in maintaining cleanliness while transferring the solution to the scattering cell add to the difficulty of using this technique. Needless to say, cleanliness in handling the polymer sample, from preparation through to dissolution, makes subsequent solution clarification easier.

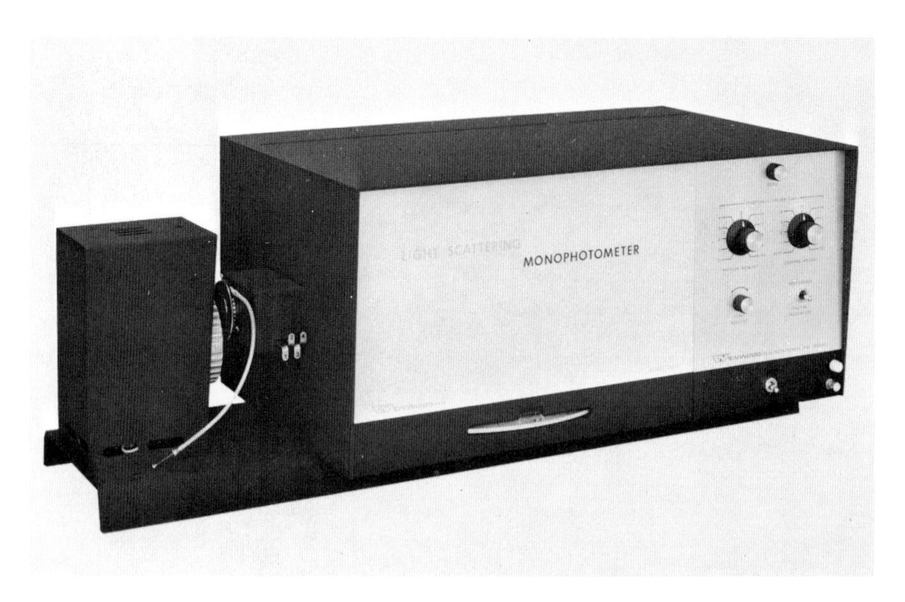

Fig. 7-9. The Brice-Phoenix light-scattering photometer (courtesy C. N. Wood Manufacturing Co.).

The clarified solutions, in the scattering cell, may be observed with the aid of a strong light beam. Dust particles can best be seen at small angles of observation, and the solutions should not exhibit any bright particles under these conditions.

Light-scattering photometers This section is based on the widely used Brice-Phoenix (Brice 1950) commercially available (Wood) light-scattering instrument and the more modern SOFICA instrument (Wippler 1954; Bausch and Lomb). From what has been said, the essential components of a photometer can be seen to be a light source, cell and holder, and detector mounted so as to view the cell over a range of angles. In the Brice instrument, these components are arranged as shown in Figs. 7-9 and 7-10. The light source S is a mercury arc lamp, used in conjunction with a monochromatizing filter F which isolates either the 546-nm green mercury line or the 436-nm blue line. (This instrument, and indeed most current commercial instruments, predates the availability of the more powerful laser as a light source.) A lens L provides an approximately parallel beam of light passing into the large sample compartment. A polarizer P_1 may be used; in the theory to follow we assume for simplicity that the light is vertically polarized (with respect to the plane of the cell and detector, which is conventionally horizontal).

The scattering cell at C is usually cylindrical except for plane entrance and exit windows for the primary light beam. (Fig. 7-10 shows an alternative form designed for observations at 45°, 90°, and 135° only.) The cell is centered on the axis of rotation of the receiver or detector R, and a photomultiplier tube powered by 700-1000

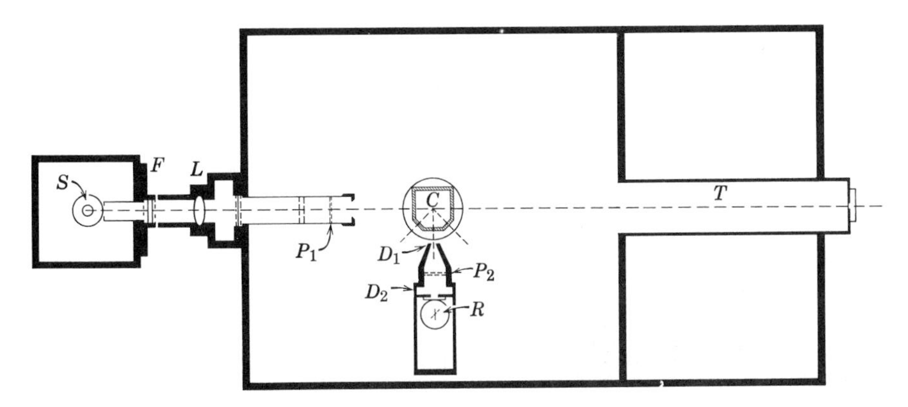

Fig. 7-10. Essential components of a light-scattering photometer (*Textbook*, Fig. 3-9).

volts from a well-regulated D.C. supply. Diaphragms D_1 and D_2 define the field of view seen by the detector, typically encompassing 2-5° of scattering angle. A second polarizer P_2 is provided, and is also normally set to pass vertically polarized light. The receiver arm can be rotated manually from outside the cell compartment. After passing through the cell, the primary light beam is caught in a trap T.

The SOFICA instrument (Bausch and Lomb; Wippler 1954) is essentially similar except for the arrangement of the cell compartment. In this photometer (Fig. 7-11*a*), a small scattering cell is immersed in a larger bath filled with a liquid having the same refractive index as the contents of the scattering cell. A prism attached to and rotating with the phototube assembly views the scattering cell from inside this bath. This arrangement effectively eliminates reflections from the surfaces of the cell. An advanced version of this instrument (Fig. 7-11*b*) offers automatic scanning and recording, compatibility with computers, and other modern features.

Since the ratio of intensities of the incident and scattered beams is typically of the order of 10^6, it is inconvenient to measure both beams with the full sensitivity of the detector. At some stage a calibration of the instrument with some material of known scattering power is expedient. Usually this is carried out in two stages: (1) A primary standard, the currently recommended one being a solution in 1*M* NaCl of 12-tungstosilicic acid, $H_4SiW_{12}O_{40}$, $M = 2879$, is used to calibrate a permanent working standard. This can be an opal glass or plastic, perhaps combined with a neutral filter to reduce the light intensity, which is permanently mounted in the instrument or replaces the scattering cell. (2) The working standard is measured from time to time during the measurement of the polymer solutions to provide the required calibration.

Differential refractometer A second quantity required in the light-scattering determination is the *specific refractive increment*, dn/dc, which is the concentration dependence of the refractive index of the polymer solution. This is a constant, so that a single determination

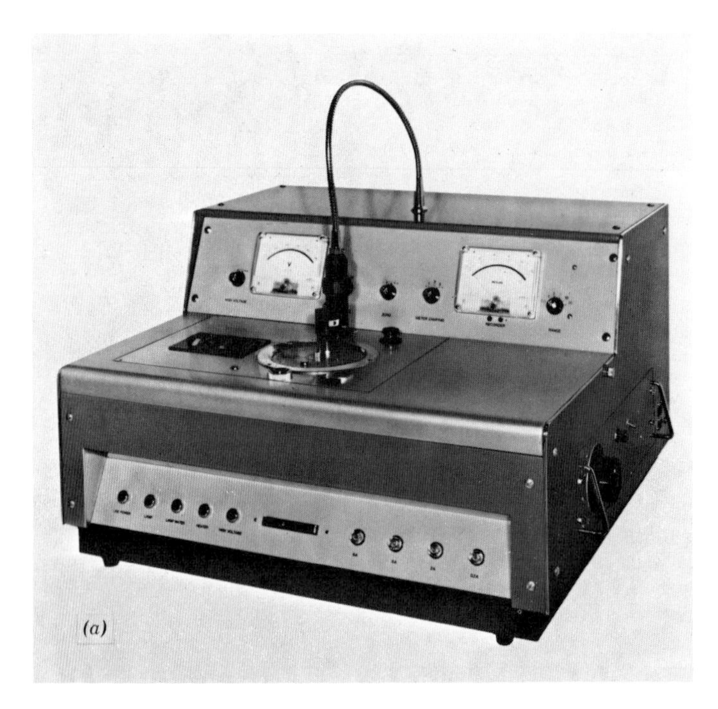

Fig. 7-11. The SOFICA light-scattering photometers: (*a*) Wippler (1954) version; (*b*) Fica 50 automated version (courtesy Bausch and Lomb).

of the difference in refractive index between a solution of known concentration and the solvent suffices for its determination. Since this difference is in the fifth or sixth decimal place of refractive index in the usual case, conventional refractomers are not sensitive enough to measure it. The use of a differential refractometer (Brice 1951) is required. This instrument (Fig. 7-12) uses a simple double-prism cell in which the deviation in direction of a light beam is directly proportional to the desired refractive-index difference.

Theory The working equation for the determination of the weight-average molecular weight by light scattering, due to Debye (1947), is

$$Kc/\Delta R_\theta = 1/\overline{M}_w P(\theta) + 2\,A_2 c \qquad (7\text{-}15)$$

Here the scattered intensity is expressed as the Rayleigh ratio, R_θ, which is the ratio of scattered-light intensity, per unit volume of scattering solution and unit solid angle at the detector, to the incident light flux per square centimeter. The Rayleigh ratio, which is a function of the scattering angle θ, has the dimensions of cm^{-1}. Here ΔR_θ denotes the Rayleigh ratio of the solution less that of the solvent, and our notation differs slightly from that in Eqs. 3-14 and

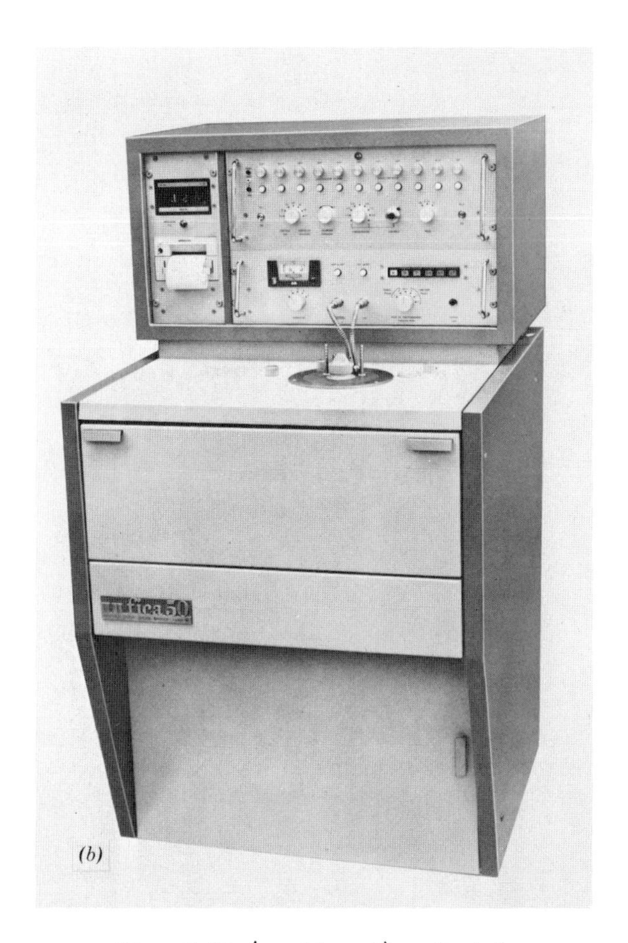

Fig. 7-11 (continued). Part *b*.

3-20 in the *Textbook*. Throughout, we simplify by assuming the use of vertically polarized light, and the absence of depolarization.

The constant K is given by

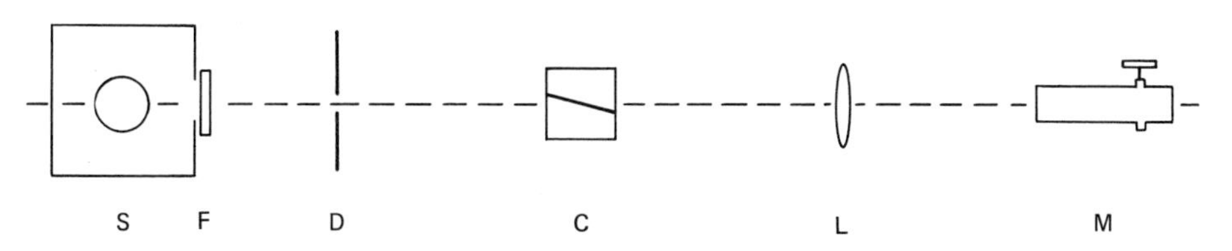

Fig. 7-12. Sketch of the optics of the Brice (1951) design differential refractometer: S, source (mercury arc lamp); F, monochromatizing filter; D, slit; C, cell; L, lens focusing image of slit onto M, microscope with traveling crosshair and scale.

TABLE 7-5. Typical data from light scattering: poly(methyl methacrylate) sample 8 (Billmeyer 1955) in 2-butanone. For explanation, see the text.

Concentration c, g/ml	Scattering Angle θ, Degrees 30	60	90	120	Reference Intensity
	A. Observed intensities, recorder chart divisions				
0	38	16.3	13.4	14.0	218
0.00060	456	205	140	134	220
0.00108	646	300	213	208	221
0.00147	770	370	266	261	220
0.00180	820	405	290	294	220
0.00208	910	450	333	339	230
	B. Intensities corrected for scattering volume				
0	19	14.1	13.4	12.1	218
0.00060	228	177.5	140	116	220
0.00108	323	260	213	180	221
0.00147	385	320	266	226	220
0.00180	410	350	290	254	220
0.00208	455	390	333	293	230
	C. Scattered intensity divided by reference intensity				
0	0.087	0.065	0.062	0.056	
0.00060	1.038	0.807	0.636	0.528	
0.00108	1.460	1.175	0.964	0.815	
0.00147	1.750	1.453	1.200	1.029	
0.00180	1.865	1.590	1.319	1.155	
0.00208	1.978	1.695	1.447	1.273	
	D. Intensity ratio less solvent reading				
0.00060	0.951	0.742	0.574	0.472	
0.00108	1.383	1.110	0.902	0.759	
0.00147	1.663	1.388	1.138	0.973	
0.00180	1.778	1.525	1.257	1.099	
0.00208	1.891	1.630	1.385	1.217	
	E. $\Delta R_\theta \times 10^4$				
0.00060	0.890	0.695	0.537	0.443	
0.00108	1.295	1.040	0.845	0.710	
0.00147	1.555	1.300	1.065	0.911	
0.00180	1.665	1.427	1.177	1.029	
0.00208	1.770	1.525	1.295	1.138	

Concentration c, g/ml	Scattering Angle θ, Degrees 30	60	90	120	Reference Intensity
	F. $c/\Delta R_\theta$				
0.00060	6.75	8.63	11.17	13.55	
0.00108	8.33	10.38	12.77	15.19	
0.00147	9.47	11.30	13.80	16.12	
0.00180	10.80	12.60	15.30	17.48	
0.00208	11.75	13.65	16.05	18.28	
	G. $\sin^2(\theta/2) + 100\,c$				
0.00060	0.127	0.310	0.560	0.810	
0.00108	0.175	0.358	0.608	0.858	
0.00147	0.214	0.397	0.647	0.897	
0.00180	0.247	0.430	0.680	0.930	
0.00208	0.275	0.458	0.708	0.958	

$$K = 2\,\pi^2 n^2 (dn/dc)^2 / N_0 \lambda^4 \qquad (7\text{-}16)$$

where n is refractive index, dn/dc is the specific refractive increment described above, N_0 is Avogadro's number, and λ is the wavelength of the incident light as measured in air.

The quantity $P(\theta)$ is a particle scattering factor, discussed further in the *Textbook* and the general references, which describes the angular dependence of the scattered light and relates it to particle size for a given model such as a sphere, random coil, or other type. In Exp. 13, we shall not be concerned with details of this function except to note that $P(\theta) = 1$ at $\theta = 0$, and that for small θ,

$$1/P(\theta) = 1 + (16\pi^2/3\lambda_s^2)\overline{s_z^2}\,\sin^2(\theta/2) \qquad (7\text{-}17)$$

where $\overline{s_z^2}$ is the z-average mean-square radius of gyration for random-coil polymers and λ_s is the wavelength of light in the solution: $\lambda_s = \lambda/n$. $P(\theta)$ can thus be considered either a correction factor to the scattered intensity to reduce it to its value at $\theta = 0$ or a means of measuring molecular size, as discussed in Sec. C.

Treatment of data The reduction of light-scattering data to yield the weight-average molecular weight, second virial coefficient, and radius of gyration, is sufficiently complex that a computer program is often written to handle the task in routine work. Here, we work out a typical example in detail, save that the data selected are limited to only a few angles of observation. They describe a sample of poly(methyl methacrylate) measured in 2-butanone, designated sample 8 in Billmeyer 1955. Appendix IV contains a statistical treatment of these data, including determination of the errors involved.

It is convenient to think of the data treatment as divided into three parts. In the first, the instrument readings are converted into values of ΔR_θ, the Rayleigh ratio for (solution less solvent). Next, the data are prepared for the extrapolations to $c = 0$ and $\theta = 0$, and these extrapolations made graphically. Finally, $\overline{M}_w$, A_2 and $(\overline{s_z^2})^{1/2}$ are calculated.

These steps are illustrated in Table 7-5. Part A gives the initial data, corrected only by applying appropriate multiplying factors for the combination of neutral filter and recorder sensitivity used. The first correction to the data is necessitated by the fact that the Rayleigh ratio is defined per unit volume of scattering liquid, whereas with the usual cell geometries the scattering volume observed is not constant, but varies inversely as sin θ. This dependence is illustrated in Fig. 7-13, and the data after multiplication by sin θ are given in Part B of the Table. Next the data are corrected for slight variations in the reference intensity, as determined by reading the secondary standard one or more times during the determination with each solution. These results are tabulated in Part C. At this point a reflection correction (Kratohvil 1966) might be applied to the data. This was omitted in the 1955 calculations. Now the solvent readings are subtracted, yielding the data given in Part D of the Table.

We are now ready to calculate ΔR_θ. A refraction correction (Billmeyer 1971) is required if the primary calibration was made using a solvent (in this case, water) with a different refractive index from that of the solvent used in the determination. The relation is

$$R_\theta = k_w(n_s/n_w)^2 \times \text{(intensity ratio)} \tag{7-18}$$

where, in this case, the subscripts s and w stand for the solvent (2-butanone) and water, respectively. The refractive indices of these liquids at $\lambda = 546$ nm, the mercury green line, are $n_s = 1.3856$ and $n_w = 1.3343$. The primary calibration, with an aqueous standard, yielded $k_w = 0.869 \times 10^{-4}$. Hence, each term in Part D is converted to ΔR_θ by multiplying by $0.869 \times 10^{-4}(1.3856/1.3343)^2 = 0.936 \times 10^{-4}$. These results are given in Part E, completing the first section of the data reduction.

The extrapolations to $c = 0$ and $\theta = 0$ required by Eq. 7-15 are usually carried out by the graphical method developed by Zimm (1948) (but see also App. IV). Use is made of the expansion of $1/P(\theta)$ for random-coil polymers, Eq. 7-17. It can be seen that $\sin^2(\theta/2)$ is

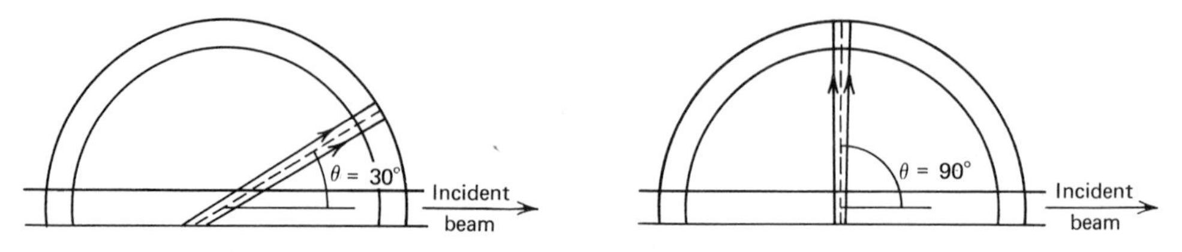

Fig. 7-13. Geometry of a light-scattering cell, showing the difference in observed volume at θ = 30° and θ = 90° (adapted from Billmeyer 1971).

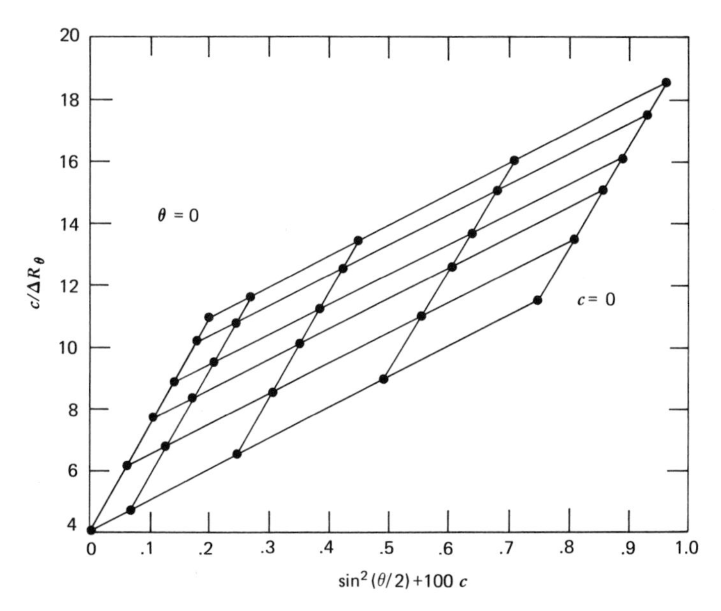

Fig. 7-14. Zimm plot for the data of Table 7-5 (Billmeyer 1955).

the appropriate variable for the extrapolation as a function of angle. Zimm's contribution was to suggest that the left-hand side of Eq. 7-19 be plotted against $\sin^2(\theta/2) + kc$, where k (not to be confused with k_w in Eq. 7-18) is an arbitrary constant selected to provide a conveniently spaced plot. We shall now develop the rectilinear grid characteristic of the Zimm plot.

For the graphical treatment, it is unnecessary to compute K at this stage. The quotient $c/\Delta R_\theta$ is calculated (Part F of Table 7-5). The quantity $\sin^2(\theta/2)$ is calculated or obtained from tables of the haversine for each scattering angle. For θ = 30°, 60°, 90°, and 120°, respectively, $\sin^2(\theta/2)$ = 0.067, 0.250, 0.500, and 0.750. The constant k is selected so that, at the highest concentration, kc is a fraction, typically 0.2-0.4, of the total range of $\sin^2(\theta/2)$, which is of course 0-1. Here k = 100 is convenient, and $\sin^2(\theta/2) + 100c$ is calculated for each data point (Part G). The Zimm plot (Fig. 7-14) is formed from corresponding data from Parts F and G (solid points in the figure). Straight lines are drawn (1) through points with varying c at constant θ, and extrapolated to $c = 0$, i.e., to the value of $\sin^2(\theta/2)$ for that angle; and (2) through points with varying θ at constant c, and extrapolated to $\sin^2(\theta/2) = 0$, i.e., to 100 c for that concentration. These sets of points are the open circles in Fig. 7-14. Straight lines are drawn through these points and extrapolated to a common intercept at the axis of ordinates. Use is made throughout of the fact that, to a good approximation in most cases, both sets of lines have constant slope and are thus parallel. The residual scatter of the experimental points in the figure is typical.

We now complete the calculation by obtaining $\bar{M}_w$ from

$$\overline{M}_w = 1/K(c/\Delta R_\theta)_{c=0,\ \theta=0} \tag{7-19}$$

A_2 from the slope of the $\theta = 0$ line, and the radius of gryation $(\overline{s_z^2})^{1/2}$ from the slope of the $c = 0$ line using Eq. 7-17. In the calculation of K, we use the conditions of the experiment, $\lambda = 546$ nm $= 5.46 \times 10^{-5}$cm, $dn/dc = 0.112$ cm^3/g, $n = 1.3856$, and $N_0 = 6.023 \times 10^{23}$. The result is $K = 0.925 \times 10^{-7}$ mole cm^2/g^2. From the Zimm plot, $(c/\Delta R_\theta)_{c=0,\ \theta=0} = 4.05$, whence $\overline{M}_w = 2{,}600{,}000$. The second virial coefficient is obtained from

$$A_2 = \frac{1}{2} K[(c/\Delta R_\theta)_{c_2} - (c/\Delta R_\theta)_{c_1}]/(c_2 - c_1) \tag{7-20}$$

Solving, $A_2 = 1.56 \times 10^{-4}$ cm^2 mole/g^2. The mean-square radius of gyration, on combining Eqs. 7-15 and 7-17, and remembering that $\lambda_s = \lambda/n$, is given by

$$\overline{s_z^2} = (3\,\lambda^2/16\,\pi^2 n^2) \times (\text{slope/intercept}) \tag{7-21}$$

where slope and intercept refer to the Zimm plot, and K cancels out. Inserting the numerical values, solving, and taking the square root, $(\overline{s_z^2})^{1/2} = 85$ nm. These results are in good agreement with those quoted in Billmeyer 1955.

Limitations of the method The light-scattering method is applicable over a wide range of molecular weights, as long as there is sufficient difference in refractive index between the polymer and the solvent. Its major limitation is the requirement that solutions be free from all extraneous scattering material. Special treatment is required for some systems, such as copolymers and very large molecules; these are mentioned and referenced in the *Textbook*.

GENERAL REFERENCES

Textbook, Chap. 3C; Billmeyer 1964*a*; McIntyre 1964.

C. *Molecular Size*

As a result of the random-coil nature of most synthetic high polymers, the concept of the size of the molecule differs significantly from that of its mass. We restrict use of the term *size* to the description of the amount of space taken up by the molecule, expressed either as a linear dimension or a volume. Since the size, so defined, of a single molecular coil varies with time as a result of conformational changes due to Brownian motion, and similarly the molecular size varies from molecule to molecule of identical mass and structure, size can be described only in terms of average properties. Two such average size parameters describing linear dimensions of polymers are the *root-mean-square end-to-end distance* $(\overline{r^2})^{1/2}$ and the *radius of gyration* $(\overline{s^2})^{1/2}$. The former is self-explanatory, and the latter is defined as the root-mean-square distance of the segments of the chain from its center of gravity. When polymer solutions are referred to, the common volume parameter is the *hydrodynamic volume*, the volume that the chain appears to occupy based on a specific property, such as the viscosity increase it imparts to the solution. For linear polymers, all these quantities are uniquely related; for example $\overline{r^2} = 6\ \overline{s^2}$. These concepts are developed simply in Billmeyer 1972, and in more detail in the *Textbook*, Chap. 2B, and the general references cited therein.

In this section we discuss only one characterization method based on measurement of molecular size, the dilute-solution viscosity experiment. We have already seen in Sec. B that light scattering yields the size parameter $(\overline{s_z^2})^{1/2}$. Unfortunately, other methods do not yield the same z-average radius of gyration, so that it is difficult to compare light-scattering results directly to others. We shall see in Sec. D that gel permeation chromatography separates molecules according to their size, not mass, but we shall treat that technique primarily as an approach to measuring the molecular-weight distribution.

Dilute-solution viscosity

Nomenclature One of the first problems to be settled in discussing the viscosity of polymer solutions is an unfortunate duplication of nomenclature. The two sets in use are given in Table 7-6, taken from the *Textbook*, Chap. 3D. The terms described as common names have the authority of long usage, and are still found in day-to-day use in most cases. Accordingly, we adopt them throughout this book. The second set was recommended for use by the International Union of Pure and Applied Chemistry as more logical. They have been adopted in some texts and to a certain extent in the literature, but are still found in the minority of cases.

A related problem arises from the units of concentration conventionally employed for viscosity-related quantities. Again, the traditional and recommended units differ, and again we adopt the traditional but still almost universally used units of g/100 ml, or g/deciliter, for c. Use of the recommended g/ml of course yields numerical values differing by 100 from those in common use, and most polymer chemists prefer to avoid this confusion by retaining the familiar magnitudes.

Table 7-6 also provides the defining equations for the quantities discussed in this section. We start with the approximate relationship that the *relative viscosity* η_r is given by the ratio of the efflux time for the solution, t, to that of the solvent, t_0. Strictly, $\eta_r = \eta/\eta_0$, where the viscosities of the solution and solvent are related to the corresponding efflux times by

$$\eta = Ctd - Ed/t^2; \quad \eta_0 = Ct_0d_0 - Ed_0/t_0^2 \tag{7-22}$$

where d is density and C and E are constants for the particular viscometer used. For dilute solutions, d and d_0 are substantially equal,

TABLE 7-6. Nomenclature of solution viscosity

Common Name	Recommended Name	Symbol and Defining Equation
Relative viscosity	Viscosity ratio	$\eta_r = \eta/\eta_0 \simeq t/t_0$
Specific viscosity	--	$\eta_{sp} = \eta_r - 1 = (\eta - \eta_0)/\eta_0 \simeq (t - t_0)/t_0$
Reduced viscosity	Viscosity number	$\eta_{red} = \eta_{sp}/c$
Inherent viscosity	Logarithmic viscosity number	$\eta_{inh} = (\ln \eta_r)/c$
Intrinsic viscosity	Limiting viscosity number	$[\eta] = (\eta_{sp}/c)_{c=0} = [(\ln \eta_r)/c]_{c=0}$

and viscometers are designed so that, for efflux times greater than 100 sec or so, the second terms are negligible.

The *specific viscosity* can be seen to be the relative increment in viscosity of the solution over that of the solvent, and the *reduced viscosity* is this quantity taken per unit concentration. This is still dependent upon c, however, so extrapolation to $c = 0$ is required. It is convenient to extrapolate not only the reduced viscosity but also the *inherent viscosity* to $c = 0$. By expanding the logarithm in the defining equation for η_{inh} and observing the behavior as $c \to 0$, it is easy to demonstrate that both quantities extrapolate to the same intercept, denoted the *intrinsic viscosity* $[\eta]$ (always written with square brackets). The units of η_{red}, η_{inh}, and $[\eta]$ can be seen to be reciprocal to those of concentration, and are dl/g, since η_r and η_{sp} are dimensionless ratios.

The families of straight lines obtained in these extrapolations, for a series of polymer samples differing only in molecular size, have slopes which are not constant but vary in a regular way, increasing as $[\eta]$ increases. They are described by no theory, but by several sets of purely empirical equations of which we use only those due to Huggins (1942)

$$\eta_{sp}/c = [\eta] + k'[\eta]^2 c \tag{7-23}$$

and Kraemer (1938)

$$(\ln \eta_r)/c = [\eta] + k''[\eta]^2 c \tag{7-24}$$

where k' and k'' are known as the Huggins and Kraemer constants, respectively. For all but unusual cases, k'' is negative, and often $k'-k'' \simeq 0.5$. Since k'' is usually smaller in absolute value than k', reflecting a smaller change with concentration in the inherent viscosity compared to the reduced viscosity, the value of the inherent viscosity at a fixed concentration, 0.2 or 0.5 g/dl, is sometimes taken as an easily obtained approximation to the intrinsic viscosity. A number of other ways of approximating $[\eta]$ from measurements at a single concentration have been described in the literature; the earliest paper appears to be Billmeyer 1949.

Many attempts have been made to read importance into values of k' or k'' or to relate them to polymer properties, but without a theoretical background, these attempts have proved of little value, and we shall not pursue them.

Viscometers In contrast to other characterization methods, viscometry requires only the simplest apparatus: one or more glass capillary viscometers, a constant-temperature bath, and a timer. Good temperature control is essential; variations of solution temperature during the experiment must be kept below 0.01-0.02°C. In most instances, the measurement of efflux times is carried out by visual observation of the passage of the liquid meniscus past two lines marked on the viscometer, at which times a stopwatch or electric timer is started and stopped. There are automatic timers (Wescan) and systems which automatically prepare dilutions from a stock solution, clean and fill a viscometer, and obtain the efflux times (Bausch and Lomb). In these instruments, photocells mounted on the viscometer actuate an electric or electronic timer. They are considerably more precise, but are expensive and have the disadvantage that they provide timing for only one viscometer at any moment. In contrast, one operator can usually operate several viscometers simultaneously, given enough timing devices. Outside of this problem, the commercial automatic devices show considerable potential, but they are not yet widely used.

There are two types of viscometer in wide use, shown in Fig. 7-15. The simpler Ostwald viscometer is a constant-volume device, whereas in the Ubbelohde instrument, the effluent from the capillary flows into a bulb separate from the main liquid reservoir, so that the viscometer operates independent of the total volume of liquid present. This allows the convenience of preparing and measuring solutions having a range of concentrations without the need of transferring them. Both types of viscometer are available commercially (Cannon) in regular (10 ml) and semi-micro (1 ml) sizes, with a variety of capillary diameters giving a selection of efflux times.

Theory There have been several theories relating the intrinsic viscosity to molecular size, but this is a recent concept in comparison to the long history of the use of dilute-solution viscosity as an empirical measure of molecular weight. Staudinger (1930) was the first to cite the viscosity of polymer solutions as evidence for their long-chain, high-molecular-weight character. He wrote, in our terminology, $\eta_{sp} = KcM$, where K is a constant which he could evaluate from measurements on oligomers of known molecular weight. His equation implies that the reduced viscosity is independent of concentration, which we now know to be in error. In addition, later work has demonstrated

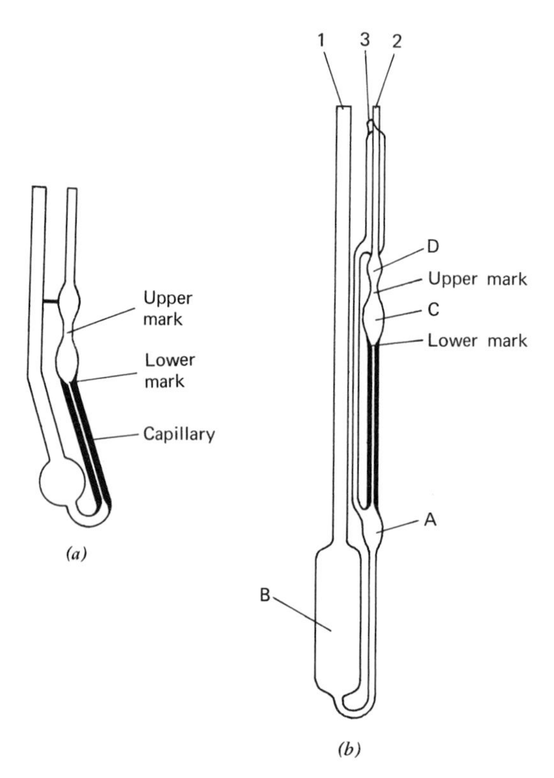

Fig. 7-15. Capillary viscometers commonly used for measurement of dilute-solution viscosity: (*a*) Cannon-Fenske; (*b*) Ubbelohde.

that Staudinger's equation must be modified by raising M to a power a, usually between 0.5 and 0.8, yielding the equation

$$[\eta] = K\overline{M}_v^a \tag{7-25}$$

which is usually attributed to Mark, Houwink, and Sakurada. Here the viscosity-average molecular weight (Schaefgen 1948)

$$\overline{M}_v = [\Sigma\ N_i M_i^{1+a} / \Sigma\ N_i M_i]^{1/a} \tag{7-26}$$

is appropriate, and it can be seen that the value of this average and its relation to others depends upon a, a measure of polymer-solvent interactions, as shown in Fig. 7-1. Since $\overline{M}_v$ is not available from other experiments (unless by calculation as described in Sec. D), one often substitutes $\overline{M}_w$ for it, as the nearest directly measurable average. Of course, for a monodisperse polymer all averages would be identical, allowing K and a to be determined with ease, but such samples are not

TABLE 7-7. Representative values of K and a for the intrinsic viscosity-molecular weight relationship (Kurata 1966)

Polymer	Solvent	T°C	$K \times 10^4$	a
Polyacrylonitrile	dimethyl formamide	25	2.4	0.75
Polycaprolactam	*m*-cresol	25	32	0.62
Polyisobutylene	cyclohexane	30	2.6	0.70
	diisobutylene	20	3.6	0.64
Poly(methyl methacrylate)	acetone	25	0.75	0.70
	benzene	25	0.55	0.76
	chloroform	20	0.60	0.79
Polystyrene	benzene	25	0.2	0.74
	2-butanone	25	3.9	0.58
	cyclohexane (Θ)	35	7.6	0.50
	tetrahydrofuran	25	1.4	0.70
	toluene	25	1.7	0.69
Poly(vinyl acetate)	acetone	25	2.1	0.68

usually available. Extensive tables of K and a are available (for example, Kurata 1966) but caution is required in using these sources since they are not selective, though reasonably comprehensive. Some values of K and a for polymers synthesized or used in the experiments in this book are given in Table 7-7.

The empirical correlation between $[\eta]$ and molecular weight is subject to several restrictions, which become understandable in light of the fundamental dependence of the intrinsic viscosity on molecular size instead of mass. Only if there is a unique relationship between size and mass can Eq. 7-25 be expected to hold. This requires that the polymer be linear, not branched; that the polymer solutions be measured in the same solvent and at the same temperature for which K and a were determined; and that the polymer be of the same chemical type as that used in determining K and a. It should be noted that, since $[\eta]$ depends upon the solvent and temperature, its specification is incomplete unless these quantities are given as well as the units. Far too often this is not done in the literature.

The most useful theory relating the intrinsic viscosity to molecular size is that of Flory (1949, 1950, 1951). The most important relation is

$$[\eta]M = \Phi_0 V_h = \Phi_0 \xi^3 (\overline{s^2})^{3/2} \qquad (7\text{-}27)$$

where V_h is the hydrodynamic volume, and Φ_0 is a universal constant with the value 2.8×10^{21} if V_h is expressed ml and $[\eta]$ in dl/g. In order to relate the product $[\eta]M$ to a more familiar size parameter such as the radius of gyration $(\overline{s^2})^{1/2}$ we must introduce another parameter, ξ, which depends upon solvent power as described below. Newman (1954) showed that, for a polydisperse system, Eq. 7-27 is correct with $M = \bar{M}_n$, and a number-average size parameter, such as $(\overline{s_n^2})^{1/2}$, results.

Perhaps the greatest value of Flory's viscosity theory lies in its interrelations with his theories of the configuration of real polymer chains and the thermodynamics of polymer solutions. While it is out of place to discuss these here (see the *Textbook*, Chap. 2), it may be recalled that the dimensions of polymer chains are determined both by short-range interactions, which can be predicted by calculation (Flory 1969), and by long-range interactions. The latter contributions depend upon the polymer-solvent interactions and are described by an expansion factor α, the ratio of the chain dimension in a given solvent and at a specified temperature to its value in the absence of such interactions (unperturbed dimensions): $\alpha = (\overline{s^2}/\overline{s_0^2})^{1/2}$, for example. The chain assumes its unperturbed dimensions at a special temperature (for a given solvent) called the Flory temperature Θ. Thermodynamic considerations relate Θ to the temperature of phase separation at infinitely high molecular weight and to the temperature at which $A_2 = 0$, described in Sec. A. The unperturbed dimensions of polymers vary in a known way with molecular weight, such that quantities of the type $\overline{s_0^2}/M$ depend only on chain structure (chemical type) and not on solvent or temperature.

Use can be made of these facts by replacing $\overline{s^2}$ by $\alpha^2\overline{s_0^2}$ in Eq. 7-27 and isolating the term in $\overline{s_0^2}/M$:

$$[\eta] = \Phi_0(\overline{s_0^2}/M)^{3/2}M^{1/2}\xi^3\alpha^3 = KM^{1/2}\xi^3\alpha^3 \qquad (7\text{-}28)$$

where $K = \Phi_0(\overline{s_0^2}/M)^{3/2}$ is a constant for a given polymer, independent of solvent, temperature, and molecular weight.

The dependence of ξ on solvent power is given by Ptitsyn (1959) as

$$\xi^3 = \xi_0^3(1 - 2.63e + 2.86e^2) \qquad (7\text{-}29)$$

where e is a parameter which can be related to other measures of solvent power, for example the exponent a in Eq. 7-25: $e = (2a - 1)/3$. The value of ξ_0, referring to the unperturbed state of the chain at $T = \Theta$, is independent of solvent power and can be incorporated into K. The result is that at the Θ temperature, where $\alpha = 1$, a is predicted to be exactly 1/2. This has been confirmed many times.

Treatment of the data Typical data for dilute-solution viscosity, including the steps in the calculation of $[\eta]$, are given in Table 7-8. The defining equations in Table 7-6 are used, and the calculations and graphing are simple and straightforward. The result is $[\eta] = 1.085$ dl/g (ethylene dichloride, 25°C), to provide the complete specification. From Eq. 7-23, the Huggins slope constant $k' = 0.33$, and from 7-24, the Kraemer constant $k'' = -0.15$. Billmeyer (1949) suggested that when k' is near 1/3, Eqs. 7-23 and 7-24 be solved simultaneously with

TABLE 7-8. Typical data from measurement of dilute-solution viscosity: poly(methyl methacrylate) in ethylene dichloride at 25°C (Billmeyer 1955)

c, g/dl	t, sec	η_r	η_{sp}	η_{red}	ln η_r	η_{inh}
0	235.6					
0.125	269.3	1.142	0.142	1.136	0.133	1.064
0.250	305.7	1.298	0.298	1.192	0.261	1.044
0.500	386.6	1.641	0.641	1.282	0.495	0.990
1.000	582.7	2.475	1.475	1.475	0.908	0.908

$k' = 1/3$ and $k'' = -1/6$ to give simple expressions allowing $[\eta]$ to be determined from η_{sp}, η_r, and c at a single concentration. It is left to the reader to derive the equations and show that they give closely the same value for $[\eta]$ as the graphical extrapolation. The inherent viscosity at 0.5 g/dl, 0.990, is obviously a poorer approximation.

The use of a series of values of $[\eta]$ and $\bar{M}_w$, for example, to obtain the constants K and a in Eq. 7-25 is straightforward and will not be illustrated.

Limitations of the method Several limitations to the empirical correlation of $[\eta]$ with M instead of molecular size were mentioned above under *Theory*. In addition, it should be noted that the viscosity of polymer solutions is dependent upon shear rate in the viscometer. For values of $[\eta]$ above about 2, this can lead to significant error.

GENERAL REFERENCES

Textbook, Chap. 3D; Kurata 1963; Lyons 1967; Moore 1967; ASTM D 2857.

D. Molecular-Weight Distribution

In this section, we discuss only the methods of measuring molecular-weight distribution which are the subject of experiments in this book, namely gel permeation chromatography (GPC), Exp. 16, and turbidimetric titration, Exp. 17. For the practicing polymer chemist, these may well not be the methods of choice, depending on equipment available and the objectives of the experiment. Reference should be made to the *Textbook*, Chap. 2E, and to Cantow (1967), Johnson (1967), and Peebles (1971) for more thorough discussion.

It was pointed out in the Introduction to this chapter that the separation which is an essential part of any study of molecular-weight distribution is often based on change in another property of the sample with molecular weight, rather than directly upon molecular weight itself. Only in ultracentrifugation, among the common methods, is the driving force for separation directly the molecular weight. Most common fractionation methods involve separation due to differences in solubility. But solubility is primarily affected by the chemical nature of the polymer and its relation to that of the solvent, and only reflects differences in molecular weight to a very minor degree. Likewise, the separation in GPC is based on differences in molecular size, and all the restrictions described in Sec. C must be adhered to before the results of this experiment can be interpreted in terms of molecular weight and its distribution.

Gel permeation chromatography

The term gel permeation chromatography was coined by John C. Moore (1964) to describe a separation technique similar to, but improved over, the method long known in the biological field as gel filtration. There has been much debate on whether either term is appropriate for describing the process, in which molecules are separated according to their hydrodynamic volume as this property influences their ability to permeate the internal pore structure of microporous "gel" particles, but both names are well established in the literature and are unlikely to be replaced.

Chromatographic processes may be defined as those in which the solute is transferred between two phases, one of which is stationary and the other moving, often traversing a long tube called a column. In GPC both phases are liquid, but in contrast to most liquid-liquid chromatography, where the two liquid phases are immiscible, the two phases in GPC are the same liquid (solvent) and are differentiated only in that the stationary phase is that part of the solvent which is inside the porous gel particles, while the mobile phase is outside.

In GPC, the transfer of solute between the two phases takes place with the driving force of diffusion, resulting from a difference in concentration of solute between the two liquid phases, and restricted by the molecule's ability (depending on its size) to penetrate the pore structure of the gel. We call this process permeation, though the term is not completely descriptive or fully accepted in this sense. Other separation processes applicable to chromatographic systems are those based on solubility differences (partition, as in gas-liquid chromatography), adsorption of a layer of molecules onto a surface (as in gas-solid chromatography), and ion exchange.

Separation media: gels The term gel is misleading in application to GPC because of its common connotation of a soft, compressible material. The crosslinked dextrans used in gel filtration do have these characteristics, but GPC gels are typically hard and incompressible, allowing the use of high pressures in the liquid transport system, up to 1000 psi or more, with correspondingly high flow rates. Moore's original gels, still widely used, consist of polystyrene crosslinked with divinyl benzene. They are polymerized in a suspension system, with large amounts of an inert solvent-nonsolvent mixture present in the monomer, selected so that the polymer when formed is on the verge of precipitation from the liquid phase. The internal pore structure results. Among several other gel materials available, porous glass appears to be second in popularity to the crosslinked polystyrenes.

The gel particles (Figs. 7-16 and 7-17) are about 100 μm in diameter. Examination of thin sections with transmission electron microscopy (Fig. 7-18) shows that the sizes of the internal pores are satis-

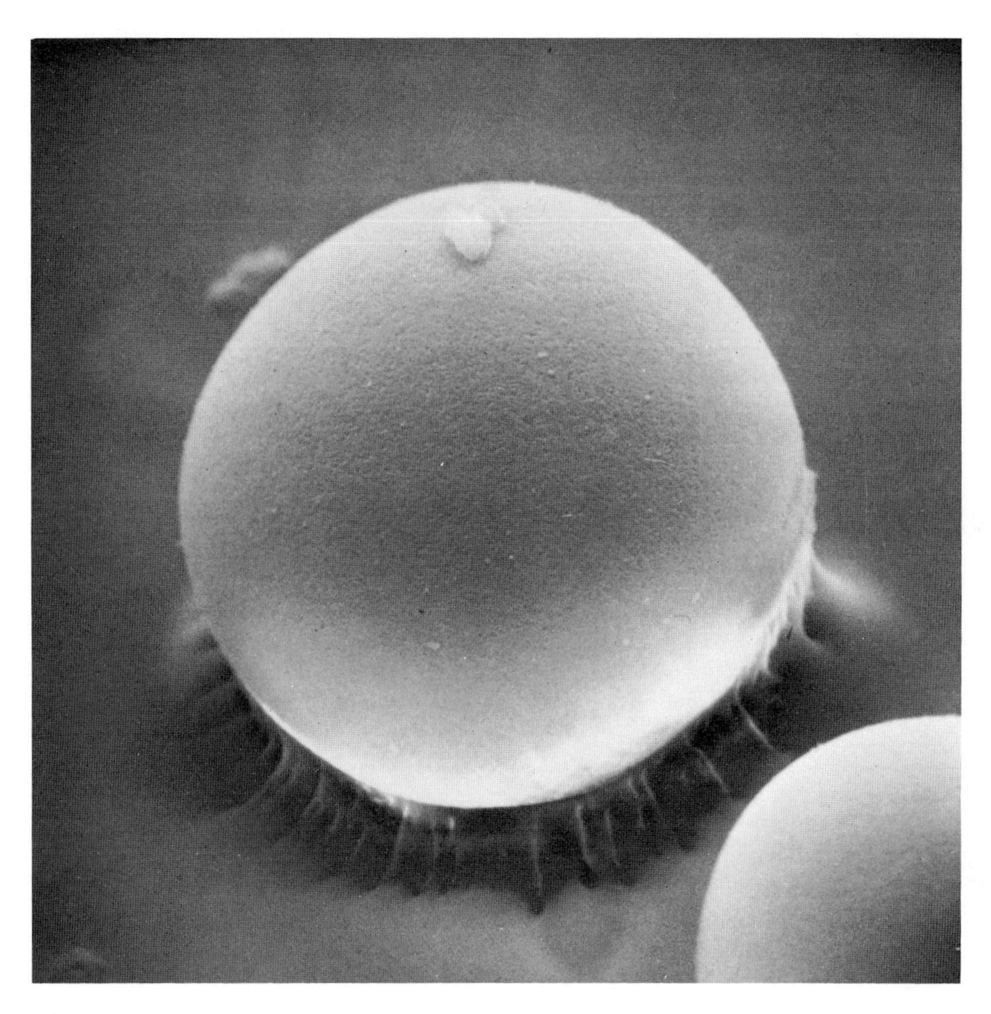

Fig. 7-16. Scanning electron micrograph of a poly(styrene-*co*-divinyl benzene) GPC gel particle (*Textbook*, Fig. 4-9, courtesy R. W. Godwin and Celanese Fibers Co.).

fyingly compatible with those calculated for polymer molecules which they are known to separate (Billmeyer 1970).

Gel permeation chromatographs The description of GPC equipment is based on the original commercial instrument (Waters), but several other liquid chromatographs capable of GPC are now available. The flow system (Fig. 7-19) consists of solvent reservoir, pump, and associated devices. The stream is split, one half going through a dummy column to provide pressure drop and into the reference side of the detector, the other through the column or columns for the separation and to the sample side of the detector. The sample is injected into the sample stream, either through a septum as is common in gas chromatography,

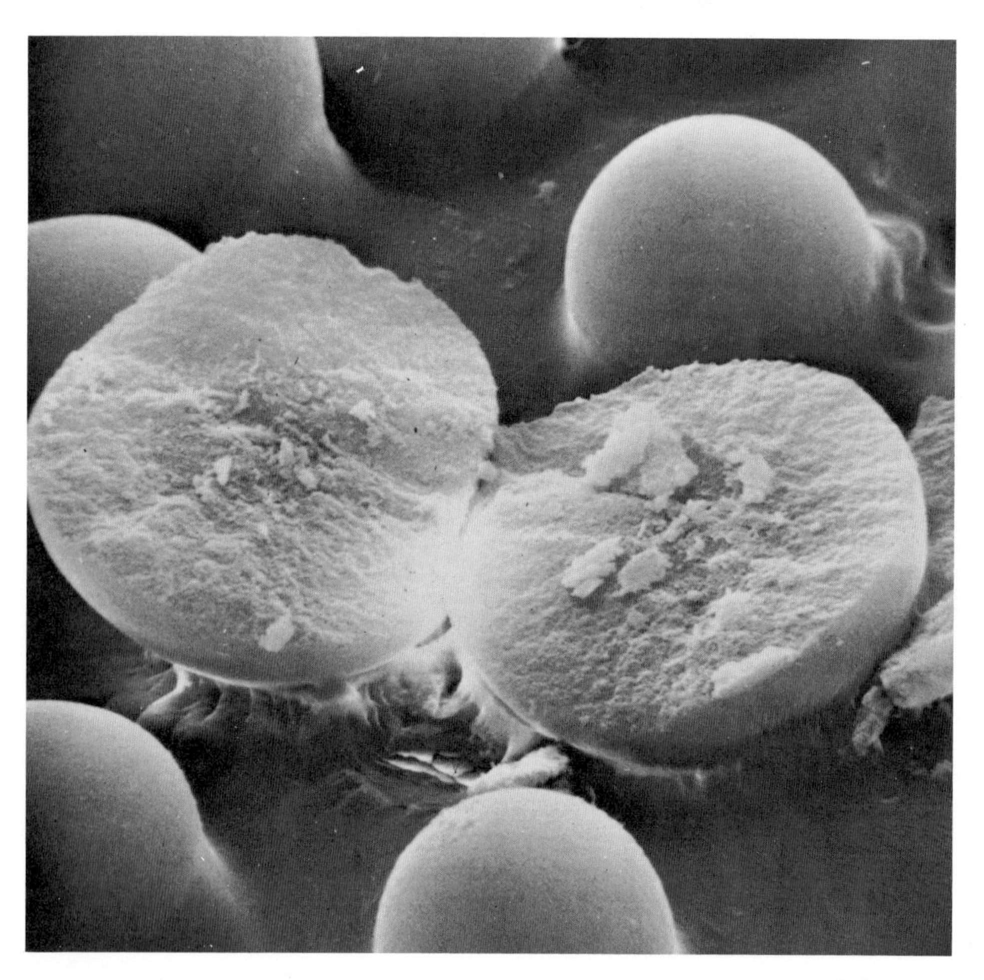

Fig. 7-17. Broken gel particle of same type as in Fig. 7-16 (from Billmeyer 1972, courtesy R. W. Godwin and Celanese Fibers Co.).

but more often with an injection loop and valve (Fig. 7-20). A differential refractometer is customarily used as the detector, but ultraviolet, infrared, or even flame-ionization detectors have also been used. A number of options are available, such as automatic injection of a series of samples for unattended operation, sample collection with a fraction collector, and recycle capability, to improve resolution by cycling that portion of the mobile phase containing the sample through the columns more than once. High-temperature operation can be carried out for polymers soluble only at elevated temperatures (this is true of all the characterization techniques described in this chapter).

GPC columns typically have a diameter of 3/8 inches and are 3-4 feet long. Three or four are commonly used in series, packed with gels having different pore sizes to insure good separation capability over

Fig. 7-18. Transmission electron micrograph of a microtomed section of a GPC gel particle like that of the preceding figures (courtesy John C. Moore).

a wide range of molecular weights. A flow rate of 1-2 ml/min is common, but the volume of the columns is such that the time for elution of all components is 2 hours or more.

Peak broadening is more serious in liquid than in gas chromatography, and has been studied extensively (Kelley 1970). It originates in part from the flow properties of the viscous mobile phase through the packed column, and in part from nonuniformities in the internal pore structure of the gel. It is customary to test GPC columns for satisfactory resolution and a low level of peak broadening by the well-known plate-count method. A monodisperse sample is injected, usually a low-molecular-weight liquid, and the plate count is calculated as indicated in Fig. 7-21. Several hundred plates per foot is required to give adequate resolution for polymer systems.

Theory Polymer molecules are separated by size in the GPC experiment because of their ability to penetrate part of the internal volume of the gel particles, that is, the stationary phase. As the sample moves

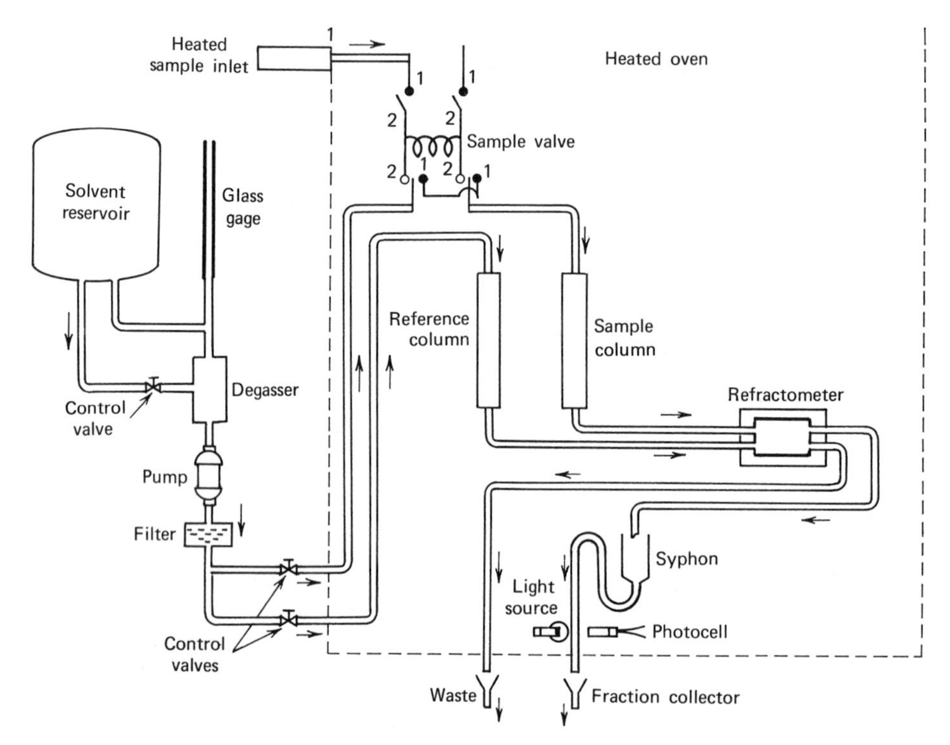

Fig. 7-19. Flow system of a gel permeation chromatograph (Maley 1965).

along the column with the mobile phase, the largest molecules are almost entirely excluded from the stationary phase, while the smallest (with a properly chosen gel) find almost all the stationary phase accessible. The smaller the molecule, the more of the stationary-phase volume is accessible to it, and the longer it stays in that phase. Small molecules thus fall behind larger ones, and are eluted from the

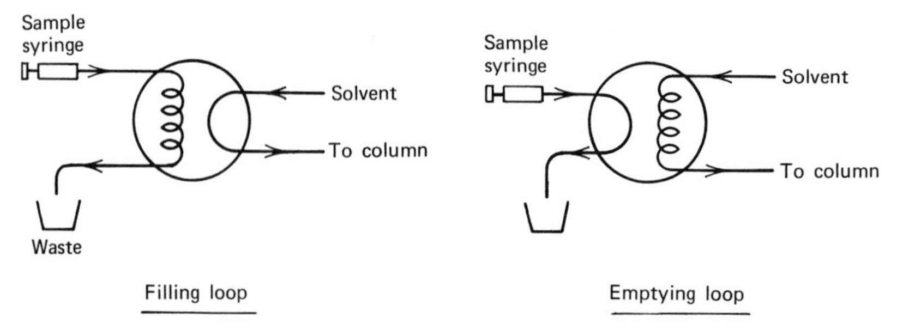

Fig. 7-20. Injection loop system commonly used for introducing the sample into a gel permeation chromatograph (Cazes 1966).

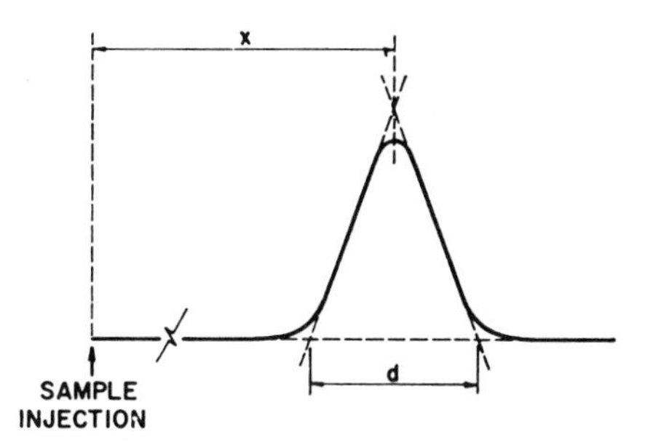

Fig. 7-21. Calculation of plate count in GPC. A monodisperse sample is injected, and the retention volume x is measured. The width d of the peak at the base line is measured, and the number of theoretical plates N is calculated as $N = 16(x/d)^2$ (Cazes 1966).

column later. For large molecules completely excluded from the gel, the retention volume V_r is equal to the interstitial or mobile-phase volume V_0, whereas for very small molecules, for which the entire stationary phase is accessible, $V_r = V_0 + V_i$ where V_i is the internal pore volume of the gel, that is, the stationary-phase volume. Intermediate species have retention times $V_r = V_0 + K_d V_i$, where K_d is the ratio of pore volume accessible to that species, to the total pore volume. Thus, K_d is a separation constant varying from 0 to 1, and in GPC all species are eluted with retention volumes between V_0 and $V_0 + V_i$; this is in contrast to partition chromatography, for example, in which the corresponding partition coefficient can have values much greater than unity.

Little use is made of this theory in the practice of GPC, however, since V_r cannot yet be predicted from molecular parameters. Instead, calibration curves are prepared by running standards of known molecular weight and narrow molecular-weight distribution, and preparing a plot (Fig. 7-22) of log molecular weight versus retention volume. By experience, it is found that this plot is often nearly linear over a range of molecular weights which, depending upon the selection of the gels used, can be quite wide.

At this point account must be taken of the true nature of the separation, which is based on hydrodynamic volume and not molecular weight. All the restrictions given in Sec. C apply when the attempt is made to substitute the latter variable for the former. Thus, a calibration made with samples of narrow-distribution linear polystyrene, in tetrahydrofuran at 25°C, will not apply to a polymer of another composition, or a branched polystyrene, or to measurements in another solvent or at another temperature. Since narrow-distribution polymers other than polystyrene are not generally available, this becomes a serious problem but one often glossed over in the literature. An approximate correction (Q-factor), which takes account of differences in molecular weight per unit chain length but ignores the effects of solvent power, is often applied and is described below. Exact treatment requires that the calibration be made directly in terms of

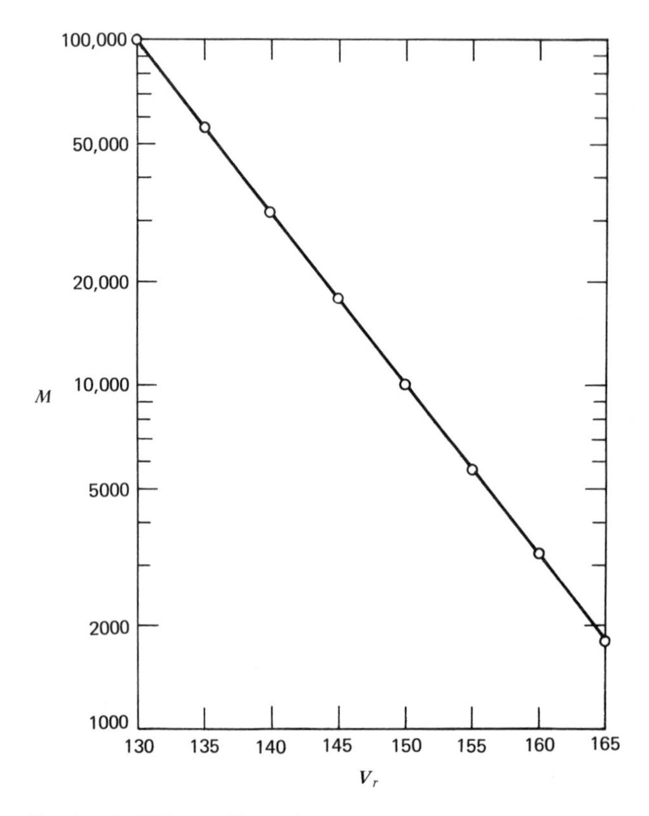

Fig. 7-22. Typical GPC calibration curve, based on data of Table 7-9.

hydrodynamic volume, using for example Eq. 7-27 (Benoit 1966, Grubisic 1967).

To do this, the calibration curve is plotted as log $[\eta]M$ versus V_r, as in Fig. 7-23. If the solvent used is one in which polystyrene is soluble, the narrow-distribution samples serve very well to provide this curve, and as the figure shows, results for a wide variety of linear and branched polymers fall on it. For linear polymers, interpretation in terms of molecular weight is straightforward: If the Mark-Houwink-Sakurada constants K and a are known, for example, log $[\eta]M$ can be written log $\overline{M}^{1+a}$ + log K, and V_r can be directly related to M. It should be noted that a new and different average molecular weight, which we call the "GPC-average," is defined by this process:

$$\overline{M}_{GPC} = \sum_{i=1}^{\infty} w_i M_i^{1+a} \Big/ \sum_{i=1}^{\infty} w_i M_i \tag{7-30}$$

The GPC-average molecular weight is equal to $\overline{M}_z$, defined in Eq. 7-31,

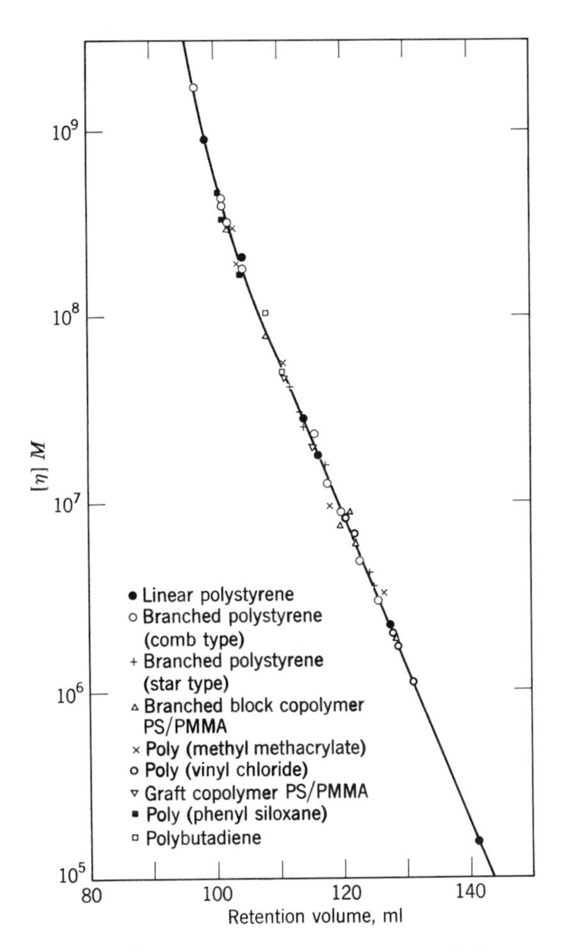

Fig. 7-23. "Universal" calibration curve for GPC, based on hydrodynamic volume (*Textbook*, Fig. 2-19, after Grubisic 1967).

if a = 1, and is somewhat greater than $\overline{M}_w$ if a = 0.5 It is shown in relation to the more-common molecular-weight averages in Fig. 7-1.

For branched polymers, one must reconcile himself to thinking solely in terms of molecular size, since the material eluting at any value of V_r consists of a mixture of species having different molecular weights and degrees of branching but constant hydrodynamic volume. When narrow-distribution samples of known molecular weight and intrinsic viscosity are not available to provide a universal calibration curve for the solvent used, more complex analyses are possible, but these are beyond the scope of this book.

Treatment of the data As indicated in Table 7-9, the basic data from GPC are recorder chart readings, proportional to amount of material, and molecular weights. These are connected through corresponding

TABLE 7-9. Typical data from gel permeation chromatography: poly-(methyl methacrylate) in tetrahydrofuran at 25°C. Calibration with narrow-distribution polystyrene samples, same conditions

V_r, ml	Recorder Divisions (N_iM_i)	Molecular Weight (M_i)	$N_i \times 10^3$	$N_iM_i^2 \times 10^{-5}$
		Sample A		
130	0.5	98,000	0.005	0.49
135	6.0	55,000	0.109	3.30
140	25.7	31,500	0.816	8.10
145	44.5	18,000	2.47	8.01
150	40.0	10,000	4.00	4.00
155	20.1	5,700	3.53	1.15
160	5.8	3,250	1.785	0.19
165	1.2	1,800	0.667	0.02
Sum	143.8		13.38	25.26
		Sample B		
	(same as Sample A below V_r = 150 ml)			
150	42.0	10,000	4.20	4.20
155	25.6	5,700	4.49	1.46
160	8.9	3,250	2.74	0.29
165	2.2	1,800	1.22	0.04
Sum	155.2		16.06	25.89

values of V_r from sample chromatogram and calibration curve, respectively; V_r serves no other purpose. In the Table, recorder readings from the differential refractometer are given in chart divisions, corrected for base line. The corresponding chromatogram is sketched in Fig. 7-24. These readings represent total concentration of polymer at that retention volume and are proportional to the quantities N_iM_i appearing in Eqs. 7-1 for $\overline{M}_n$, 7-14 for $\overline{M}_w$, and elsewhere. By the use of corresponding values of M_i from the calibration curve, it is easy to calculate N_i and $N_iM_i^2$, as tabulated, and to obtain the sums necessary to calculate $\overline{M}_w$ and $\overline{M}_n$ from the equations indicated. One can also find the necessary data to calculate $\overline{M}_v$ and $\overline{M}_{GPC}$ from Eqs. 7-26 and 7-30, if a is known, and the z-average molecular weight

$$\overline{M}_z = \sum_{i=1}^{\infty} N_iM_i^3 \Big/ \sum_{i=1}^{\infty} N_iM_i^2 = \sum_{i=1}^{\infty} w_iM_i^2 \Big/ \sum_{i=1}^{\infty} w_iM_i \qquad (7\text{-}31)$$

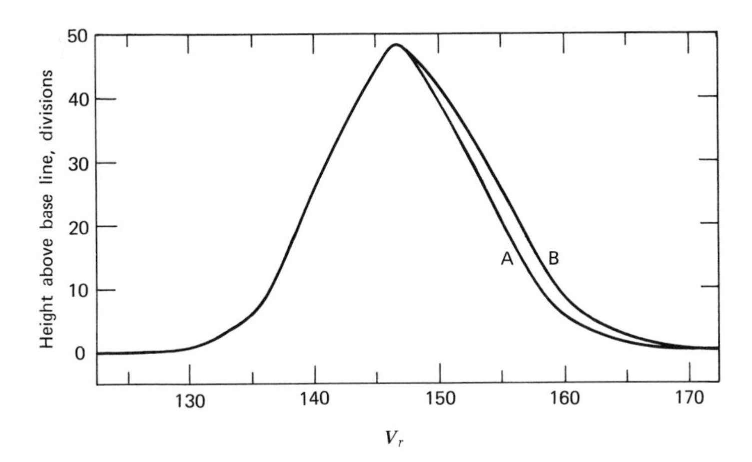

Fig. 7-24. GPC chromatogram of a poly(methyl methacrylate) in tetrahydrofuran at 25°C, data of Table 7-9.

For Sample A, $\overline{M}_w$ = 17,570, $\overline{M}_n$ = 10,750, and $\overline{M}_w/\overline{M}_n$ = 1.63.

The data for Sample B were added to show the effect of a small increment in the amount of low-molecular-weight material present. The calculated values of $\overline{M}_w$ = 16,680, $\overline{M}_n$ = 9,660, and $\overline{M}_w/\overline{M}_n$ = 1.75 show that $\overline{M}_n$ changes much more than $\overline{M}_w$ and the distribution is effectively broadened, even though in each case the last material in the sample eluted at V_r = 165 ml.

These data are totally uncorrected for the fact that a polystyrene calibration was used with a poly(methyl methacrylate) sample. For relative work this is often satisfactory, but it must be remembered that the quoted molecular weights are not correct, and even the ratio $\overline{M}_w/\overline{M}_n$ is in error (Nichols 1971). The first approximation to correcting the data is to consider the fact that, neglecting differences in polymer-solvent interactions, equal numbers of repeat units in the chain should make the same contribution to its hydrodynamic volume. This leads to the concept of the "*Q* factor," the weight in daltons per angstrom of chain length. Considering one repeat unit, of length 2.54 A in these polymers, *Q* (polystyrene) = 104/2.54 = 40.9, and *Q* (poly(methyl methacrylate)) = 100/2.54 = 39.4. The polystyrene calibration underestimates the molecular weight (at a given value of V_r) of the lighter (lower *Q*) poly(methyl mechacrylate), so we postulate that, for example, it should have at V_r = 150 ml a molecular weight of 10,000 × (40.9/39.4) = 10,400. In fact, polymer-solvent interactions are such that this presumed correction is actually in the wrong direction: a calibration based on samples of poly(methyl methacrylate) showed that at V_r = 150 ml, *M* = 8,600. A *Q*-factor of 47.6 would be needed to account for this. We do not recommend use of the *Q*-factor approach. Use of a universal calibration would be appropriate.

Accepting relative results, the data in Table 7-9 can be used to provide graphs of the molecular-weight distribution. Two types are useful: The first is the *cumulative* distribution curve, obtained by

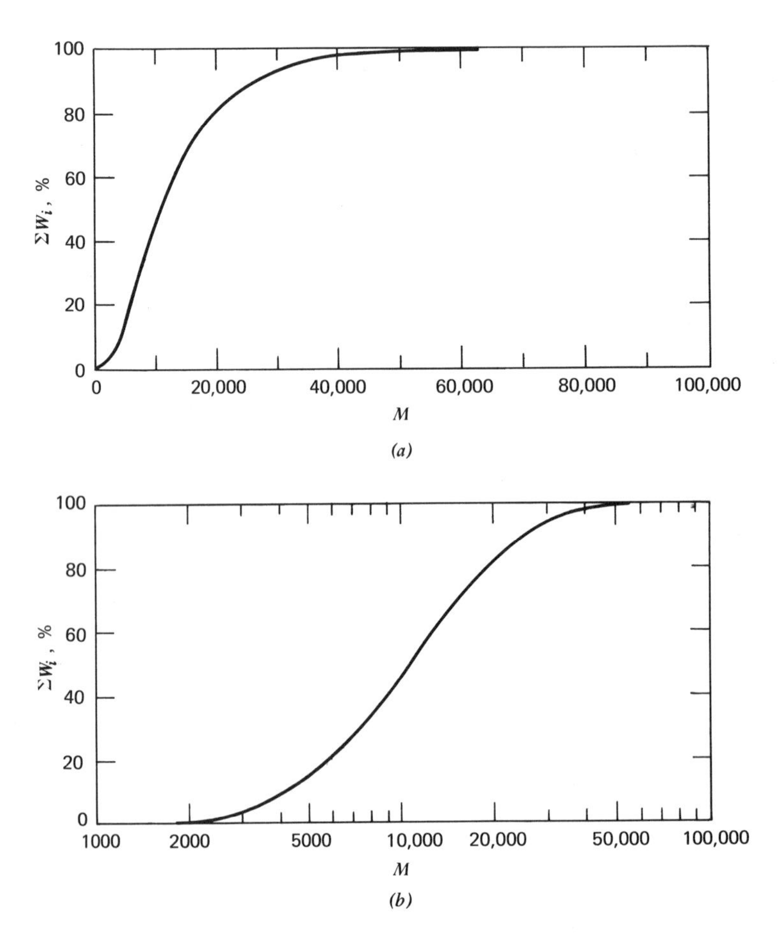

Fig. 7-25. Cumulative molecular-weight distribution curve of poly-(methyl methacrylate), calculated from the data in Tables 7-9 and 7-10: (*a*) linear plot; (*b*) plot against log *M*.

accumulating and then normalizing the recorder-chart readings in the second column of the table; this calculation is carried out in Table 7-10. The resulting data are plotted against *M* (Fig. 7-25*a*), or as is useful for broader distributions, against log *M* (Fig. 7-25*b*). In considering the logarithmic plot, it is useful to note that

$$w(\log M)\ d(\log M) = 2.3\ M\ w(M)\ dM\ /\ 2.3\ M = w(M)\ dM \qquad (7\text{-}32)$$

The second useful curve is the *differential* distribution, obtained by differentiating the smooth curve drawn through the points of Fig. 7-25*a*, as is shown in Fig. 7-26. An approximate relative curve can be obtained by omitting the smoothing and plotting the recorder read-

TABLE 7-10. Calculation of molecular-weight distribution from GPC data of Table 7-9

M_i	Recorder Divisions	Same, Normalized	Sum up to given M_i
1,800	1.2	0.8	0.8
3,250	5.8	4.0	4.8
5,700	20.1	14.0	18.8
10,000	40.0	27.9	46.7
18,000	44.5	30.9	77.6
31,500	25.7	17.9	95.5
55,000	6.0	4.2	99.7
98,000	0.5	0.3	100.0
Sum	143.8	100.0	

ings directly against M. Note that the chromatogram itself (Fig. 7-24) is not unlike the differential distribution curve in general features, differing primarily in that the scale of V_r is not linear but approximately logarithmic in M (Fig. 7-27).

Limitations of the method As has been emphasized above, many of the limitations of the GPC method stem from the fact that the separation is based upon molecular size, yet we usually wish to interpret the results in terms of molecular weight. For linear homopolymers, this leads to difficulties in absolute calibration techniques, which are partially overcome by making the calibration in terms of hydrodynamic

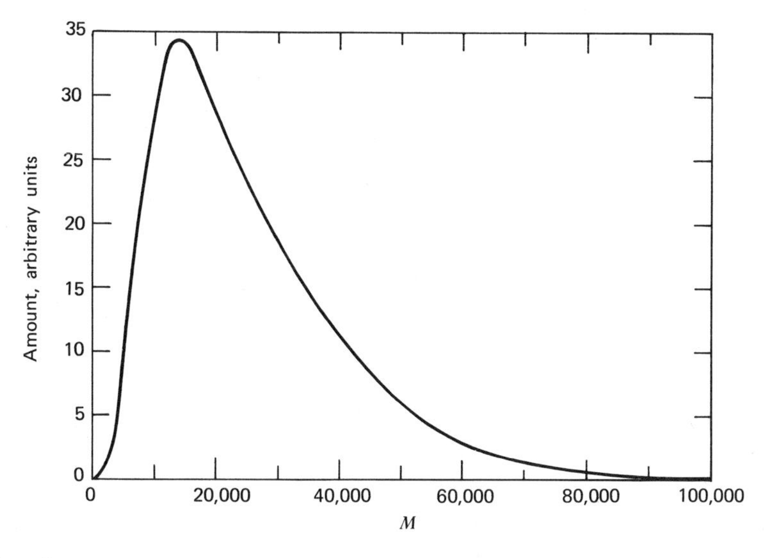

Fig. 7-26. Differential molecular-weight distribution corresponding to Fig. 7-25*a*.

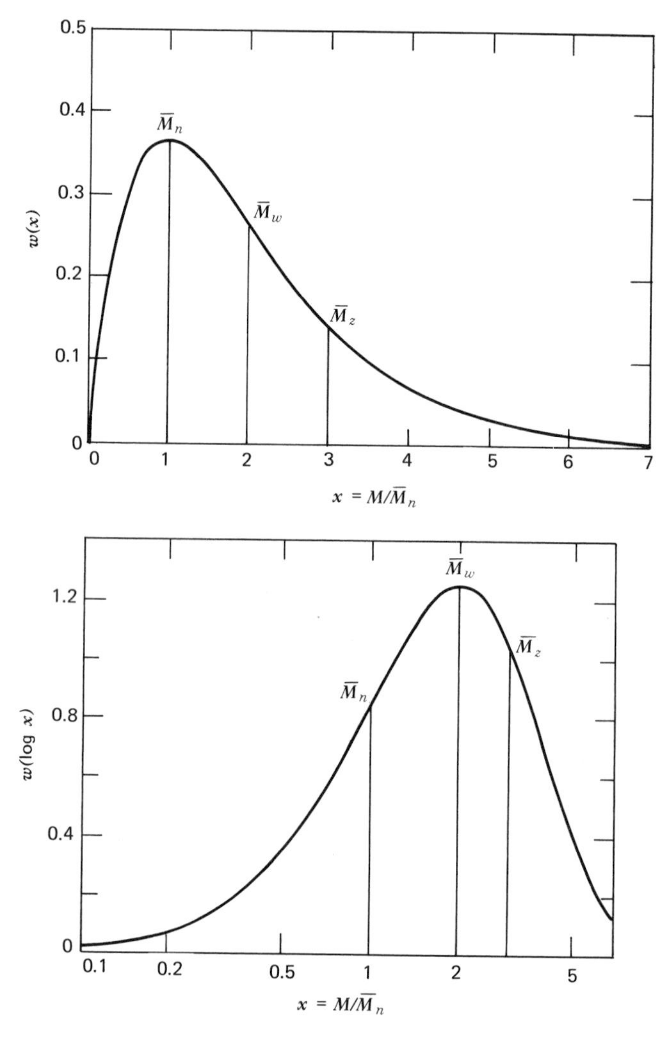

Fig. 7-27. "Most probable" molecular-weight distribution curve and the corresponding GPC chromatogram, assuming a calibration curve linear in log M. Note the relative positions of the various molecular-weight averages (courtesy A. R. Shultz).

volume. The problem of branched polymers can be simplified by use of the universal calibration technique, but no complete interpretation in terms of molecular-weight distribution is possible. Copolymers present a different problem in that (a) V_r is determined at least in part by chemical composition, which influences molecular size because of polymer-solvent interactions, and (b) refractive index, and thus detector response per unit concentration, also depend upon chemical composition. Similarly, if the refractive index of a homopolymer depends upon molecular weight, as it inevitably must at sufficiently low molecular weight, detector response is no longer constant over the entire chromatogram and a separate calibration must be made. Other detectors can be used, however, including those based on infrared or ultraviolet absorption or flame ionization. Useful information about the chemical composition of the eluate can be obtained by combining results obtained with two different detectors, especially if one is sensitive to an absorption band specific to one of the species in a copolymer. Finally, with existing columns, peak broadening can seriously affect the distribution curve for samples with M_w/M_n less than about 2. Empirical corrections for this effect have been discussed by Duerksen (1968).

Turbidimetric titration

Turbidimetric titration is an analytical fractionation technique, in which polymer is precipitated continuously by decreasing its solubility in the solvent. Originally (Morey 1945) this was done by titrating a nonsolvent into the polymer solution, but the technique of cooling a solution to cause precipitation is also used (Taylor 1962). The amount of polymer precipitated is measured by the turbidity of the solution.

The key to the success of turbidimetric titration lies in locating a suitable solvent-precipitant system, in which the precipitated phase remains stably suspended, rather than settling rapidly as is usually desirable in other fractionation methods. Since turbidity is uniquely related to concentration only at constant size and mass of the scattering particles, these parameters should be maintained constant throughout the titration, new particles forming as more polymer is precipitated, rather than existing particles growing or agglomerating. In addition, the particle size should be relatively independent of the molecular weight and the solvent composition. Finally, since the turbidity depends on the difference in refractive index between the two phases, the solvent and nonsolvent must have similar refractive indices to minimize the change in the refractive index of their mixture as the titration proceeds.

These are formidable requirements, yet they have been met in many cases. A recent review article (Giesekus 1967) lists over 30 major types of polymers which have been successfully studied by turbidimetric titration, and in addition many copolymer systems (random, block, and graft). Giesekus also states, however, that finding a successful solvent-precipitant system for a new polymer requires on the average about 100 preliminary tests, at least two weeks' work. Thus, this is scarcely the method of choice for obtaining information rapidly about a previously unstudied polymer.

Instrumentation Usually, turbidity itself is not measured in this experiment, but scattering at a convenient angle. Equipment ranges from conventional or modified light-scattering photometers to analytical spectrophotometers. Most often, scattered-light intensity is measured at 90° angle. It is convenient but not essential to measure the transmitted light intensity also in this case. A typical apparatus (Hall 1959) is shown in block diagram in Fig. 7-28. A further improvement allows measurement also at 45° and 135° to determine the dissym-

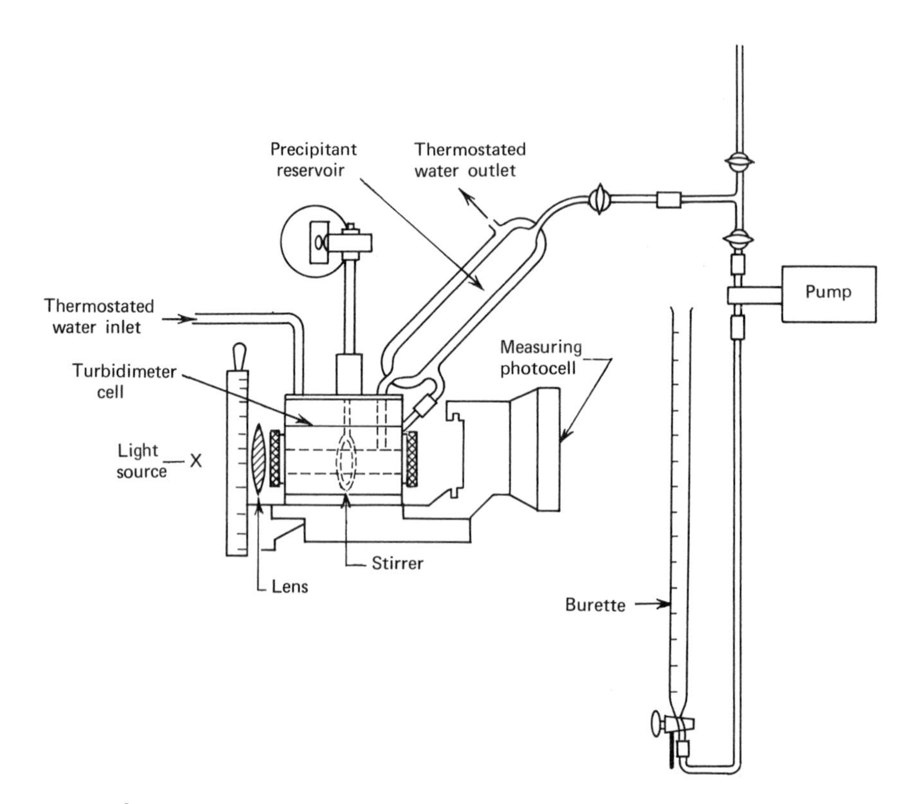

Fig. 7-28. Essential features of an apparatus for turbidimetric titration (Hall 1959).

metry of the scattered light. The importance of this measurement is indicated below.

A common feature of turbidimetric titration is a pumping device to inject the precipitant uniformly and continuously into the scattering cell. Syringe-type burets, with the plunger motor-driven at constant speed, are normally used. The cell itself may be open (variable volume) or closed (constant volume). Efficient agitation to insure good mixing is mandatory. Automatic recording of the turbidity as a function of some variable related to solvent-nonsolvent composition is usually carried out.

Theory Interpretation of the results of turbidimetric titration can be made at several levels of sophistication. The simplest procedure is to compare curves of observed turbidity (perhaps expressed as a fraction of the maximum turbidity attained) as a function of percent precipitant, and look for differences. More meaningful evaluations can be made if certain corrections, described below, are applied to the experimental data. The result, in principle, is a curve of weight fraction precipitated as a function of precipitant concentration. Ideally, one would then wish to convert this to an integral molecular-weight distribution curve by calibration of the scale of precipitant concentration. This can be done only approximately, as will be described.

The several correction factors are summarized by Giesekus (1967) and are only enumerated here:

1. Because of the continuous change in the volume of liquid under investigation, precipitant concentration is not linearly related to the volume of nonsolvent pumped in. The modifying functions are simple if it can be assumed that there is no volume change on mixing, but differ for the cases of closed and open cells. Similar considerations apply to the total polymer concentration (in both phases) at any time.

2. Determination of the concentration of precipitated polymer requires a number of approximations. By proper cell design, it is possible to relate the ratio of 90°-scattered to transmitted light to the desired turbidity, assuming that the size of the precipitated particles remains constant. This can be checked by periodic measurement of the dissymmetry, and Giesekus shows that there is a range of dissymmetries over which the effect of changing particle size is negligible. Change in relative refractive index of the precipitated particles and the surrounding medium must be taken into account. Even for homopolymers, where the polymer refractive index is invariant, this can be done exactly only if the polymer-solvent ratio and the solvent composition in the precipitated particles are known at all stages in the titration. This is not usually the case, and Giesekus recommends reliance on empirical corrections (Allen 1960). For copolymers, where there is a difference in refractive index between the monomers, the situation is further complicated to the point where no quantitative interpretation is feasible.

The difficulty in converting turbidimetric titration curves to molecular-weight distributions is that the molecular weight being precipitated at any solvent-nonsolvent composition depends on the molecular weights and amounts of all the other species present, as discussed in the *Textbook*, Chap. 2D. Thus, simple calibration of the molecular-weight scale, by noting the point of, say, 50% precipitation of sharp fractions of known molecular weight, leads to gross systematic errors. This is clearly demonstrated in Fig. 7-29, where the precipitation of one fraction from a polystyrene sample is seen to be around 60% complete at a solvent composition where precipitation of the parent polymer is just beginning. No theory has been developed to correct for this solubilization of any species by the presence of others, and it seems unlikely that a useful exact theory can be developed in the near future. It is advisable, therefore, to consider this strictly an empirical method, and it is so treated in Exp. 17.

Treatment of data Typical data for the turbidimetric titration of two samples of polystyrene in 0.005% solution in toluene are given in Table 7-11. In these experiments, methanol was added at a constant rate by means of an infusion pump to cause the precipitation of the polymer. The nonsolvent concentration is expressed as volume percent. Scattered-light intensity was measured in a conventional photometer with a stirred cell, and the table lists the increment in scattering, corrected for change in solution volume, over that with no precipitate present. The polymers were NBS 705 and NBS 706 (App. I), the former having a narrow ($\overline{M}_w/\overline{M}_n \simeq 1.05$) and the latter a relatively broad ($\overline{M}_w/\overline{M}_n \simeq 2.9$) distribution of molecular weights. Figure 7-30 shows

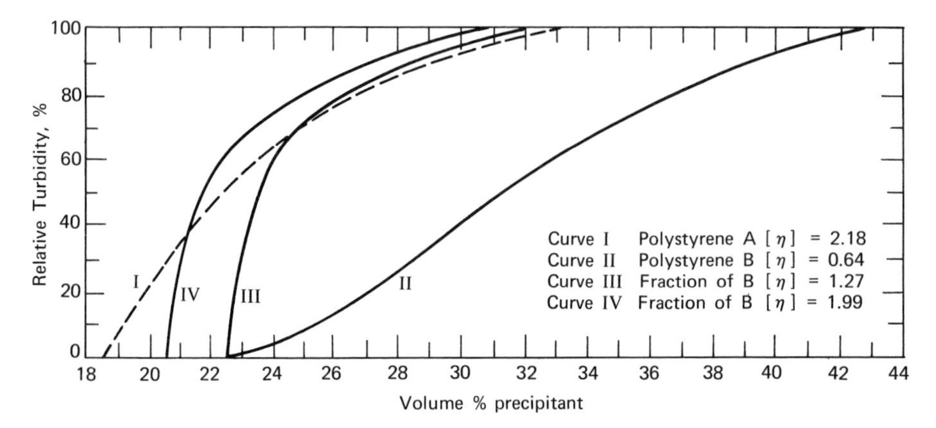

Fig. 7-29. Results of turbidimetric titration of two polystyrene samples and two fractions of one of them (Hall 1959). Note that the higher-molecular-weight fraction of polystyrene B, when isolated, is about 60% precipitated (curve IV) at a solvent composition at which the whole polymer containing this fraction is still just completely soluble (curve II).

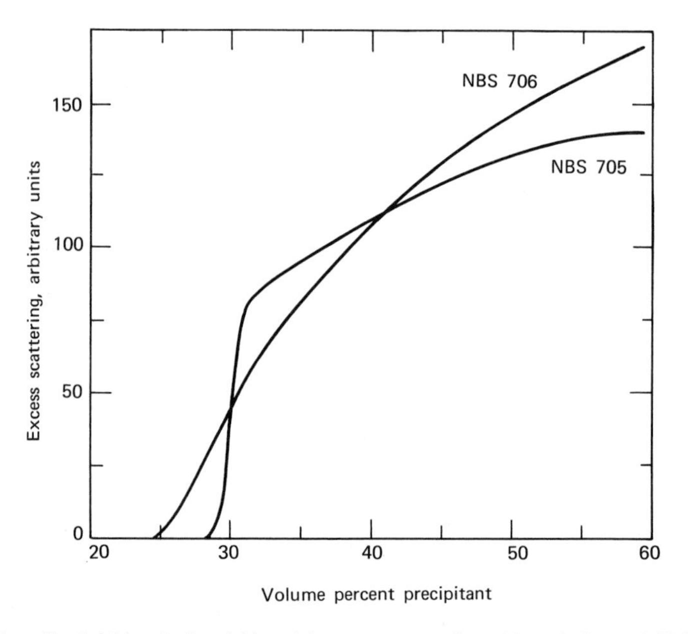

Fig. 7-30. Turbidimetric titration curves for the data of Table 7-11.

TABLE 7-11. Data for the turbidimetric titration of polystyrene in toluene with methanol as the precipitant

Volume percent precipitant	Increment in scattered intensity, arbitrary units, for	
	NBS 705	NBS 706
24.7	0	0
28.3	0	27.5
31.5	80.2	61.0
34.5	93.8	78.8
37.2	101	94.3
39.7	110	107
42.0	116	118
44.2	120	126
46.2	124	134
48.0	128	140
49.8	131	146
51.3	133	150
52.8	136	155
54.8	137	158
55.6	138	163
57.0	139	164
58.0	140	166
59.1	139	170

the curves resulting from these data.

Limitations of the method As stated above, the turbidimetric titration experiment is best considered empirical, since no theory exists allowing the data to be converted to exact molecular-weight distributions. The experiment is most useful if applied to a polymer for which a suitable solvent-nonsolvent system has previously been found. A number of recent papers, listed below, testify to the usefulness of the technique under these conditions.

GENERAL REFERENCES

Gel permeation chromatography: Cazes 1966, 1970; Altgelt 1967, 1968; Johnson 1970; Bly 1971.
Turbidimetric titration: Giesekus 1967; Cornet 1968; Rayner 1969; Peaker 1970; Springer 1970.

E. Summary

Table 7-12 summarizes some of the characteristics of the methods described in this chapter which are pertinent to the selection of a method for the molecular characterization of a polymer. There are many factors pertinent to this choice, and not all of them can be summarized in tabular form; some are discussed below.

The data in the table are of necessity only approximate. It should, for example, be clear that the range of molecular weights covered by any method will depend upon a number of factors peculiar to the samples under study; thus ranges can only be stated approximately. By low equipment cost is meant under $1,000; by moderate, up to $6,000; and by high, $10,000 and up. A long measurement time means several hours, including special sample preparation; moderate, about 1 hour; and short, a few minutes. A large sample size means of the order of a gram; moderate, 100 mg; small, a few milligrams.

It will be recalled that end-group analysis is limited by the fact that not all polymers contain end groups that can be detected and measured quantitatively by physical or chemical methods. In this case, the upper limit of applicability depends upon the method used, and the stated limit of 25,000 is only an order of magnitude.

The major limitation of membrane osmometry was seen to be the diffusion of low-molecular-weight species through the membrane, limiting the applicability of the method to samples containing small amounts of such species. This criterion cannot be related specifically to $\bar{M}_n$, and the lower limit on this quantity also depends upon the type of membrane and the conditioning procedure, but it is likely to be near the value quoted.

As experience with vapor-phase osmometry grows, most users come to the realization that many of the difficulties experienced with it stem from deficiencies in the samples and not the method. Unlike other methods for measuring $\bar{M}_n$, vapor-phase osmometry counts all molecules and thus is unusually (but correctly) sensitive to small amounts of low-molecular-weight impurities such as residual monomer, solvents, traces of moisture, etc. As is always the case, the measurement is no better than the sample being tested.

As far as experimental techniques are concerned, light scattering probably requires more care than any other, particularly in solution clarification and the demonstration that this step has been carried out completely. The method is applicable over a wide range of molecular weight, however, and the suggested limits can be far exceeded on both ends of the range in favorable circumstances.

TABLE 7-12. *Summary of widely used methods for the determination of molecular weight and its distribution*

Method	Value Determined	Range of M Covered	Equipment Cost	Measurement Time	Sample Size
End-group analysis	$\overline{M}_n$	Up to 25,000	Low	Moderate	Large
Membrane osmometry	$\overline{M}_n$	20,000 - 500,000	Moderate	Moderate	Moderate
Vapor-phase osmometry	$\overline{M}_n$	Up to 25,000	Moderate	Short	Small
Light scattering	$\overline{M}_w$	10,000 - 10,000,000	Moderate to high	Long	Large
Solution viscosity	$\overline{M}_v$	20,000 - 1,000,000	Low	Short	Moderate
GPC	Distribution	Up to 1,000,000	High	Long	Small

Both solution-viscosity measurement and GPC are limited with respect to molecular weight because they are based on considerations of molecular size instead, as has been discussed in detail in Secs. C and D. This leads to serious problems in the interpretation of the data in familiar terms. Despite this, both techniques are widely used, the one for its experimental simplicity and the other for the wealth of information it gives, despite difficulties in interpretation, on a topic that is otherwise very difficult to study.

All of the techniques discussed in this chapter require that the polymer be soluble, and this gives rise to some common difficulties. A suitable solvent must be found, calling into play the considerations discussed in Chap. 6A and the *Textbook*, Chap. 2A. This is especially difficult for crystalline polymers, and operation at elevated temperatures is sometimes required. Additional difficulties result when it is necessary to control polymer-solvent interactions, for example to eliminate association or to facilitate extrapolation to $c = 0$.

Certain polymer characteristics lead to special problems with some methods. The presence of gel or microgel poses serious problems in light scattering, but is of virtually no consequence in methods for determining $\overline{M}_n$. Branched polymers can be handled by any of the absolute methods for determining molecular weight, but require special consideration in methods based on molecular size. Copolymers require special treatment in light scattering and GPC, because of the variation of refractive index and molecular size with chemical composition. Polyelectrolytes, likewise, require special treatment but can be measured by most of the techniques described (*Textbook*, Chap. 3F).

While the recitation of these restrictions and limitations may give the impression that these methods are not valuable, this is far from the case. What is implied is that none of them represents a cut-and-dried analytical technique, for which samples can be submitted without careful consideration of every step of the process (Strazielle 1971). Most chemists not trained in polymer characterization are not familiar with either the requirements for satisfactory results or the interpretation of these results in terms of molecular structure. To impart at least a little of such training is one of the objectives of this book.

BIBLIOGRAPHY

Allen 1960. P. E. M. Allen, R. Hardy, J. R. Maser and P. Molyneux, "Applications and Limitations of Turbidimetric Titration as a Method of Characterizing Solutions of Polymers and Mixtures of Polymers," *Makromol. Chem. 39*, 52-66 (1960).

Altgelt 1967. K. H. Altgelt and J. C. Moore, "Gel Permeation Chromatography," Chap. B.4 in Manfred J. R. Cantow, ed., *Polymer Fractionation,* Academic Press, New York, 1967.

Altgelt 1968. K. H. Altgelt, "Theory and Mechanics of Gel Permeation Chromatography," pp. 3-46 in J. Calvin Giddings and Roy A. Keller, eds., *Advances in Chromatography*, Vol. 7, Marcel Dekker, New York, 1968.

ASTM D 2857. Standard Method of Test for Dilute-Solution Viscosity of Polymers. ASTM Designation: D 2857. American Society for Testing and Materials, Philadelphia, Pennsylvania.

Armstrong 1968. Jerold L. Armstrong, "Critical Evaluation of Commercially Available Hi-Speed Membrane Osmometers," pp. 51-55 in Donald McIntyre, ed., *Characterization of Macromolecular Structure,* National Academy of Sciences Publication No. 1573, Washington, D.C., 1968.

Bausch and Lomb. Bausch and Lomb Company, Rochester, New York 14602.

Benoit 1966. Henri Benoit, Zlatka Grubisic, Paul Rempp, Danielle Decker, and Jean-Georges Zilliox, "Study by Liquid-Phase Chromatography of Linear and Branched Polystyrenes of Known Structure" (in French), *J. Chim. Phys. 63,* 1507-1514 (1966).

Billmeyer 1949. Fred W. Billmeyer, Jr., "Methods for Estimating Intrinsic Viscosity," *J. Polymer Sci. 4,* 83-86 (1949).

Billmeyer 1955. F. W. Billmeyer, Jr., and C. B. de Than, "Dissymmetry of Molecular Light Scattering in Polymethyl Methacrylate," *J. Am. Chem. Soc. 77,* 4763-4767 (1955).

Billmeyer 1964a. Fred W. Billmeyer, Jr., "Principles of Light Scattering," Chap. 56 in I. M. Kolthoff and Philip J. Elving, eds., with the assistance of Ernest B. Sandell, *Treatise on Analytical Chemistry,* Part I, Vol. 5, Interscience Div., John Wiley and Sons, New York, 1964.

Billmeyer 1964b. F. W. Billmeyer, Jr. and V. Kokle, "The Molecular Structure of Polyethylene. XV. Comparison of Number-Average Molecular Weights by Various Methods," *J. Am. Chem. Soc. 86,* 3544-3546 (1964).

Billmeyer 1965. Fred W. Billmeyer, Jr., "Characterization of Molecular Weight Distributions in High Polymers," *J. Polymer Sci. C8,* 161-178 (1965).

Billmeyer 1970. Fred W. Billmeyer, Jr., and K. H. Altgelt, "The Sizes of Polymer Molecules and the GPC Separation," *Separation Science 5*, 393-402 (1970).

Billmeyer 1971. Fred W. Billmeyer, Jr., Harold I. Levine, and Peter J. Livesey, "The Refraction Correction in Light Scattering," *J. Colloid Interface Sci. 35*, 204-214 (1971).

Billmeyer 1972. Fred W. Billmeyer, Jr., *Synthetic Polymers*, Doubleday and Co., New York, 1972.

Bly 1971. Donald D. Bly *et al.*, "Proposed Recommended Practice for Gel Permeation Chromatography Terms and Relationships," *J. Polymer Sci. B9*, 401-411 (1971).

Brice 1950. B. A. Brice, M. Halwer, and R. Speiser, "Photoelectric Light-Scattering Photometer for Determining High Molecular Weights," *J. Opt. Soc. Am. 40*, 768-778 (1950).

Brice 1951. B. A. Brice and M. Halwer, "A Differential Refractometer," *J. Opt. Soc. Am. 41*, 1033-1037 (1951).

Cannon. Cannon Instrument Co., State College, Pennsylvania 16801.

Cantow 1967. Manfred J. R. Cantow, ed., *Polymer Fractionation*, Academic Press, New York, 1967.

Cazes 1966. Jack Cazes, "Topics in Chemical Instrumentation. XXIX. Gel Permeation Chromatography," *J. Chem. Educ. 43*, "Part One" A567-A582 and "Part Two" A625-A642 (1966).

Cazes 1970. Jack Cazes, "Current Trends in Gel Permeation Chromatography 1970," *J. Chem Educ. 47*, "Part One: Theory and Experiment" A461-A471 and "Part Two: Methodology" A505-A514 (1970).

Chiang 1964. R. Chiang "Characterization of High Polymers in Solutions - with Emphasis on Techniques at Elevated Temperatures," Chap. XII in Bacon Ke, ed., *Newer Methods of Polymer Characterization*, Interscience Div., John Wiley and Sons, New York, 1964.

Coleman. Coleman Instruments Div., Perkin-Elmer Corp., Maywood, Illinois 60153.

Coll 1968. Hans Coll and F. H. Stross, "Determination of Molecular Weights by Equilibrium Osmotic-Pressure Measurements," pp. 10-27 in Donald McIntyre, ed., *Characterization of Macromolecular Structure*, National Academy of Sciences Publication No. 1573, Washington, D.C., 1968.

Cornet 1968. C. F. Cornet, "Turbidity as a Tool for the Characterization of Molecular Weight Distribution," *Polymer 9*, 7-14 (1968).

Debye 1947. P. Debye, "Molecular-Weight Determination by Light Scattering," *J. Phys. & Coll. Chem. 51*, 18-32 (1947).

Duerksen 1968. J. H. Duerksen and A. E. Hamilec, "Polymer Reactors and Molecular Weight Distribution. Part III. Gel Permeation Chromatography, Methods of Correcting for Imperfect Resolution," *J. Polymer Sci. C21*, 83-103 (1968).

Elias 1968. Hans-Georg Elias, "Dynamic Osmometry," pp. 28-50 in Donald McIntyre, ed., *Characterization of Macromolecular Structure*, National Academy of Sciences Publication No. 1573, Washington, D.C., 1968.

Flory 1949. Paul J. Flory, "The Configuration of Real Polymer Chains," *J. Chem. Phys. 17*, 303-310 (1949).

Flory 1950. Paul J. Flory and Thomas G Fox, Jr., "Molecular Configuration and Thermodynamic Parameters from Intrinsic Viscosities," *J. Polymer Sci. 5*, 745-747 (1950).

Flory 1951. P. J. Flory and T. G Fox, Jr., "Treatment of Intrinsic Viscosities," *J. Am. Chem. Soc. 73,* 1904-1908 (1951).
Flory 1953. Paul J. Flory, *Principles of Polymer Chemistry,* Cornell University Press, Ithaca, New York, 1953.
Flory 1969. Paul J. Flory, *Statistical Mechanics of Chain Molecules,* Interscience Div., John Wiley and Sons, New York, 1969.
Giesekus 1967. Hanswalter Giesekus, "Turbidimetric Titration," Chap. C.1 in Manfred J. R. Cantow, ed., *Polymer Fractionation,* Academic Press, New York, 1967.
Grubisic 1967. Z. Grubisic, P. Rempp, and H. Benoit, "A Universal Calibration for Gel Permeation Chromatography," *J. Polymer Sci. B5,* 753-759 (1967).
Hall 1959. R. W. Hall, "The Fractionation of High Polymers," Chap. 2 in P. W. Allen, ed., *Techniques of Polymer Characterization,* Butterworths, London, 1959.
Hallikainen. Hallikainen Instruments, 750 National Court, Richmond, California 94804
Hewlett-Packard. Hewlett-Packard Corp., Avondale, Pennsylvania 19311.
Huggins 1942. Maurice L. Huggins, "The Viscosity of Dilute Solutions of Long-Chain Molecules. IV. Dependence on Concentration," *J. Am. Chem. Soc. 64,* 2716-2718 (1942).
Johnson 1967. Julian F. Johnson, Manfred J. R. Cantow, and Roger S. Porter, "Fractionation," pp. 231-260 in Herman F. Mark, Norman G. Gaylord, and Norbert M. Bikales, eds., *Encyclopedia of Polymer Science and Technology,* Vol. 7, Interscience Div., John Wiley and Sons, New York, 1967.
Johnson 1970. J. F. Johnson and R. S. Porter, "Gel Permeation Chromatography," Chap. 4 in A. D. Jenkins, ed., *Progress in Polymer Science,* Vol. 2, Pergamon Press, New York, 1970.
Kelley 1970. Richard N. Kelley and Fred W. Billmeyer, Jr., "A Review of Peak Broadening in Gel Chromatography," *Separation Science 5*, 291-316 (1970).
Kraemer 1938. Elmer O. Kraemer, "Molecular Weights of Cellulose and Cellulose Derivatives," *Ind. Eng. Chem. 30,* 1200-1203 (1938)
Kratohvil 1966. Josip P. Kratohvil, "Calibration of Light Scattering Instruments. IV. Corrections for Reflection Effects," *J. Colloid Interface Sci. 21,* 498-512 (1966).
Krigbaum 1967. W. R. Krigbaum and R.-J. Roe, "Measurement of Osmotic Pressure," Chap. 79 in I. M. Kolthoff and Philip J. Elving, eds., with the assistance of Ernest B. Sandell, *Treatise on Analytical Chemistry,* Part I, Vol. 7, Interscience Div., John Wiley and Sons, New York, 1967.
Kurata 1963. M. Kurata and W. H. Stockmayer, "Intrinsic Viscosities and Unperturbed Dimensions of Long Chain Molecules," *Fortschr. Hochpolym. Forsch. (Advances in Polymer Science) 3,* 196-312 (1963).
Kurata 1966. Michio Kurata, Masamichi Iwawa, and Kensuke Kamada, "Viscosity-Molecular Weight Relationships and Unperturbed Dimensions of Linear Chain Molecules," pp. IV-1 to IV-72 in J. Brandrup and E. H. Immergut, eds., with the collaboration of H.-G. Elias, *Polymer Handbook*, Interscience Div., John Wiley and Sons, New York, 1966.
Lyons 1967. John W. Lyons, "Measurement of Viscosity," Chap. 83 in I. M. Kolthoff and Philip J. Elving, eds., with the assistance of Ernest B. Sandell, *Treatise on Analytical Chemistry,* Part I, Vol. 7, Interscience Div., John Wiley and Sons, New York, 1967.

Maley 1965. L. E. Maley, "Application of Gel Permeation Chromatography to High and Low Molecular Weight Polymers," *J. Polymer Sci. C8*, 253-268 (1965).

McCaffery 1970. Edward M. McCaffery, *Laboratory Preparation for Macromolecular Chemistry*, McGraw-Hill Book Co., New York, 1970.

McIntyre 1964. D. McIntyre and F. Gornick, eds., *Light Scattering from Dilute Polymer Solutions*, Gordon and Breach Science Publishers, New York, 1964.

Melabs. Melabs, Inc., Scientific Instruments Dept., 3300 Hillview Ave., Palo Alto, California 94304. (Reiff osmometers are now manufactured by Wescan.)

Moore 1964. J. C. Moore, "Gel Permeation Chromatography. I. A New Method for Molecular Weight Distribution of High Polymers," *J. Polymer Sci. A2*, 835-843 (1964).

Moore 1967. W. R. Moore, "Viscosities of Dilute Polymer Solutions," Chap. 1 in A. D. Jenkins, ed., *Progress in Polymer Science*, Vol. 1, Pergamon Press, New York, 1967.

Morawetz 1965. Herbert Morawetz, *Macromolecules in Solution*, Interscience Div., John Wiley and Sons, New York, 1965.

Morey 1945. D. R. Morey and J. W. Tamblyn, "The Determination of Molecular Weight Distribution in High Polymers By Means of Solubility Limits," *J. Applied Phys. 16*, 419-424 (1945).

Newman 1954. Seymour Newman, William R. Krigbaum, Claude Laugier, and Paul J. Florv. "Molecular Dimensions in Relation to Intrinsic Viscosities," *J. Polymer Sci. 14*, 451-462 (1954).

Nichols 1971. Edgar Nichols, "Molecular Weight Averages from Gel Permeation Chromatography Employing the Universal Calibration Method," *Polymer Preprints* (ACS Div. Polymer Chem.) *12* (2), 828-834 (1971).

Overton 1971. J. R. Overton, "Determination of Osmotic Pressure," pp. 309-346 in Arnold Weissberger and Bryant W. Rossiter, eds., *Determination of Thermodynamic and Surface Properties*, Vol. 1 of A. Weissberger, ed., *Techniques of Chemistry*, 3rd ed., Part 5, Interscience Div., John Wiley and Sons, New York, 1971.

Pasternak 1962. R. A. Pasternak, P. Brady, and H. C. Ehrmantraut, "Apparatus for the Rapid Determination of Molecular Weight," *Dechema Monograph 44*, 205-207 (1962).

Peaker 1970. F. W. Peaker and M. G. Rayner, "Molecular Weight Distributions from Light Scattering Studies of Polystyrene Solutions During Phase Separation," *European Polymer J. 6*, 107-119 (1970).

Peebles 1971. Leighton H. Peebles, Jr., *Molecular Weight Distributions in Polymers*, Interscience Div., John Wiley and Sons, New York, 1971.

Ptitsyn 1959. O. B. Ptitsyn and Yu. E. Eizner, "Hydrodynamics of Polymer Solutions. II. Hydrodynamic Properties of Macromolecules in Active Solvents;" in Russian, *Tech. Fiz. 29*, 1117-1134 (1959); in English, *Soviet Phys.--Tech. Phys. 4*, 1020-1036 (1960).

Rayner 1969. M. G. Rayner, "Molecular Weight Distributions from Turbidimetric Titration and Related Techniques: Theoretical Evidence for a New Calibration Procedure," *Polymer 10*, 827-32 (1969).

Reiff 1959. Theodore R. Reiff and Marvin J. Yiengst, "Rapid Automatic Semimicro Colloid Osmometer," *J. Lab. Clin. Med. 53*, 291-298 (1959).

Rolfson 1964. F. B. Rolfson and Hans Coll, "Automatic Osmometer for Determination of Number Average Molecular Weights of Polymers," *Anal. Chem. 36*, 888-894 (1964).

Schaefgen 1948. John R. Schaefgen and Paul J. Flory, "Synthesis of Multichain Polymers and Investigation of their Viscosities," *J. Am. Chem. Soc. 70*, 2709-2718 (1948).

Springer 1970. J. Springer, K. Ueberreiter, and W. Weinle, "Experiments for the Determination of Molecular Weight Distribution by Turbidimetric Measurements." *European Polymer J. 6*, 87-95 (1970).

Staudinger 1930. H. Staudinger and W. Heuer, "Highly Polymerized Compounds. XXXIII. A Relation Between the Viscosity and the Molecular Weight of Polystyrenes" (in German), *Ber. 63B*, 222-234 (1930).

Steele 1963. R. E. Steele, W. E. Walker, D. E. Burge, and H. C. Ehrmantraut, paper presented at the Pittsburgh Conference on Analytical Chemistry, March, 1963.

Strazielle 1971. Claude Strazielle and Henri Benoit, "Molecular Characterization of Commercial Polymers," *Pure Appl. Chem. 26*, 451-479 (1971).

Tanford 1961. Charles Tanford, *Physical Chemistry of Macromolecules*, John Wiley and Sons, New York, 1961.

Taylor 1962. W. C. Taylor and L. H. Tung, "Turbidimetric Determination of the Molecular Weight Distribution of Linear Polyethylene," *SPE Trans. 2*, 119-121 (1962).

Wachter 1969. Alfred H. Wachter and Wilhelm Simon, "Molecular Weight Determination of Polystyrene Standards by Vapor Pressure Osmometry," *Anal. Chem. 41*, 90-94 (1969).

Waters. Waters Associates, Inc., 61 Fountain Street, Framingham, Massachusetts 01701.

Wippler 1954. C. Wippler and G. Scheibling, "Description of an Apparatus for the Study of Light Scattering" (in French), *J. Chim. Phys. 51*, 201-205 (1954).

Wescan. Wescan Instruments, Inc., P. O. Box 525, Cupertino, California 95014.

Wood. C. N. Wood Manufacturing Company, Newtown, Pennsylvania 18940.

Zimm 1948. Bruno H. Zimm, "Apparatus and Methods for Measurement and Interpretation of the Angular Variation of Light Scattering: Preliminary Results on Polystyrene Solutions," *J. Chem. Phys. 16*, 1099-1116 (1948).

8

Polymer Morphology

Morphology is the science of form and structure, and polymer morphology is the study of the arrangement of polymer molecules into crystalline and amorphous regions, the form and structure of these regions, and the manner in which they are organized, if at all, into larger and more complex units.

Our knowledge of the morphology of polymers is limited almost entirely to the behavior of the crystalline regions of crystallizable polymers, and the bulk of this chapter necessarily deals with that topic. Papers are beginning to appear, however, suggesting that solid amorphous polymers do not have the completely disordered structure of randomly coiled and entangled chains that was once attributed to them. Electron-microscope observations now suggest that varying degrees of molecular order exist in amorphous polymers, the amorphous regions of semicrystalline polymers, and very likely in polymer melts also (Yeh 1967, 1972*a*, *b*; Bonart 1968; Beecher 1969; Siegmann 1970; Klement 1971, 1972). It should be realized that in all probability a complete spectrum of structures exists in polymers, ranging from total disorder through various degrees and kinds of one- and two-dimensional order, up to that high degree of three-dimensional order we associate with crystallinity.

The key to crystallinity lies in the regular packing of atoms and molecules, and the key to achieving this regularity lies in control during synthesis of the regularity of the configuration of the polymer chain. As pointed out in the *Textbook*, Chap. 5B, crystallizable polymers consist of those with essentially linear chains in which substituents or side chain groups are either (a) small enough to fit into an orderly arrangement without disrupting it, or (b) disposed regularly and symmetrically along the chain. Considering polyethylene as a starting point, polymers falling in the first group include poly(vinyl alcohol) and poly(vinyl fluoride), and in the second, isotactic or syndiotactic polymers on the one hand and regularly repeating condensation polymers such as the polyesters and polyamides on the other. (For the concepts of tacticity, see the *Textbook*, Chap. 5A.) Such atactic polymers as polystyrene and poly(methyl methacrylate), with bulky side groups and irregular configurations, do not crystallize under any circumstances.

Even among crystallizable polymers, the ease with which crystallization takes place varies considerably from one polymer to another. When a molten crystallizable polymer is cooled below its melting point T_m, crystallization starts at a number of nucleation sites in the fused mass. As the temperature is lowered, the rate of crystallization increases, goes through a maximum, and decreases to zero at the glass-transition temperature T_g. (This behavior is described in Chap. 9B and the *Textbook*, Chap. 5E; for the glass-transition temperature, see Chap. 9A and the *Textbook*, Chap. 6D.) Polymers which crystallize rapidly, such as polyethylene and polypropylene, always do so no matter how rapidly one attempts to cool them, but polymers which crystallize more slowly, including poly(ethylene terephthalate) and isotactic polystyrene and poly(methyl methacrylate), can be quenched to below T_g in the amorphous state. These materials then crystallize when warmed to temperatures between T_g and T_m.

The rate and extent of crystallization in a given polymer determine not only the total amount of crystallinity but the sizes and shapes of the crystalline regions and of larger morphological units. All these are intimately related to the physical and mechanical behavior of the material. While these relations between structure and properties of polymers are dealt with in Chap. 10, it should already be apparent that crystallinity and morphology strongly influence polymer behavior and performance. Some of the unique mechanical and physical properties of such engineering plastics as linear polyethylene, polypropylene, polyoxymethylene, the nylons, and polytetrafluoroethylene result from their two-phase crystalline-amorphous nature in which rigid crystalline regions are interconnected by flexible amorphous or disordered regions. Higher crystallinity leads to polymers which are denser, stiffer, harder, tougher, and more resistant to solvents, while the amorphous domains add softness, flexibility, and ease of processing below melt temperatures.

In the following section, we review briefly the features of the morphology of crystalline polymers, discussed also in the *Textbook*, Chap. 5, and in the literature cited. We then examine the two major techniques used in the study of morphology, x-ray diffraction and microscopy.

A. *Morphology of Crystalline Polymers*

Because of the importance of x-ray (and electron) diffraction in determining the structures of crystals, we make considerable reference to results obtained with this technique, in advance of the discussion of methods in Sec. B. A complication in polymer systems compared to those of low-molecular-weight substances arises at once, in that polymer x-ray photographs, such as the powder diagram in Fig. 8-1, show the presence of both liquid-like and crystalline-like diffraction patterns. The sharp diffraction rings are characteristic of regions of three-dimensional order at least several hundred A in size, while the more diffuse features are associated with the presence of disordered regions similar to liquids or polymer melts. The presence of both types of pattern from the same sample was used, historically, to argue that polymers were semicrystalline, consisting of distinct crystalline

Fig. 8-1. X-ray powder diffraction pattern showing sharp crystal rings and diffuse amorphous bands. Compression-molded linear polyethylene (courtesy P. H. Geil).

regions surrounded by an amorphous phase in a composite two-phase structure known as the fringed-micelle model (*Textbook*, Fig. 5-12). It now appears more fruitful, however, to consider some polymers, especially highly crystalline ones, as one-phase systems of varying degrees of order from point to point, and from sample to sample (*Textbook*, Chap. 5D, Geil 1963).

Molecular packing

The production of an x-ray diffraction pattern like that in Fig. 8-1 requires that there be a regular, repeating arrangement of individual atoms extending in all three dimensions for at least several hundred A; if some relaxation is permitted in the requirement of perfection, even larger distances in one or more directions may be required. To achieve this demands the existence of a unit cell containing portions of, usually, several polymer chains, and this in turn places stringent requirements on the regularity of the conformation (Fig. 8-2) as well as the configuration of at least those sequences of the chain falling within a given crystal.

Fig. 8-2. An example of a regular conformation of a polymer chain as it occurs in the crystal lattice: the helical conformation of polypropylene, in side and end views (Natta 1960).

Here and throughout, as in the *Textbook*, Chap. 5A, we use the term *configuration* to describe spatial arrangements of atoms which can be interconverted only by breaking and reforming chemical bonds, such as *cis* and *trans* isomerization, or isotactic, atactic and syndiotactic arrangements. *Conformation* is used to describe arrangements which could be interconverted by rotations about single bonds, such as *trans* and *gauche* arrangements in a carbon-carbon chain. In solid polymers above T_g, application of heat (as in annealing) or stress (as in drawing or orientation) can lead to rearrangement of chain conformations.

Polymers crystallize with a wide variety of unit cells, many of which are described by Geil (1963, 1966). A comprehensive treatment is in preparation (Wunderlich 1973). One feature common to all and required by the long-chain nature of polymers is an anisotropy resulting from differences between the chemical-bond forces along the chains and the intermolecular forces between them. This anisotropy provides the possibility of great enhancement of physical properties in one direction by orienting the polymer molecules such that their chains extend in the direction of maximum expected stress. This orientation is taken advantage of, for example, in the drawing of fibers. X-ray diffraction experiments with fibers utilize the symmetry provided by such orientation, and provide most of the information we have about the atomic and molecular arrangements within polymer crystals. An example is the diffractometer scan of linear polyethylene in Fig. 8-3, such a scan being a plot of diffracted intensity versus angle along what would correspond to one of the radii in the powder diagram of Fig. 8-1. Indicated on the figure are the portions of the scattering attributed to crystalline and amorphous regions in the polymer, and various background contributions.

From the study of results such as these, plus the generous application of prior knowledge obtained from model compounds, such as bond lengths and angles, the structures of unit cells have been deduced. In polyethylene, segments of the molecules are packed in the crystal in a manner identical to that of the orthorhombic unit cell previously known for the short-chain normal paraffins. In the polyethylene unit cell (Fig. 8-4), the molecule is in the fully extended planar zig-zag conformation. In the paraffins, the c-axis repeat distance (in the chain direction) is the total length of the molecule, but in polyethylene, the c-axis repeat distance is taken as the 2.54 A between corresponding carbon atoms two apart, or one zig-zag.

It is interesting to note that, in the early history of polymer science, those familiar with crystals of low-molecular-weight substances failed to realize that a chain molecule could, by virtue of

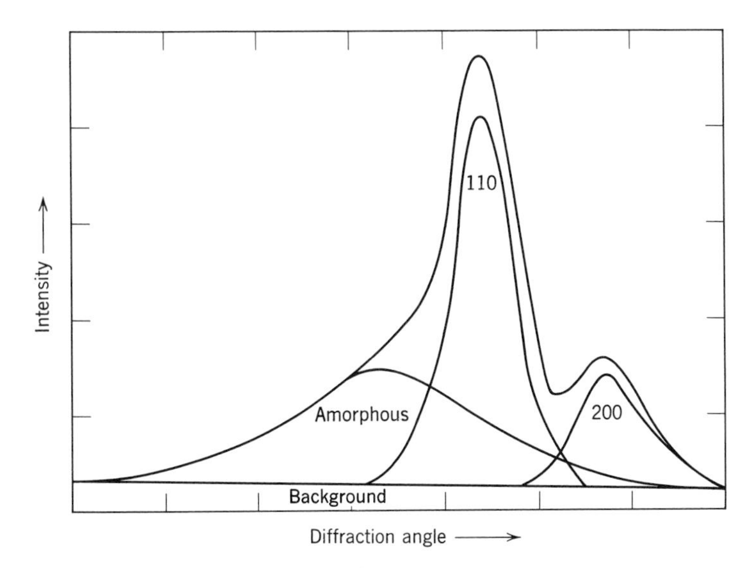

Fig. 8-3. Diffractometer scan of linear polyethylene, showing resolution into contributions from the (110) and (200) crystalline reflections, the amorphous peak, and the background (*Textbook*, Fig. 5-22).

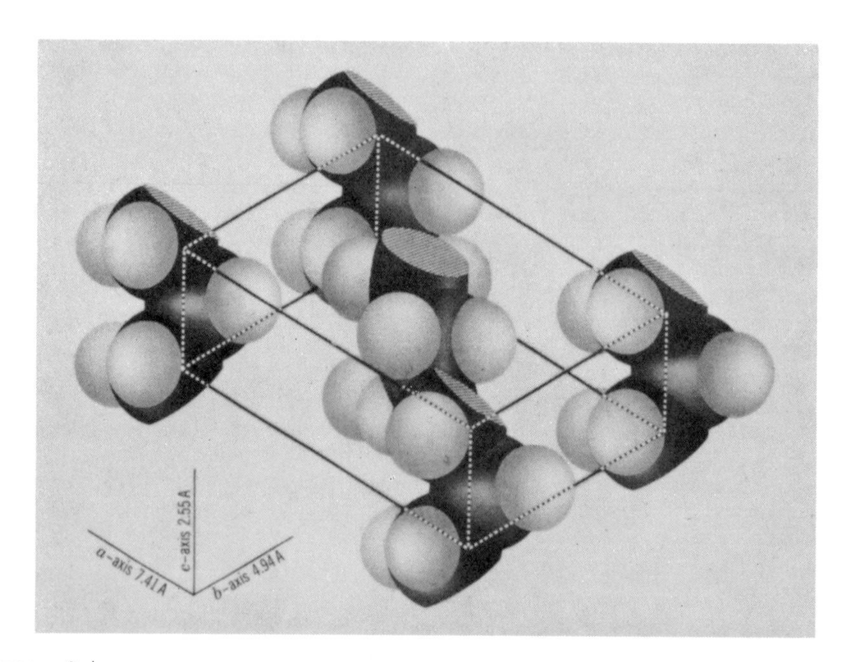

Fig. 8-4. Arrangement of chains in the unit cell of polyethylene (*Textbook*, Fig. 5-23, from Geil 1963).

a regular repeating structure, run through many consecutive unit cells. Thinking that an entire molecule must be accommodated within a single unit cell, they argued that the long-chain macromolecular hypothesis was in error. The pioneer work of Herman Mark and K. H. Meyer in the early 1930's in demonstrating the principles outlined above in the unit cells of such polymers as cellulose did much to bring about the acceptance of this hypothesis, now so familiar.

In the cases of most crystalline polymers, as in the single crystals discussed below, the sequence of unit cells containing portions of a single molecule persists only for 100 A or so. Subsequent segments of the molecule may be incorporated into a different crystal or another part of the same crystal, with an intercrystalline, less-ordered segment of variable length in between. Only in rare instances of extended-chain crystals, produced at high pressures and temperatures, is the same conformation maintained throughout the entire length of the chain (Geil 1964, Wunderlich 1968). The thickness of the crystal in the chain direction is here equal to the molecular chain length. These crystals represent the closest approach discovered to date to the thermodynamically most stable crystal form in polymers.

Polymer single crystals

Single crystals (*Textbook*, Chap. 5C) of many different polymers can be grown from dilute solution. Most polymer single crystals have the same general appearance (Fig. 8-5*a*), being thin platelets or lamellae, often diamond-shaped, about 100 A thick and 1-10 μm in lateral dimensions. Many examples are shown in Geil's book (1963). The size, shape, and regularity of the crystals depend on their growth conditions, the important variables being solvent, temperature, and concentration. The thickness of the lamellae is independent of chain length, but depends on the crystallization temperature, and can be altered by subsequent annealing treatments.

Although polymer single crystals are too small to be examined by x-ray diffraction, most electron microscopes can be adjusted to provide electron-diffraction patterns (Sec. C) from selected areas of the specimen as small as one square micrometer. Such patterns (Fig. 8-5*b*), obtained with polymer single crystals, have not only confirmed the fact that they are single crystals, but unequivocally show that the chain axis is perpendicular to the plane of the lamella. Since the molecular chains are many times longer than the lamellar thickness, it follows that they must fold through 180° at the surface and re-enter the crystal (Keller 1957). Whether this folding is regular or not is still a debatable question (*Textbook*, Chap. 5C and references cited therein), but chain-folded growth has been established as a basic and general mechanism of polymer crystallization; the only widespread exceptions are thought to be materials of very low crystallinity, such as poly(vinyl chloride), where the fringed-micelle model may be appropriate, and the extended-chain crystals mentioned above.

Chain folding introduces a new structural parameter, the fold length or lamellar thickness, depending as noted above primarily on the degree of supercooling and thus the crystallization temperature. In the well-explored system of single crystals of linear polyethylene grown from xylene solution, the fold length ranges between 90 and 170 A.

Bulk polymers

When polymers are crystallized from the melt, two structural features are seen so often that they appear to be quite general and characteristic (*Textbook*, Chap. 5D). The first of these is the *lamella*. In fact, in typical cases of polymers with high crystallinity, lamellae are the only conspicuous feature of the electron micrographs (Fig. 8-6), and there is no direct evidence for the presence of substantial amorphous regions. Photographs like this have done much to favor the replacement of the fringed-micelle concept with that of the folded-chain lamella as the basic structural element in highly crystalline polymers in the bulk.

During crystallization from the melt, and to some extent occurring also in crystallization from concentrated solutions, parts of a molecule may be incorporated into more than one lamella. These "tie molecules" are thought to bind the lamellae together and contribute greatly to the mechanical strength properties and the ductility of bulk polymers.

The second characteristic feature of the morphology of bulk polymers is the *spherulite* (Fig. 8-7). This is a radially symmetrical structure formed by the cooperative growth of crystallites outward from a nucleus in three dimensions. In many cases, the crystallites

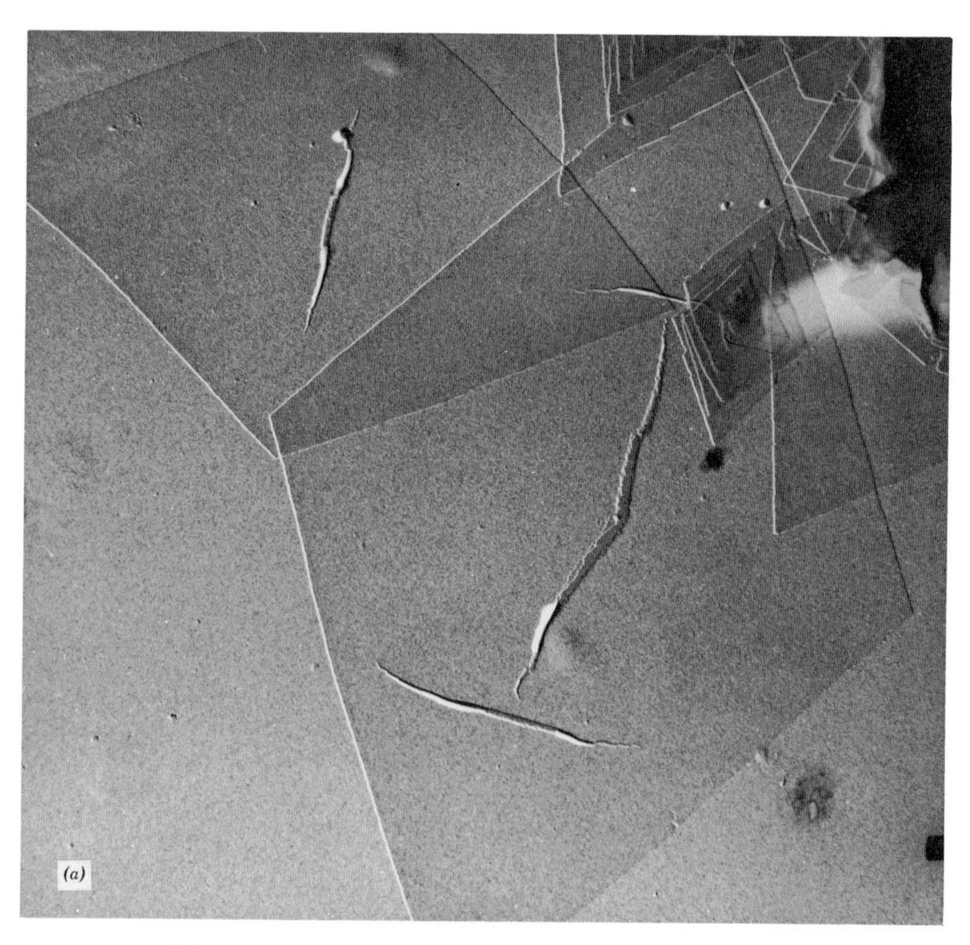

Fig. 8-5. (*a*) Electron micrograph of a single crystal of linear polyethylene (*Textbook*, Fig. 5-9, from Reneker 1960); and (*b*) electron diffraction pattern from a similar crystal (courtesy P. H. Geil).

appear to be ribbon-like lamellae or fibrils with width of the order of 1 μm and thickness in the range 100-500 A. Occasionally, the lamellae undergo a regular twist for reasons not yet known (Fig. 8-8). In all polymer spherulites studied to date, the chain axis is approximately tangential to the radial direction; this is consistent with the chain-folding model. In the polarizing microscope, spherulites have a characteristic Maltese-cross extinction pattern (Fig. 8-7), with extinction occurring along radii parallel to the optic axes of the polarizer and analyzer. A simple demonstration experiment on the growth and observation of spherulites was described by Billmeyer (1960).

Spherulite size and structure also has an influence on the properties of polymers, particularly in regard to failure mechanisms (*Textbook*, Chap. 7C).

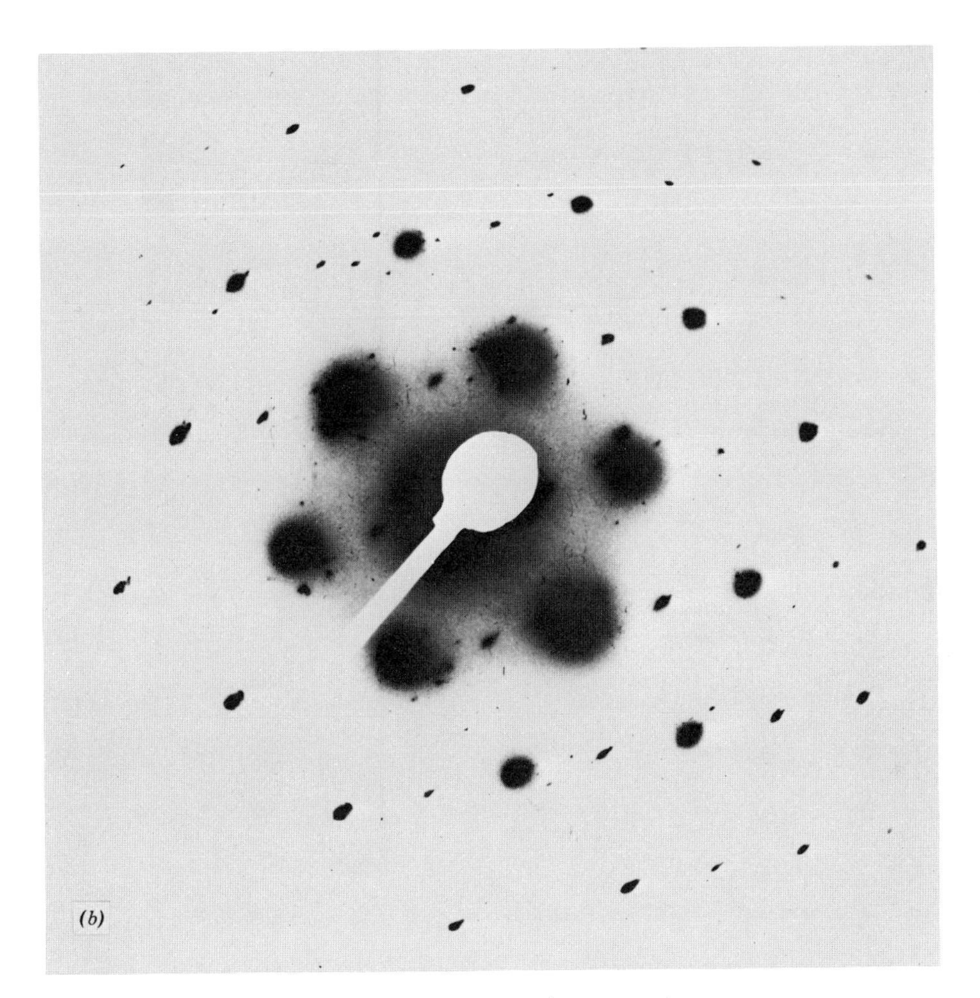

Fig. 8-5 (continued). Part *b*.

GENERAL REFERENCES

Textbook, Chap. 5; Tobolsky 1960, Chap. 2; Geil 1963, 1966; Hoffman 1964; Mandelkern 1964; Meares 1965, Chaps. 2 and 4; Peterlin 1965, 1967; Miller 1966, Chap. 10; Sharples 1966, Chap. 2; Cottham 1968; Keller 1968; Wunderlich, 1969, 1973; Rodriguez 1970, Chap. 3; Rees 1971; Williams 1971, Chap. 6.

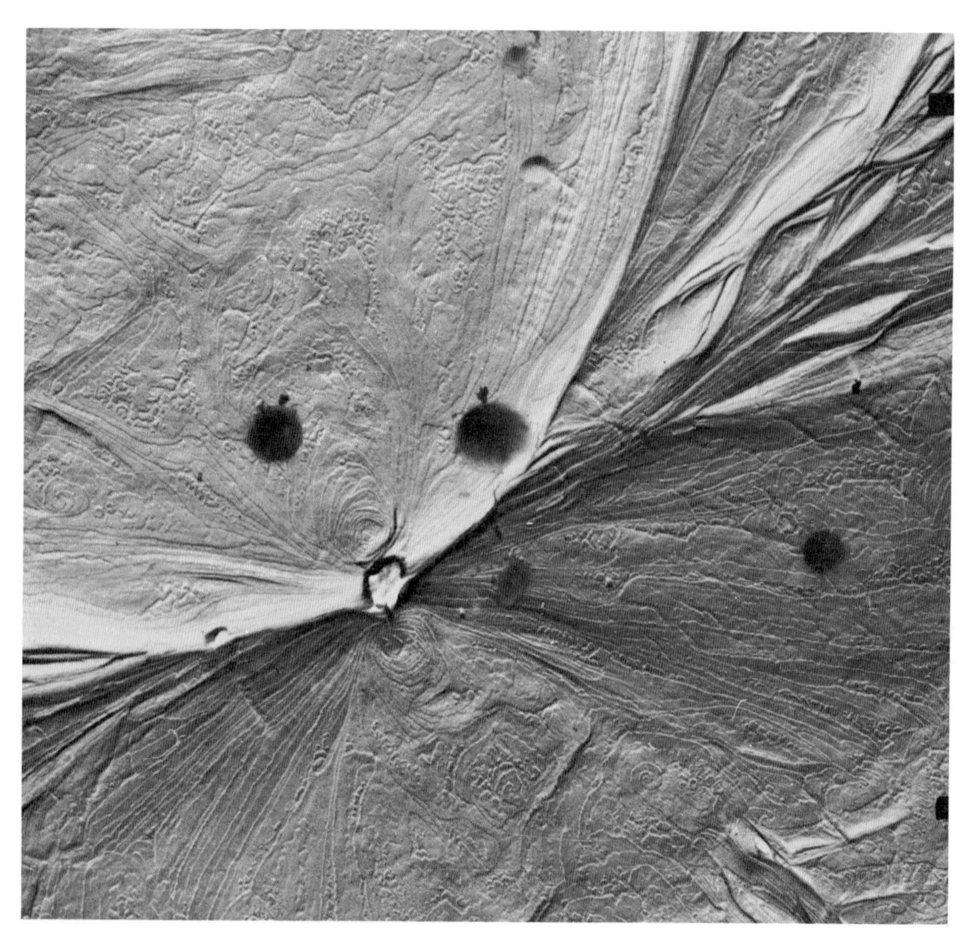

Fig. 8-6. Electron micrograph of the surface of melt-crystallized polyoxymethylene (Geil 1960).

B. *X-Ray Diffraction*

Principle

When a crystalline material is exposed to a beam of x-rays, the three-dimensional array of atoms in the crystal scatters the electromagnetic radiation in such a way that the scattered waves from different atoms reinforce each other only in certain directions; elsewhere, destructive interference occurs and no scattering is observed. These directions are characteristic of the orientation of the crystal with respect to the beam and of the interatomic separations in the material. They are governed by Bragg's Law,

Fig. 8-7. Spherulites of isotactic polypropylene as seen in the light microscope with crossed polarizers. The two types of birefringence result from growth of some spherulites from hexagonal crystals, and others from those with a monoclinic unit cell (courtesy P. H. Geil).

$$n\lambda = 2d_{hkl} \sin \theta \tag{8-1}$$

where n is an integer denoting the order of the diffraction, λ is the wavelength of the x-rays, θ is the angle of incidence and reflection, and d_{hkl} is the spacing between planes of atoms causing the reflection, designated by a set of three small integers (Miller indices) h, k, and l.

It can be shown that Eq. 8-1 is over-specified and in the general case no diffraction is observed. To generate diffraction maxima from the planes with a given spacing d, either λ or θ must be varied. In the usual modern experiment, it is θ that is varied, in one of two ways. If the sample is ground to a fine powder, the x-ray beam interacts with many tiny crystals, which take up all possible orientations. Now diffraction from a single set of planes takes place at all angles having a fixed relation to the direction of the beam, that is, along the surface of a cone with half angle 2θ around the beam direction. A

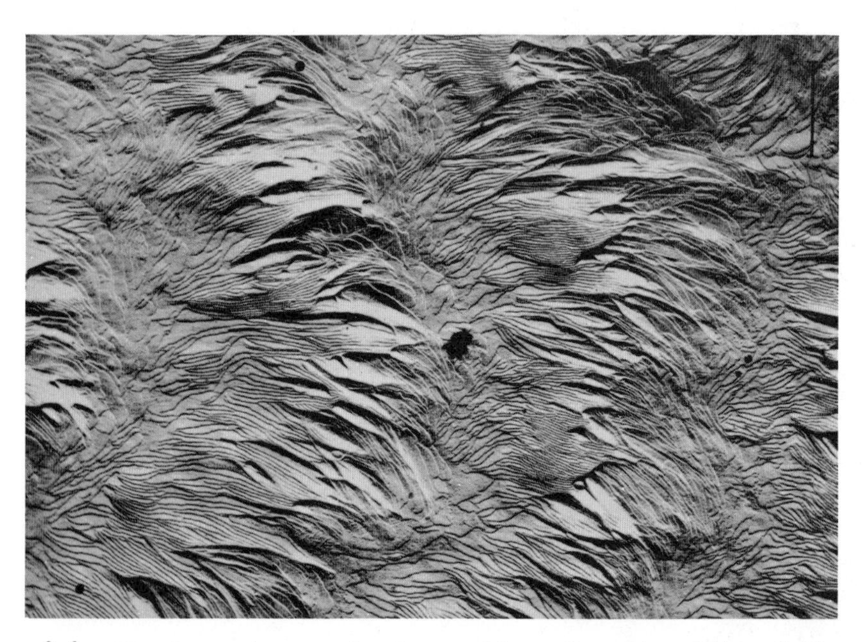

Fig. 8-8. Electron micrograph of a portion of a spherulite of linear polyethylene, showing twisted lamellae (*Textbook*, Fig. 5-16, from Geil 1963; photograph by E. W. Fischer).

flat film placed normal to the beam cuts this cone in a circle, while a cylindrical film with axis perpendicular to the beam intercepts arcs of such circles to produce the concentric rings shown in Fig. 8-1.

If the sample is a single crystal or a highly oriented fiber, placed so that one crystallographic axis is perpendicular to the beam, and rotated about that axis, each set of planes produces a discrete spot, as shown in Fig. 8-9. From the positions and intensities of these spots are determined both the orientation of the crystal and information about its structure, as explained in part below. Other experiments are possible, including those in which λ is varied by the use of polychromatic rather than monochromatic x-rays (Laue photographs, now largely of historical interest), but the fiber pattern is most often used for polymeric materials.

The locations of the diffracted beams serve primarily to determine the d-spacings, through the Bragg equation. The intensity of any spot is determined by a structure factor taking account of phase differences in the diffraction process in a complicated way. The total diffracted intensity turns out to be proportional to the number of electrons per unit volume in the sample. Materials composed of heavy atoms diffract x-rays much more strongly than polymers, and generally yield more complete information more rapidly. Hydrogen atoms cannot be located at all in x-ray diffraction experiments; their locations must be inferred from bond lengths and angles. They can be detected by electron or neutron scattering, however.

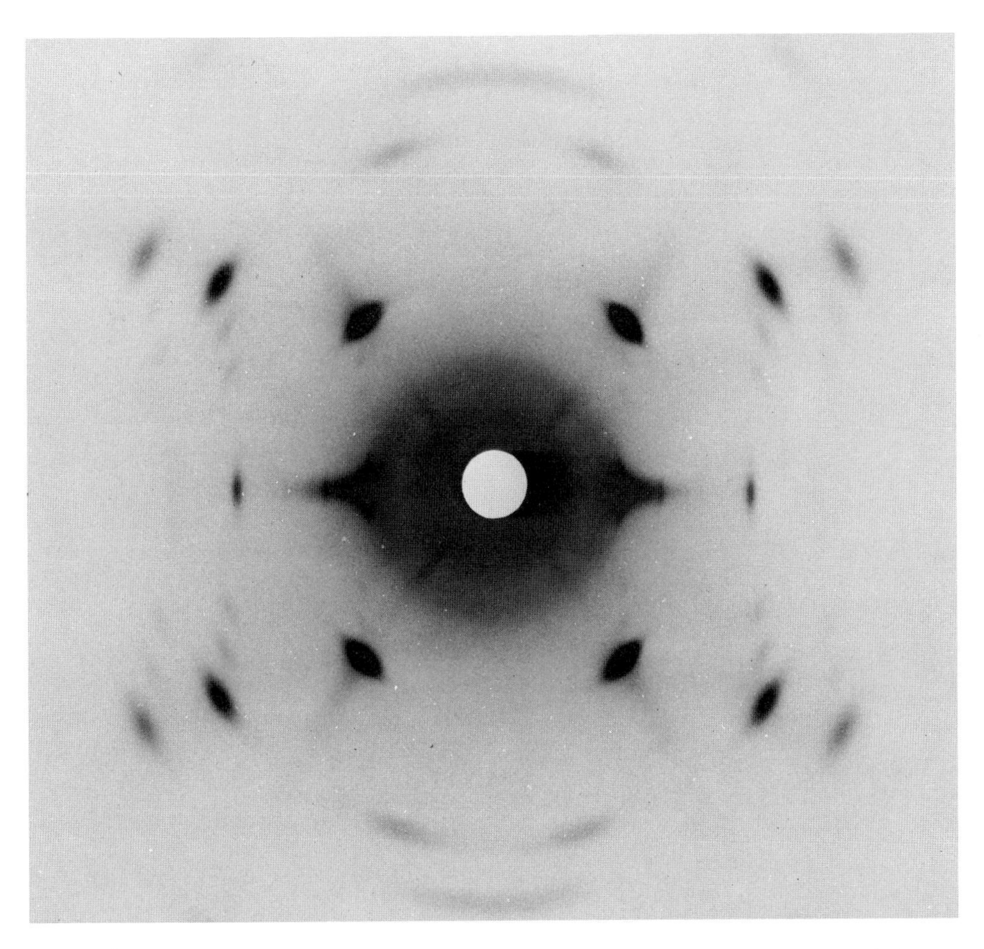

Fig. 8-9. X-ray fiber diffraction pattern of polyoxymethylene (*Textbook*, Fig. 5-26*b*, courtesy E. S. Clark).

As mentioned in Sec. A, an amorphous material, randomly oriented, produces diffuse rather than sharp rings in both the powder and fiber photographs. When such a material is stretched, some alignment of the chains in the direction of stretch takes place, and the ring becomes more intense along the equator (perpendicular to this direction) and may even break up into spots. The average spacing of the chains can be determined from the Bragg equation, and a measure of the degree of orientation can be calculated from the intensities in the various directions around the ring (Alexander 1969, Chap. 4).

Apparatus

The general features of an x-ray diffraction unit are shown in Fig. 8-10. X-rays are generated by the impact of electrons on a metal

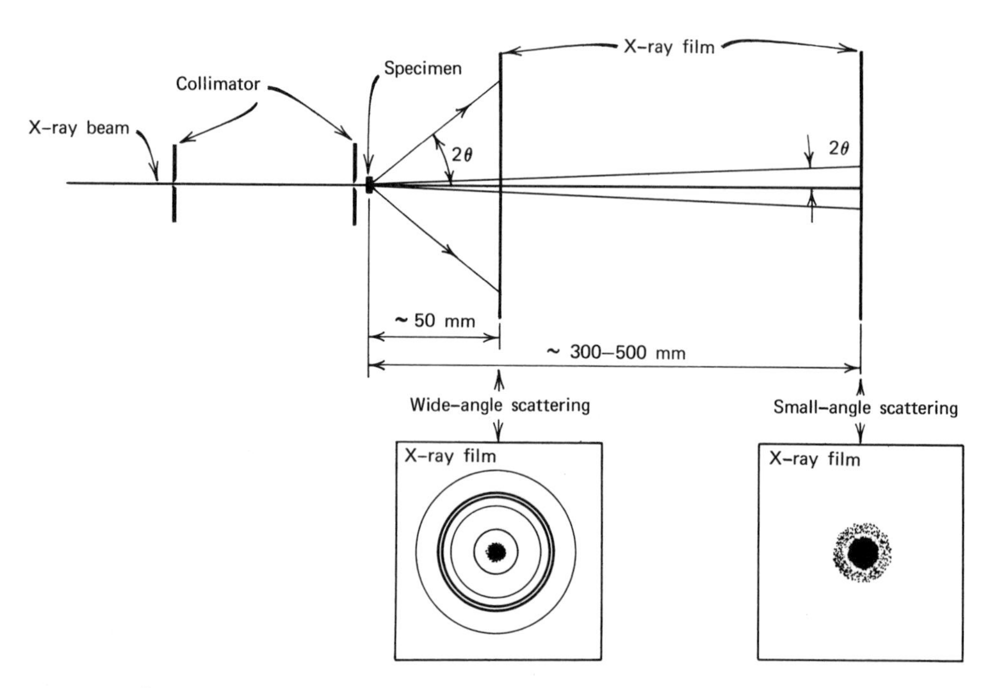

Fig. 8-10 Essential features of an x-ray diffraction instrument.

target in an evacuated tube; many metals can be used but copper provides radiation with a good balance of wavelength and penetrating power for organic materials. A nickel filter is used to isolate the Cu Kα line, λ = 1.54 A. The beam is collimated with suitable slits or pinholes, passes through and is diffracted by the sample, and is captured by absorption in a small lead stop.

The diffracted x-rays may be recorded photographically, as described before, or by the use of one of several types of x-ray-sensitive detectors followed by electronic amplification. Most often a proportional counter is used. This device detects and counts the separate x-ray photons in a diffracted beam. In this case, a recorded pattern of intensity versus 2θ is produced, like that of Fig. 8-3. Each of these methods has distinct advantages and disadvantages. The photographic technique records the entire diffraction pattern at once, even though long exposures may be required. It is excellent for locating diffraction maxima initially, but poor for the production of quantitative data, since photometry and tedious calibration are required for accuracy. The diffractometer counter provides quantitative results at once, but spots may be missed unless their presence is known so that the conditions of the scan can be set appropriately. Complete mapping of the entire diffraction pattern by diffractometer is too time consuming for practical application.

There are two regions of x-ray diffraction, both highly useful for polymers, from which data are obtained: wide angles (from a few degrees in the Bragg diffraction angle 2θ up to 30-50°) and small angles

Fig. 8-11. Philips table-top x-ray diffraction generator (courtesy Philips).

(from a few seconds of arc up to a degree or so). The following paragraphs describe the results obtained in these two regions, but some comments on the apparatus requirements are pertinent.

Equipment for wide-angle x-ray diffraction is widely used, and little need be said about it. A unit that is particularly suitable for student use is the Philips Table Top Diffraction Generator (Fig. 8-11) (Cooke 1967).

A practical small-angle camera for general use is shown in Fig. 8-12. The smaller the Bragg angle, the more stringent the requirements on resolution in the instrument. For small-angle work, much finer collimators must be used to define the x-ray beam, with corresponding loss in intensity. Alignment is critical, and in addition the equipment must be evacuated to eliminate air scattering, which is intense enough to mask the sample diffraction pattern at small angles. To aid in separating the primary and diffracted beams, the film or detector is placed farther from the sample than in wide-angle work (typically, 30 versus 5 cm). The exposure or counting time must be increased by the square of this ratio. With older equipment and weakly scattering materials, it is not unusual for photographic exposures to last several days, but with modern equipment one to a few hours usually suffices (Statton 1964, Alexander 1969).

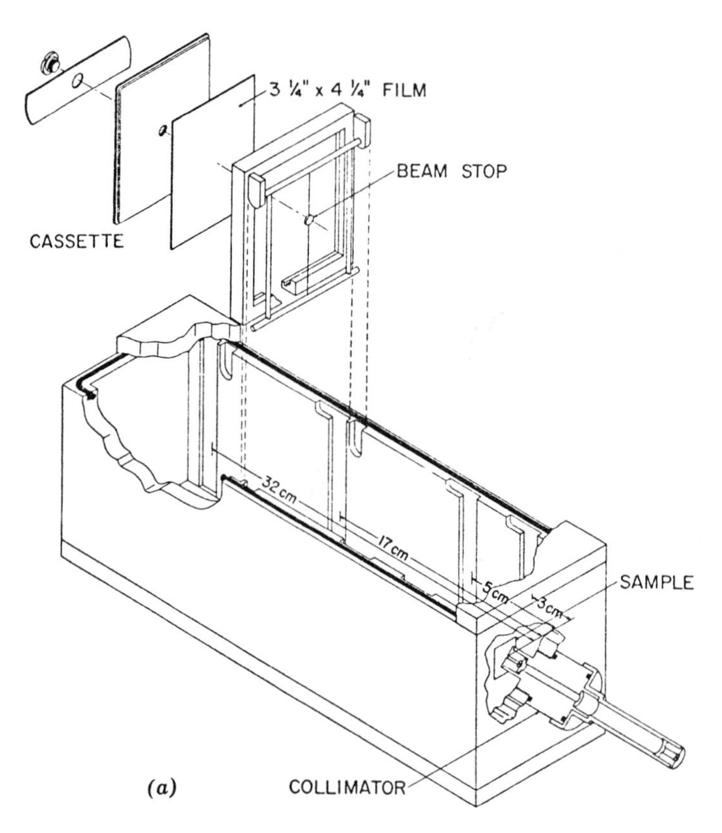

Fig. 8-12. Small-angle x-ray diffraction camera designed by W. O. Statton: (*a*) sketch (courtesy Warhus); (*b*) shown in place on General Electric XRD-3 x-ray diffraction unit, with wide-angle camera to the left.

Results of wide-angle diffraction

The major results of wide-angle x-ray diffraction of chief interest to polymer scientists are information on the relative amounts of crystalline and amorphous materials, the size, orientation and perfection of crystals, the nature of ordering in the system, and of course the chain arrangement in the crystals.

Degree of crystallinity There are several ways of measuring the degree of crystallinity of polymers, some of which are discussed by Kavesh (1969) and Wunderlich (1973). In the x-ray experiment, the total amounts of crystalline and amorphous scattering are measured, as the sum of the areas underneath all the crystalline peaks and underneath the amorphous ring, respectively. The degree of crystallinity is taken as the ratio of crystalline to total scattering. Generally, a

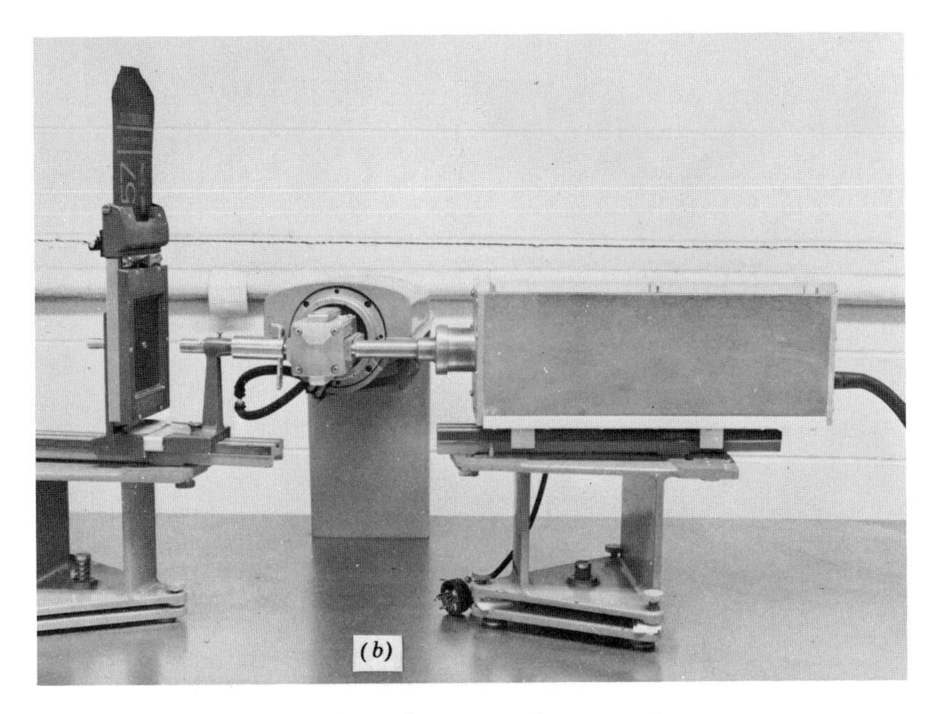

Fig. 8-12 (continued). Part *b*.

powdered sample is used in order to avoid problems which might arise from orientation. A typical diffractometer scan showing the separation into amorphous and crystalline scattering is shown in Fig. 8-3.

Size and perfection of crystals As the crystals exposed to x-rays become either smaller or less perfect, the diffracted peaks or lines become broader. Unfortunately, it is not possible in most cases to differentiate between these two causes of line broadening for polymer samples (Thielke 1964), unless two or more orders of reflection [for example, from planes with Miller indices (110) and (220)] can be observed. In the diffraction patterns of polymers, this is very rarely the case. In the absence of multiple orders, it is virtually fruitless to attempt to assign the origins of line broadening to imperfections or small crystallite size, unless some other information allowing a decision is at hand. If one assumes a model including only one of these parameters, then the measurement can be interpreted in terms of the model. Thus, the fringed-micelle model assumed only small crystallite size, and line broadening (corrected for contributions from the instrument itself) gave sizes in the range of 50 to several hundred A. The folded-chain model, however, is usually considered to include both small size (lamellar thickness) and imperfect crystals (Hosemann 1967).

Orientation When a sample is increasingly oriented, the rings of a powder diagram first break up into arcs, which then narrow to the

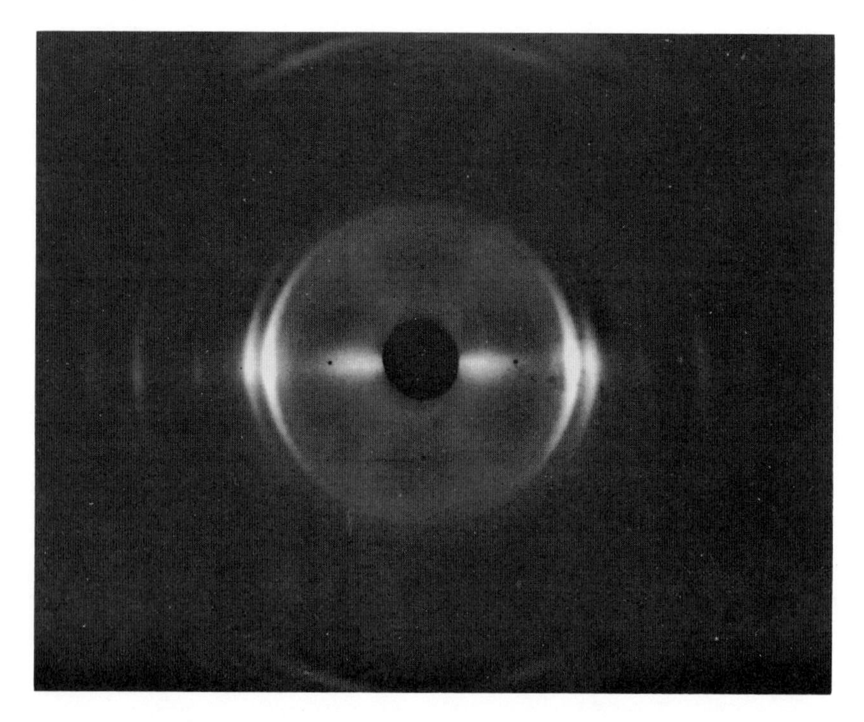

Fig. 8-13. X-ray fiber diagram showing the arcs and spots resulting from partial orientation; compare Fig. 8-1, characteristic of completely unoriented material, and Fig. 8-9, of a highly oriented sample.

relatively sharp spots seen in a fiber pattern. An example is shown in Fig. 8-13. Similar effects in the amorphous scattering ring were mentioned above. In each case, the usual result of orientation is that the molecular chains are aligned parallel to the direction of stretch (called the *meridian*; the direction perpendicular to this is the *equator*). Orientation is a complicated phenomenon, and all the morphological changes accompanying it are not yet understood (*Textbook*, Chap. 5F).

Order and packing The equatorial reflections in a wide-angle fiber diagram provide information about the packing between chains, the regularity and long-range lateral order in the crystal, and the symmetry of the unit cell. Information about order along the chains is obtained from examination of reflections along the meridian.

Many polymers, including most of those with the isotactic configuration, crystallize in a helical conformation, an example being polypropylene. Here, successive chain bonds are in the *gauche* conformation, so that the methyl groups stick out at 120° intervals and it takes three repeat units of the chain to produce one turn of the helix (Fig. 8-2); this is therefore termed a 3:1 helix. The x-ray diagram from such a helix shows strong reflections along the meridian at every third possible position (layer line), with intervening ones being absent. By examining this intensity pattern, one can determine the pitch of the helix, sometimes its hand (left or right), and the number of chain repeat units per turn.

Some amorphous materials appear to contain molecules in helical conformations, packed in such a way that two amorphous rings are observed. One corresponds to the van der Waals contact distance between atoms of 4-5 A (the usual amorphous ring), and the other a center-to-center interhelix distance of 10-12 A. This can be seen in polystyrene and poly(methyl pentene-1), among others.

Results of small-angle diffraction

Small-angle x-ray diffraction provides information on the external dimensions of such morphological domains as lamellae, spherulites (if small enough), separated phases, and voids, and in some cases provides evidence on their shapes. These results are independent of the internal arrangement or order in such regions. The small-angle experiment allows spacings characteristic of these features in the range 50-200 A to be resolved easily, and with special cameras (Kratky or Bonse-Hart), resolution up to several thousand A is possible.

An obvious application of small-angle x-ray diffraction is the determination of lamellar thickness, since this is in the range around 100 A (Blais 1967). More generally, the presence of two phases in the applicable size range and with different electron densities can be detected regardless of their origin. They may be crystallites in a largely amorphous matrix, corresponding roughly to the fringed-micelle theory, or two amorphous phases such as those found after microphase separation in block copolymers, for example those of styrene and butadiene. Voids can be treated similarly.

In such cases, it is the average center-to-center spacing of the domains and not their size which is determined. To obtain the latter, additional information is required, such as the assumption that the scattering units are close packed, or that they have a specific shape (e.g., spherical) and are present in a known volume concentration, or their direct observation, say by electron microscopy. In dilute solutions or suspensions, one can make use of the intensity of small-angle x-ray scattering, analogous to the light-scattering experiment described in Chap. 7B, to obtain a measure of particle mass, or with density known, of particle size. This approach has also been applied to solid two-phase systems such as styrene-butadiene block copolymers (McIntyre 1970). In this case, however, Bragg's equation is not applicable because of the complexity of the interference patterns (Guinier 1955).

In general, information on the shapes of domains cannot be obtained unless the sample is oriented. Then the spacing may become different in the meridional and equatorial directions, and some inferences regarding shape can be drawn. The dependence of intensity on scattering angle in the "tail" of the scattering curve can yield information on the surface-to-volume ratio of the dispersed phase, assuming that the electron densities of the two phases are known and making some additional hypotheses concerning the regularity of their spacing (Guinier 1955, Brumberger 1968, Alexander 1969, Chap. 5).

GENERAL REFERENCES

Textbook, Chaps. 4C, 5; Guinier 1955, 1963; Statton 1959, 1964, 1967; Tanford 1961, Chap. 2; Hosemann 1962; Posner 1962; Geil 1963; Liebhafsky 1964; Brumberger 1968; Alexander 1969; Jeffery 1971; Wilkes 1971.

C. *Electron Microscopy*

The application of electron microscopy has played an important role in our understanding of polymer morphology, and holds promise of providing opportunity for even greater contributions in the future. It is appropriate to cover the particular areas in which this technique can be applied to polymer samples.

Principle

The principle of electron microscopy is entirely analogous to that of visible-light microscopy. The wavelength, of course, is that associated with the electron beam through the application of de Broglie's principle of the wave-particle duality of electrons, and electrostatic or electromagnetic lenses replace the usual glass ones. Since the reader is no doubt familiar with the working of a conventional microscope, we proceed at once to the apparatus of electron microscopy.

Apparatus

The essential features of a transmission electron microscope are shown in Fig. 8-14. An electron gun delivers a beam of electrons, accelerated by a potential of the order of 100,000 V, through a system of condensing lenses onto the specimen. This takes place in a chamber or column evacuated to a pressure of 10^{-5} torr or less, because of the strong scattering of electrons by air. While the need to operate under high vacuum is a serious detriment to the use of the electron microscope with biological materials, which must be completely dried and free from all volatiles, it is of little consequence in the study of synthetic polymers.

The electron beam penetrates the specimen, with some of the electrons being diverted by scattering. The primary source of image contrast is differences in the fraction of scattered electrons from one part of the specimen to another. The sample thickness is limited to a few thousand A for the usual instrument with an accelerating voltage of 100,000 V. Higher voltages produce greater penetrating power but lower contrast, and *vice versa.*

Not all electron microscopes are suitable for use by relatively unskilled personnel. The JEM 30 or JEM 50 "Superscope," manufactured by Japan Electron Optics Laboratory Co. (JEOL) and shown in Fig. 8-15 has simple design and is easy to operate, making it particularly suitable for student use in Exp. 19. It uses a beam voltage of 30,000-50,000 V, and has a resolution of 100-150 A.

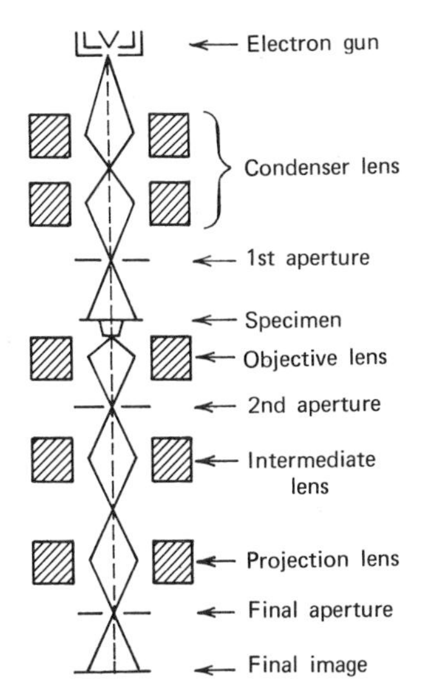

Fig. 8-14. Essential features of a transmission electron microscope (Baker 1971).

Specimen preparation and damage

The preparation of specimens thin enough to be suitable for electron microscopy can be extremely difficult. Microtoming is an obvious technique, but is is difficult to be sure that the specimen has not been deformed in the process. Replication and shadowing by standard electron-microscopy techniques have been widely used, especially to study the morphology of fracture surfaces. These and other sample-preparation techniques are described by Scott (1959), Hamm (1960), Bassett (1961), Bradley (1961), Haine (1961), Newman (1962), and Geil (1963).

In optical microscopy, it is common to enhance contrast by staining transparent tissue with dyes which preferentially adsorb on or react with certain structures. Staining is also practiced in electron microscopy, with the obvious requirement that the stain must be a powerful electron scatterer instead of an absorber for visible light (Kassenbeck 1967; Kato 1967*a*,*b*; Keskkula 1967). Heavy-atom compounds, such as bromine or osmium tetroxide, are satisfactory, but attack only sites of unsaturation. Materials known as negative stains are sometimes used with particulate samples; they coat the surrounding areas but leave the specimen uncovered. Phosphotungstic and phosphomolybdic acids are examples. Solvent etching is another useful preparation technique (Riew 1971; Wilska 1971), but care must be taken not to modify the surface of interest.

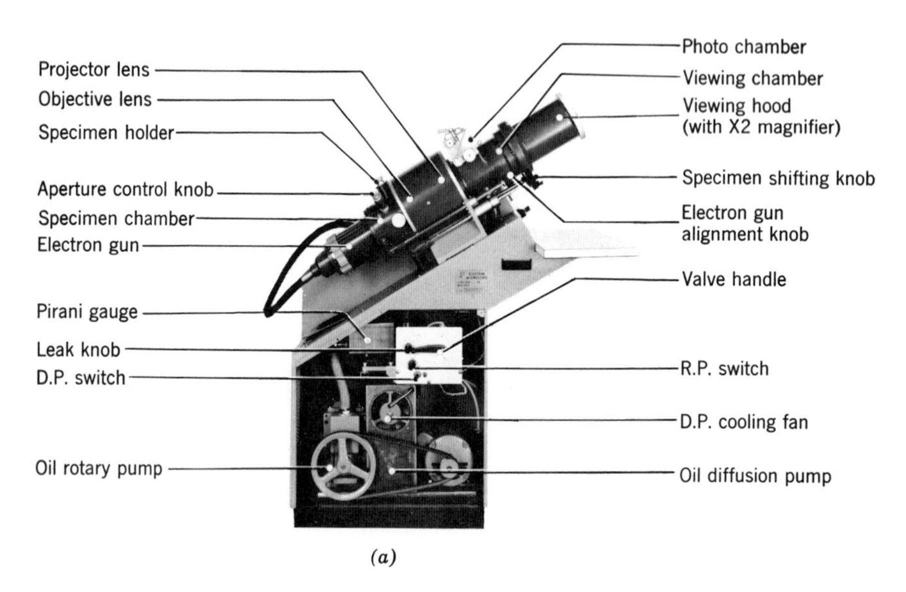

Fig. 8-15. Photograph (*a*) and diagram (*b*) of the JEM 50B electron microscope (courtesy JEOL).

The scattering of electrons by the specimen, essential to the formation of an image in the electron microscope, leads to the generation of heat. Temperatures of the order of 100-200°C can easily be reached, and as high as 300°C has been observed (Hamm 1960, Hall 1966). Exposure to these temperatures may not only melt, but also degrade the polymer. To minimize damage to the sample, it is necessary to operate at high accelerating voltage and low beam current, to cool the sample with a cold stage, and to minimize the time of observation.

The sample is supported on a grid or screen, typically about 200 mesh, made of copper or stainless steel, and covered with a thin (*ca.* 200 A) transparent film or substrate. A collodion film is often used, on which a layer of amorphous carbon has been deposited by evaporating graphite in an arc. The specimen is placed on this substrate, and may be shadowed with a heavy metal such as chromium or platinum, to increase contrast (Fig. 8-16).

Resolving power

The resolving power of any microscope is related to the wavelength of the radiation and the numerical aperture *NA* by the well-known relation

$$\text{resolving power} = \lambda/(2\,NA) \tag{8-2}$$

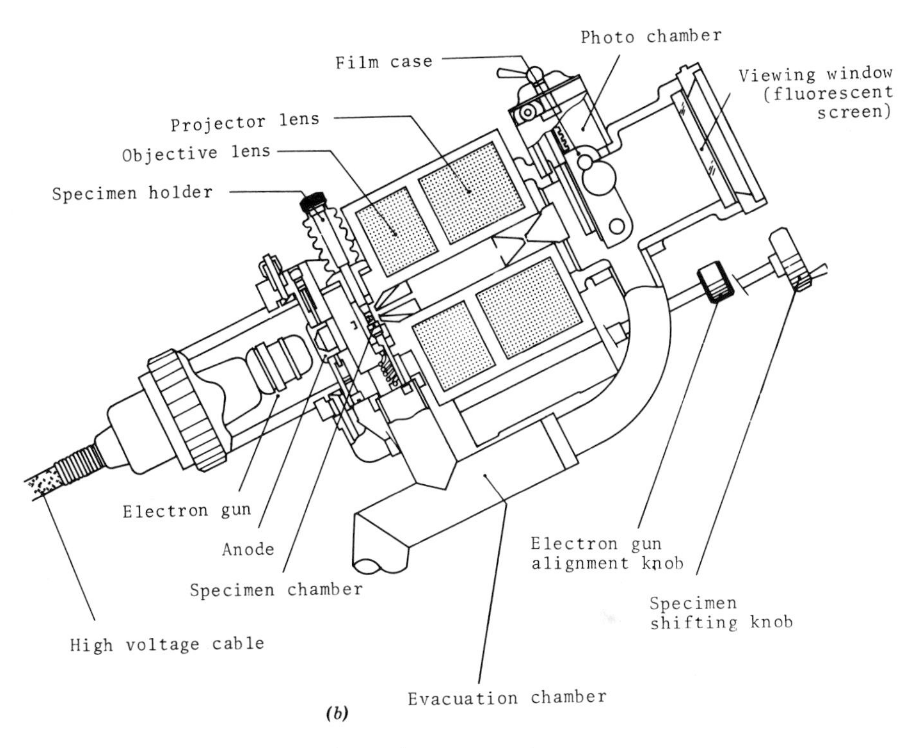

Fig. 8-15 (continued). Part *b*.

where $NA = n \sin(\theta/2)$, n is the refractive index of the medium between specimen and lens (in this case, vacuum, n = 1.000), and θ is the angle of acceptance of the lens. The wavelength of the electron beam is given by de Broglie's relation

$$\lambda = h/mv \tag{8-3}$$

where h is Planck's constant and mv the momentum of the electron. This can be related to the accelerating voltage, and substitution of numerical values yields the approximate equation

$$\lambda \simeq 12/V^{1/2} \tag{8-4}$$

where λ is in A and V in volts. Thus, at $V = 100{,}000$, the electron wavelength is about 0.04 A.

From this consideration alone, one might expect that the electron microscope should have a resolving power of a fraction of an Angstrom unit. It is not possible to achieve high numerical apertures, however, because of lens aberrations which cannot be controlled nearly as effectively as in the light microscope (Cosslett 1968). The best resolution routinely achieved in electron microscopy is of the order of 2-5 A.

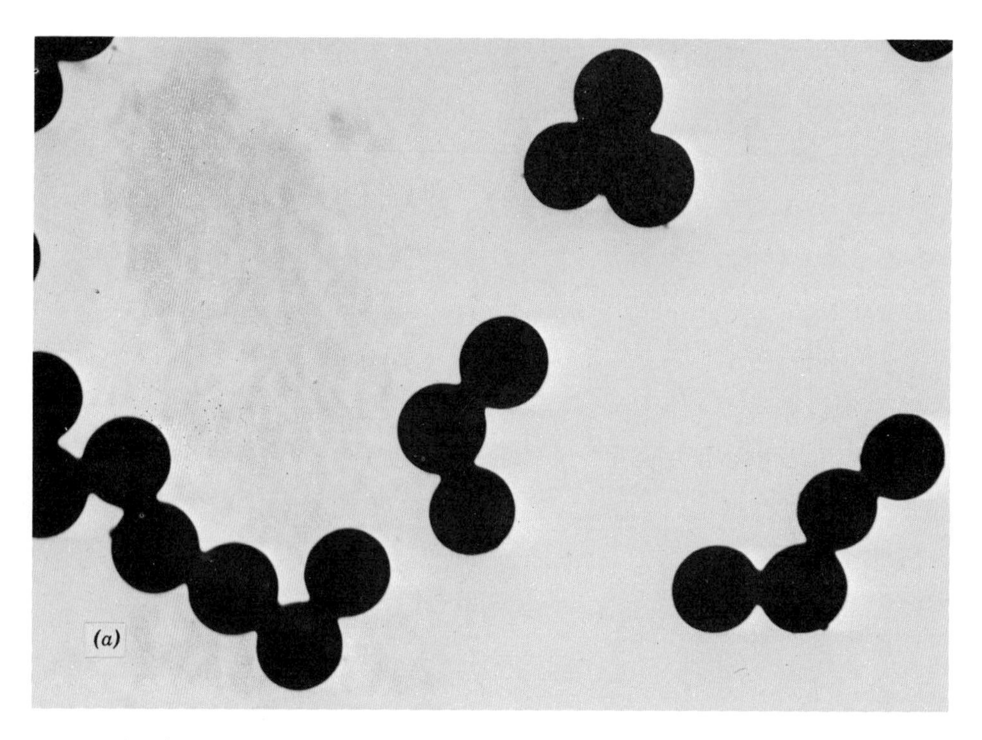

Fig. 8-16. Electron micrographs of (*a*) unshadowed and (*b*) shadowed poly(vinyl chloride) latex particles.

To utilize this resolving power, the image must be magnified so that points 2-5 A apart become approximately 0.1 mm apart, this being the resolving power of the unaided eye. This requires magnifications of 200,000-500,000, and anything beyond this is useless or "empty" magnification, providing no additional detail. The magnification is determined by photographing a diffraction grating with known spacing of the ruled lines, or a monodisperse polystyrene latex of known particle diameter.

Scanning electron microscopy

In the scanning electron microscope (Fig. 8-17), the electron beam is made very fine and scanned across the subject in synchronism with the beam in a cathode-ray tube, much as is done in television. As the beam hits the specimen, secondary electrons are produced by scattering. These are collected and produce a signal used to modulate the intensity of the beam in the cathode-ray tube. The region in the sample from which scattered electrons are gathered is 50-100 A in diameter, and resolution is limited to this value, but the nature of the scanning beam is such as to produce a very great depth of field, much in contrast to that in the light microscope or the transmission electron microscope. Images from the scanning electron microscope have in con-

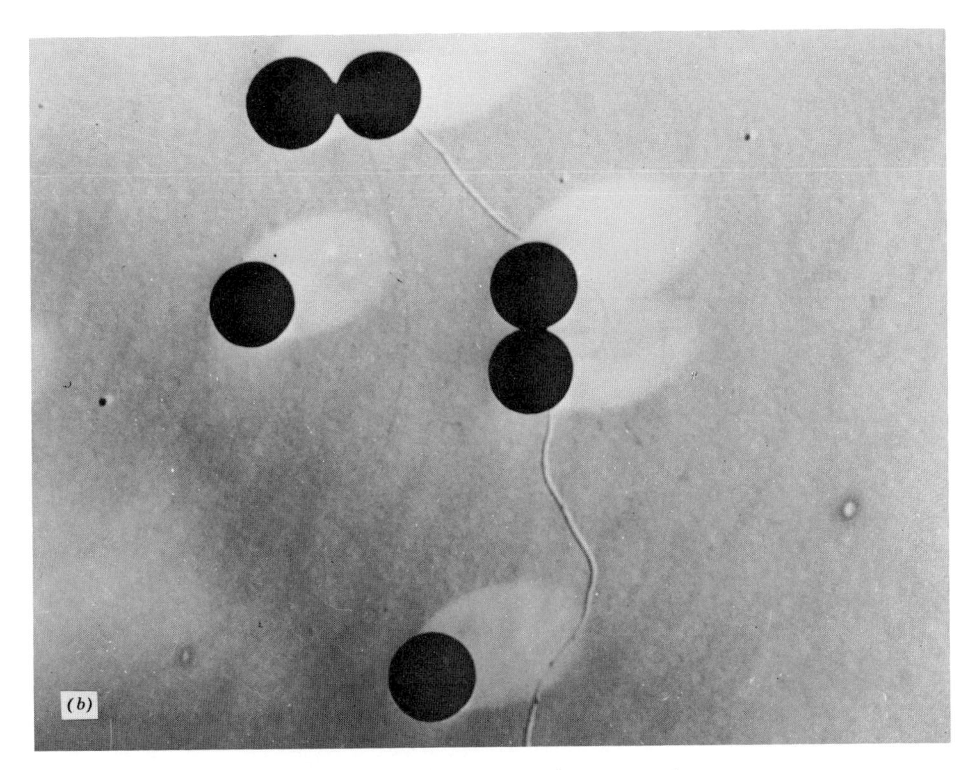

Fig. 8-16 (continued). Part *b*.

sequence a very natural, almost three-dimensional appearance. To date, this relatively new technique has contributed more to the study of biological specimens and of surface phenomena than it has to the elucidation of polymer morphology, but it is a highly useful supplementary and complementary technique to the others discussed in this section (Thornton 1968, Kimoto 1969, Baker 1971).

Electron diffraction

Most modern electron microscopes allow the production of electron-diffraction patterns by a simple readjustment of the lenses. The patterns are much like those from x-ray diffraction but offer some advantages. The short-wavelength electron beam reacts more intensely with the specimen, so that photographic exposure times are shorter (a few seconds instead of an hour or so) and more diffraction maxima can be observed. The shorter wavelength also restricts line broadening to features smaller than 50-100 A.

A typical electron-diffraction pattern from a polymer single crystal is shown in Fig. 8-5*b*. The patterns obtained from polycrystalline material resemble powder patterns and like them can be used to deduce

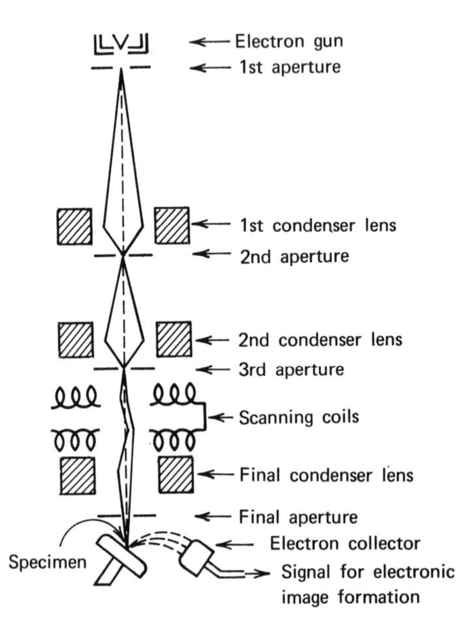

Fig. 8-17. Principle of the scanning electron microscope; compare Fig. 8-14 (Baker 1971).

the lattice spacings of the crystals. The major difficulty with electron diffraction, as with electron microscopy, is the possibility of damage of the specimen by the electron beam (Halliday 1961, Fischer 1964; Hall 1966, Chap. 8).

GENERAL REFERENCES

Textbook, Chap. 4F; Wyckoff 1958; Hamm 1960; Haine 1961; Kay 1961; Wischnitzer 1962; Heidenreich 1964; Geil 1966; Hall 1966; Sjöstrand 1967; Hayat 1970; Swift 1970.

D. *Optical Microscopy*

Obtaining the maximum information about the morphology of polymers requires the use of a variety of techniques of optical as well as electron microscopy. The major aspects of optical microscopy of interest to the polymer scientist (in addition to polarized light microscopy for the observation of spherulites, discussed in Sec. A) are interference and phase-contrast microscopy and the observation of birefringence.

While the electron microscope is an essential tool for the study of polymer morphology, its cost and relative difficulty of manipulation represent real shortcomings which are not characteristic of the light microscope. Light microscopy is often a method of choice if the size of the object is large enough (resolution is at best around 2000 A and useful magnification seldom exceeds 500-1000) and adequate contrast can be obtained. Techniques for growing large polymer single crystals have been developed (Wunderlich 1962, Geil 1963) and lamellae up to 1 mm diameter have been grown. Contrast can be provided by phase-contrast or interference techniques. Difference in optical thickness or path length (physical path length or specimen thickness times refractive index) are accentuated with the phase-contrast microscope, and large areas of constant optical thickness are shown well in the interference microscope. The latter instrument also allows quantitative determination of optical thickness (Sullivan 1964), and the presence of differences in thickness can be strikingly revealed in color.

Interference microscopy

The basis of interferometry is the division of a beam of light into two or more separate but coherent rays which are later combined. In the application of interferometry to polymers in the optical microscope, a polymer crystal is inserted into one of the rays, while the other passes through the medium nearby. The difference in optical thickness due to the crystal gives rise to a phase difference between the rays, and this is converted to an intensity difference by their interference. In this way the crystal is made visible, and in favorable instances its thickness can be determined with high precision.

A schematic diagram of an interference microscope is given in Fig. 8-18. Plane-polarized light passes through the condenser and is divided into two beams by a birefringent plate. One beam passes through the sample and the other alongside it. The two are recombined by a second birefringent plate at the objective, and a quarter-wave plate produces plane polarized light again, but the plane of polarization has now been rotated as described below. The amount of rotation is determined using an adjustable analyzer. If white light is used, differences in optical thickness result in different colors, which change as the analyzer is rotated.

The plane of vibration of the linearly polarized beam emerging from the quarter-wave plate depends on the phase difference δ between the interfering rays, and is measured by the angle α between the optic axes of the quarter-wave plate and the analyzer:

$$2\alpha = \delta = (2\pi/\lambda)(n_2 - n_1)\,\ell \qquad (8\text{-}5)$$

where λ is the wavelength of the light (here monochromatic for convenience), n_1 and n_2 are the refractive indices of the sample and surrounding medium, and ℓ is the specimen thickness. If the refractive indices are known, ℓ can be calculated.

Phase-contrast microscopy

Not unlike the interference microscope, the phase-contrast instrument converts differences in optical thickness to differences in in-

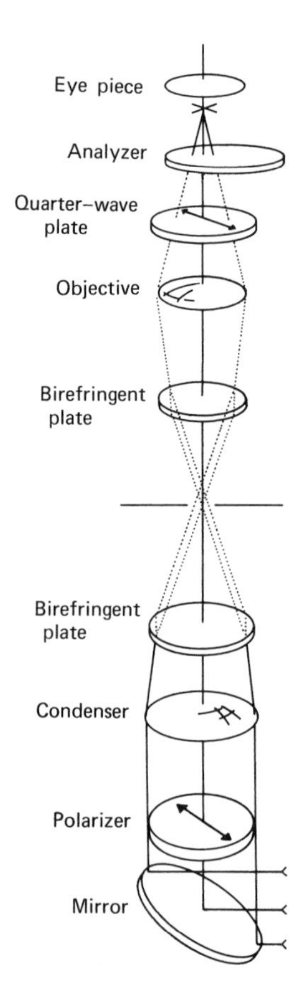

Fig. 8-18. Schematic diagram of an interference microscope (courtesy A. O. Baker).

tensity, thus making visible details of specimens which do not absorb light (the usual source of contrast in the optical microscope). The phase-contrast microscope is more convenient to operate than the interference microscope, but of course lacks the ability of the latter to provide quantitative information on optical thickness. It can, however, be used as a null instrument for the determination of refractive index (Newman 1962): A specimen differing in refractive index by less than 0.002-0.004 from its surrounding medium will not be visible in the phase-contrast microscope.

Birefringence

Birefringence, usually but not necessarily observed in the optical microscope, results when a material is optically anisotropic, that is, when its refractive index depends upon direction. It is observed as the ability of the material to rotate the plane of polarized light and is calculated as the difference in refractive index in two selected perpendicular directions. Completely amorphous polymers without any strain exhibit no birefringence, but both orientation and crystallization lead to its presence (*Textbook*, Chap. 5F). The refractive indices and the birefringence can be measured as described below, and in a variety of other ways discussed by Bauer (1960), Brauer (1962), Forziate (1962), and Newman (1962).

When a crystalline polymer is heated, the birefringence disappears gradually as the crystallites melt. (In contrast to low-molecular-weight substances, crystalline melting in polymers takes place over a range of temperatures, as discussed in Chap. 9B.) The point of disappearance of the last trace of birefringence, usually observed with the hot-stage microscope (Fig. 8-19) using crossed polarizers (Exp. 20), is taken as the crystalline melting point. As it is a measure of the highest melting point of any of the crystallites, it is somewhat higher than the crystalline melting points determined by other techniques. Any birefringence arising from strain or orientation in the amorphous regions disappears at the glass-transition temperature.

Birefringence is a useful technique for measuring orientation (Forziate 1962, Newman 1962, Stein 1964, Clough 1967, Wilkes 1971). It results because the polarizability, and hence the refractive index, of a polymer is different in directions along the chain and perpendicular to it. Birefringence occurs as the molecular chains are aligned in the process of orientation, the magnitude and sign of the effect being determined by the chemical nature of the polymer. Spherulites exhibit birefringence, as discussed in Sec. A, and occasionally the growth of more than one type of spherulite, with different birefringence, is observed (Fig. 8-7).

Orientation

If orientation is considered to occur only in one dimension (the simplest case, and an obvious simplification), birefringence and several related phenomena (infrared dichroism, polarization of fluorescence, etc.) measure the quantity $\overline{\cos^2\theta}$, where the bar indicates averaging and θ is the angle between the molecular chain direction and that of the orienting force, such as the fiber stretch axis. It is convenient to introduce an orientation function

$$f = (3\,\overline{\cos^2\theta} - 1)/2 \qquad (8\text{-}6)$$

which has the value 0 for a completely unoriented or isotropic material for which $\overline{\cos^2\theta} = 1/3$, and 1 for a perfectly oriented fiber, for which $\overline{\cos^2\theta} = 1$. In the case where all the molecular chains lie in the plane perpendicular to the fiber axis, $\theta = \pi/2$, $\overline{\cos^2\theta} = 0$, and $f = -1/2$.

The orientation function is useful in relating the mechanical properties of polymers to their structure (Stein 1964, Maeda 1970, Roe 1970, Samuels 1970).

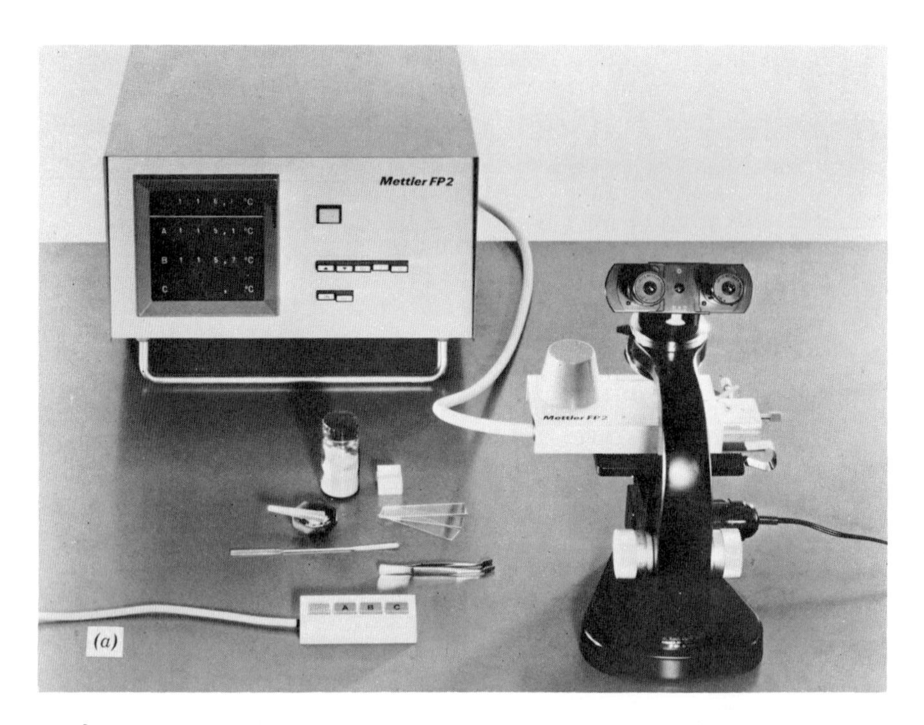

Fig. 8-19. The hot-stage optical microscope (courtesy Mettler): (*a*) overall view; (*b*) detail of stage with cover open.

Measurement of birefringence and orientation

In the simple case of uniaxial orientation discussed above, the birefringence Δ is given as the difference in refractive index n_2 in the direction of the orientation and the average refractive index n_1 in the direction perpendicular to the orientation: $\Delta = n_2 - n_1$. A relation between Δ and the orientation function f (Eq. 8-6) can be derived from the Lorentz-Lorenz equation (Eq. 5-5) (Stein 1964).

The polarizing microscope (Exp. 20) can be used to provide a measure of Δ through the phase shift δ between the two rays, much as was done in the interference microscope, using Eq. 8-5. The phase shift is measured with the compensator described in Fig. 8-20.

GENERAL REFERENCES

Textbook, Chap. 4F; Schaeffer 1953; Chamot 1958; Needham 1958; Richards 1959; Jelley 1960; Setlow 1962, Chap. 12; McCrone 1965; Martin 1966; Ross 1967; Meinecke 1968; Wilchinski 1968; Wahlstrom 1969; Hallimond 1970; Hartshorn 1970; Loveland 1970; Richardson 1971.

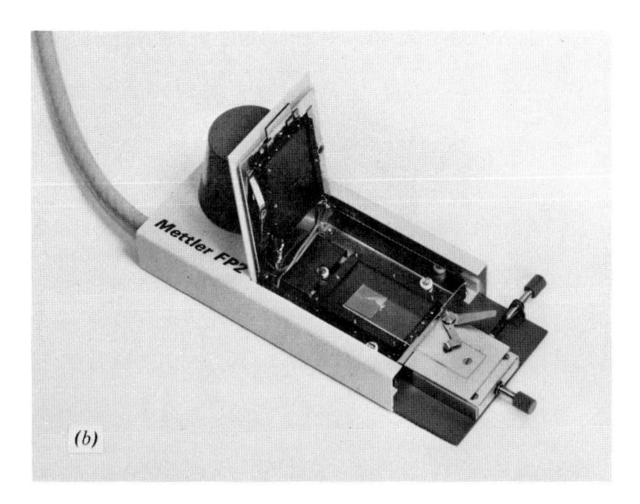

Fig. 8-19 (continued). Part *b*.

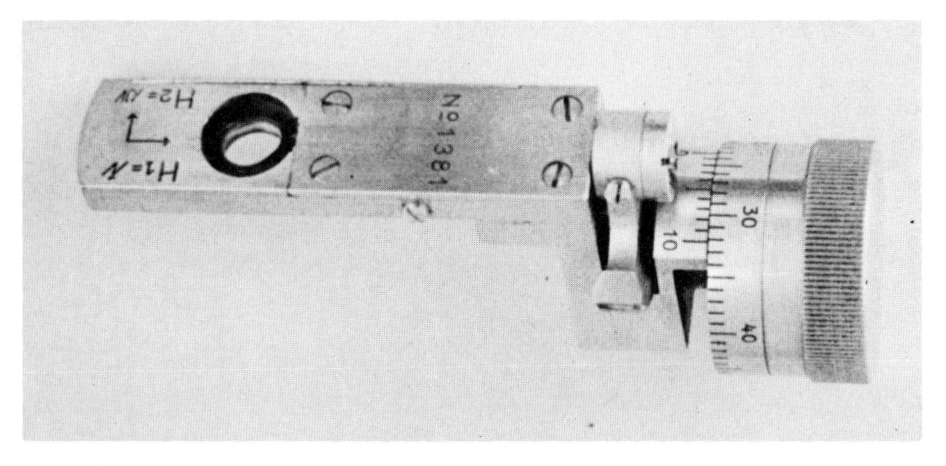

Fig. 8-20. Berek compensator for measuring phase shift in the polarizing microscope.

BIBLIOGRAPHY

Alexander 1969. Leroy E. Alexander, *X-ray Diffraction Methods in Polymer Science*, John Wiley and Sons, New York, 1969.

A. O. Baker. A. O. Baker *Interference Microscope (Manual)*, American Optical Company, Buffalo, New York.

Baker 1971. F. L. Baker and L. H. Princen, "Scanning Electron Microscopy," pp. 498-507 in Herman F. Mark, Norman G. Gaylord, and Norbert M. Bikales, eds., *Encyclopedia of Polymer Science and Technology*, Vol. 15, Interscience Div., John Wiley and Sons, New York, 1971.

Bassett 1961. D. C. Bassett, "Surface Detachment from Polyethylene Crystals," *Phil. Mag. 6*, 1053-1056 (1961).

Bauer 1960. N. Bauer, K. Fajans, and S. Z. Lewin, "Refractometry," pp. 1139-1282 in Arnold Weissberger, ed., *Physical Methods of Organic Chemistry*, Vol. 1 of A. Weissberger, ed., *Technique of Organic Chemistry*, 3rd ed., Part 2, Interscience Div., John Wiley and Sons, New York, 1960.

Beecher 1969. James F. Beecher, L. Marker, R. D. Bradford, and S. L. Aggarwal, "Morphology and Mechanical Behavior of Block Polymers," *J. Polymer Sci. C26*, 117-134 (1969).

Billmeyer 1960. F. W. Billmeyer, Jr., P. H. Geil, and K. R. Van der Weg, "Growth and Observation of Spherulites in Polyethylene: A High Polymer Demonstration," *J. Chem. Educ. 37*, 460-461 (1960).

Blais 1967. J. J. B. P. Blais and R. St. John Manley, "X-ray Long Periods in Bulk Crystalline Polymers," *J. Macromol. Sci.--Phys. B1*, 525-566 (1967).

Bonart 1968. R. Bonart, "X-ray Investigations Concerning the Physical Structure of Cross-linking in Segmented Urethane Elastomers," *J. Macromol. Sci.--Phys. B2*, 115-188 (1968).

Bradley 1961. D. E. Bradley, "Replica and Shadowing Techniques," Chap. 5 in Desmond Kay and V. E. Cosslett, eds., *Techniques for Electron Microscopy*, Blackwell Scientific Publications, Oxford, England, 1961

Brauer 1962. G. M. Brauer and E. Horowitz, "Systematic Procedures," Chap. 1 in Gordon M. Kline, ed., *Analytical Chemistry of Polymers*, Part III, *Identification Procedures and Chemical Analysis*, Interscience Div., John Wiley and Sons, New York, 1962.

Brumberger 1968. H. Brumberger, "Determination of Specific Surfaces by Small-Angle X-ray Scattering Methods--A Brief Review," pp. 76-85 in H. van Olphen and William Parrish, eds., *X-ray and Electron Methods of Analysis*, Plenum Press, New York, 1968.

Chamot 1958. E. M. Chamot and C. W. Mason, *Handbook of Chemical Microscopy*, 3rd ed., John Wiley and Sons, New York, 1958.

Clough 1967. S. Clough, M. B. Rhodes, and R. S. Stein, "The Transmission of Light by Films of Crystalline Polymers," *J. Polymer Sci. C18*, 1-32 (1967).

Cooke 1967. S. L. Cooke, Jr., "X-ray Diffraction in the Chemistry Curriculum," *Norelco Reporter 14* (1), 19-21 (Jan.-Mar., 1967).

Cosslett 1968. V. Ellis Cosslett, "High Voltage Electron Microscopy," *Physics Today 21* (7), 23-31 (July, 1968).

Cottham 1968. Leonard Cottham, "The Characterization of Crystallinity in Addition Polymers," Chap. 8 in Derek A. Smith, ed., *Addition Polymers: Formation and Characterization*, Plenum Press, New York, 1968.

Fischer 1964. E. W. Fischer, "Electron Diffraction," Chap. VII in Bacon Ke, ed., *Newer Methods of Polymer Characterization*, Interscience Div., John Wiley and Sons, New York, 1964.

Forziati 1962. A. F. Forziati, "Optical Methods," Chap. III in Gordon M. Kline, ed., *Analytical Chemistry of Polymers,* Part II, Interscience Div., John Wiley and Sons, New York, 1962.

Geil 1960. P. H. Geil, "Morphology of an Acetal Resin," *J. Polymer Sci. 43,* 65-74 (1960).

Geil 1963. Phillip H. Geil, *Polymer Single Crystals,* Interscience Div., John Wiley and Sons, New York, 1963.

Geil 1964. Phillip H. Geil, Franklin R. Anderson, Bernhard Wunderlich, and Tamio Arakawa, "Morphology of Polyethylene Crystallized from the Melt Under Pressure," *J. Polymer Sci. A2,* 3707-3720 (1964).

Geil 1966. Phillip H. Geil, "Electron Microscopy," pp. 662-669 in Herman F. Mark, Norman G. Gaylord, and Norbert M. Bikales, eds., *Encyclopedia of Polymer Science and Technology*, Vol. 5, Interscience Div., John Wiley and Sons, New York, 1966.

Guinier 1955. André Guinier and Gerard Fournet, *Small Angle Scattering of X-rays,* John Wiley and Sons, New York, 1955.

Guinier 1963. A. Guinier, *X-ray Diffraction in Crystals, Imperfect Crystals, and Amorphous Bodies*, W. H. Freeman and Co., San Francisco, 1963.

Haine 1961. M. E. Haine and V. E. Cosslett, *The Electron Microscope, The Present State of the Art,* E. and F. N. Spon Ltd., London, 1961.

Hall 1966. Cecil E. Hall, *Introduction to Electron Microscopy,* 2nd ed., McGraw-Hill Book Co., New York, 1966.

Halliday 1961. J. S. Halliday and R. Phillips, "Electron Diffraction," Chap. 12 in Desmond Kay and V. E. Cosslett, eds., *Techniques for Electron Microscopy,* Blackwell Scientific Publications, Oxford, England, 1961.

Hallimond 1970. Arthur F. Hallimond, *The Polarizing Microscope,* 3rd ed., Vickers Ltd., York, England, 1970.

Hamm 1960. F. A. Hamm, "Electron Microscopy," pp. 1561-1640 in Arnold Weissberger, ed., *Physical Methods of Organic Chemistry,* Vol. 1 of A. Weissberger, ed., *Technique of Organic Chemistry,* 3rd ed., Part 2, Interscience Div., John Wiley and Sons, New York, 1960.

Hartshorne 1970. N. H. Hartshorne and A. Stuart, *Crystals and the Polarizing Microscope,* 4th ed., American Elsevier, New York, 1970.

Hayat 1970. M. Arif Hayat, *Principles and Techniques of Electron Microscopy, Biological Applications,* Van Nostrand-Reinhold, New York, 1970.

Heidenreich 1964. Robert D. Heidenreich, *Fundamentals of Transmission Electron Microscopy,* Interscience Div., John Wiley and Sons, New York, 1964.

Hoffman 1964. J. D. Hoffman, "Theoretical Aspects of Polymer Crystallization with Chain Folds and Bulk Polymers," *SPE Trans. 4* (4), 1-48 (Oct., 1964).

Hosemann 1962. R. Hosemann and S. N. Bagchi, *Direct Analysis of Diffraction by Matter,* North Holland Publishing Co., Amsterdam, 1962.

Hosemann 1967. R. Hosemann, "Molecular and Supramolecular Paracrystalline Structure of Linear Synthetic High Polymers," *J. Polymer Sci. C20,* 1-17 (1967).

Jeffrey 1971. J. W. Jeffrey, *Methods in X-Ray Crystallography,* Academic Press, New York, 1971.

Jelley 1960. Edwin E. Jelley, "Light Microscopy," pp. 1347-1474 in Arnold Weissberger, ed., *Physical Methods of Organic Chemistry,*

Vol. 1 of A. Weissberger, ed., *Technique of Organic Chemistry*, 3rd ed., Part 2, Interscience Div., John Wiley and Sons, New York, 1960.

JEOL. Japan Electron Optics Laboratory Co., Ltd., Medford, Massachusetts 02155.

Kassenbeck 1967. P. Kassenbeck, "Application of Histochemical Techniques in Electron Microscopy as Applied to High Polymers," *J. Polymer Sci. C20*, 49-57 (1967).

Kato 1967a. Koichi Kato, "The Osmium Tetroxide Procedure for Light and Electron Microscopy of ABS Plastics," *Polymer Eng. Sci. 7*, 38-39 (1967).

Kato 1967b. Koichi Kato, "ABS Mouldings for Electroplating--An Electron Microscopy Study," *Polymer 8*, 33-39 (1967).

Kavesh 1969. S. Kavesh and J. M. Schultz, "Meaning and Measurement of Crystallinity in Polymers: A Review," *Polymer Eng. Sci. 9* (6), 452-460 (1969).

Kay 1961. Desmond Kay and V. E. Cosslett, *Techniques for Electron Microscopy*, Blackwell Scientific Publications, Oxford, England, 1961.

Keller 1957. A. Keller, "A Note on Single Crystals in Polymers: Evidence for a Folded Chain Configuration," *Phil. Mag. 8* (2), 1171-1175 (1957).

Keller 1968. A. Keller, "Polymer Crystals," *Rep. Prog. Physics 31*, 623-704 (1968).

Keskkula 1967. Henno Keskkula and P. A. Taylor, "Microstructure of Some Rubber-Reinforced Polystyrenes," *J. Applied Polymer Sci. 11*, 2361-2372 (1967).

Kimoto 1969. S. Kimoto and J. C. Russ, "The Characteristics and Applications of the Scanning Electron Microscope," *Am. Scientist 57* (1), 112-133 (1969).

Klement 1971. J. J. Klement and P. H. Geil, "Deformation and Annealing Behavior. I Polyethylene Terephthalate Films," *J. Macromol. Sci.--Phys. B5*, 505-534 (1971).

Klement 1972. J. J. Klement and P. H. Geil, "Deformation and Annealing Behavior. III Thin Films of Polycarbonate, Isotactic Polymethylmethacrylate and Isotactic Polystyrene," *J. Macromol. Sci.--Phys. B6*, 31-56 (1972).

Liebhafsky 1964. H. A. Liebhafsky, H. G. Pfeiffer, and E. H. Winslow, "X-ray Methods: Absorption, Diffraction, and Emission," Chap. 60 in I. M. Kolthoff and Philip J. Elving, eds., with the assistance of Ernest B. Sandell, *Treatise on Analytical Chemistry*, Part I, Vol. 5, Interscience Div., John Wiley and Sons, New York, 1964.

Loveland 1970. R. P. Loveland, *Photomicrography*, Interscience Div., John Wiley and Sons, New York, 1970.

Maeda 1970. Matsuo Maeda, Sadao Hibi, Fumikizo Itoh, Shunji Namura, Tatsuro Kawaguchi, and Hiromichi Kawai, "General Description of Mechanical Anisotropy of Semicrystalline Polymers in Relation to Orientation of Structural Units Composed of Crystalline and Noncrystalline Materials," *J. Polymer Sci. A-2 8*, 1303-1322 (1970).

Mandelkern 1964. Leo Mandelkern, *Crystallization of Polymers*, McGraw-Hill Publishing Co., New York, 1964.

Martin 1966. L. C. Martin, *The Theory of the Microscope*, Blackie, London, 1966.

McCrone 1965. Walter C. McCrone and Lucy B. McCrone, "Chemical Microscopy," Chap. 72 in I. M. Kolthoff and Philip J. Elving, eds., with the assistance of Ernest B. Sandell, *Treatise on Analytical Chemistry,* Part I, Vol. 6, Interscience Div., John Wiley and Sons, New York, 1965.
McIntyre 1970. D. McIntyre and E. Campos-Lopez, "Small Angle X-ray Scattering of Tri-Block Polymers," pp. 19-30 in S. L. Aggarwal, ed., *Block Polymers,* Plenum Press, New York, 1970.
Meares 1965. Patrick Meares, *Polymers: Structure and Properties,* D. Van Nostrand Co., Princeton, New Jersey, 1965.
Meinecke 1968. E. Meinecke, "Optical Properties," pp. 525-548 in Herman F. Mark, Norman G. Gaylord, and Norbert M. Bikales, eds., *Encyclopedia of Polymer Science and Technology,* Vol. 9, Interscience Div., John Wiley and Sons, New York, 1968.
Mettler. Mettler Instrument Corp., Hightstown, New Jersey 08520.
Miller 1969. M. L. Miller, *The Structure of Polymers,* Reinhold Publishing Corp., New York, 1966.
Natta 1960. G. Natta and P. Corradini, "Structure and Properties of Isotactic Polypropylene," *Nuovo Cimiento, Supp. 15,* 40-51 (1960).
Needham 1958. G. H. Needham, *Practical Use of the Microscope,* C. Thomas Co., Springfield, Ohio, 1958.
Newman 1962. Sanford B. Newman, "Microscopy," Chap. III in Gordon M. Kline, ed., *Analytical Chemistry of Polymers,* Part III, *Identification and Chemical Analysis,* Interscience Div., John Wiley and Sons, New York, 1962.
Peterlin 1965. A. Peterlin, "Crystalline Character in Polymers," *J. Polymer Sci. C9,* 61-89 (1965).
Peterlin 1967. A. Peterlin, "The Role of Chain Folding in Fibers," pp. 283-340 in H. F. Mark, S. M. Atlas, and E. Cernia, eds., *Man-Made Fibers, Science and Technology,* Interscience Div., John Wiley and Sons, New York, 1967.
Philips. Philips Electronic Instruments, Southfield, Michigan 48075.
Posner 1962. Aaron S. Posner, "X-ray Diffraction," Chap. II in Gordon M. Kline, ed., *Analytical Chemistry of Polymers,* Part II, Interscience Div., John Wiley and Sons, New York, 1962.
Rees 1971. D. V. Rees and D. C. Bassett, "The Texture of Crystalline Polymers: A Brief Review," *J. Material Sci. 6,* 1021-1035 (1971).
Reneker 1960. D. H. Reneker and P. H. Geil, "Morphology of Polymer Single Crystals," *J. Applied Phys. 31,* 1916-1925 (1960).
Richards 1959. Oscar W. Richards, "Measurement with Phase and Interference Microscopes," pp. 6-28 in *Symposium on Microscopy,* ASTM Special Technical Publication No. 257, American Society for Testing and Materials, Philadelphia, Pennsylvania, 1959.
Richardson 1971. James H. Richardson, *Optical Microscopy for the Material Sciences,* Marcel Dekker, New York, 1971.
Riew 1971. C. K. Riew and R. W. Smith, "Modified Osmium Tetroxide Stain for the Microscopy of Rubber-Toughened Resins," *J. Polymer Sci. A-1 9,* 2739-2744 (1971).
Rodriguez 1970. Ferdinand Rodriguez, *Principles of Polymer Systems,* McGraw-Hill Book Co., New York, 1970.
Roe 1970. Ryong-Joon Roe, "Methods of Description of Orientation in Polymers," *J. Polymer Sci. A-2 8,* 1187-1194 (1970).
Ross 1967. K. F. A. Ross, *Phase Contrast and Interference Microscopy for Cell Biologists,* E. Arnold, London, 1967.

Samuels 1970. Robert J. Samuels, "Quantitative Structural Characterization of Mechanical Properties of Isotactic Polypropylene," *J. Macromol. Sci.--Phys. B4,* 701-759 (1970).

Schaeffer 1953. Harold F. Schaeffer, *Microscopy for Chemists,* D. Van Nostrand Co., New York, 1953.

Scott 1959. Robert G. Scott, "The Structure of Synthetic Fibers," pp. 121-131 in *Symposium on Microscopy,* ASTM Special Technical Publication No. 257, American Society for Testing and Materials, Philadelphia, Pennsylvania, 1959.

Setlow 1962. Richard B. Setlow and Ernest C. Pollard, *Molecular Biophysics,* Addison-Wesley, New York, 1962.

Sharples 1966. Allan Sharples, *Introduction to Polymer Crystallization,* Edward Arnold, London, 1966.

Siegmann 1970. A. Siegmann and P. H. Geil, "Crystallization of Polycarbonate from the Glassy State. Part I. Thin Films Cast from Solution,"*J. Macromol. Sci.--Phys. B4,* 239-272 (1970).

Sjöstrand 1967. Fritiof S. Sjöstrand, *Electron Microscopy of Cells and Tissues,* Vol. I, *Instrumentation and Techniques,* Academic Press, New York, 1967.

Statton 1959. W. O. Statton, "The Use of X-ray Diffraction and Scattering in Characterization of Polymer Structure," pp. 242-256 in *International Symposium on Plastics Testing and Standardization,* ASTM Special Technical Publication No. 247, American Society for Testing and Materials, Philadelphia, Pennsylvania, 1959.

Statton 1964. W. O. Statton, "Small-Angle X-ray Studies of Polymers," Chap. VI in Bacon Ke, ed., *Newer Methods of Polymer Characterization,* Interscience Div., John Wiley and Sons, New York, 1964.

Statton 1967. W. O. Statton, "The Meaning of Crystallinity when Judged by X-rays," *J. Polymer Sci. C18,* 33-50 (1967).

Stein 1964. R. S. Stein, "Optical Methods of Characterizing High Polymers," Chap. IV in Bacon Ke, ed., *Newer Methods of Polymer Characterization,* Interscience Div., John Wiley and Sons, New York, 1964.

Sullivan 1964. P. Sullivan and B. Wunderlich, "The Interference Microscopy of Crystalline Linear High Polymers," *SPE Trans. 4* (2), 2-8 (April, 1964).

Swift 1970. H. F. Swift, *Electron Microscopes,* Barnes and Noble, New York, 1970.

Tanford 1961. Charles Tanford, *Physical Chemistry of Macromolecules,* John Wiley and Sons, New York, 1961.

Thielke 1964. H. G. Thielke and F. W. Billmeyer, Jr., "Origins of X-ray Line Broadening in Polyethylene Single Crystals," *J. Polymer Sci. A 2,* 2947-2950 (1964).

Thornton 1968. P. R. Thornton, *Scanning Electron Microscopy,* Chapman and Hall, London, 1968.

Tobolsky 1960. Arthur V. Tobolsky, *Properties and Structure of Polymers,* John Wiley and Sons, New York, 1960.

Wahlstrom 1969. Ernest E. Wahlstrom, *Optical Crystallography,* 4th ed., Interscience Div., John Wiley and Sons, New York, 1969.

Warhus. William H. Warhus, Wilmington, Delaware 19803.

Wilchinski 1968. Zigmond W. Wilchinski, "Orientation," pp. 624-648 in Herman F. Mark, Norman G. Gaylord and Norbert M. Bikales, eds., *Encyclopedia of Polymer Science and Technology,* Vol. 9, Interscience Div., John Wiley and Sons, New York, 1968.

Wilkes 1971. G. L. Wilkes, "The Measurement of Molecular Orientation in Polymeric Solids," *Fortschr. Hochpolym. Forsch. (Advances in Polymer Sci.) 8*, 91-136 (1971).

Williams 1971. David J. Williams, *Polymer Science and Engineering*, Prentice-Hall, Englewood Cliffs, New Jersey, 1971.

Wilska 1971. Seppo Wilska and Vuorikemia Oy, "Solvent Etching: New Techniques for Electron Microscope Studies of Dry Paint Films," *J. Paint Technol. 43* (555), 65-74 (April, 1971).

Wischnitzer 1962. Saul Wischnitzer, *Introduction to Electron Microscopy*, Pergamon Press, New York, 1962.

Wunderlich 1962. B. Wunderlich and P. Sullivan, "Interference Microscopy of Crystalline High Polymers. Determination of the Thickness of Single Crystals," *J. Polymer Sci. 56*, 19-25 (1962).

Wunderlich 1968. Bernhard Wunderlich and Louis Melillo, "Morphology and Growth of Extended Chain Crystals of Polyethylene," *Makromol. Chem. 118*, 250-264 (1968).

Wunderlich 1969. Bernhard Wunderlich, *Crystalline High Polymers: Molecular Structure and Thermodynamics*, American Chemical Society, Washington, D.C., 1969.

Wunderlich 1973. Bernhard Wunderlich, *Crystals of Linear Macromolecules*, Academic Press, New York, in preparation for publication in 1973.

Wyckoff 1958. Ralph W. G. Wyckoff, *The World of the Electron Microscope*, Yale University Press, New Haven, Connecticut, 1958.

Yeh 1967. G. S. Y. Yeh and P. H. Geil, "Crystallization of Polyethylene Terephthalate from the Glassy Amorphous State," *J. Macromol. Sci.--Phys. B1*, 235-249 (1967).

Yeh 1972a. G. S. Y. Yeh, "Order in Amorphous Polystyrenes as Revealed by Electron Diffraction and Diffraction Microscopy," *J. Macromol. Sci.--Phys. B6*, 451-464 (1972).

Yeh 1972b. G. S. Y. Yeh, "A Structure Model for the Amorphous State of Polymers: Folded Chain Fringed Micellar Grain Model," *J. Macromol. Sci.--Phys. B6*, 465-478 (1972).

9

Thermal Behavior

One of the most important properties of a high polymer is its thermal behavior. Knowledge of this behavior is essential not only for the selection of proper processing and fabrication conditions, but also for the full characterization of the material's physical and mechanical properties, and for the selection of appropriate end uses. The temperature-dependent properties of polymers undergo their major changes at one of two transition points: For crystalline polymers, the one of greater importance is the crystalline melting point, while for amorphous polymers, it is the glass-transition temperature.

The importance of the thermal properties of polymers was discussed also in Chap. 6, where emphasis was placed on practical considerations and on simple methods of evaluating thermal behavior. Here, we consider in more detail methods for measuring the important thermal properties, and their dependence upon molecular-structure parameters. The discussion is limited in the main to those properties which are determined in Exps. 21-24, namely the glass-transition temperature, the crystalline melting point, specific heat and enthalpy, thermal conductivity, and thermal degradation.

A. The Glass-Transition Temperature

Although it is exhibited by many low-molecular-weight compounds, the glass transition is usually studied in polymers, and may be unfamiliar to those approaching polymer science for the first time. Perhaps the simplest of the many definitions of the phenomenon is that the glass-transition temperature, T_g, is that temperature below which the polymer is glassy and above which it is rubbery. These descriptions apply strictly only to amorphous polymers, and in the presence of substantial crystallinity, the change of properties at the glass transition (which is practically confined to the amorphous regions) may be obscured. For simplicity, we refer only to amorphous polymers in this section.

The polymer scientist is concerned with the molecular interpretation of T_g as the temperature of the onset of large-scale motion of molecular chain segments. At very low temperatures, near the absolute zero, chain atoms undergo only low-amplitude vibratory motion around fixed positions. As the temperature is raised, both the amplitude and the cooperative nature of these vibrations among neighboring atoms increase, until at a well-defined transition at temperature T_g, segmental motion becomes possible, and the material becomes leathery or rubbery. Above T_g, the chain segments can undergo cooperative rotational, translational, and diffusional motions, and as the temperature is raised sufficiently (to, say, T_g + 100°C) the material behaves like a liquid, of course having very high viscosity.

To obtain a better understanding of the glass transition, consider two well-known polymers, polystyrene and natural rubber (*cis*-polyisoprene). At room temperature, polystyrene is brittle (in the glassy state), while the rubber exhibits typical elastomeric properties. But when rubber is cooled below its T_g (-72°C), as by immersion in liquid nitrogen, it becomes both rigid and quite brittle. Tapping it on a hard surface or hitting it with a hammer shatters the piece by brittle or glassy fracture. On the other hand, when polystyrene is heated above T_g = 100°C, it is no longer brittle but has typical if poor elastomer properties. This simple experiment serves to demonstrate both the nature of the glass transition and its value as a reference temperature for comparing the physical and mechanical behavior of materials (Billmeyer 1969). Another simple demonstration of the glass transition, using glucose pentaacetate, is described by Koleski (1966).

Measurement of T_g

The glass-transition temperature can be measured in a variety of ways, not all of which yield the same value. This results from the kinetic, rather than thermodynamic, nature of the transition; as discussed below, T_g depends upon the heating rate of the experiment and the thermal history of the specimen.

It will be recalled that in a first-order thermodynamic transition, such as a crystalline melting point, there are discontinuities in such properties as heat content and specific volume (first derivatives of the Gibbs free energy), associated with the heat of fusion and the volume change on fusion. In a second-order transition, such changes do not exist, but the second derivatives of the free energy--heat capacity and volume expansion coefficient--change abruptly. These changes are also characteristic of the glass transition and are often used to determine T_g, and in fact in some older books this phenomenon is referred to as a second-order or apparent second-order transition, but it is not a thermodynamic transition because of the rate effects mentioned above.

These rate effects are manifested in the dependence of T_g on the preceding thermal history of the specimen (Fig. 9-1). As the theories to be described suggest, the polymer is not in thermodynamic equilibrium at T_g when cooled at any finite rate. On subsequent heating after rapid cooling, the volume-temperature curve may in fact lose its simple two-straight-line form and become quite complicated. It is usually possible, however, to cool the specimen slowly enough (say, around 0.3°C/min) that a further decrease in cooling rate changes T_g only a negligible amount. Experiments of this sort are easily carried

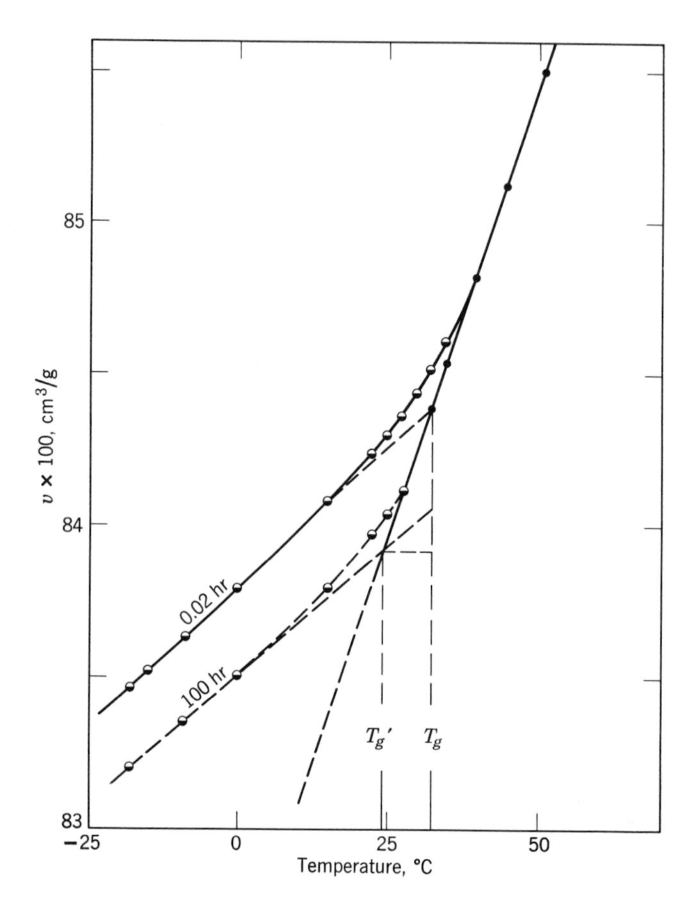

Fig. 9-1. Influence of thermal history on the glass transition. Specific volume of poly(vinyl acetate), measured after cooling quickly from well above T_g: ●, equilibrium values; ⊖, measured 0.02 hr and 100 hr after cooling, as indicated (Ferry 1961).

out in a simple glass dilatometer (Fig. 9-2) (Bekkedahl 1949, Danusso 1959, Bareš 1972).

In addition to measurement of heat capacity, specific volume, or density as a function of temperature, the glass transition can be located by many different tests (Table 9-1). Among these are differential thermal analysis (Sec. C and Exp. 21), the temperature dependence of refractive index or dielectric constant or loss (Exp. 28), dynamic mechanical properties (Exp. 27), impact modulus, infrared or nuclear magnetic absorption spectra, and numerous industrial methods based on softening point (ASTM D 1525), hardening point (ASTM D 1053), heat deflection temperature (ASTM D 648), hardness (ASTM D 785), elastic modulus, or viscosity. Some of these methods are dynamic in the sense that temperature is varied at a constant rate until some change in property such as softening or deflection occurs; these may yield end-

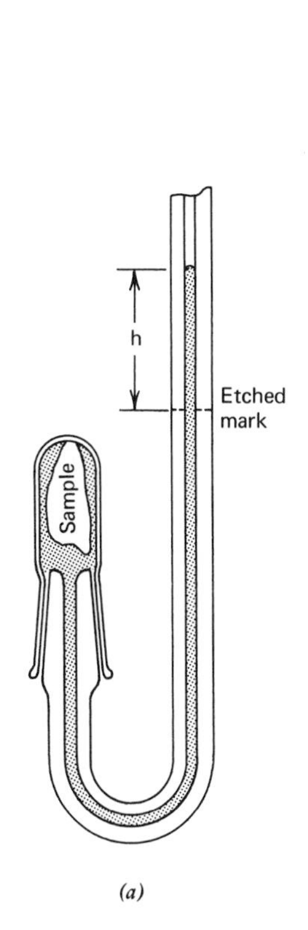

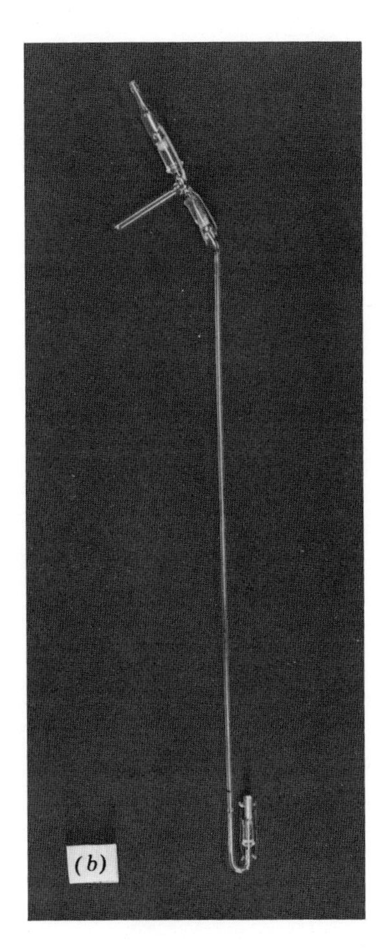

Fig. 9-2. Volume dilatometer with mercury as the confining liquid: (*a*) sketch of the bulb and mercury column; (*b*) photograph showing devices for filling with mercury under vacuum.

point temperatures 10-20°C higher than T_g as determined by other methods.

The glass-transition temperatures of a number of polymers are listed in Table 9-2.

Theories of the glass transition

The combination of some characteristics of thermodynamic transitions with obvious rate effects has led to much confusion regarding the origins of the glass transition, and there are three apparently opposing but not mutually contradictory views in current consideration.

TABLE 9-1. A partial list of tests for the glass-transition temperature

Property varying	Method	Rate of Experiment	Equipment Cost
Tests defined by equilibrium thermodynamics			
Volume, V, or dV/dT	Dilatometry (linear or volume)	Medium	Low
Refractive index	Refractometry	Medium	Medium
--	β-ray absorption	Slow	Medium
Compressibility $(dV/dP)_T$	Dilatometry (volume)	Slow	Medium
Specific heat	Calorimetry (DSC, Sec. C, Exp. 22)	Fast	High
--	Thermal analysis (Sec. C, Exp. 21)	Fast	High
Tests defined by dynamic or transport theories			
Vibrational energy levels	Infrared spectroscopy	Medium	High
Proton environment	Nuclear magnetic resonance	Slow	High
Refractive index	Stress birefringence	Fast	Medium
Dipole moment	Dielectric constant and loss (Exp. 28)	Slow	Medium
Viscosity, elastic modulus	Creep and stress relaxation	Slow	Medium
Mechanical energy absorption	Dynamic mechanical properties (Exp. 27)	Slow	High
Tests related to end-use properties			
Failure mode	Impact resistance (ASTM D 746)	Fast	Medium
Viscosity	Softening point (ASTM D 1525)	Slow	Medium

Property varying	Method	Rate of Experiment	Equipment Cost
Viscosity	Heat deflection (ASTM D 648)	Slow	Medium
Viscosity	Hardness (ASTM D 785)	Slow	Medium

Equilibrium theory In the theory of Gibbs (1956, 1958) and DiMarzio (1958, 1959) (*Textbook*, Chap 6D), it is concluded that the observed glass transition is the result of kinetic manifestations of the approach to a true equilibrium thermodynamic transition. At infinitely long times, a second-order thermodynamic transition could be achieved under equilibrium conditions. The approach to the transition is viewed as the change with temperature of the configurational entropy of the material. As the temperature is lowered, the number of states available to the polymer decreases and the rate of approach to equilibrium also decreases. At equilibrium, the configurational entropy becomes zero at T_g.

Relaxation or hole theory The theory of Hirai (1958, 1959) and Eyring regards the vitrification process as a reaction involving the passage of kinetic units (for example, chain segments in which all the atoms move cooperatively) from one energy state to another. This requires a "hole" or empty space (see the concept of free volume, below) for the unit to move into, and the creation of this hole requires both a hole energy needed to overcome the cohesive forces of the surrounding molecules and an activation energy to overcome the potential barrier associated with the rearrangement, via an activated state. All holes are characterized by a single mean volume. This theory allows a description of the approach to equilibrium. As a rubbery liquid is cooled, the glass-transition temperature, defined as the temperature of half-freezing of the hole equilibrium, depends only on the rate of cooling since there is ample molecular motion in the liquid for equilibrium to be achieved. In contrast, as a glass is heated T_g depends not only on the heating rate but also on the thermal history of the sample since equilibrium cannot be achieved below T_g. Glasses which have been cooled differently represent different starting materials, which have different enthalpies and different time-dependent heat capacities in the glass-transition range (Wunderlich 1964*a*).

Free-volume theory (Kaelble 1969) No discussion of the glass transition can neglect the theory of Fox (1950) and Flory based on a free-volume model. The glass transition occurs when the free or unoccupied volume in the material reaches a constant value, and does not decrease further as the material is cooled below T_g. The remaining or critical free volume is supposed to remain frozen in the glassy state; the residual small volume-expansion coefficient observed below T_g is believed to have the same origin as the thermal-expansion coefficient of a crystalline solid.

The fraction of free volume f can be written

TABLE 9-2. *Glass transition and crystalline melting temperatures of some common polymers*

Polymer	T_g, °C	T_m, °C
Poly(α-methylstyrene)	175	--
Polycarbonate	150	220 (267)
Polymethacrylonitrile	120	--
Poly(acrylic acid)	106	--
Poly(methyl methacrylate) (syndiotactic)	105	>200
Polyacrylonitrile	104	317
Polystyrene	100	--
Poly(vinyl chloride)	83	220-240
Poly(ethylene terephthalate)	69	267
Poly(ethyl methacrylate)	65	--
Polycaproamide (6 nylon)	50	225 (215)
Poly(hexamethylene adipamide) (66 nylon)	50	265
Poly(ω-aminoundecanoic acid) (11 nylon)	46	194
Poly(methyl methacrylate) (isotactic)	45	160
Poly(hexamethylene sebacamide) (610 nylon)	40	227
Poly(vinyl acetate)	29	--
Poly(*n*-butyl methacrylate)	22	--
Poly(methyl acrylate)	9	--
Poly(vinylidene chloride)	-17	190-198
Polypropylene	-19	176
Poly(ethyl acrylate)	-22	--

Polymer	T_g, °C	T_m, °C
Poly(butyl acrylate)	-55	--
Polyoxymethylene	-68	180-200
Polyisoprene (*cis*)	-73	28 (36)
Polyethylene (branched)	-80	105-125
Polyethylene (linear)	-80	137
Polybutadiene (syndiotactic)	-85	154
Poly(dimethyl siloxane)	-123	--

$$f = V_f/(V_f + V_0) = 0.025 + 0.00048\ (T-T_g) \qquad (9\text{-}1)$$

(where V_f is free volume and V_0 is occupied volume) according to the "iso-free volume" hypothesis (Fox 1950). The critical free volume is evaluated as 0.025 by the WLF equation (Williams 1955; see the *Textbook*, Chap. 6D); it seems to be roughly a constant for the majority of polymers. The relation of this increment in volume to the volume-expansion coefficients above and below T_g and the location of the pseudo-equilibrium value of T_g at infinite time, T_∞, are shown in Fig. 9-3. The figure also illustrates an alternative approach by Simha (1962) and Boyer, in which the critical free volume is measured by extrapolating the liquid line to the absolute zero of temperature, and is about 0.113 at T_g.

Theories of the glass transition are reviewed in an excellent article by Eisenberg (1970).

Effect of molecular structure on T_g

The basis for considering the effects of molecular-structure variables on the glass transition is the onset, as T_g is approached, of cooperative motion of groups of atoms larger than, say, the monomer. As the temperature increases, these groups become larger until above T_g the entire polymer coil is the elastic unit, as required by the kinetic theory of rubber elasticity.

Thus, any molecular parameter affecting chain mobility can be expected to influence T_g. Among the variables to be considered are the chain microstructure, including chemical type of monomer units, copolymerization effects, and tacticity; molecular architecture variables such as molecular weight and distribution, branching, and crosslinking; and the presence of low-molecular-weight compounds such as plasticizers and diluents.

As discussed also in the *Textbook*, Chap. 7B, the major parameters influencing T_g are chain stiffness, bulkiness and flexibility of side

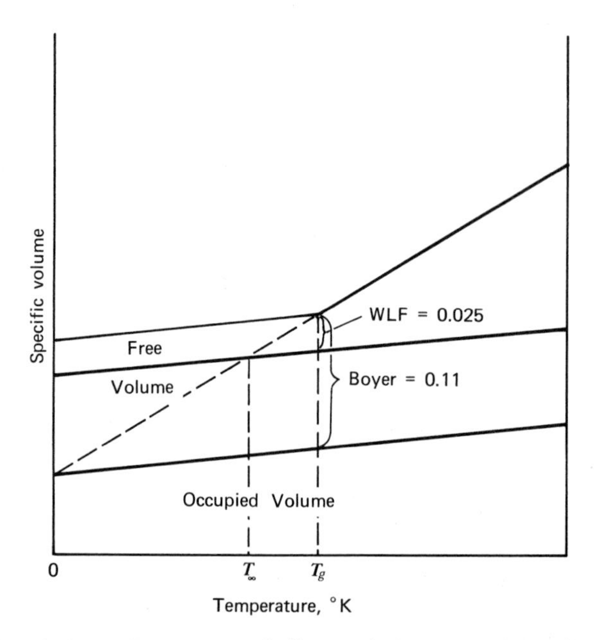

Fig. 9-3. Specific volume around T_g as interpreted by the free-volume theory. See the text for discussion.

chains, and the polarity of the chain, expressed by the cohesive-energy density or solubility parameter (Chap. 6B). For example, atactic polypropylene has a solubility parameter $\delta_2 \simeq 7$-8 and $T_g = -19°C$, whereas polyacrylonitrile has $\delta_2 = 15.4$ and $T_g = 104°C$. The influence of polar substituents is to increase chain cohesion and T_g; compare polybutadiene, $T_g = -85°C$, and poly(vinyl chloride), $T_g = 83°C$. The addition of rigid, bulky side groups decreases the flexibility of the chain and increases T_g; compare polystyrene, 100°C, poly(α-methylstyrene), 175°C, and polyacenaphthalene, where chain mobility is severely hindered, 264°C. An increase in chain symmetry, however, lowers T_g; compare poly(vinyl chloride), 83°C, and poly-(vinylidene chloride), -17°C.

The effect of substituent side chains is illustrated by comparing values of T_g for the lower alkyl acrylates and methacrylates (Table 9-3). Since the side chains are fairly flexible, they have little effect within each series on steric hindrance to rotation, but they force the main chains farther apart, increasing the free volume and reducing T_g. Adding the methyl group in going from the acrylates to the methacrylates does increase the steric hindrance, however, and T_g increases correspondingly. Quantitative treatment of these influences by the free-volume theory has shown reasonable agreement with experiment.

The influences on the glass transition of molecular weight, branching, crosslinking, copolymerization and diluents or plasticizers can also be accounted for by the free-volume theory. Quantitative relations can be derived with the assumptions that chain ends contribute more free volume to the system than do similar segments along the chain, that the volumes of the components are additive, and that both components contribute to the thermal expansion of the polymer by their glassy-state expansion coefficients below T_g and their rubbery-state coefficients above T_g. For example, the influence of molecular weight can be expressed as

$$T_g = T_{g\infty} - K/\overline{M}_n \tag{9-2}$$

where $T_{g\infty}$ is the glass transition temperature for polymer of infinite molecular weight and K is a constant, both being characteristic of a given polymer type. For polystyrene (Fox 1954), $T_{g\infty}$ = 100°C and K = 10^5.

TABLE 9-3. Glass-transition temperatures of some acrylate and methacrylate polymers (Billmeyer 1969)

Acrylate	T_g, °C	*Methacrylate*	T_g, °C
Methyl	-9	Methyl	105
Ethyl	-22	Ethyl	65
n-Butyl	-56	*n*-Butyl	22
2-Ethyl hexyl	-70	*n*-Hexyl	-5

Amorphous random copolymers exhibit a single glass-transition temperature which lies between the values of T_g for the two homopolymers. The best-known formula (Gordon 1952) among several for predicting this glass temperature is

$$T_g = (w_1 T_{g1} + Kw_2 T_{g2})/(w_1 + Kw_2) \tag{9-3}$$

where T_{g_1} and T_{g_2} refer to the two homopolymers present in weight fractions w_1 and w_2 (where $w_1 + w_2 = 1$), and K is given by

$$K = (\alpha_r - \alpha_g)_2/(\alpha_r - \alpha_g)_1 \tag{9-4}$$

where the α's are the thermal-expansion coefficients of the homopolymers, the subscript r referring to the rubbery state and g to the glassy state. It is often better to determine K experimentally, but the equation fits experimental data very well (Illers 1963). In contrast to this behavior, block and graft copolymers exhibit two glass transitions if the blocks are long enough to cause phase separation.

Equation 9-3 is also valid for polymer-diluent systems; since the value of T_g for a low-molecular-weight diluent or plasticizer is very low, T_g is always decreased.

In accord with the predictions of the free-volume theory, branching increases free volume and decreases T_g, whereas crosslinking decreases free volume and increases T_g. The extent to which tacticity influences T_g varies with the type of polymer. The lower value of T_g for isotactic poly(methyl methacrylate) compared to the syndiotactic polymer is explained in terms of easier rotation about chain bonds in the isotactic structure. Similarly, increasing the syndiotacticity of poly-(vinyl chloride) raises T_g substantially.

In crystalline polymers, the crystallites tend to reinforce or stiffen the structure, increasing T_g proportional to the amount of crystallinity. There is also a well-known but approximate proportionality between T_g and the crystalline melting point T_m, as exhibited in Fig. 9-4. This is not unexpected, since the same cohesive energy factors must exist in both the crystalline and amorphous regions, and influence the two transitions in similar ways.

The effects of structural variables on T_g are summarized in Table 9-4.

TABLE 9-4. Summary of structural factors affecting T_g

Factors favoring increase in T_g	Factors favoring decrease in T_g
Main-chain rigidity	Main-chain flexibility
Increased polarity	Increased symmetry
Bulky or rigid side chains	Flexible side chains
Increased molecular weight	Addition of diluents or plasticizers
Increased cohesive-energy density	Increased tacticity
Crosslinking	Branching

GENERAL REFERENCES

Textbook, Chaps. 6D and 7B; Kauzmann 1948; Bueche 1962, Chaps. 4, 5; Tobolsky 1962, Chap. 2; Boyer 1963; Gordon 1965; Meares 1965, Chap. 10; Miller 1966; Shen 1966; Eisenberg 1970; Ferry 1971, Chap. 11; Williams 1971, Chap. 7.

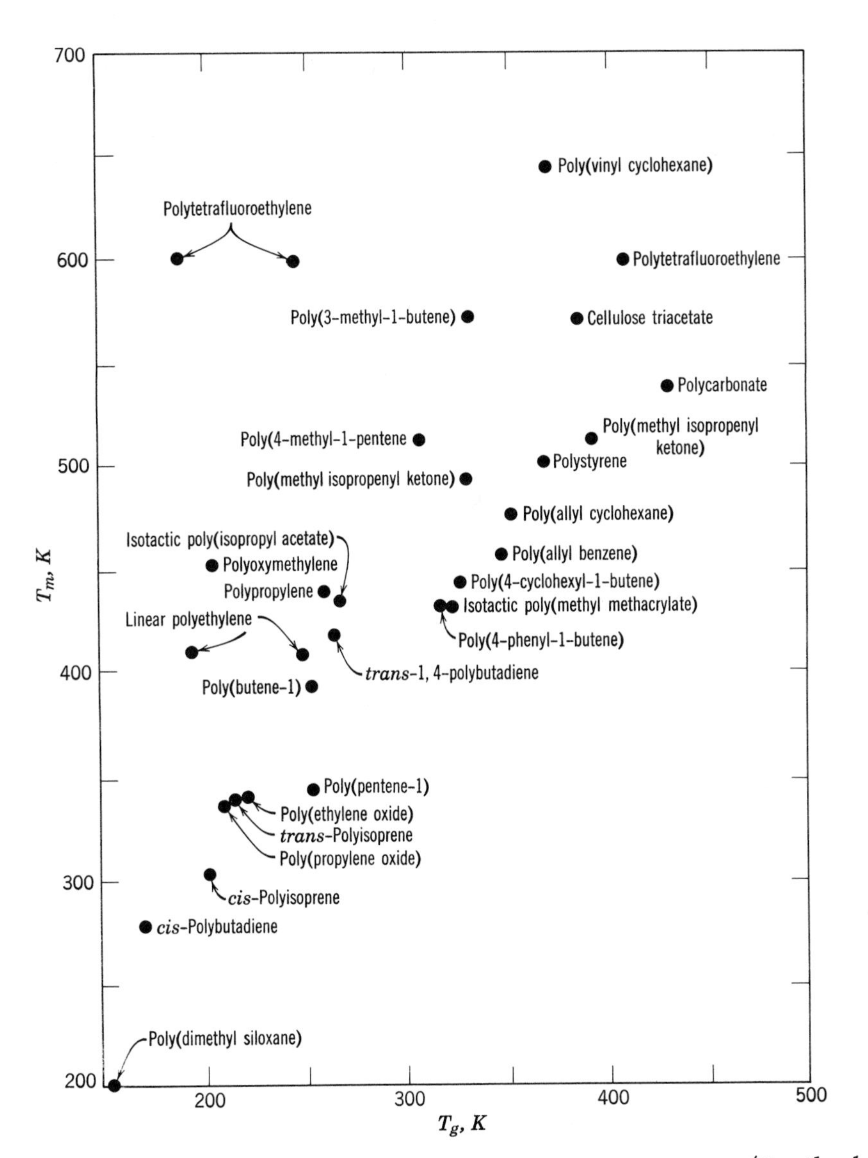

Fig. 9-4. Relation between T_m and T_g for various polymers (*Textbook*, Fig. 7-4, from Boyer 1963).

B. *The Crystalline Melting Point*

For crystalline polymers, the melting point constitutes the most important thermal transition. The property changes at T_m are far more drastic than those at T_g, particularly if the polymer is highly crystalline. These changes are characteristic of a thermodynamic first-order transition, and include a heat of fusion and discontinuous changes in heat capacity, volume or density, refractive index and birefringence, and transparency. Any of these can be used to detect T_m; we discuss in this chapter only methods based on thermal measurements. Since the crystalline melting point was discussed extensively in connection with morphology in Chap. 8, the treatment here is somewhat abbreviated; see also the *Textbook*, Chap. 5E.

The appearance and disappearance of crystallinity in polymers shows behavior different from that in low-molecular-weight substances in some respects. Crystalline melting occurs, for example, over a rather broad temperature range (Fig. 9-5), since the exact melting point of each crystalline region depends upon both its size and its perfection, larger and more perfect crystallites having higher melting points. The value of T_m is usually taken to be the point of melting of the highest melting crystallites, that is, the point of disappearance of the last traces of crystallinity. This definition is adhered to, for example, in the determination of T_m by optical microscopy described in Chap. 8D.

Since crystal perfection and crystallite size are influenced by the rate of crystallization, T_m depends to some extent upon the thermal history of the specimen, and it is of interest to examine the crystallization process briefly. The rate of polymer crystallization increases as the temperature is dropped below T_m (more properly, the melting point of the largest and most perfect crystals that can be

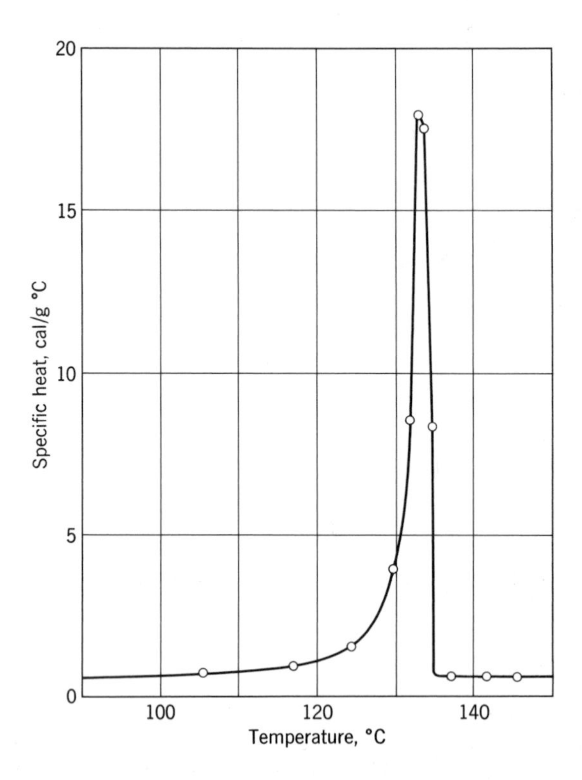

Fig. 9-5. Specific heat as a function of temperature over the crystalline melting range in linear polyethylene (*Textbook*, Fig. 5-21, from Wunderlich 1957).

formed by slow crystallization or annealing, that is, the equilibrium melting point). At some temperature a maximum rate will occur, which may or may not be realizable experimentally; at lower temperatures, the rate again decreases and reaches zero at T_g. This behavior is shown in Fig. 9-6 for natural rubber, one of several common polymers for which the entire curve is experimentally accessible. In cases of this sort, it is also possible to quench the polymer in the amorphous state to below T_g; crystallization does not take place until the temperature is subsequently raised above that point. If the temperature is increased continuously, crystals will first form and then melt. The thermal effects accompanying these changes are shown for poly(ethylene terephthalate) in Fig. 9-7. For many other polymers, notably polyethylene and polypropylene, the entire crystallization curve cannot be observed, the rate of crystallization being so great that the material cannot be quenched to the amorphous state. The dependence of the transition temperatures on rates of cooling and heating have been discussed by Bekkedahl (1941), Wood (1946), Hellmuth (1965), Collins (1966), and Jaffe (1967, 1969).

Values of T_m for some common polymers are given in Table 9-2.

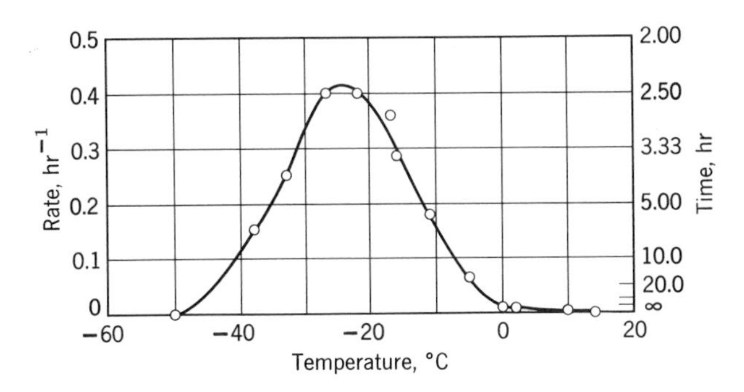

Fig. 9-6. Rate of crystallization of natural rubber as a function of temperature (*Textbook*, Fig. 5-18, from Wood 1946).

GENERAL REFERENCES

Textbook, Chap. 5; Gordon 1965; Meares 1965, Chaps. 2 and 4; Rodriguez 1970, Chap. 3; Williams 1971, Chap. 6.

C. *Thermal Analysis and Calorimetry*

Differential thermal analysis

Differential thermal analysis (DTA) is a technique for detecting the thermal effects accompanying physical or chemical changes in a sample as its temperature is varied through a region of transition or reaction, by means of programmed heating or cooling. The instrumentation involved is described by means of the following figures: Figure 9-8 shows the widely used Du Pont DTA instrument, while Fig. 9-9 gives its basic components in block diagram. The cell is sketched in detail

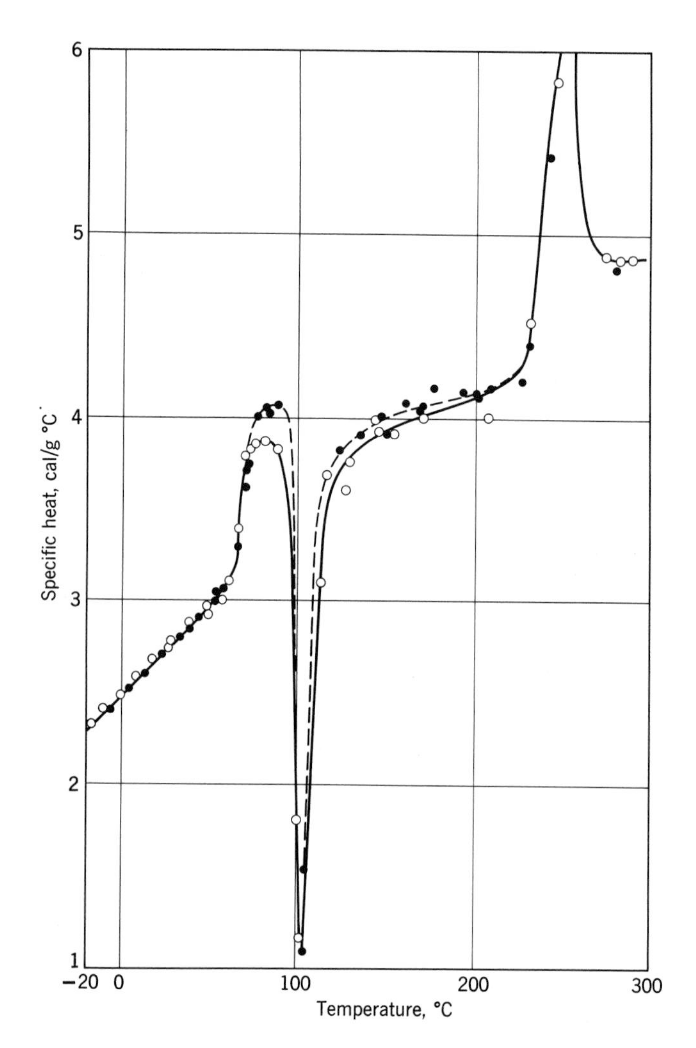

Fig. 9-7. Specific heat as a function of increasing temperature for quenched (amorphous) poly(ethylene terephthalate) (*Textbook*, Fig. 4-6, from Smith 1956).

in Fig. 9-10. Its essential features include the sample S, a reference material R, and a chamber contained in a heating block with electrical heating element H. Thermocouples placed in the centers of the sample and reference measure their temperature and also the temperature difference between them. This temperature difference depends in a complicated way on the density, thermal conductivity, specific heat, and thermal diffusivity of sample and reference, and it is often not possible to separate the effects of these individual factors from one another and from the influence of the geometry of the system.

Typical DTA experiment In a typical DTA experiment, the temperature of the heating block is programmed to increase linearly with time. The reference material is selected to be thermally inert except for slight changes in heat capacity with temperature; such materials as fused quartz, porcelain, glass beads, or MgO are used. At the start of the experiment, the block, sample, and reference are all at the same temperature. As the block temperature rises, the temperatures registered by the thermocouples at the centers of sample and reference rise more slowly due to the finite heat capacities of these materials. Soon a steady state is reached, however, in which the temperatures all rise uniformly with time, as indicated in Fig. 9-11*a*, and the temperature difference between sample and reference is zero, as indicated in Fig. 9-11*b*.

As the temperature reaches the glass-transition temperature of the sample, its heat capacity changes abruptly (but as seen in Sec. A its enthalpy does not change). The sample now absorbs more heat because of its higher heat capacity, and its temperature lag behind the reference is increased. Its temperature therefore changes more slowly than that of the reference, with the results indicated in the figure.

If the polymer sample is crystallizable but in the quenched amorphous state, crystallization can now take place. The heat of fusion is evolved, the sample temperature rises, and the DTA trace (Fig. 9-11*b*) shows an exothermic peak at T_c, the temperature at which the rate of crystallization is a maximum. On further heating the crystallites melt, heat is absorbed, and an endothermic peak occurs at T_m, here defined as the temperature at which the rate of melting is highest; note that this is probably lower than T_m as defined in Sec. B.

Finally, other changes in the sample may occur. Oxidation or crosslinking (at T_{ox}) are exothermic, while decomposition (at T_d) is endothermic. Of course, not all samples exhibit all of the aforementioned phenomena. Some typical DTA traces for common polymers are shown in Fig. 9-12.

Factors influencing DTA To obtain accurate results by DTA, it is essential to achieve uniform temperature throughout sample and reference and to operate under steady-state conditions. Ideally, sample and surroundings should be in temperature equilibrium at all times, and though this obviously cannot be achieved, it is necessary to work toward this objective to avoid deviations in the base line of temperature difference and asymmetry of the peaks. Some of the factors leading to such aberrations are mismatch in heat capacities between sample and reference, poor heat transfer, poor sample packing or improper particle size, lack of symmetry in the geometry, and effects arising from the presence of diluents. These factors have their origin in several parameters of the experiment which are discussed in the following paragraphs.

The *sample holder assembly* or heating block must be made so that uniform heat transfer to both sample and reference takes place, without any fluctuations that might be mistaken for transitions in the sample. To obtain a reproducible heat flux which is relatively little affected by changes in the sample properties, a steep temperature gradient must be set up between the block and the sample; this is favored by the use of small samples and by keeping the sample and reference close together. Considerations of resolution and sensitivity cannot be neglected,

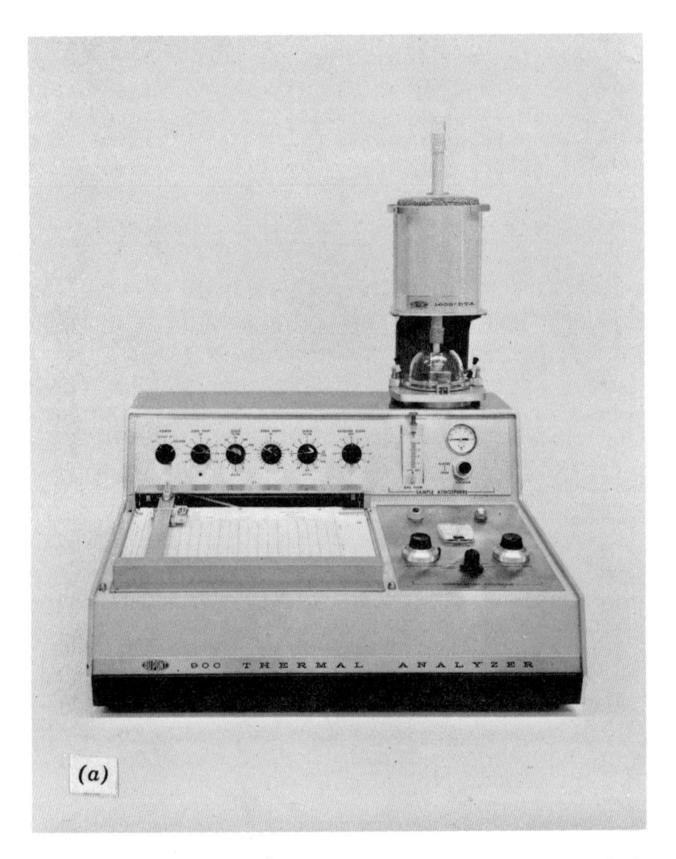

Fig. 9-8. The Du Pont DTA apparatus: (*a*) Model 900; (*b*) Model 990, showing DTA cell at the left (courtesy Du Pont).

however, and the solution is often not simple. A comprehensive review of the design of heating blocks and sample holders is found in Wunderlich (1971).

The selection of *amount of sample and diluents* is both important and complex. Diluents in this context are inert materials (often identical to the reference) mixed with the sample to match its thermal conductivity and diffusivity to that of the reference. In general, as indicated above, small sample sizes and small amounts of diluent are preferred. A compromise must be made, however, since sensitivity increases with sample size but resolution decreases. In practice, samples of 25-50 mg are satisfactory when measuring large heats of transition but too small for the observation of glass transitions. Melting peaks can be sharpened by the use of smaller samples and lower heating rates. Peaks due to chemical reaction can be sharpened by increasing the heating rate (Berkelhamer 1945). Maximum sensitivity and resolution are obtained by minimizing the sample size and increasing the efficiency of heat transfer from the sample to its thermocouple. This can be accomplished by careful design of the sample holder and thermocouple and location of the thermocouple.

Fig. 9-8 (continued). Part *b*.

In addition to amount of sample, the *size and packing of the sample particles* also influence the temperature measurement (Schwenker 1964). Too small particle size increases the surface area and shifts the transition peaks to lower temperatures. In addition, small particles often undergo reorganization, causing thermocouple movement leading to artifacts in the temperature measured (Smothers 1966). Since thermal conductivity and diffusivity are influenced by sample density, tighter packing increases heat conduction, reduces the chance of peak spreading, and improves reproducibility. Packing is especially important in cases where gaseous products are evolved or where samples are studied under a gaseous atmosphere.

The *reference material* should be chosen to give as small a temperature difference as possible between sample and reference by balancing their heat capacities. Though commercial equipment is designed to permit electronic compensation of a constant base line slope, resulting from mismatch in heat capacities, it is advantageous to reduce the mismatch as much as possible by judicious selection of reference material.

Ideally, the *selection of the thermocouple providing the system temperature reading* should make no difference, but in practice there are some significant differences. If the block thermocouple is used, the slight lag between sample and block, which changes magnitude at a sample transition, can cause a shift in observed peak temperature which depends upon heating rate (Wunderlich 1964*b*). Use of the reference thermocouple does not solve the problem since a similar shift can occur. This would disappear if the sample thermocouple were used, but as noted before the position and possible movement of this thermocouple are critical. Additionally, the sample thermocouple characteristics

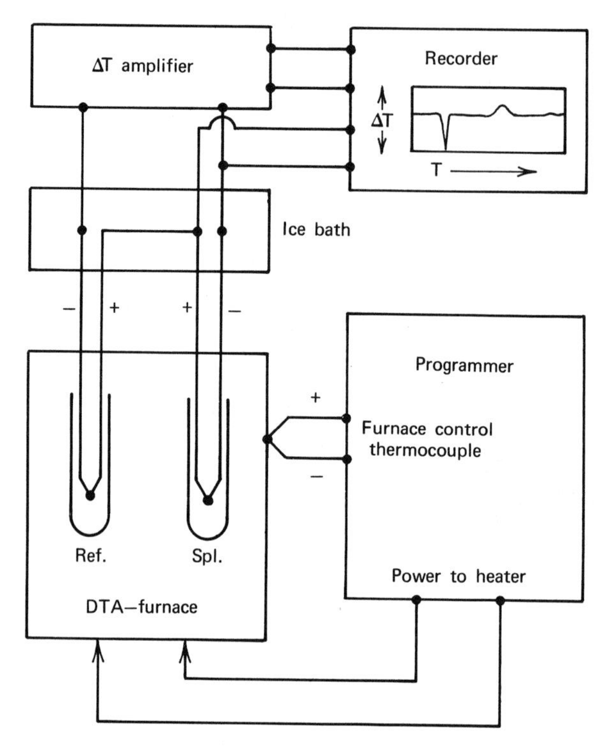

Fig. 9-9. Sketch of the essential components of a DTA apparatus (Wunderlich 1971).

may change as a result of reactions with the sample. These and other factors are considered in detail by Smyth (1951), Barrall (1962), and David (1966).

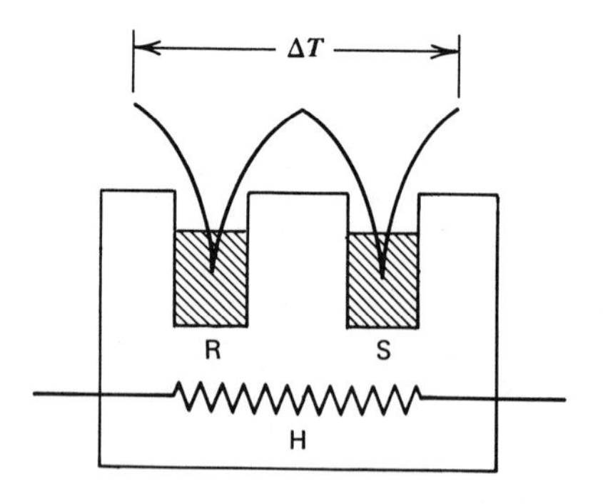

Fig. 9-10. Sketch of a DTA cell.

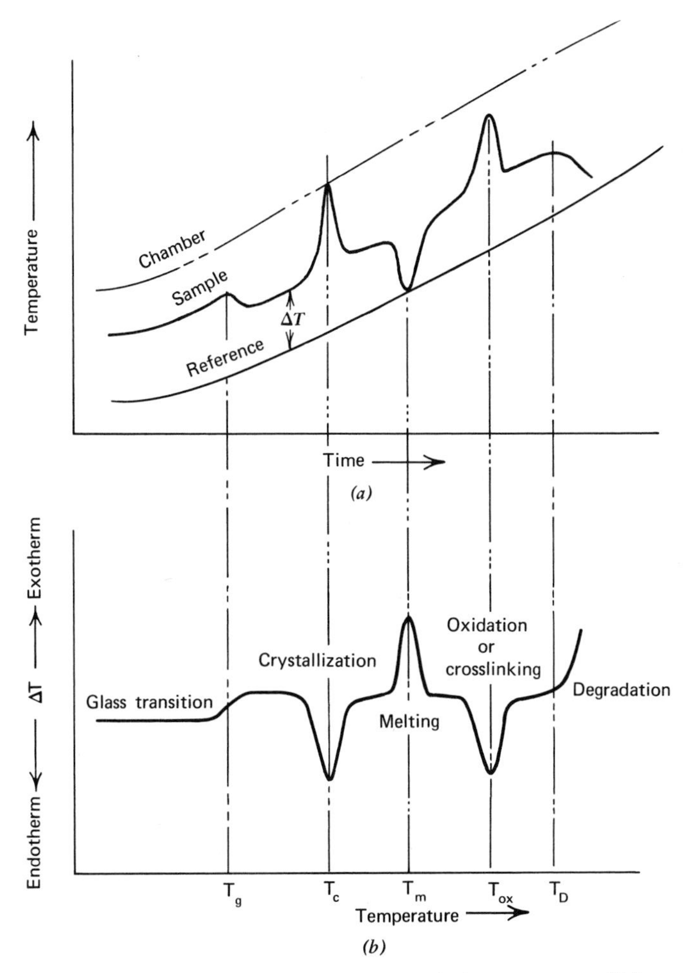

Fig. 9-11. Dependence of DTA signals on (*a*) time and (*b*) temperature.

It is obvious from the foregoing that the *heating rate* is an important parameter of the experiment, and is closely interrelated with sample size, thermocouple placement, and other variables. To avoid shifts in peak temperatures with heating rate, the sample thermocouple should provide the temperature base, otherwise T_g tends to increase slightly with heating rate, and T_c and T_d significantly. T_m is little affected in this respect, but of course lower heating rates measure values of T_m closer to the equilibrium value, as the result of annealing the sample during the DTA experiment. Unless the heating rate is linear, slight shifts in base line can occur which may be mistaken for sample transitions.

Various types of DTA phenomena are summarized in Table 9-5, while factors of importance influencing the several types of determination by DTA are listed in Tables 9-6 through 9-8.

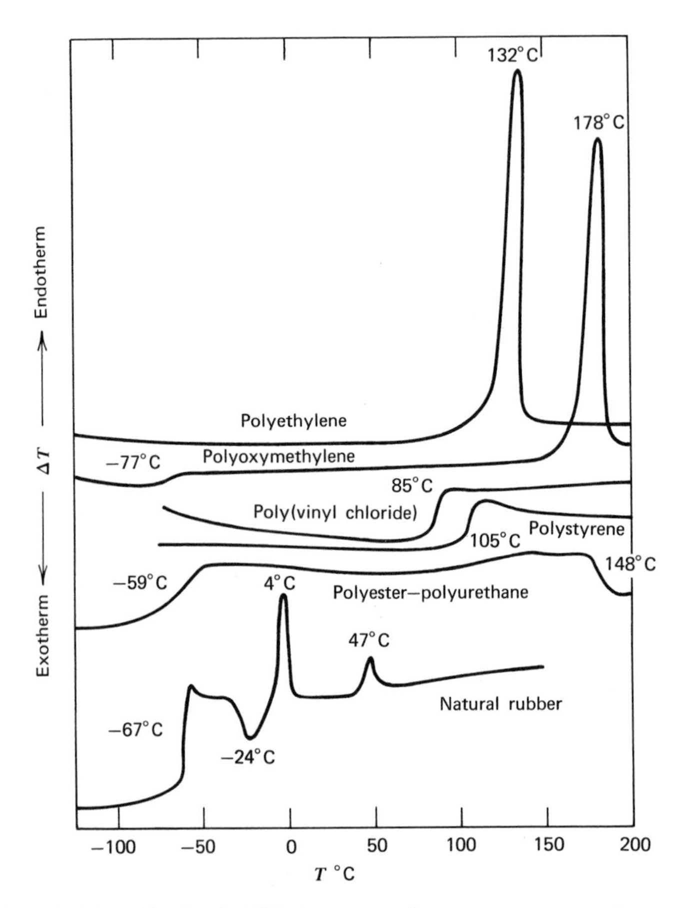

Fig. 9-12. Typical DTA traces of some common polymers.

Differential scanning calorimetry

The basic differences between DTA and differential scanning calorimetry (DSC) lie in the design of the heating system and the mode of operation of the instrument. In DSC, the sample and reference are heated separately by individually controlled elements, as the sketch of Fig. 9-13 shows. The power to these heaters is adjusted continuously in response to any thermal effects in the sample, so as to maintain sample and reference at identical temperatures. The differential power required to achieve this condition is recorded as the ordinate on an x-y recorder, with the (programmed) temperature of the system as the abscissa.

Fig. 9-14 shows the widely-used Perkin-Elmer differential scanning calorimeter, for which a block diagram is given in Fig. 9-15. By keeping the thermal mass of the sample and reference holders and all other thermal resistances in the system to a minimum, and by using high-gain electronic amplification, the response time of the instru-

TABLE 9-5. Classification of DTA phenomena

Origin	Type	Nature
Transitions	Kinetic or thermodynamic	Glass transition (curve shift, not peak)
	Thermodynamic	Solid-solid: crystallization (exothermic) phase change (either)
		Solid-liquid: crystallization (exothermic) melting (endothermic)
		Solid-gas: sublimation (endothermic)
		Liquid-gas: boiling (endothermic)
Chemical changes	Thermodynamic	Volatilization (endothermic)
	Kinetic	Reaction (either)

ment is kept short enough that the readout is a nearly instantaneous measure of the energy transferred to or from the sample holder. The sample and reference temperatures are maintained identical while following a selected program of heating or cooling at constant rate.

Relation of DTA and DSC In classical DTA, the sample forms a major part of the thermal conduction path. Since its thermal conductivity changes in a way that is generally unknown during a transition, the proportionality between temperature differences and energy changes is also unknown. This makes the conversion of peak areas to energies uncertain and relegates DTA to an essentially qualitative category. In

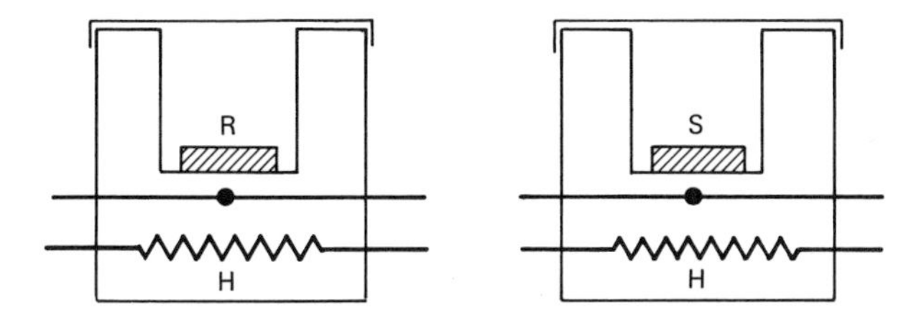

Fig. 9-13. Sketch of the essential elements of the Perkin-Elmer DSC cell.

TABLE 9-6. Factors affecting measurement of T_g by DTA

Origin	Variable	Change or Component	Result
Programming rate	Heating	Faster	Higher T_g
		Slower	Lower T_g, sharper inflection
	Cooling	Faster	Higher T_g
		Slower	Lower T_g, sharper inflection (best mode)
Sample	Concentration	Greater	Enhances sensitivity; dilution generally undesirable
	Amount	More	Increases sensitivity but may broaden transition and reduce resolution, especially at high heating rates
Instrument	Temperature gradient	Sample	Can broaden transition and reduce resolution, especially at high rates
		Block	Little effect if sample thermocouple provides temperature base, otherwise temperature errors
	Component size	Block	Large block tends to reduce sensitivity
		Thermocouple	Large thermocouple conducts heat, reduces sensitivity, leads to temperature errors
	Thermocouple providing system temperature	Sample	Most accurate mode, must be centered
		Block	Least accurate
		Reference	Approximates sample temperature but higher when heating, lower when cooling

Origin	Variable	Change or Component	Result
	Imbalance[a]	Sample *vs.* reference	Poor baseline, distorted inflection, possible temperature error
	Program	Nonlinear	Poor balance, distorted inflection, possible temperature error

[a]Similar results from imbalances in heat capacities, sample and reference size, heating, and thermocouple location.

contrast, in DSC the peak area is a true measure of the electrical energy input required to maintain the sample and reference temperatures equal, independent of the instrument's thermal constants or any changes in the thermal behavior of the sample. The calibration constant relating DSC peak areas to calories is known and constant, permitting quantitative analysis.

It should be pointed out, however, that it is not difficult, with proper design and operating conditions, to make any DTA apparatus perform functions similar to those of DSC. This was done by Boersma (1955), who removed the thermocouples from the interior of the sample and reference, locating them close by but external to these materials. This eliminates the dependence of the thermal conductivity on changes in the sample, but the temperature dependence remains. This type of cell (Figs. 9-16 and 9-17) is used in commercial DSC cells for use with DTA equipment. To make reliable calorimetric measurements with this design requires calibration with a standard having a transition temperature close to that of the unknown, and particular care must be taken that the thermal conductivity remains the same from one experiment to the next.

Heat flow equations In this section we review the mathematics of DSC, where the temperature gradient within the sample can be neglected without serious error, following the treatment of Müller (1960), Adam (1963), and Martin (1963), as outlined by Wunderlich (1971). The more complex treatment for DTA, where a temperature gradient exists within the sample, and the temperature difference depends on the thermal conductivity of the sample as well as its changes in heat capacity and enthalpy, is beyond the scope of this book but is treated by Garn (1965), Smothers (1966), and Wunderlich (1971).

In DSC, the heat flow into the sample holder can be approximated by

$$dQ/dT = K(T_b - T) \qquad (9\text{-}5)$$

where K is the thermal conductivity of the thermal-resistance layer around the sample, assumed to be dependent on geometry but independent

TABLE 9-7. Factors affecting measurement of T_m *and* T_c *in DTA*

Origin	Variable	Change or Component	Result
Programming rate	Heating	Faster	T_m: Slightly lower temperature (small effect) T_c: Higher temperature; less crystallinity, thus smaller peak
		Slower	T_m: Smaller peak; more annealing, closer approach to equilibrium T_m T_c: Crystallization (from glass) at lower temperature; more crystallinity and larger peak
	Cooling	Faster	T_c: Crystallization at lower temperature; less crystallinity, lower peak
		Slower	T_c: Higher T_c, more crystallinity, larger peak
Instrumental	Sample	Concentration	Undiluted sample preferred (unless sample movement a problem); smaller size for same amount; larger peaks, better conductivity gives lower gradient
		Packing	Tight gives less movement, more sample, better thermal gradient, narrower peaks
	Temperature gradient	Sample	Gradient within a sample causes poor resolution, peak distortion, possible temperature error
		Block	Gradient in block causes errors if block thermocouple furnishes system temperature; otherwise little effect

Origin	Variable	Change or Component	Result
	Sample size	Small	Ideal; easier to balance with reference
		Large	Self-heating gives appearance of supercooling; tendency to poor thermal gradients; poor resolution, multiple peaks
	Cell size	Large	Conducts heat, reducing peak size; helps prevent self-heating or cooling. (Small is ideal.)
	Thermocouple size	Large	Conducts heat, reducing peak size and giving temperature errors. (Small is ideal.)
	Thermocouple providing system temperature	Sample	If well centered, best for accuracy
		Block	Least accurate
		Reference	Approximates sample temperature, but higher when heating, lower when cooling
	Imbalance[a]	Sample *vs.* reference	Poor baseline, skewed peaks, temperature errors
	Program linearity	Nonlinear	Poor balance; skewed peaks; temperature errors; kinetic studies not possible. T_c more sensitive.

[a]Similar results from imbalance in heat capacities, sample and reference size, heating, and thermocouple location.

of the temperature, T is is the sample temperature, and T_b the programmed block temperature:

$$T_b = T_0 + qt \tag{9-6}$$

TABLE 9-8. Factors affecting measurement of reaction parameters in DTA

Origin	Variable	Change or Component	Result
Programming rate	Heating	Fast	Shifts reactions to higher temperatures; gives larger, sharper peaks
		Slow	Also needed since rate is part of reaction data
Instrumental	Sample	Concentration	High concentration can cause poor gas release and self-heating if reaction is exothermic. Small surface reduces effects of atmosphere
		Packing	Tight preferred for less movement, more sample, better gradient, unless gas release interfered with
	Temperature gradient	Sample	Gradient within sample causes poor resolution, peak distortion, possible temperature error
		Block	Causes temperature error if block thermocouple furnishes system temperature; otherwise little effect
	Sample size	Large	Self-heating may change rate of exothermic reaction, interfere with gas release, give poor gradient and resolution. (Small is ideal.)
	Cell size	Large	Absorbs heat, reduces peak size in exothermic reactions, opposite in endothermic, but helps stabilize temperature fluctuations. (Small is ideal.)

Origin	Variable	Change or Component	Result
	Thermocouple size	Large	Conducts heat, reducing peak size and giving temperature error. (Small is ideal.)
	Thermocouple providing system temperature	Sample	If well centered, best for accuracy
		Block	Least accurate
		Reference	Approximates sample temperature but somewhat higher
	Imbalance[a]	Sample *vs.* reference	Poor baseline, skewed peaks, temperature errors
	Program linearity	Nonlinear	Poor balance, skewed peaks, temperature errors, kinetics not possible

[a]Similar results from imbalance in heat capacities, sample and reference size, heating, and thermocouple location.

where q is the heating rate. An equation similar to 9-5 can be written for the reference. The heat absorbed on heating a sample with constant heat capacity* C_p between temperatures T_0 and T is, by definition,

$$Q = C_p(T - T_0) \tag{9-7}$$

From these concepts can be derived the basic equation for DSC (Gray 1968, Baxter 1969),

$$\Delta T = qC_p/K \tag{9-8}$$

where the symbol Δ refers to the difference (reference minus sample).

*Here we differ from the *Textbook* by using the symbol C_p for heat capacity and c_p for specific heat: If m is the mass of the sample in grams, $C_p = mc_p$.

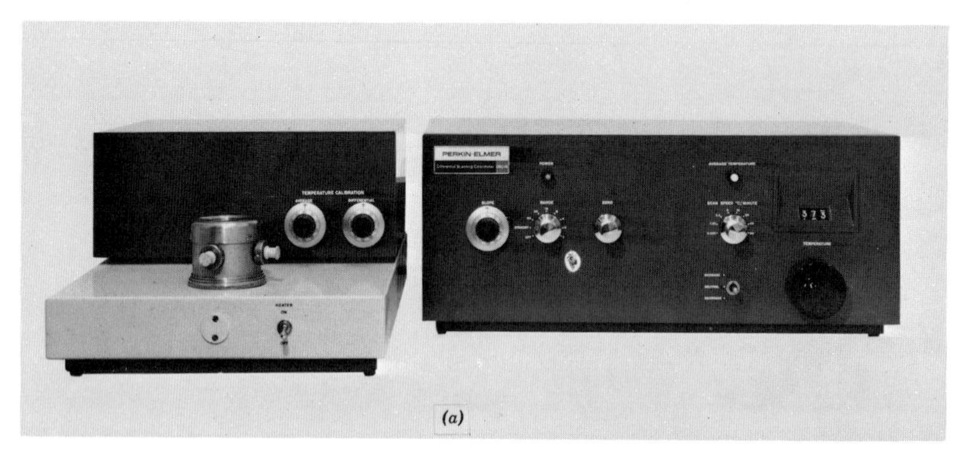

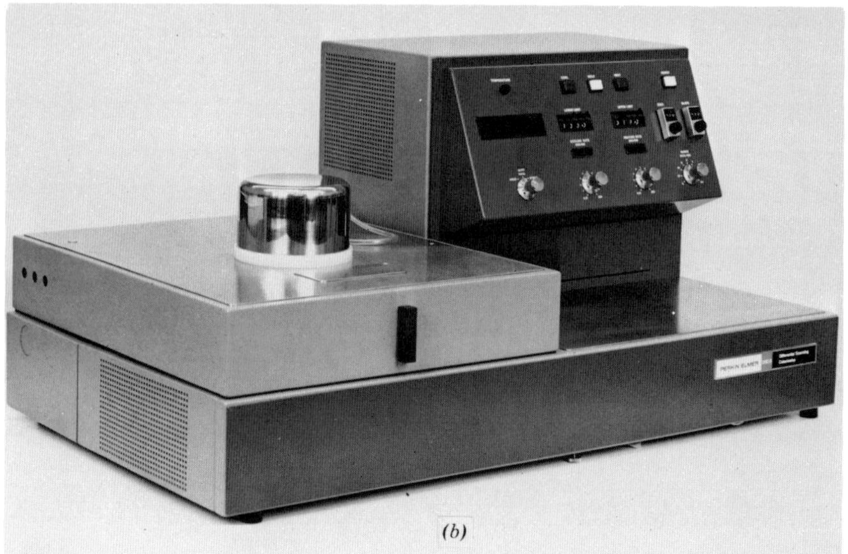

Fig. 9-14. The Perkin-Elmer differential scanning calorimeter: (*a*) Model DSC 1B; (*b*) Model DSC 2 (courtesy Perkin-Elmer).

Heat capacity In measuring heat capacity by DSC, it is generally assumed that changes occur sufficiently slowly on heating that a steady state is maintained. Then q is constant for sample and reference, and

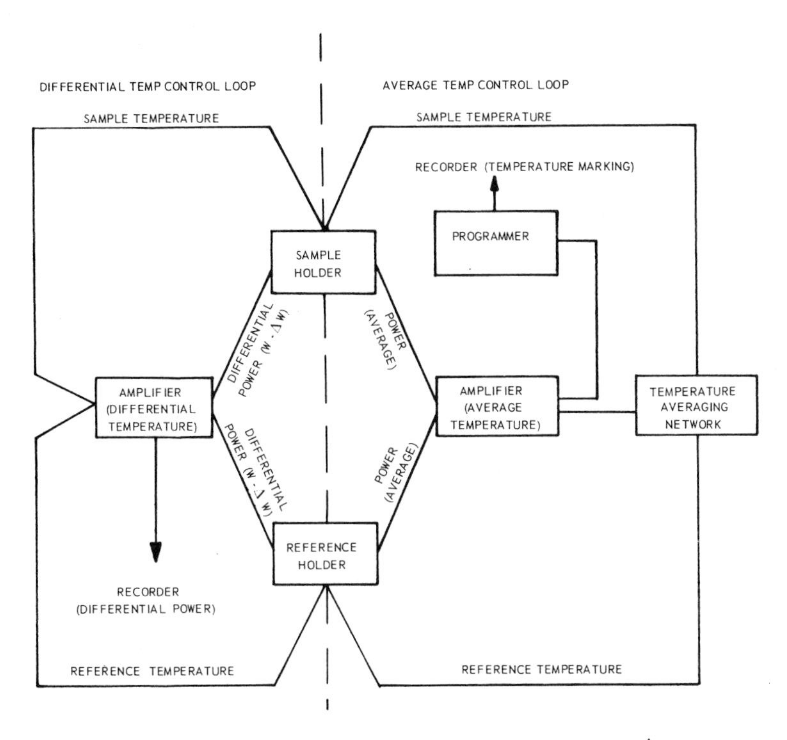

Fig. 9-15. Block diagram of the Perkin-Elmer DSC 1B (courtesy Perkin-Elmer).

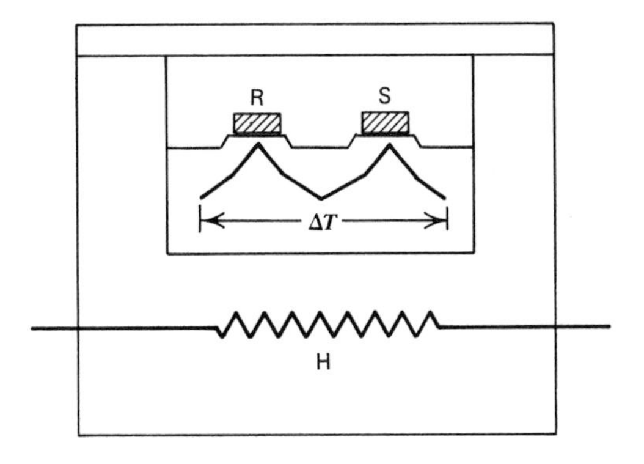

Fig. 9-16. Sketch of the Boersma-type DSC cell as used in the Du Pont DSC.

Fig. 9-17. The Du Pont cell as used in the Du Pont 990 DTA (courtesy Du Pont).

Eq. 9-8 applies. If this is not so, a simple correction can be made. Specific-heat measurements can be obtained with an accuracy of about ±2% over a wide temperature range (at least -100 to 600°C) by DSC.

Typical data were obtained with a Boersma DSC cell on the Du Pont DTA: A calibration run at 80°C with 17.6 mg zinc (c_p = 0.0928 cal/g, handbook value at 80°C) made at a heating rate of 10°C/min yielded ΔT = 0.12°C. Rearranging that part of Eq. 9-8 referring to the sample alone, since the same reference and heating rate are used throughout, yields

$$K = qmc_p/\Delta T = 10 \times 0.0176 \times 0.0928/0.12 = 0.136 \text{ cal/min} \qquad (9\text{-}8a)$$

A measurement with 6.8 mg polystyrene, also at 80°C, gave ΔT = 0.162°C. Again rearranging Eq. 9-8,

$$c_p = K\Delta T/qm = 0.136 \times 0.162/10 \times 0.0068 = 0.324 \text{ cal/g} \qquad (9\text{-}8b)$$

The literature value (Boundy 1952, p. 67) is 0.341 cal/g.

Enthalpy The calculation of enthalpy from DSC differs depending upon whether the transition is slow enough that a steady-state condition is maintained (Fig. 9-18*a*) or so rapid that the change of the sample temperature is halted until sufficient heat is supplied (Fig. 9-18*b*). We consider only the slow transition, and further assume that there is no difference in heating rates for sample and reference, so that the DSC trace returns to the original base line after the transition. In this case,

$$\Delta H = \int_{T_i}^{T_f} C_p \, dT = \int_{T_i}^{T_f} (K\Delta T/q)\, dT \tag{9-9}$$

where the subscripts i and f refer to the initial and final temperatures of the transition. The more complicated cases are treated by Adam (1963) and Gray (1968) among others. The effect of a change in K on peak area and shape for a sharp transition is illustrated for DTA and DSC in Fig. 9-19. In DSC, peak area is independent of K since

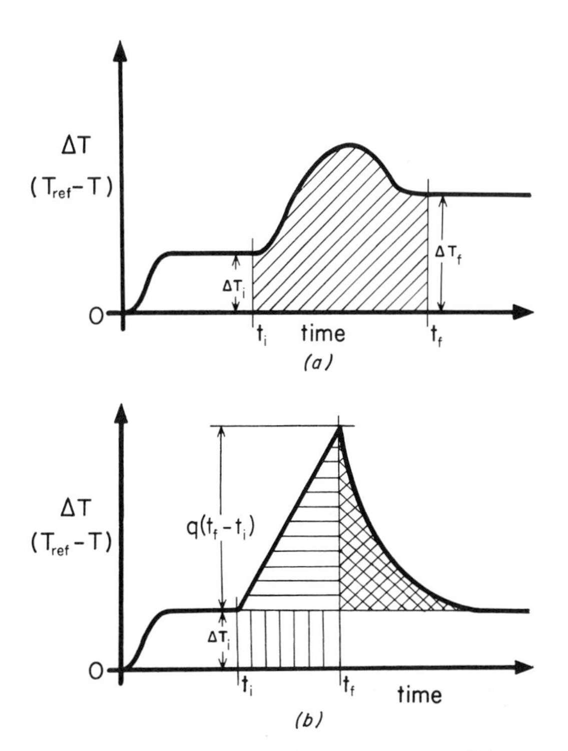

Fig. 9-18. DSC traces typical of (*a*) slow and (*b*) rapid transitions involving a change in enthalpy (Wunderlich 1971).

this constant enters only into the time lag between the temperature of the sample and its holder. Resolution improves, however, as K increases. In DTA the latter is also true but peak area decreases as K increases.

Typical enthalpy data, obtained with 8.6 mg polyoxymethylene in a Boersma-type DSC cell on the Du Pont DTA, are shown in Fig. 9-20. Here we wish to evaluate the enthalpy per gram of polymer, $\Delta H/m$. The values of $K = 0.446$, read from a calibration curve (not shown) for $T = 178°C$, the peak of the transition, and $q = 10°C/min$, are constant and can be taken out of the integral in Eq. 9-9. Taking $T_i = 120°C$ and $T_f = 190°C$, the integral is evaluated as area (sq. in.) × (scale factors for ordinate and abscissa, deg/in.). The peak area was 1.91 sq. in. and the scale factor for T was 20°C/in. and for ΔT, 0.278°C/in. Thus,

$$\Delta H/m = (K/mq) \int_{T_i}^{T_f} \Delta T \, dT$$

$$= (0.446/0.086 \times 10)(1.91 \times 0.278 \times 20) \qquad (9\text{-}9a)$$

$$= 55.1 \text{ cal/g}$$

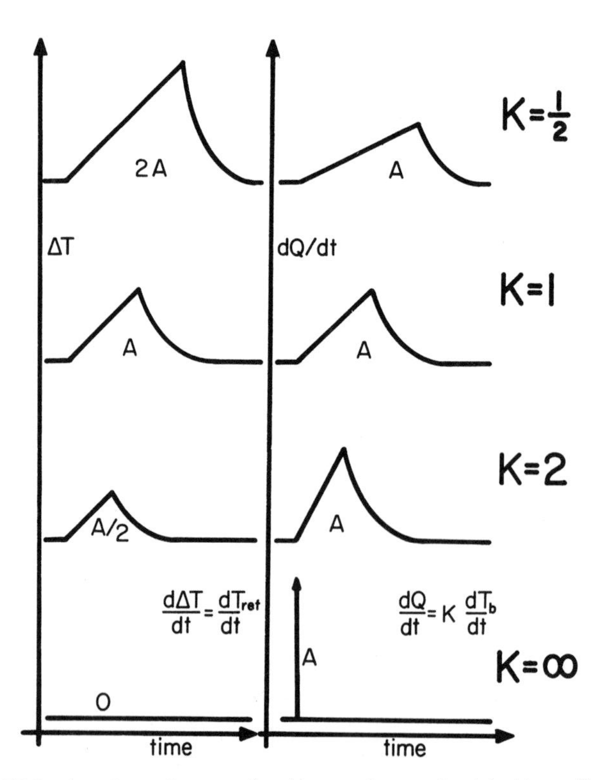

Fig. 9-19. Effect of a change in thermal conductivity, K, on DTA and DSC traces (Wunderlich 1971).

The literature value for polyoxymethylene (Inoue 1961) is 53.0 cal/g.

GENERAL REFERENCES

Textbook, Chap. 4E, and specifically:

DTA: Kissinger 1962; Manley 1963; Ke 1964; Wendlandt 1964, 1972; Barrall 1966; David 1966; Smothers 1966; Porter 1968; Garn 1969; Schwenker 1969; Mackenzie 1970*a*, Chaps. 2 and 4; Wunderlich 1971.
DSC: Wendlandt 1964, 1972; Garn 1965; Barrall 1966; Smothers 1966; Ford 1967; Gray 1968; Baxter 1969; Mackenzie 1970*a*,*b*; Wunderlich 1970, 1971.

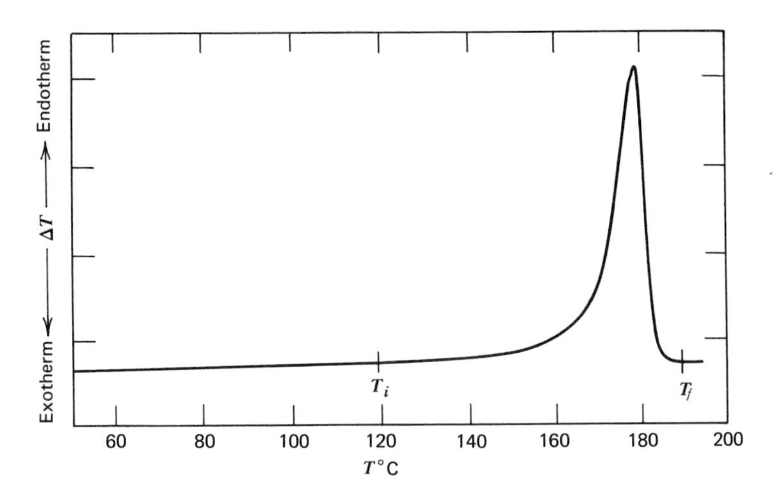

Fig. 9-20. DSC trace of the melting transition of polyoxymethylene.

D. *Thermal Conductivity*

A knowledge of the thermal conductivity of a material is essential for any application involving the conversion, exchange, or transport of thermal energy in that material. The importance of thermal conductivity in DTA and DSC was demonstrated in Sec. C.

The thermal conductivity K is defined as the ratio of the heat flow across unit area of a surface to the negative of the temperature gradient in the direction of flow:

$$K = (dQ/dt)/(dT/dx) \tag{9-10}$$

Methods for measuring K thus require the establishment of a known temperature distribution and measurement of the heat flux. The measurements can be made at the steady state and under transient conditions. Only the former are considered here; transient methods are reviewed by Harmathy 1964, Goldfein 1965, Anderson 1966*a*, Eiermann 1966*a*, Steere

1966, and Powell 1969.

Several steady-state methods are in wide use, based on different cell geometries (ASTM C 177, Kline 1961, Schröder 1963). Schröder's method, using the Colora Thermoconductometer (Dynatech) (Fig. 9-21), is used in Exp. 23. In this instrument (Hansen 1972), the thermal conductivity of a sample is determined by comparing the time required to vaporize a given amount of a known liquid when heated by conduction through the sample and through a reference material. In the apparatus (Fig. 9-22) a pure liquid is boiled in A, heating a silver plate S_1, and returned after condensation to A. An upper vessel, B, contains a liquid boiling 10-20°C lower than that of the liquid in A. A second silver plate S_2 forms the bottom of B, and the sample is fitted between the two plates at P. Vapor from liquid boiling in B is condensed into a graduated vessel M.

Since the plates S_1 and S_2 are maintained at the constant boiling points of the liquids in A and B, a fixed temperature gradient $\Delta T = T_A - T_B$ is imposed on the sample of thickness ℓ and area A. When the steady state is reached, the time t necessary to distill a fixed

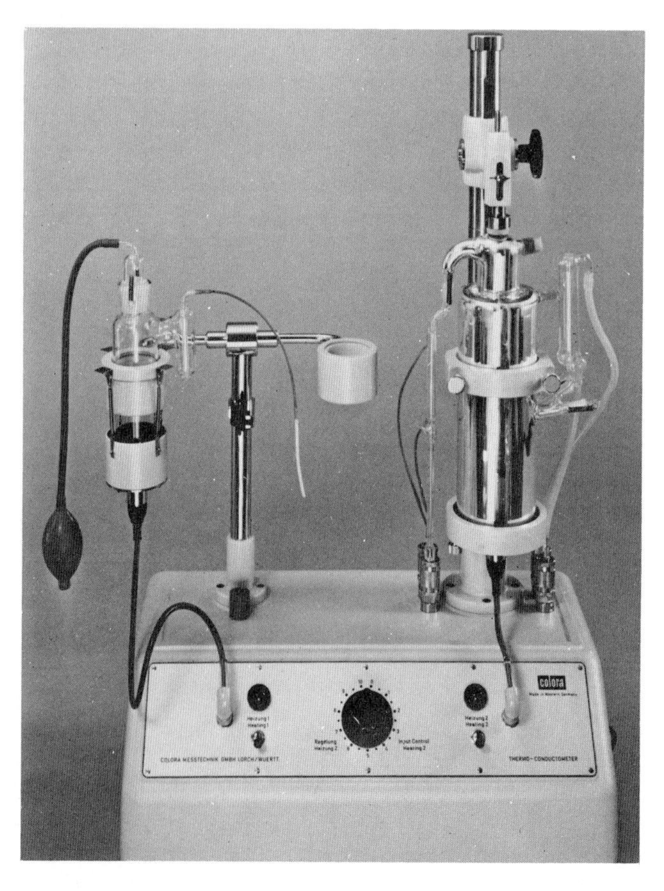

Fig. 9-21. The Colora Thermoconductometer (courtesy Dynatech).

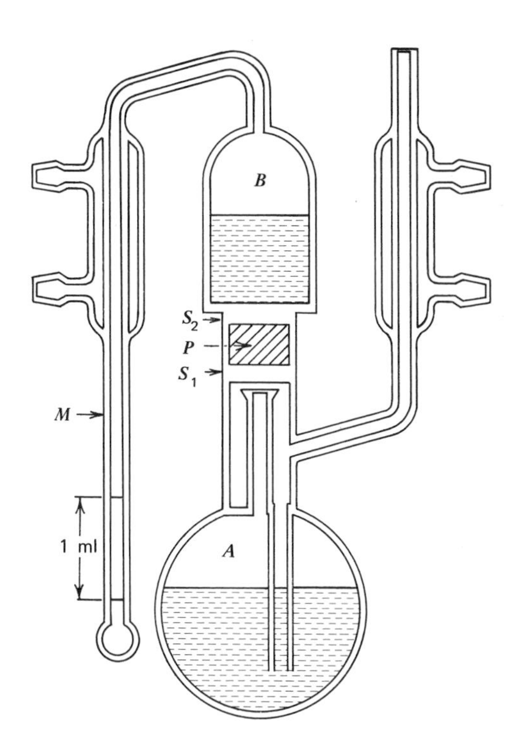

Fig. 9-22. Sketch of the essential parts of the Colora apparatus (courtesy Dynatech).

amount (1 ml) of the liquid in B into M is measured. If the heat of vaporization of this amount of liquid is ΔH_v,

$$K = \ell \Delta H_v / A t \Delta T \tag{9-11}$$

where ℓ and A are the thickness and area of the sample, respectively. In practice, one takes as a calibration parameter the thermal resistance $R = t\Delta T/\Delta H_v$ in sec °C/cal, whence

$$K = \ell / RA \tag{9-12}$$

R is first determined for a standard material of known thermal conductivity, for a range of sample thicknesses yielding times encompassing those anticipated for the polymer sample. R is plotted versus t, and values read off for the observed distillation times for polymer samples, again at several values of ℓ. Substitution into Eq. 9-12 completes the calculation.

Typical data are given in Fig. 9-23, which shows the calibration curve of R versus t for three samples of glass with the indicated thicknesses. From the curve are read R = 133 for a polystyrene sample

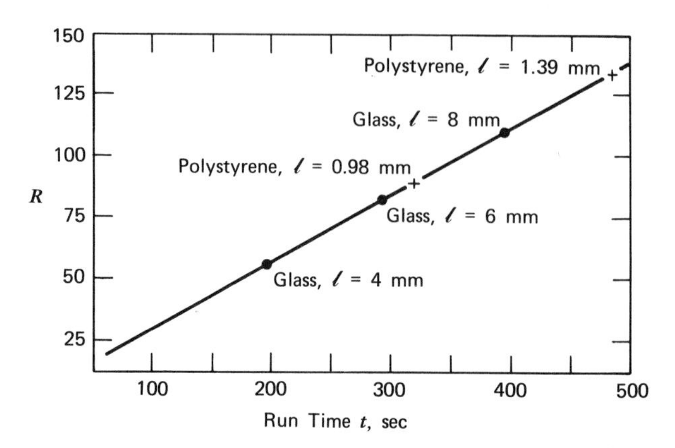

Fig. 9-23. Calibration and data curve for measurement of thermal conductivity of polystyrene. The solvent pair was acetone and pentane, and the measurement temperature was the mean of their boiling points, 46.2°C.

for which t = 491 sec, ℓ = 1.39 mm, and A = 2.545 cm^2; and R = 88.5 for a second sample for which t = 319 sec, ℓ = 0.98 mm, and A = 2.502 cm^2. Calculated values of K for the polystyrene are 4.11 and 4.43 × 10^{-4} cal/cm sec °C, compared to the literature value of 4.0 × 10^{-4} (Knappe 1969).

Effect of polymer molecular structure on thermal conductivity

The theory of thermal conductivity was described for linear amorphous polymers by Hansen (1965). For crystalline polymers, the contributions of the crystalline and amorphous regions are assumed to be additive; the thermal conductivity of a crystalline polymer is greater than that of an amorphous material, but its dependence on temperature is the same (Eiermann 1963, Hansen 1965, Sheldon 1965). Hansen (1965) showed that K is proportional to the square root of $\bar{M}_w$. It increases with crosslinking (Ueberreiter 1951), and increases in the direction of orientation but decreases in the perpendicular direction as the result of molecular alignment (Eiermann 1961*b*, 1962; Hellwege 1963; Hansen 1972). The effect of diluents is equivalent to that of lowering the molecular weight (Hansen 1966).

GENERAL REFERENCES

Tye 1969; Kline 1970; Knappe 1971.

E. *Thermogravimetric Analysis*

As usually practiced, thermogravimetric analysis (TGA) is a dynamic technique in which the weight loss of a sample is measured continuously while its temperature is increased at a constant rate. Alternatively, weight loss can be measured as a function of time at constant temperature. The main use of TGA in application to polymers has been in studies of thermal decomposition and stability. A DTA trace provides knowledge of the temperatures at which a thermal event occurs, and TGA tells whether the event is accompanied by weight loss. The use of gas chromatography or mass spectrometry to analyze the effluent gases provides positive identification.

In a typical thermogram (Fig. 9-24), change in weight is plotted as a function of temperature. A small initial weight loss, such as that from w to w_0, generally results from desorption of solvent. If it occurs near 100°C, this is usually assumed to be loss of water. In the example in the figure, extensive decomposition starts at T_1, with a weight loss $w_0 - w_1$. Between T_2 and T_3 another stable phase exists, then further decomposition occurs. Some typical TGA traces are shown in Fig. 9-25.

In many thermograms, phenomena occur so closely spaced that it is difficult to assign appropriate temperatures. This is more easily done with reference to the differential curve of rate of weight change *vs.* temperature, also shown in Fig. 9-24. The area under the curve provides the change in weight. Such curves can be obtained automatically on some instruments.

Factors affecting TGA (Wendlandt 1964, Chap. 2, Anderson 1966*b*, Doyle 1966) Increasing the heating rate in TGA increases the apparent decomposition temperature, and too high a rate may cause a two-step process to be seen as one because equilibrium conditions are not reached. The presence of a self-generated gaseous atmosphere can obscure details of the process, and it is often advantageous to supply a small flow of inert gas through the furnace chamber. The shape of the sample container may be important during volatilization or when a

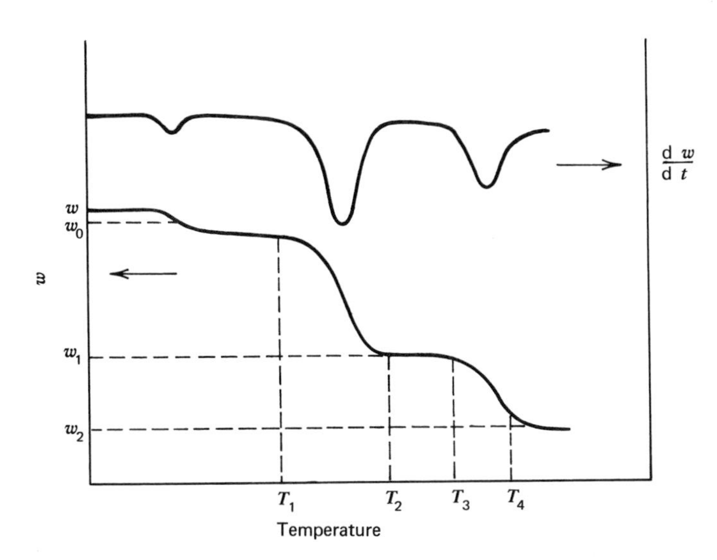

Fig. 9-24. Sketch of a typical TGA thermogram.

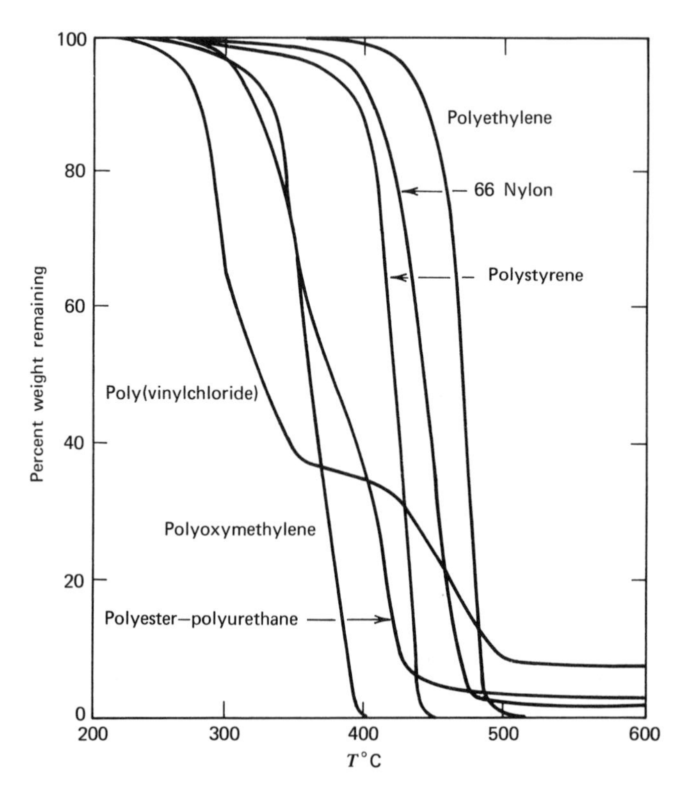

Fig. 9-25. TGA thermograms of some common polymers. Heating rate, 10°C/min, helium atmosphere.

large volume of gas is evolved, since these processes depend upon the surface area of the sample and the way it changes with time.

Large particles of sample, or thick sections, may lead to inefficient heat transfer and temperature gradients, either from the applied heat or from heat of reaction. Problems with the diffusion of volatiles may also arise. In general, decreasing the particle size lowers the temperature of both onset and completion of thermal decomposition. The effect of solubility of gases in the sample is pronounced and difficult to measure or eliminate.

Kinetic parameters The shape of the TGA curve is a function of the kinetics of the decomposition reaction, specifically the order of the reaction and the activation energy and frequency factor in the Arrhenius equation. These parameters can be determined from the thermogram in several ways, as outlined by Freeman (1958), Doyle (1966), and Reich (1971).

Instrumentation Numerous TGA units have been described by Wendlandt (1964, Chap. III), Anderson (1966*b*), and Reich (1971). The most widely used measuring unit is the Cahn electrobalance; this is described in Exp. 24 and used in the Perkin-Elmer TGA unit (Fig. 9-26). The Du Pont design (Fig. 9-27) is based on the same principle.

Applications TGA is useful for studying, in addition to simple stability and decomposition in air or an inert atmosphere, such phenomena as solid-state reactions; determination of moisture, volatiles, and ash; absorption, adsorption and desorption; rates and latent heats of evaporation and sublimation; oxidative degradation; plasticizer volatility; dehydration and hygroscopicity; extent of cure in condensation polymers; composition of filled polymers and composites; and identification by means of characteristic thermograms.

GENERAL REFERENCES

Wendlandt 1964; Anderson 1966*b*; Doyle 1966; Reich 1967, 1971

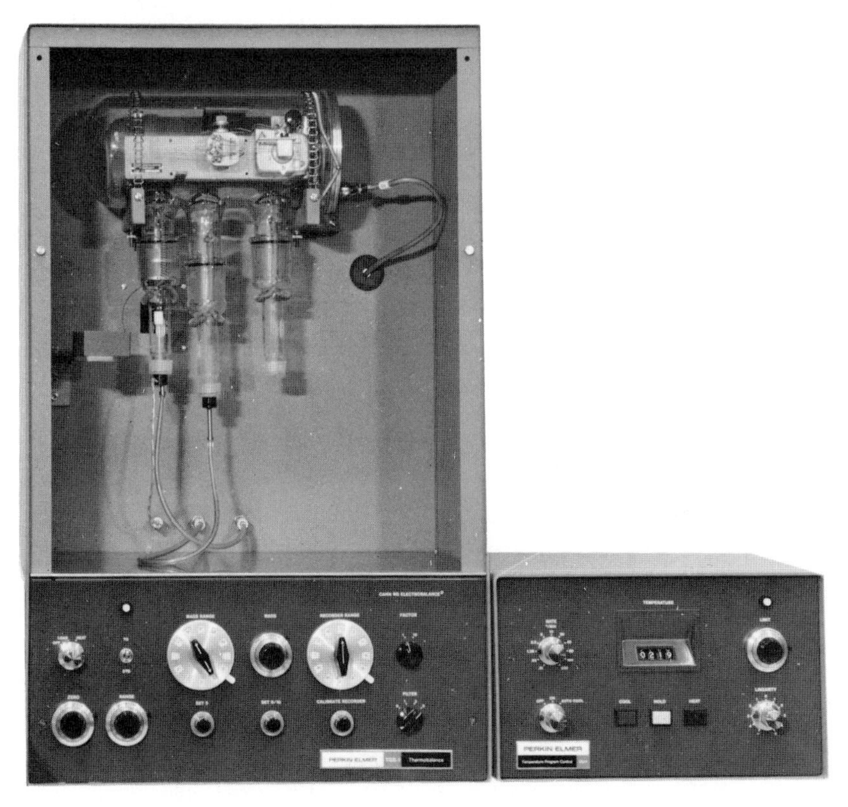

Fig. 9-26. The Perkin-Elmer TGA accessory for the DSC-1B apparatus: left, TSG-1 Cahn-type Thermobalance; right, UU-1 Temperature Program Control (courtesy Perkin-Elmer).

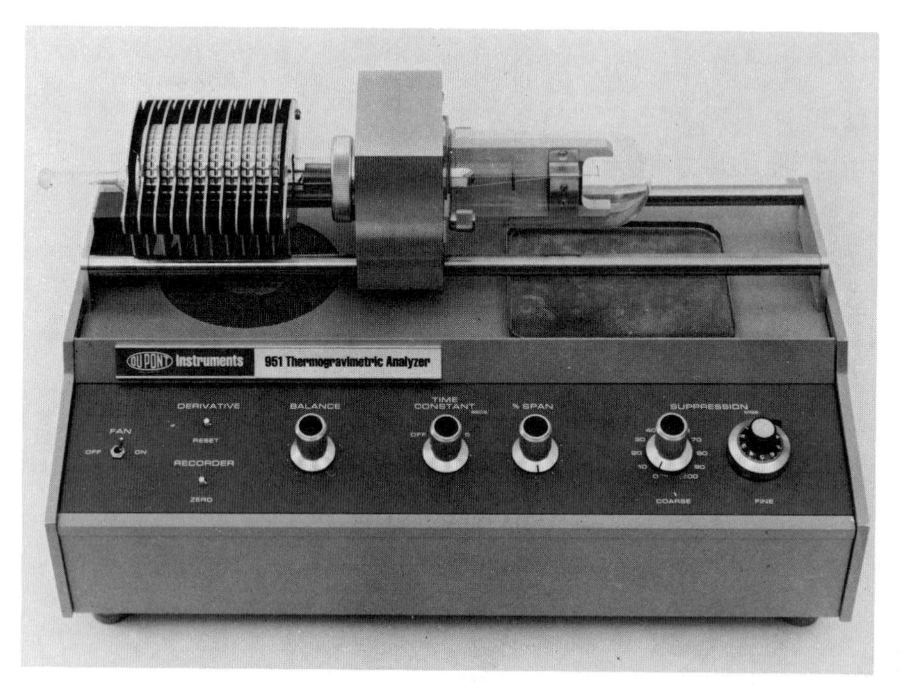

Fig. 9-27. The Du Pont 951 Thermogravimetric Analyzer attachment for the 990 DTA unit (courtesy Du Pont).

BIBLIOGRAPHY

Adam 1963. G. Adam and F. H. Müller, "Theory of the Dynamic Measurement of Heat Quantities by the Differential Thermal Analysis Principle," *Kolloid Z.--Z. Polymere 192*, 29-34 (1963).

Anderson 1966a. D. R. Anderson, "Thermal Conductivity of Polymers," *Chem. Revs. 66*, 677-690 (1966).

Anderson 1966b. Hugh C. Anderson, "Instrumentation, Techniques and Applications of Thermogravimetry," Chap. 3 in Philip E. Slade, Jr., and Lloyd T. Jenkins, eds., *Thermal Analysis*, Vol. 1 of *Techniques and Methods of Polymer Evaluation*, Marcel Dekker, New York, 1966.

ASTM C 177. *Standard Method of Test for Thermal Conductivity of Materials by Means of the Guarded Hot Plate*. ASTM Designation: C 177. American Society for Testing and Materials, Philadelphia, Pennsylvania 19103.

ASTM D 648. *Standard Method of Test for Deflection Temperature of Plastics Under Load*. ASTM Designation: D 648. American Society for Testing and Materials, Philadelphia, Pennsylvania 19103.

ASTM D 746. *Standard Method of Test for Brittleness Temperature of Plastics and Elastomers by Impact*. ASTM Designation: D 746. American Society for Testing and Materials, Philadelphia, Pennsylvania 19103.

ASTM D 785. Standard Method of Test for Rockwell Hardness of Plastics and Electrical Insulating Materials. ASTM Designation: D 785. American Society for Testing and Materials, Philadelphia, Pennsylvania 19103.

ASTM D 1525. Standard Method of Test for Vicat Softening Point of Plastics. ASTM Designation: D 1525. American Society for Testing and Materials, Philadelphia, Pennsylvania 19103.

Bareš 1972. J. Bareš, "Use of Computer for Polymer Characterization, Viscoelasticity: WLF Equation, Viscoelastic Spectra, H and L, and Moduli, Dilatometry," *Ing. Chim. Ital.*, Suppl. to *Chim. e Ind. (Milan) 8*, 27-33 (1972).

Barrall 1962. Edward M. Barrall, II, and L. B. Rogers, "Differential Thermal Analysis of Organic Compounds. Effects of Diluting Agents," *Anal. Chem. 34*, 1106-1111 (1962).

Barrall 1966. E. M. Barrall, II, and Julian F. Johnson, "Instrumentation, Techniques and Applications of Differential Thermal Analysis," Chap. 1 in Philip E. Slade, Jr., and Lloyd T. Jenkins, eds., *Thermal Analysis*, Vol. 1 of *Techniques and Methods of Polymer Evaluation*, Marcel Dekker, New York, 1966.

Baxter 1969. R. A. Baxter, "A Scanning Microcalorimetry Cell Based on a Thermoelectric Disc--Theory and Applications," pp. 65-84 in R. F. Schwenker and P. D. Garn, eds., *Thermal Analysis*, Vol. 1, *Instrumentation, Organic Materials and Polymers*, Academic Press, New York, 1969.

Beatty 1950. K. O. Beatty, Jr., A. A. Armstrong, Jr., and E. M. Schoenborn, "Thermal Conductivity of Homogeneous Materials. Determination by an Unsteady State Method," *Ind. Eng. Chem. 42*, 1527-1532 (1950).

Bekkedahl 1941. Norman Bekkedahl and Lawrence A. Wood, "Influence of the Temperature of Crystallization on the Melting of Crystalline Rubber," *J. Chem. Phys. 9*, 193 (1941).

Bekkedahl 1949. Norman Bekkedahl, "Volume Dilatometry," *J. Res. Natl. Bur. Stds. 43*, 145-156 (1969).

Berkelhamer 1945. I. L. H. Berkelhamer and S. S. Peil, "Differential Thermal Analysis," *Mine and Quarry Eng. 10*, 221-225, 273-279 (1945).

Bier 1965. Gerhard Bier, "PVC, An Old and New Plastic. Part II. Vinyl Chloride Polymers" (in German), *Kunststoffe 55*, 694-700 (1965).

Billmeyer 1969. Fred W. Billmeyer, Jr., "Molecular Structure and Polymer Properties," *J. Paint Technol. 42*, 1-16 (1969).

Boersma 1955. S. L. Boersma, "A Theory of Differential Thermal Analysis and New Methods of Measurement and Interpretation," *J. Am. Ceramic Soc. 38*, 281-284 (1955).

Boundy 1952. Ray H. Boundy and Raymond F. Boyer, eds., *Styrene, Its Polymers, Copolymers and Derivatives*, Reinhold Publishing Corp., New York, 1952.

Boyer 1963. Raymond F. Boyer, "The Relation of Transition Temperatures to Chemical Structure in High Polymers," *Rubber Chem. Tech. 36*, 1303-1421 (1963).

Bueche 1962. F. Bueche, *Physical Properties of Polymers*, Interscience Div., John Wiley and Sons, New York, 1962.

Collins 1966. Edward A. Collins and Lyle A. Chandler, "Temperature and Rate Effects on Crystalline Transitions in Cis-1,4-Polybutadiene as Measured by DTA," *Rubber Chem. Tech. 39*, 193-205 (1966).

Danusso 1959. F. Danusso, G. Moraglio, W. Ghilia, L. Motta, and G. Talamini, "Volumetric and Dilatometric Property of Olefin Polymers" (in Italian), *Chim. e Ind. (Milan) 41,* 748-757 (1959).

David 1966. J. J. David, "Transition Temperatures by Differential Thermal Analysis," Chap. 2 in Philip E. Slade, Jr., and Lloyd T. Jenkins, eds., *Thermal Analysis,* Vol. 1 of *Techniques and Methods of Polymer Evaluation,* Marcel Dekker, New York, 1966.

Di Marzio 1958. E. A. Di Marzio and J. H. Gibbs, "Chain Stiffness and the Lattice Theory of Polymer Phases," *J. Chem. Phys. 28,* 807-813 (1958).

Di Marzio 1959. E. A. Di Marzio and J. H. Gibbs, "Glass Temperatures of Copolymers," *J. Polymer Sci. 40,* 121-131 (1959).

Doyle 1966. Charles D. Doyle, "Quantitative Calculations in Thermogravimetric Analysis," Chap. 4 in Philip E. Slade, Jr., and Lloyd T. Jenkins, eds., *Thermal Analysis,* Vol. 1 of *Techniques and Methods of Polymer Evaluation,* Marcel Dekker, New York, 1966.

Du Pont. Instrument Products Division, E. I. du Pont de Nemours and Co., Wilmington, Delaware 19898.

Dynatech. Dynatech Corporation, Cambridge, Massachusetts 02139.

Eiermann 1961a. K. Eiermann, K. H. Hellwege and W. Knappe, "Quasi-stationary State Measurement of the Specific Heat of Plastics in the Temperature Range -180°C to +90°C" (in German), *Kolloid Z. 174,* 134-142 (1961).

Eiermann 1961b. K. Eiermann, "Thermal Conductivity of Plastics and Its Dependence on Structure, Temperature and Pre-Treatment," *Kunststoffe 51,* 512-517 (1961).

Eiermann 1962. K. Eiermann and K. H. Hellwege, "Thermal Conductivity of High Polymers from -180°C to 90°C," *J. Polymer Sci. 57,* 99-106 (1962).

Eiermann 1963. K. Eiermann, "Thermal Conductivity of High Polymers," *J. Polymer Sci. C6,* 157-165 (1963).

Eisenberg 1970. A. Eisenberg and M. Shen, "Recent Advances in Glass Transitions in Polymers," *Rubber Chem. Tech. 43,* 156-170 (1970).

Ferry 1961. John D. Ferry, *Viscoelastic Properties of Polymers,* John Wiley and Sons, New York, 1961.

Ferry 1970. John D. Ferry, *Viscoelastic Properties of Polymers,* 2nd ed., John Wiley and Sons, New York, 1970.

Ford 1967. R. W. Ford and R. A. Scott, "Influence of Thermal History on Fusion Curves of Polyethylene," *J. Applied Polymer Sci. 11,* 2325-2330 (1967).

Fox 1950. T. G. Fox and P. J. Flory, "Second Order Transition Temperatures and Related Properties of Polystyrene. I. Influence of Molecular Weight," *J. Applied Phys. 21,* 581-591 (1950).

Fox 1954. T. G Fox and P. J. Flory, "The Glass Temperature and Related Properties of Polystyrene. Influence of Molecular Weight," *J. Polymer Sci. 14,* 315-319 (1954).

Freeman 1958. E. S. Freeman and B. Carroll, "The Application of Thermoanalytical Techniques to Reaction Kinetics. Thermogravimetric Evaluation of the Kinetics of the Decomposition of Calcium Oxalate Monohydrate," *J. Phys. Chem. 62,* 394-397 (1958).

Garn 1965. P. D. Garn, *Thermoanalytical Methods of Investigation,* Academic Press, New York, 1965.

Garn 1965. Paul D. Garn, "Thermoanalytical Techniques," Chap. 7 in Raymond R. Myers and J. S. Long, eds., *Treatise on Coatings,* Vol. 2, Part I, *Characterization of Coatings,* Marcel Dekker, New York, 1969.
Gibbs 1956. Julian H. Gibbs, "Nature of the Glass Transition in Polymers," *J. Chem. Phys. 25,* 185-186 (1956).
Gibbs 1958. Julian H. Gibbs and Edmund A. Di Marzio, "Nature of the Glass Transition and the Glassy State," *J. Chem. Phys. 28,* 373-383 (1958).
Goldfein 1965. S. Goldfein and J. Calderon, "Apparatus for Determining Thermal Conductivity of Insulation Materials," *J. Applied Polymer Sci. 9,* 2985-2991 (1965).
Gordon 1952. Manfred Gordon and James S. Taylor, "Ideal Copolymers and the Second-Order Transitions of Synthetic Polymers. I. Non-Crystalline Copolymers," *J. Applied Chem. 2,* 493-500 (1952).
Gordon 1965. M. Gordon, "Thermal Properties of High Polymers," Chap. 5 in P. D. Ritchie, ed., *Physics of Plastics,* D. Van Nostrand Co., Princeton, New Jersey, 1965.
Gray 1968. Allan P. Gray, "A Simple Generalized Theory for the Analysis of Dynamic Thermal Measurement," pp. 209-218 in Roger S. Porter and Julian F. Johnson, eds., *Analytical Calorimetry,* Plenum Press, New York, 1968.
Hansen 1965. David Hansen and Cheng C. Ho, "Thermal Conductivity of High Polymers," *J. Polymer Sci. A 3,* 659-670 (1965).
Hansen 1966. David Hansen, R. C. Kantayya and C. C. Ho, "Thermal Conductivity of High Polymers--The Influence of Molecular Weight," *Polymer Eng. Sci. 6,* 260-262 (1966).
Hansen 1972. D. Hansen and G. A. Bernier, "Thermal Conductivity of Polyethylene: The Effects of Crystal Size, Density, and Orientation on the Thermal Conductivity," *Polymer Eng. Sci. 12,* 204-208 (1972).
Harmathy 1964. T. Z. Harmathy, "Variable-State Methods of Measuring the Thermal Properties of Solids," *J. Applied Phys. 35,* 1190-1200 (1964).
Hellmuth 1965. E. Hellmuth and Bernhard Wunderlich, "Superheating of Linear High Polymer Polyethylene Crystals," *J. Applied Phys. 36,* 3039-3044 (1965).
Hellwege 1963. K. H. Hellwege, J. Hennig and W. Knappe, "Anisotropy of the Thermal Expansion and Thermal Conductivity in Amorphous High Polymers that are Stretched Along One Axis," *Kolloid Z.--Z. Polymere 188,* 121-127 (1963).
Hirai 1958. N. Hirai and H. Eyring, "Bulk Viscosity of Liquids," *J. Applied Phys. 29,* 810-816 (1958).
Hirai 1959. N. Hirai and H. Eyring, "Bulk Viscosity of Polymeric Systems," *J. Polymer Sci. 37,* 51-70 (1959).
Illers 1963. Karl-Heinz Illers, "The Glass Transition Temperatures of Copolymers" (in German), *Kolloid Z.--Z. Polymere 190,* 16-33 (1963).
Inoue 1961. M. Inoue, "Heat of Fusion and Interaction of Polyoxymethylene with Diluents,"*J. Polymer Sci. 51,* S18-S20 (1961).
Jaffe 1957. M. Jaffe and B. Wunderlich, "Melting of Polyoxymethylene," *Kolloid Z.--Z. Polymere 216-217,* 203-216 (1967).
Jaffe 1969. M. Jaffe and B. Wunderlich, "Superheating of Extended-Chain Polymer Crystals," pp. 387-403 in R. F. Schwenker and

P. D. Garn, eds., *Thermal Analysis*, Vol. 1, *Instrumentation, Organic Materials, and Polymers*, Academic Press, New York, 1969.

Kaelble 1969. D. H. Kaelble, "Free Volume and Polymer Rheology," Chap. 5 in Frederick R. Eirich, ed., *Rheology--Theory and Applications*, Vol. 5, Academic Press, New York, 1969.

Kauzmann 1948. W. Kauzmann, "The Nature of the Glassy State and the Behavior of Liquids at Low Temperature," *Chem. Revs. 43*, 219-256 (1948).

Ke 1964. Bacon Ke, "Differential Thermal Analysis," Chap. IX in Bacon Ke, ed., *Newer Methods of Polymer Characterization*, Interscience Div., John Wiley and Sons, New York, 1964.

Kissinger 1962. H. E. Kissinger and S. B. Newman, "Differential Thermal Analysis," Chap. 4 in Gordon M. Kline, ed., *Analytical Chemistry of Polymers*, Part II, *Analysis of Molecular Structure and Chemical Groups*, Interscience Div., John Wiley and Sons, New York, 1962.

Kline 1961. Donald E. Kline, "Thermal Conductivity Studies of Polymers," *J. Polymer Sci. 50*, 441-450 (1961).

Kline 1970. Donald E. Kline and David Hansen, "Thermal Conductivity of Polymers," Chap. 5 in Philip E. Slade, Jr., and Lloyd T. Jenkins, eds., *Thermal Characterization Techniques*, Vol. 2, Marcel Dekker, New York, 1970.

Knappe 1969. W. Knappe, P. Lohe, and R. Wutschig, "Chain Structure and Thermal Conductivity of Amorphous Linear Polymers" (in German), *Angew. Makromol. Chem. 7*, 181-193 (1969).

Knappe 1971. W. Knappe, "Thermal Conductivity of Polymers" (in German), *Fortschr. Hochpolym. Forsch. (Advances in Polymer Sci.) 7*, 477-535 (1971).

Koleske 1966. Joseph V. Koleske and Joseph A. Faucher, "Demonstration of the Glass Transition," *J. Chem. Educ. 43*, 254-256 (1966).

Mackenzie 1970a. R. C. Mackenzie, *Differential Thermal Analysis*, Vol. 1, *Fundamental Aspects*, Academic Press, New York (1970).

Mackenzie 1970b. R. C. Mackenzie and B. D. Mitchell, "Instrumentation," Chap. 3 in R. C. Mackenzie, ed., *Differential Thermal Analysis*, Vol. 1, *Fundamental Aspects*, Academic Press, New York, 1970.

Manley 1963. T. R. Manley, "Differential Thermal Analysis and Its Application to Polymer Science," pp. 175-197 in *Techniques of Polymer Science*, S.C.I. Monograph No. 17, Gordon and Breach Science Publishers, New York, 1963.

Martin 1963. H. Martin and F. H. Müller, "Changes with Crystalline Portions of Polymers Through Deformation," *Kolloid Z.--Z. Polymere 192*, 19-23 (1963).

Meares 1965. Patrick Meares, *Polymer Structure and Bulk Properties*, D. Van Nostrand Co., Princeton, New Jersey, 1965.

Miller 1966. M. L. Miller, *The Structure of Polymers*, Reinhold Publishing Corp., New York, 1966.

Müller 1960. F. H. Müller and H. Martin, "Calorimeter for Measuring the Specific Heat of Small Amounts of Mixed Materials" (in German), *Kolloid Z. 172*, 97-116 (1960).

Perkin-Elmer. Perkin-Elmer Corporation, Instruments Division, Norwalk, Connecticut 06852.

Porter 1968. Roger S. Porter and Julian F. Johnson, *Analytical Calorimetry*, Plenum Press, New York, 1968.

Powell 1969. R. W. Powell, "Thermal Conductivity Determinations by Thermal Comparator Method," Chap. 6 in R. P. Tye, ed., *Thermal Conductivity*, Vol. 2, Academic Press, New York, 1969.

Reich 1967. Leo Reich and David W. Levi, "Dynamic Thermogravimetric Analysis in Polymer Degradation," *Macromol. Revs. 1*, 173-275 (1967).

Reich 1971. Leo Reich and David W. Levi, "Thermogravimetric Analysis," pp. 1-41 in Herman F. Mark, Norman G. Gaylord, and Norbert M. Bikales, eds., *Encyclopedia of Polymer Science and Technology*, Vol. 14, Interscience Div., John Wiley and Sons, New York, 1971.

Rodriguez 1970. Ferdinand Rodriguez, *Principles of Polymer Systems*, McGraw-Hill Book Co., New York, 1970.

Schröder 1963. J. Schröder, "Apparatus for Determining the Thermal Conductivity of Solids in the Temperature Range from 20 to 200°C," *Rev. Sci. Inst. 34*, 615-621 (1963).

Schwenker 1964. Robert F. Schwenker, Jr. and Rober K. Zuccarello, "Differential Thermal Analysis of Synthetic Fibers," *J. Polymer Sci. C6*, 1-16 (1964).

Schwenker 1969. Robert F. Schwenker, Jr. and Paul D. Garn, eds., *Thermal Analysis*, Vol. 1, *Instrumentation, Organic Materials and Polymers*; Vol. 2, *Inorganic Materials and Physical Chemistry*, Academic Press, New York, 1969.

Sheldon 1965. R. P. Sheldon and K. Lane, "Thermal Conductivities of Polymers--II. Polyethylene," *Polymer 6*, 205-212 (1965).

Shen 1966. M. C. Shen and A. Eisenberg, "Glass Transitions in Polymers," Chap. 9 in H. Reiss, ed., *Progress in Solid State Chemistry*, Vol. 3, Pergamon Press, New York, 1966. (Reprinted in *Rubber Chem. Tech. 43*, 95-155 (1970).

Simha 1962. Robert Simha and Ray F. Boyer, "On a General Relation Involving the Glass Temperature and Coefficients of Expansion of Polymers," *J. Chem. Phys. 37*, 1003-1007 (1962).

Smith 1956. Carl W. Smith and Malcolm Dole, "Specific Heat of Synthetic High Polymers. VII. Polyethylene Terephthalate," *J. Polymer Sci. 20*, 37-56 (1956).

Smothers 1966. W. J. Smothers and Y. Chiang, *Handbook of Differential Thermal Analysis*, Chemical Publishing Co., New York, 1966.

Smyth 1951. H. T. Smyth, "Temperature Distribution During Mineral Inversion and Its Significance in Differential Thermal Analysis," *J. Am. Ceramic Soc. 34*, 221-224 (1951).

Steere 1966. Robin C. Steere, "Thermal Properties of Thin-Film Polymers by Transient Heating," *J. Applied Phys. 37*, 3338-3344 (1966).

Tobolsky 1960. Arthur V. Tobolsky, *Properties and Structure of Polymers*, John Wiley and Sons, New York, 1960.

Tye 1969. R. P. Tye, ed., *Thermal Conductivity*, Academic Press, New York, 1969.

Ueberreiter 1951. K. Ueberreiter and S. Nens, "Specific Heat, Specific Volume, Temperature and Heat Conductivity of High Polymers, Bistyrene and High Polymer Styrene," *Kolloid Z. 123*, 92-99 (1951).

Wendlandt 1964. Wesley Wm. Wendlandt, *Thermal Methods of Analysis*, Vol. XIX in P. J. Elving and I. M. Kolthoff, eds., *Chemical Analysis*, Interscience Div., John Wiley and Sons, New York, 1964.

Wendlandt 1972. W. W. Wendlandt, "Topics in Chemical Instrumentation. LXVII. Thermal Analysis Techniques: Part II. Differential Thermal Analysis and Differential Scanning Calorimetry," *J. Chem. Educ. 49*, A624-A635, A671-A680 (1972).

Williams 1955. Malcolm L. Williams, Robert F. Landel, and John D. Ferry, "The Temperature Dependence of Relaxation Mechanisms in Amorphous Polymers and Other Glass Forming Liquids," *J. Am. Chem. Soc. 77*, 3701-3707 (1955).

Williams 1971. David J. Williams, *Polymer Science and Engineering,* Prentice-Hall, Inc., Englewood Cliffs, New Jersey, 1971.

Wood 1946. Lawrence A. Wood and Norman Bekkedahl, "Crystallization of Unvulcanized Rubber at Different Temperatures," *J. Applied Phys. 17,* 362-375 (1946).

Wunderlich 1957. Bernhard Wunderlich and Malcolm Dole, "Specific Heat of Synthetic High Polymers. VIII. Low Pressure Polyethylene," *J. Polymer Sci. 24,* 201-213 (1957).

Wunderlich 1964a. Bernhard Wunderlich, David M. Bodily, and Mark H. Kaplan, "Theory and Measurement of the Glass-Transformation Interval of Polystyrene," *J. Applied Phys. 35,* 95-102 (1964).

Wunderlich 1964b. Bernhard Wunderlich and David M. Bodily, "Dynamic Differential Thermal Analysis of the Glass Transition Interval," *J. Polymer Sci. C6,* 137-148 (1964).

Wunderlich 1970. B. Wunderlich and H. Baur, "Heat Capacities of Linear High Polymers," *Fortschr. Hochpolym. Forsch. (Advances in Polymer Sci.) 7,* 151-368 (1970).

Wunderlich 1971. Bernhard Wunderlich, "Differential Thermal Analysis," Chap. VIII in Arnold Weissberger and Bryant W. Rossiter, eds., *Physical Methods of Chemistry,* Vol. 1 of A. Weissberger, ed., *Techniques of Chemistry,* Part 5, Interscience Div., John Wiley and Sons, New York, 1971.

10

Structure-Property Relationships

The physical and mechanical properties of high polymers depend primarily on their molecular weight and molecular structure. Many properties, for example tensile strength, elongation, modulus or stiffness, hardness, and processability show a characteristic relationship to molecular weight, and a certain minimum molecular weight must be attained to realize the useful or optimum properties of a polymer. The effect of T_m, T_g, and the increase of viscosity with molecular weight shown in Fig. 10-1 serve to define, in terms of the variables molecular weight and temperature, regions in which the properties of typical plastics, rubbers, viscous liquids, and other forms of polymeric materials may be found. An increase in molecular weight leads to improvement in many physical and mechanical properties, but this is achieved at the expense, for example, of ease of processability. Thus, in practice, one must almost always compromise between optimum properties on the one hand and optimum processing on the other. To achieve such compromises requires that the polymer scientist synthesize materials to within fairly specific molecular-weight limits, as well as select appropriate molecular structures.

As amplified in the *Textbook*, Chaps. 1A and 7, the most important variables determining the state of a polymer are the nature and magnitude of the restraints on the motion of its chain atoms, that is, its molecular motion. Kinds of restraints and their effect on properties are illustrated in Fig. 10-2, in which increasing restriction to molecular motion occurs as one moves either down or to the right, or both.

Perhaps the properties most often associated with long-chain polymers are those characteristic of the amorphous state. These properties include low elastic modulus, high extensibility, and the whole range of viscoelastic behavior. The molecular basis for this behavior is related to the randomly coiled conformation of long chains in the amorphous state. Crystallinity, if present, stiffens the material and makes it harder, tougher, higher melting, and more resistant to solvents, decreases its extensibility, and enhances its mechanical strength. The ductility of crystalline polymers, for example, depends to a great extent upon the size, perfection, and organization of the

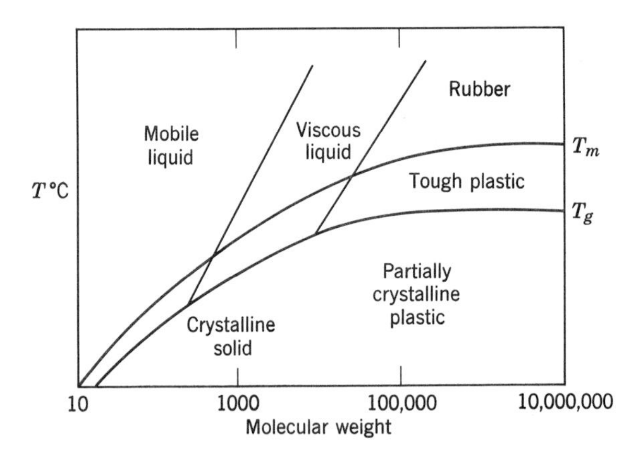

Fig. 10-1. Approximate relations among molecular weight, T_m, T_g, and polymer properties (*Textbook*, Fig. 7-9).

crystals. By synthesizing polymers whose architecture embodies the proper arrangement of crystalline and amorphous domains, the polymer chemist can achieve a balance of engineering properties on which the outstanding performance of the material depends.

In considering structure-property relationships, it is instructive to classify polymers into one of four regimes shown on the volume-temperature plot in Fig. 10-3. Every polymer is a mixture of these states in varying amounts, but with a preponderance of one (with the exception of the liquid state). This means that a homogeneous system

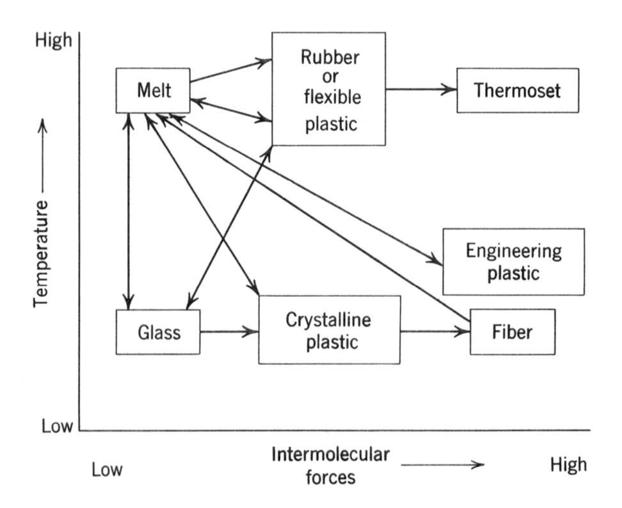

Fig. 10-2. The interrelation of the states of bulk polymers. The arrows indicate the directions in which changes from one state to another can take place (*Textbook*, Fig. 1-4, from Billmeyer 1968).

is seldom achieved. The properties of most polymers can be described in terms of the amounts of these species that they contain.

While molecular structure can be elucidated by such classical techniques as ultraviolet and infrared spectroscopy, nuclear magnetic resonance spectroscopy, etc., detailed consideration of these techniques is beyond the scope of this book. We have, however, considered x-ray diffraction (Chap. 8B) and its use in determining the amorphous-crystalline ratio, the nature of the crystalline region, and the effect of orientation in both the amorphous and crystalline regions. Furthermore, we have seen how morphology can be observed directly by optical and electron microscopy (Chaps. 8C and 8D) and how measurement of thermal properties (Chap. 9) can be used to delineate the four regions in addition to providing detailed information on the crystalline state.

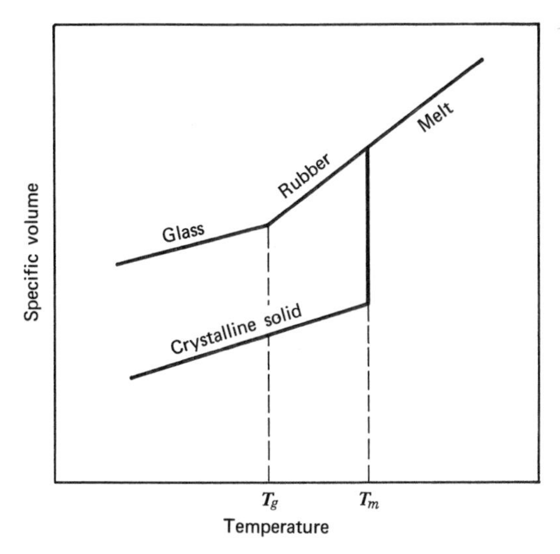

Fig. 10-3. Regimes of bulk polymers in terms of volume and temperature.

In this chapter we turn our attention to the behavior of polymers in bulk as a function of molecular structure, as determined by melt and concentrated-solution viscosity, dynamic electrical and mechanical behavior, stress-strain measurements, diffusion, and swelling. Each of these is considered only to the extent necessary to understand and interpret the related experiments, Exps. 25-31. This behavior, except for swelling and diffusion, can be unified in terms of stimuli applied to the polymer material and its response, as determined by its properties. As summarized in Fig. 10-4, the stimulus in (for example) the bulk-viscosity experiment is a shear strain applied at a constant rate. What one observes is the response to this strain in the form of stress in the material, related to the strain through the property viscosity. The stimuli, material properties, and responses for the cases of mechanical and electrical behavior are also shown in the figure.

The stimulus may be thought of as disturbing the thermodynamic equilibrium of the system, and the response shows the route by which

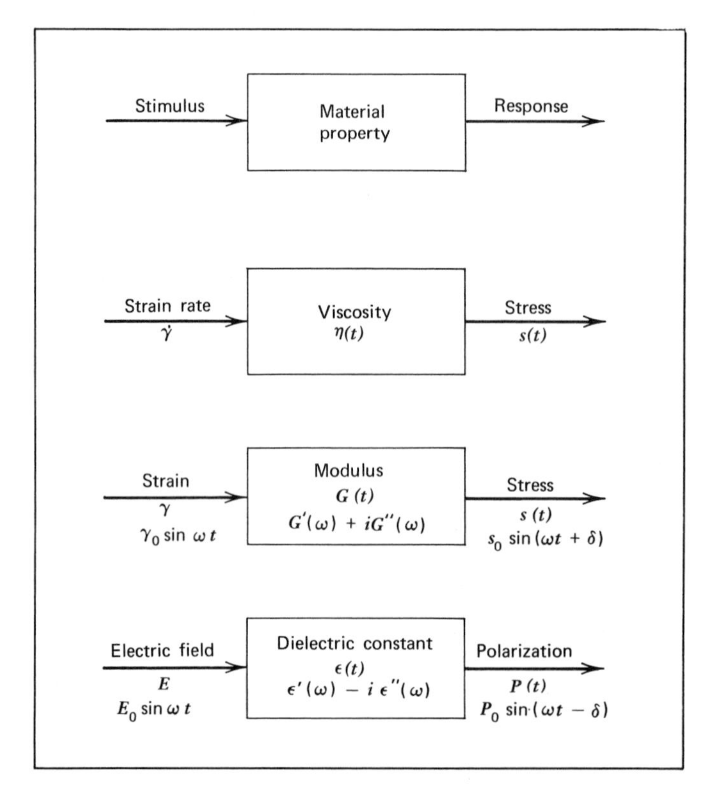

Fig. 10-4. Stimuli, responses, and material properties for the linear viscoelastic behavior of polymers.

equilibrium is again approached. In general, the response always occurs later than the stimulus; this delay gives rise to the phenomenon of *relaxation*, in which the response continues after the stimulus is removed. This behavior is characteristic of polymers.

For quite small stimuli, the relation between the magnitudes of the stimulus and response may be linear. In this case *Boltzmann's superposition principle* (*Textbook*, p. 199) applies, stating that the total response to several stimuli is the sum of the responses to each one applied separately. This situation holds well for dynamic electrical and mechanical behavior, but not at all in the usual measurement of stress-strain properties, where the deformations are large and lead to such phenomena as yielding, permanent distortion, and failure.

A. *Bulk Viscosity*

Polymeric substances, as commonly prepared, are composed of macromolecules similar in configuration but varying over a wide range of molecular weight. The configurations of the various types of macromolecules range from linear and flexible through branched and partially crosslinked to completely crosslinked systems, in which the molecular weight has increased without limit. Heavily crosslinked polymers may be mechanically dispersed in liquids, producing systems whose flow properties are dependent on the distribution of the sizes and shapes of the particles, and on particle-particle and particle-solvent interaction forces. For these dispersions, the internal nature of the dispersed particles is of little significance in determining the flow properties of the mixture. For linear and flexible, branched, or lightly crosslinked macromolecules in solution, however, the structure of the macromolecule is of great importance with respect to flow properties. Such polymer molecules may be regarded as bundles or coils through which solvent may drain, giving rise to internal particle-solvent dynamic interaction. The segments of an individual molecule are in constant agitation, so that the shape and conformation of each random coil is constantly varying. At any fixed temperature, however, a most probable extension of the individual molecule exists (*Textbook*, Chap. 2B), so that each coil may be regarded as a three-dimensional spring with a force of compression or extension arising when the equilibrium conformation is disturbed. In laminar shear, such a coil moves in the flow field with a rotational frequency equal to $\dot{\gamma}/2$, where $\dot{\gamma}$ is the rate of shear. In this rotation, segments are subject to alternate extension and compression so that the coil is constrained to vibrate with a frequency dependent on the rate of shear (Bueche 1962).

On the basis of coil structures of the type described, the flow of solvent-polymer systems has been analyzed, with various assumptions as to models of the coil-solvent interaction, the spring-like nature of the coil, and the effect of polydispersity. Such calculations have been successful in explaining the flow properties of dilute solutions in terms of the nature of the dissolved polymer molecule and the solvent. The study of dilute solutions has led to interesting and useful knowledge of polymer properties, with excellent agreement between theory and experiment, as amplified in Chap. 7C.

When the molecular weight of a polymer reaches a critical value in solutions of fixed concentration, or when the concentration of a particular polymer exceeds a critical range, flow characteristics can no longer be predicted from the information gained in dilute solution studies. While the theory of concentrated solutions has not developed rapidly for the quantitative interpretation of flow properties, data obtained from flow studies may be of great value in polymer process control and fabrication, and in the development of new polymeric materials. It is surprising that comparatively little study has been made of the concentrated systems, when it is considered that the industrial use of polymers, as in the spinning and coating fields, often involves concentrated solutions or melts (but for a recent review, see Osaki 1971).

Terms and definitions

Consider a thin layer of polymer between two parallel plates, one of which is moved with respect to the other. This requires the application of a force, the *shear stress* s, which is proportional to the rate of change of flow rate with distance through the layer in the direction normal to the plates. This rate of change is $\dot{\gamma}$, the time rate of change of shear strain, that is, the *shear rate*. The proportionality constant is the *viscosity*, η:

$$s = \eta\dot{\gamma} \tag{10-1}$$

The material is said to be *Newtonian* if η is independent of $\dot{\gamma}$, and *non-Newtonian* if it is not. In the latter case, the viscosity at specified values of s and $\dot{\gamma}$ is termed the *apparent viscosity* η_a.

Viscous behavior of polymers can be described by several types of flow curves, one of which is shown in Fig. 10-5. Here curve *N* represents a Newtonian material, the slope of the straight line being the viscosity, η; curve *P* represents a material for which the viscosity *decreases* with increasing shear rate, that is, a *shear-rate thinning* material; curve *D* represents a material for which the viscosity *increases* with increasing shear rate, that is, a *shear-rate thickening* material; and curve *B* represents a material which behaves like a Newtonian material after a critical shear stress, the so-called *yield stress*, $s = Y$, is exceeded. For many materials the viscosity is also dependent on time of shear as well as the shear-rate history and the time of rest (Bauer 1967) but a discussion of these effects goes beyond the scope of this book.

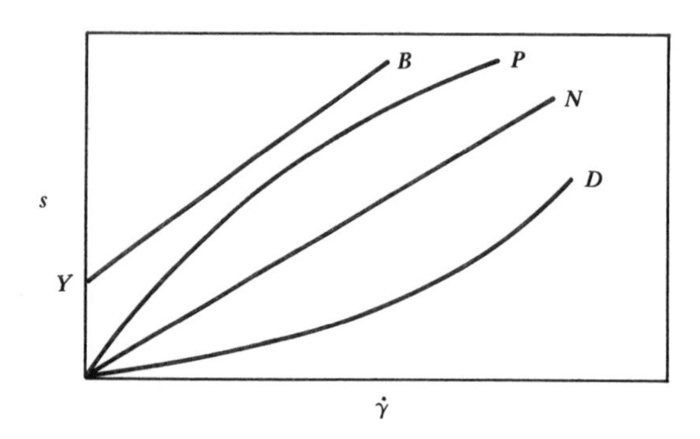

Fig. 10-5. Types of viscosity behavior demonstrated by polymer melts or concentrated solutions: *N*, a Newtonian material; *P*, shear-rate thinning; *D*, shear-rate thickening; *B*, Newtonian viscosity after a critical yield stress *Y* is exceeded.

When a concentrated polymer solution or a melt is subjected to shear, the flow curve is often non-linear. At low rates of shear a constant viscosity, η_0, is exhibited. At some critical shear rate, the shear rate increases more rapidly than the shear stress. After this second regime of flow, in which shear thinning takes place, the shear rate again becomes linear with shear stress, and an upper regime of flow ensues, with a viscosity designated η_∞. This behavior (Fig. 10-6) is one of the striking features of the flow of concentrated systems and melts of all kinds. It is shown for such diverse dispersions as clay suspensions (Rebinder 1954), polyisobutylene in decalin (Brodnyan 1957), lubricating oils (Philippoff 1936, 1958), aluminum soaps in toluene (Weber 1956), and polyethylene (Porter 1960), for example. These and many subsequent investigations have established the

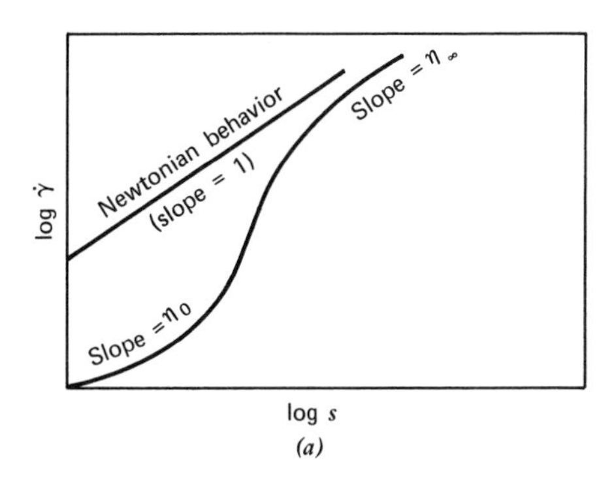

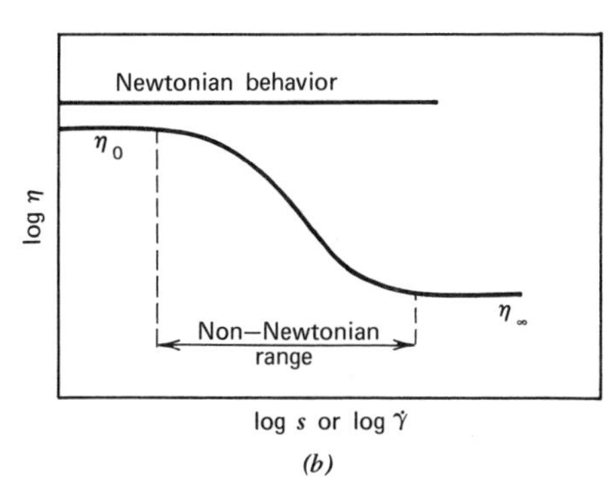

Fig. 10-6. Upper and lower Newtonian viscosity behavior in terms of (*a*) log shear rate versus log shear stress and (*b*) log apparent viscosity versus either log shear rate or log shear stress.

generality of the phenomenon. Systems which exhibit the two regimes of viscosity are also generally capable of storing energy when subjected to laminar shear. This capacity is evidenced by the appearance of *normal stresses*, that is, stresses in the direction perpendicular to the direction of the applied shear. These are commonly observed in several ways, including climbing phenomena on rotating elements. The stored energy may be held in the form of the energy of non-equilibrium orientation, of shear strain arising from particle deformation, or of dissociation of polymer molecules. All of these forms of stored energy vary in time of relaxation, according to the viscosity of the system and the nature of the storing capacity. Stored energy of orientation may be dissipated rapidly in capillary viscometry by swelling and loss of velocity of the extrudate. The relaxation of shear stress is generally much more rapid than the relaxation of the stored shear-generated energy. Solutions which are referred to as *viscoelastic* are capable of storing energy as recoverable shear strain, but the forms of energy storage in flowing systems are clearly not limited to this. When the energy storage is reversible, normal forces may be expected, but these will not be shown when the stored energy is dissipated as heat. Methods of assessing the normal stresses in flow have only recently been developed, and quantitative data are not available for most polymer systems.

Concentration and molecular-weight dependence Flow behavior in the non-Newtonian region shown in Fig. 10-6 can often be described by an empirical power-law equation of the type

$$s = K\,\dot{\gamma}^{n} \qquad (10\text{-}2)$$

where the exponent n is called the non-Newtonian flow index. This type of equation is widely used, even though it may fit the data only over a narrow range of shear rate. Fortunately, it often describes flow data under polymer processing conditions quite well.

The empirical approach has been applied also to the limiting zero-shear viscosity, η_0, in order to describe its dependence on concentration of polymer c and molecular weight M. Ferry (1953), Fox (1955), and others have shown that

$$\eta_0 = K' c^{n'} \tag{10-3}$$

where n' often has a value near 5 for polymers with molecular weight above a critical value M_c. Bueche (1952, 1953, 1956, 1959, 1960) and Fox (1956) have shown that, below M_c, η_0 is proportional to $\bar{M}_w$ and above it, to $\bar{M}_w^{3.4}$, as depicted in Fig. 10-7. These two dependences can be combined in the equation, valid above $M = M_c$

$$\eta_0 = K'' c^5 \bar{M}_w^{3.4} \tag{10-4}$$

Many plots of the type of Fig. 10-7 have appeared in the literature (Berry 1968, Chinai 1965, Porter 1966*a*). Similar breaks have been found in plots of log zero-shear viscosity versus log concentration (Onogi 1963, Fox 1965, Hayahara 1967, Lyons 1970). The critical molecular weight M_c or the concentration at the break have been attributed to the "entanglement transition"--the point above which the viscosity is strongly influenced by polymer entanglement, polymer association, or polymer-polymer interaction. On the other hand, some workers (Chinai 1965, Pezzin 1966) have not observed an abrupt change

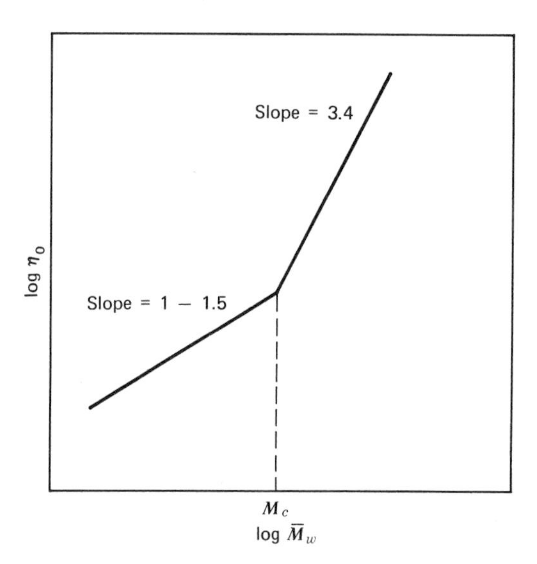

Fig. 10-7. Flow behavior of polymers above and below the critical molecular weight at which the power-law exponent changes.

in the slope of the log zero-shear viscosity - log concentration curve, and slopes above the transition considerably greater than 5 (as high as 50) have been reported (Fujita 1961, 1962; Hayahara 1967). The observation of a power near 5 is quite general, however, and limited theoretical substantiation has been developed (Onogi 1963, Berry 1968).

Melt viscosity measurement

Rheometers There are many different instruments for the measurement of melt viscosity, but the capillary viscometer is by far the most common and is the only one we describe. In this instrument the polymer melt is forced from a reservoir through a capillary of known length and diameter, either at constant flow rate (shear rate $\dot{\gamma}$) or at constant load (stress s) by a plunger activated by either mechanical or pneumatic means.

In the constant-load design, of which the extrusion plastometer measuring "melt index" (ASTM D 1238) and the CIL rheometer (Mullowney) are examples, a fixed load is applied to the top of the plunger, and the output rate is measured.

The constant-rate rheometer provides results more readily interpreted in terms of fundamental quantities, and is used in Exp. 25. The widely used instrument forming the basis for this experiment is the Instron Capillary Rheometer, an accessory to the standard Instron tensile testing machine. In this instrument, the crosshead drive of the testing machine advances the plunger at constant rate and the load cell measures the force on the plunger. The precision-bore capillary is made from tungsten carbide or stainless steel. Temperatures from just above ambient to 340°C can be obtained, and steady-state temperature is reached within 2-3 min after loading the instrument. A self-contained capillary rheometer of similar design is shown in Fig. 10-8.

The shear rate at the wall of the capillary, $\dot{\gamma}_w$, is proportional to the volume flow rate of polymer, and inversely proportional to the cube of the capillary diameter, d_c. In terms of instrument variables, the flow rate is given as $\pi d_p^2 v$, where d_p is the diameter of the plunger and v is the velocity of the crosshead. If both diameters are measured in inches and v in in./min (any other unit of linear dimensions is correct), the flow rate in sec^{-1} is given by

$$\dot{\gamma}_w = (2\, d_p^2/15\, d_c^3)v \qquad (10\text{-}5)$$

where the factor in parentheses is a constant for a given instrument and capillary. The shear stress at the wall is proportional to the pressure drop across the capillary and its diameter, and inversely proportional to its length, ℓ. In terms of instrument variables,

$$s_w = (d_c/\pi d_p^2 \ell)F \qquad (10\text{-}6)$$

where F is the force on the plunger. Again, the factor in parentheses is an instrument constant.

Fig. 10-8. The Instron capillary rheometer (courtesy Instron).

Typical data Typical melt-flow data, obtained on the Instron Rheometer using linear polyethylene, are shown in Table 10-1 and Fig. 10-9. For this example, d_p = 0.375 in., ℓ = 0.993 in., and d_c = 0.0516 in. Substituting into the above equations and converting to cgs units, one obtains $\dot{\gamma}$ = 148.5 v sec^{-1} and s_w = 7885 F dynes/cm^2 when v is in in./min and F in pounds. The apparent viscosity η_a is calculated by Eq. 10-1 as $\eta_a = s_w/\dot{\gamma}_w$. In the figure, points at intermediate rates of shear are plotted in addition to those corresponding to the tabulated data.

Corrections to the data A number of corrections can be made to the melt-viscosity data in order to obtain closer approximations to absolute values: The *Rabinowitsch correction* (Rabinowitsch 1929) compensates for non-Newtonian behavior in the calculation of the shear rate. The true shear rate is given in this correction by

TABLE 10-1. Melt viscosity data for Marlex 6035 linear polyethylene

	T°C	Crosshead speed, in./min			
		0.02	0.2	2	20
$\dot{\gamma}$, sec^{-1}	all	2.97	29.7	297	2970
F, lbs	160	14.3	60	193	440
	170	15.5	56.2	176	415
	180	12.8	50.5	164	435
	190	9.0	40	140	405
s_w, dynes/cm^2	160	113	473	1520	3470
	170	122	443	1390	3270
	180	101	398	1290	3430
	190	71	315	1100	3190
η_a, poise	160	38.0	15.9	5.12	1.17
	170	41.1	14.9	4.67	1.10
	180	34.0	13.4	4.35	1.16
	190	24.6	10.9	3.83	1.11

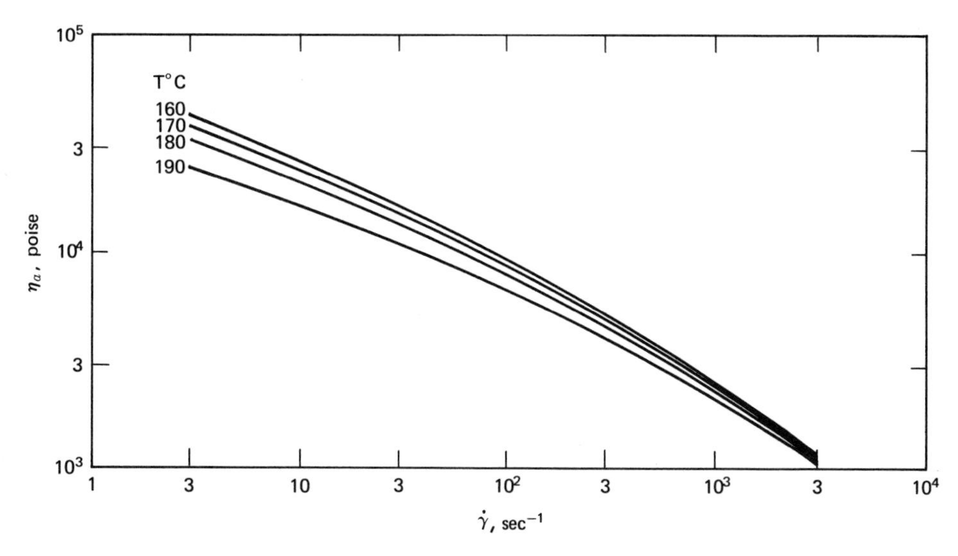

Fig. 10-9. Melt viscosity as a function of shear rate and temperature for Marlex 6035 linear polyethylene.

$$\dot{\gamma}_{t,w} = \dot{\gamma}_w(3n + 1)/4n \qquad (10\text{-}7)$$

where n is the non-Newtonian flow index defined in Eq. 10-2. This correction is of the order of 2-5%.

The *entrance correction* is significant only if the ratio of length to diameter of the capillary is small, and is determined experimentally by taking measurements with capillaries of varying length (Bagley 1957).

A correction due to *pressure drop across the plunger barrel* can be significant for short, large-diameter capillaries (Metzger 1965). These and several other corrections are described by Van Wazer (1963).

Other information from flow measurements Other useful observations during melt-flow determinations include measurement of the swelling of the polymer stream as it is extruded from the capillary, and qualitative observation of such surface characteristics as melt fracture (Tordella 1969, Rudin 1970) and change in gloss. Die swelling, the ratio of the diameter of the extrudate to that of the capillary, is related to the elasticity of the melt (Lenk 1968, Mendelson 1968, Brydson 1970). It generally increases with molecular weight, the presence of long-chain branching, and breadth of the molecular-weight distribution (Zabusky 1964, Bagley 1969, Brydson 1970, Williams 1971).

Concentrated-solution viscosity measurement

Since viscosities and operating temperatures are lower, equipment for measuring concentrated-solution viscosity is generally less complicated than melt rheometers. Three rotational viscometers are described, any of which can be used in Exp. 26.

Cone-and-plate viscometer The cone-and-plate design has a significant advantage over a capillary viscometer in that in the former the shear rate is constant for small cone angles, whereas in the latter the shear rate varies from center to wall of the capillary. The Ferranti-Shirley cone-and-plate viscometer (Fig. 10-10) has a stationary flat lower plate and a rotating cone driven by a variable-speed motor through a gear train. Shear rate is proportional to cone speed, and is calculated simply as the angular velocity of the cone (radians/sec) divided by the cone angle in radians. Shear stress is calculated from the torque of the spring, G, as

$$s = 3G/2\pi r^3 \qquad (10\text{-}8)$$

where G is in dyne cm and r is the radius of the cone in cm.

Typical data obtained with the Ferranti-Shirley viscometer for solutions of poly(vinyl chloride) in cyclohexanone are given in Table 10-2. The resulting viscosities (plus intermediate values not tabulated) are plotted as a function of shear rate in Fig. 10-11, and extrapolated values of η_0 are plotted according to Eq. 10-3 in Fig. 10-12. For the instrument used, r = 2.0 cm, the cone angle is 0.0055 radians, and scale divisions are converted to torque by means of a calibration constant of 1310 dyne cm/div, determined by experiments with a standard oil of known viscosity.

In Fig. 10-11, the break in the viscosity-concentration curve at a critical concentration is clearly seen. For the sample used, the slopes below and above the break, respectively, are 1.35 and 4.14.

Coaxial-cylinder viscometer In a coaxial-cylinder (couette) type viscometer, the fluid is contained in a narrow gap between two coaxial cylinders, either of which can be rotated. The Haake Rotovisco (Fig. 10-13) is such an instrument, in which the inner cylinder is connected through a spring to a drive system. The operating principles of the Haake and Ferranti-Shirley instruments are similar. The shear rate in this instrument is simply calculated from the radii of the inner and outer cylinders:

$$\dot{\gamma} = 2\, r_i r_o\, \Omega/(1/r_i^2 + 1/r_o^2) \qquad (10\text{-}9)$$

where Ω is the angular velocity of the inner cylinder.

Typical data obtained with this instrument, using the same solutions of poly(vinyl chloride) described in Table 10-2 and Figs. 10-11 and 10-12, are plotted as a function of shear rate in Fig. 10-14. The

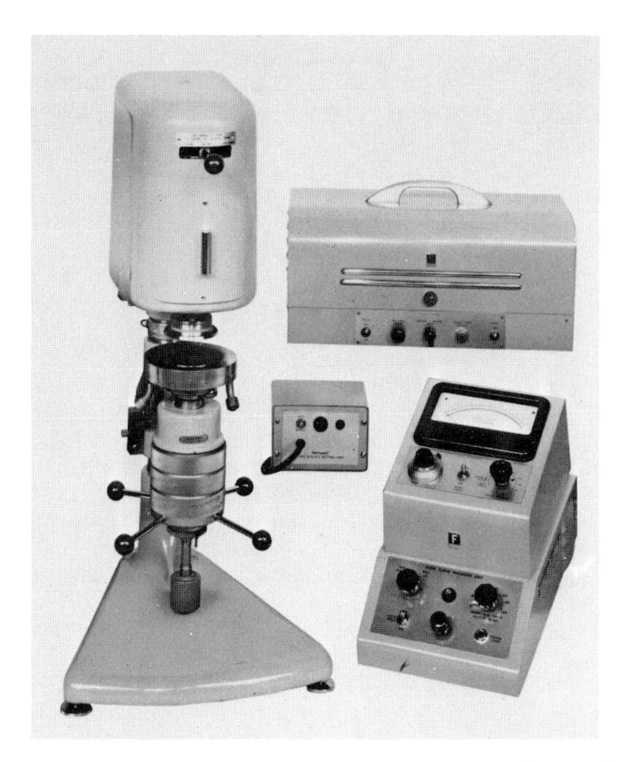

Fig. 10-10. The Ferranti-Shirley cone-and-plate viscometer (courtesy Ferranti).

TABLE 10-2. Solution viscosity of poly(vinyl chloride) (Geon 103) in cyclohexanone, 25°C, as measured on the Ferranti-Shirley cone-and-plate viscometer

Shear rate, sec^{-1}	Viscosity, cp, at concentration, g/ml				
	3	5	7	10	12
169	2.0	4.0	10.2	40.9	102.3
337	2.0	4.0	10.2	40.9	92.1
648	2.0	4.0	10.2	35.8	79.3
1012	2.0	4.0	10.2	34.1	72.6
1350	2.0	4.0	10.2	33.2	69.1
1687	2.0	4.0	10.2	31.7	65.5

plot of log η_0 versus log c for these data is quite similar to that of Fig. 10-12, the slopes below and above the critical concentration being 1.65 and 4.43, respectively.

Rotating-cylinder viscometer A modification of the couette-type viscometer leading to a very simple design is to eliminate the outer cylinder entirely. The inner (rotating) cylinder is simply immersed in a relatively large container of the fluid to be tested. The Brookfield viscometer (Fig. 10-15) and the similar Ferranti portable viscometer are of this type.

The Brookfield instrument is one of the most widely used and popular industrial instruments for measuring viscosity because of its portability, simplicity of operation, and low cost. Data are obtained rapidly (less than 5 min) and a wide range of viscosities is covered (0-20,000 poise) by the use of different size cylinders. The shear rate and shear stress are not readily calculated (Van Wazer 1963), but the simple approximation that the shear rate is approximately 0.2 times the revolutions per minute of the cylinder is useful. Conversion of dial readings to viscosities is normally made by the use of factors supplied by the manufacturer for specific combinations of speed and spindle.

Data obtained with the Brookfield viscometer for the same set of poly(vinyl chloride) solutions are shown in Fig. 10-16. The remarks of the preceding paragraph concerning the Haake Rotovisco and Ferranti-Shirley data apply. The slopes below and above the break in the log η_0 versus log c curve were 1.71 and 4.47, respectively.

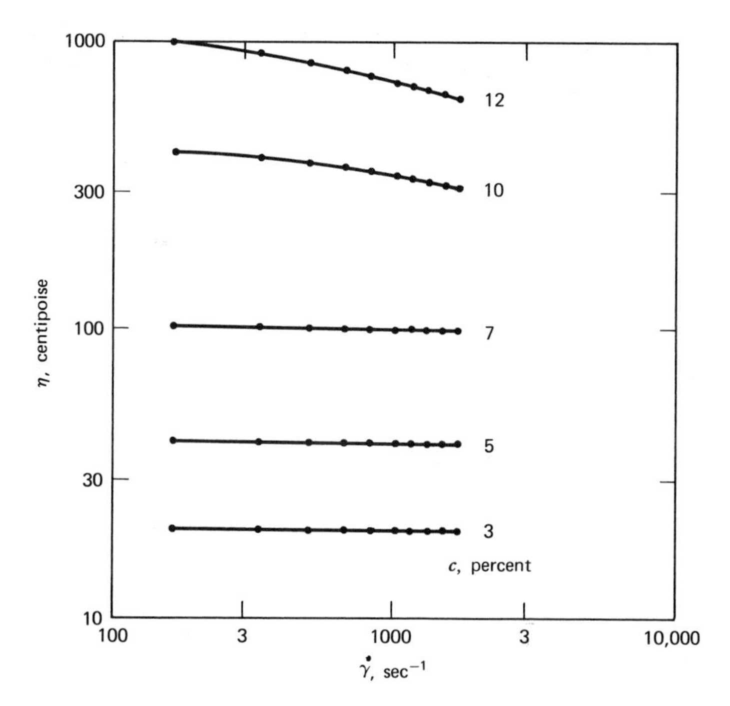

Fig. 10-11. Concentrated-solution viscosity of Geon 103EP poly(vinyl chloride) solutions in cyclohexanone, as a function of shear rate and concentration, determined with the Ferranti-Shirley cone-and-plate viscometer.

Relation of flow to temperature and molecular parameters

Temperature Classically, the decrease in the bulk viscosity of polymers with increasing temperature is described by an equation of the Arrhenius type:

$$\eta = A \exp(-\Delta E/RT) \tag{10-10}$$

where A is a frequency factor depending on shear rate, shear stress, and molecular structure, and ΔE is the activation energy for viscous flow. The equation holds only over small temperature ranges, however, and in the non-Newtonian region the apparent activation energy depends upon whether viscosity at constant shear rate or constant shear stress is considered. For many systems, ΔE decreases with increasing shear rate, and its value at constant shear rate is less than at constant shear stress (Porter 1966*b*). The two apparent activation energies are equal in the Newtonian region (Collins 1970).

The temperature dependence of polymer melt viscosities can also be expressed in terms of the free volume (Chap. 9A) (Doolittle 1951, 1952):

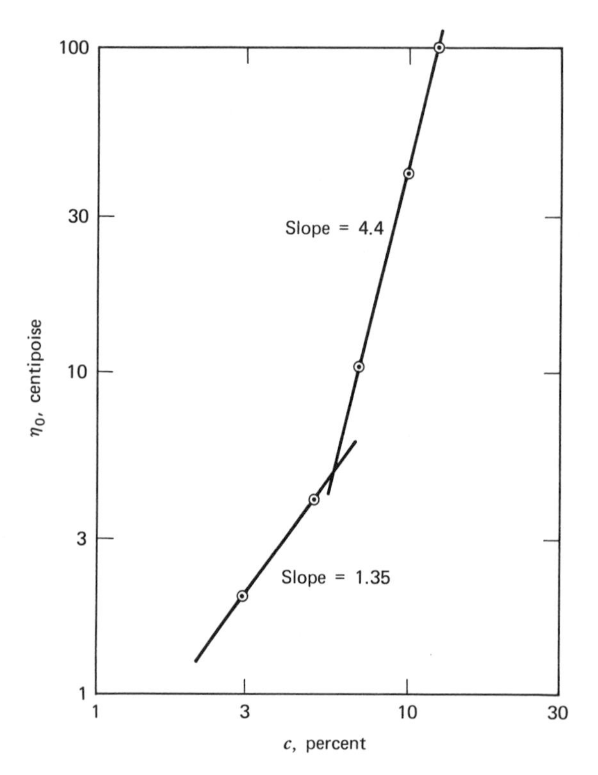

Fig. 10-12. Zero-shear viscosity-concentration plot for the data of Fig. 10-11 and Table 10-2.

$$\eta = A \exp(BV_0/V_f) \qquad (10\text{-}11)$$

where the free volume V_f is the difference between the specific volume of the melt and its value V_0 extrapolated to the absolute zero of temperature without phase change. For polymer melts, the Doolittle equation often holds over a wider temperature range than the Arrhenius equation.

It has been known for almost a hundred years that most of the change in viscosity of a material with temperature is associated with the concurrent change in its volume. This relationship is applied to polymers in the WLF equation (Williams 1955) (Chap. 9A),

$$\log(\eta_T/\eta_{T_g}) = 17.44(T - T_g)/(51.6 + T - T_g) \qquad (10\text{-}12)$$

The WLF equation holds up to about $T_g + 100°$; above this, the Arrhenius equation usually represents the data better.

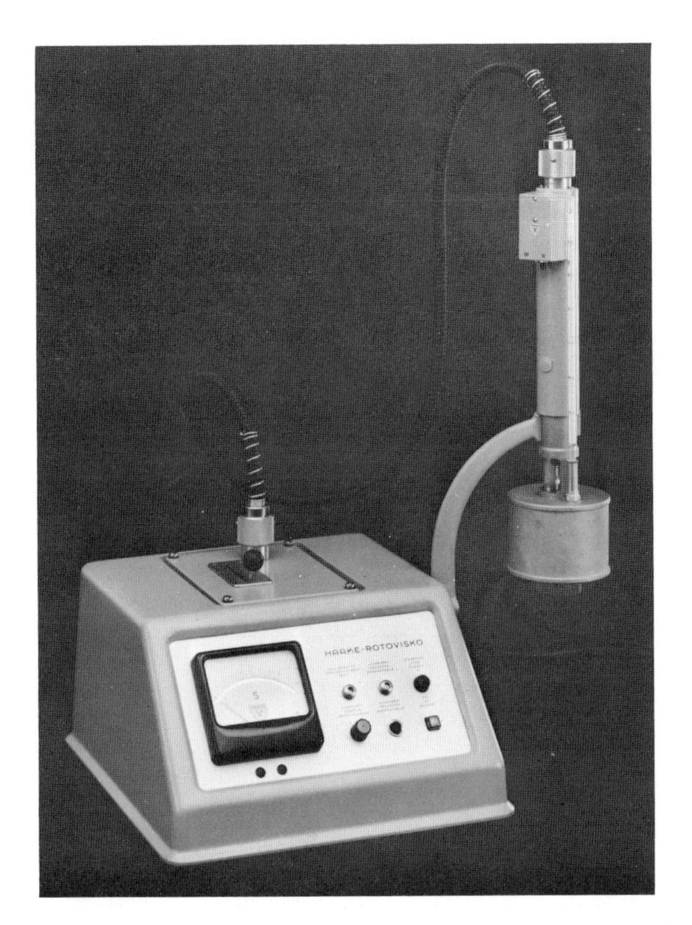

Fig. 10-13. The Haake Rotovisco couette-type viscometer (courtesy Haake).

It has also been known for many years that the ability to express the viscosity of a wide variety of polymer solutions and melts by a single universal equation permits the application of the *time-temperature equivalence principle* (*Textbook*, p. 199). This principle allows superposition of flow curves to produce a master curve from which flow behavior as a function of time can be predicted from data obtained in relatively short-time experiments at various temperatures (Leaderman 1958, Tobolsky 1958, Mendelson 1968).

For low shear stress, Eqs. 10-4 and 10-12 can be combined to yield an equation of the type

$$\log \eta = 3.4 \log \overline{M}_w + 5 \log c - 17.44(T-T_g)/(51.6 + T-T_g) + k \tag{10-13}$$

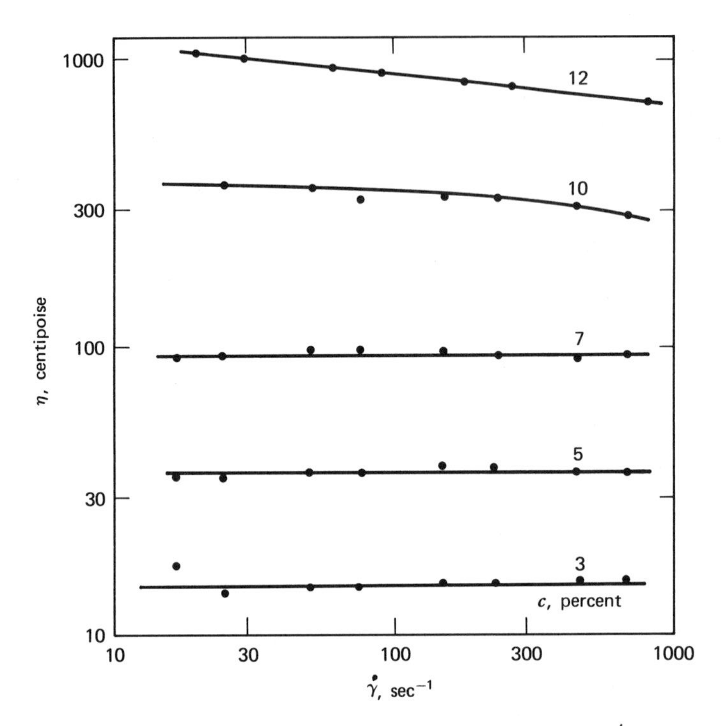

Fig. 10-14. Concentrated-solution viscosities of poly(vinyl chloride), as determined with the Haake Rotovisco; see Fig. 10-11.

Further discussion of free-volume theories of polymer flow is given by Ferry (1961, 1970), West (1967), Kaelble (1969), and Ward (1971).

Molecular weight The most widely used relationship between flow properties and molecular weight is the empirical observation expressed by Eq. 10-4, namely that the zero-shear viscosity is proportional to $\bar{M}_w^{3.4}$ above the critical value of molecular weight (which corresponds to a chain length of about 600 carbon atoms for many polymers). Application of this relation yields only relative values of $\bar{M}_w$ unless polymers with known molecular weight are available for calibration purposes, just as is the case with dilute-solution viscosity measurement (Chap. 7C).

Bueche (1958) has suggested that absolute values of molecular weight can be obtained from the log η - log $\dot{\gamma}$ flow curve. His procedure involves matching the experimental flow curve for the polymer being investigated to a standard curve prepared from his theory but modified empirically. The point of match allows evaluation of a parameter interpreted as a characteristic vibrational frequency of the polymer coil. With fundamental constants, knowledge of this parameter allows calculation of the molecular weight. Graessley (1967) has developed a molecular theory which successfully predicts the shape of Bueche's empirical master curve. Tests of Bueche's theory have not been extensive or conclusive to date.

The effect of molecular weight on the flow curve for polymers of the same type and molecular-weight distribution is illustrated schematically in Fig. 10-17. As the molecular weight decreases, the onset of shear-rate dependence shifts to progressively higher rates of shear as indicated by the arrows in the figure. In addition, of course, the magnitude of the limiting zero-shear viscosity also decreases. The high-molecular-weight portion of the distribution influences the low-shear behavior, while the high-shear behavior is affected by the low-molecular-weight fractions in the sample. Data of this type have been reported in the literature for polyethylene (Mendelson 1970), polystyrene (Stratton 1966), poly(vinyl chloride) (Collins 1970), poly(dimethyl siloxane) (Johnson 1961), and polybutadiene (Gruver 1964).

Molecular-weight distribution Much effort has been applied in obtaining structural information, such as that on molecular-weight distribution and chain branching, from the flow curve (Edelmann 1954, 1960; Schurz 1959, 1970; Kirschke 1962, 1969; Chinai 1964). The de-

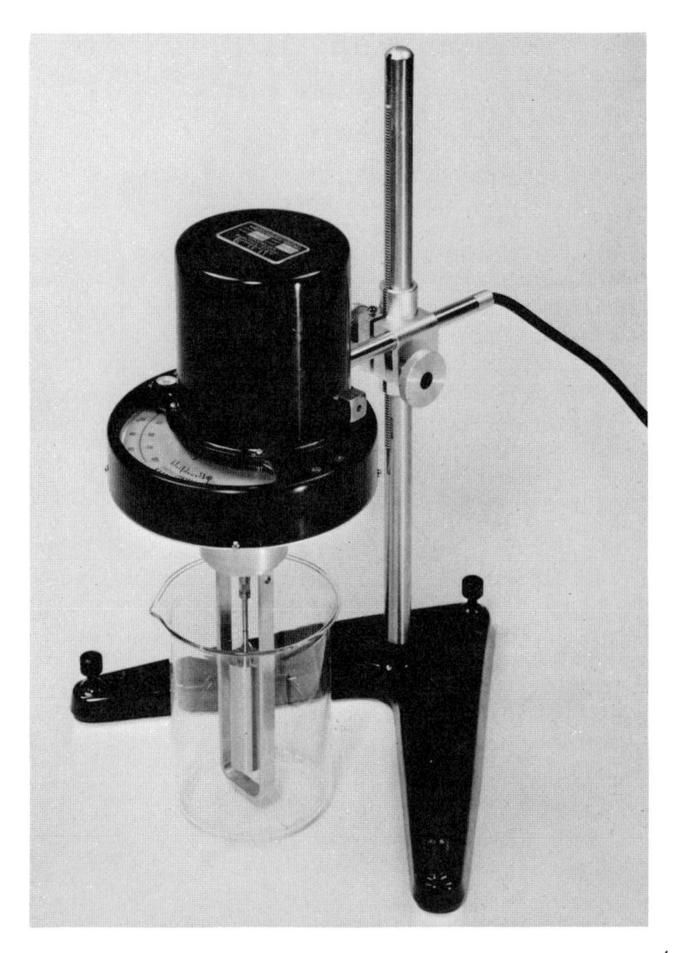

Fig. 10-15. The Brookfield rotating-cylinder viscometer (courtesy Brookfield).

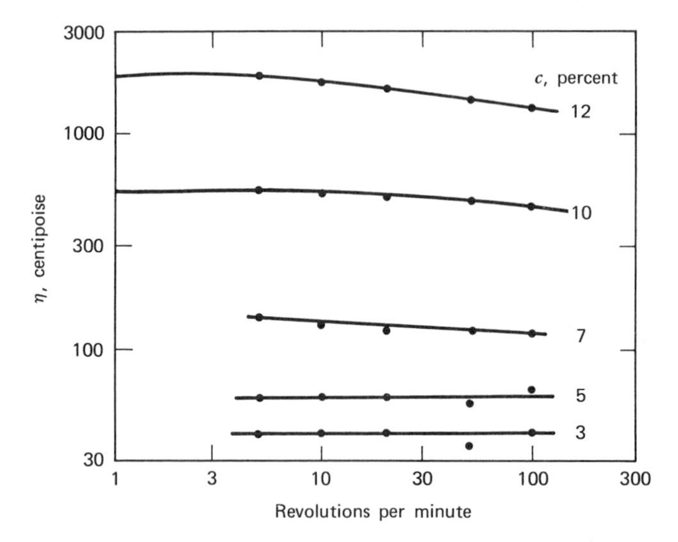

Fig. 10-16. Concentrated-solution viscosities of poly(vinyl chloride), as determined with the Brookfield viscometer; see Fig. 10-11.

pendence of the flow curve on distribution breadth for linear polymers with the same weight-average molecular weight is illustrated schematically in Fig. 10-18. Polymers with broad distributions show a more pronounced dependence of viscosity on the shear rate, the onset of the non-Newtonian region occurring at lower rates of shear in comparison to similar polymers with narrower distributions.

From the standpoint of processing, it should be clear that a polymer with a broad molecular-weight distribution would have a lower viscosity and hence be easier to process, producing higher output for given processing conditions. Unfortunately, advantage cannot always be taken of this fact because changes in the mechanical properties of the material occur as well. For example, as the distribution is broadened, tensile strength decreases while elongation increases. Where this is not objectionable, the direction to easier processing is clear.

The increased sensitivity of the melt viscosity to shear rate for broader molecular-weight distributions has been demonstrated by numerous workers (Rudd 1960; Mills 1961; Ferguson 1964; Porter 1964; Stabler 1967). Some success has been achieved in developing a molecular theory to account for this behavior (Graessley 1965, 1967). There have not been many attempts, however, to extend the limiting zero-shear-rate dependence of the viscosity on the molecular weight and molecular-weight distribution to higher shear rates, that is, to the non-Newtonian region.

While the manner in which polymer melt viscosity transforms from Newtonian to non-Newtonian as a function of molecular weight, molecular-weight distribution and interchain entanglement has not been established with certainty, the data seem to favor the behavior shown schematically

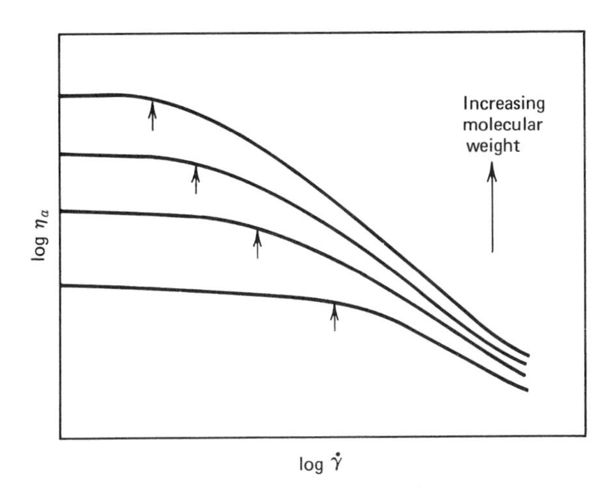

Fig. 10-17. The effect of varying the molecular weight on the melt flow curves of polymers of the same chemical type and molecular-weight distribution. Arrows indicate the point of deviation from Newtonian behavior.

in Fig. 10-19 (Collins 1965; Onogi 1966, 1968; Stratton 1966; Hayahara 1968). As the shear rate is increased, the viscosity decreases, and as molecular weight increases, the viscosity ultimately becomes independent of shear rate. The critical molecular weight for entanglement, M_c, may also shift to higher molecular weights with increasing shear rate, as required by Graessley's theories.

Ferguson (1964) has shown that narrowing the molecular-weight distribution leads to a greater dependence of viscosity on temperature and lower melt elasticity.

Chain branching The effect of long-chain branching on melt and concentrated-solution viscosity is even more complex and less well understood than the dependences discussed previously. Qualitatively, long-chain branching appears to have the same effect on the viscosity-shear rate curve as broadening the molecular-weight distribution (Schaefgen 1948; Schreiber 1962; Kraus 1965; Graessley 1968). For molecular weights below M_c, a branched polymer has a lower viscosity than a linear polymer of the same weight-average molecular weight. On the other hand, for molecular weights above M_c, a branched polymer has a higher viscosity at low shear rates (Moore 1959, Long 1964), but becomes more non-Newtonian with increasing shear rate and has a lower viscosity than the corresponding linear material at high shear rates. Because of this complex behavior and because the effects of branching cannot easily be separated from the effects of polydispersity, it has not been possible to derive information about branching from rheological measurements alone.

For a recent review, see Nagasawa 1972.

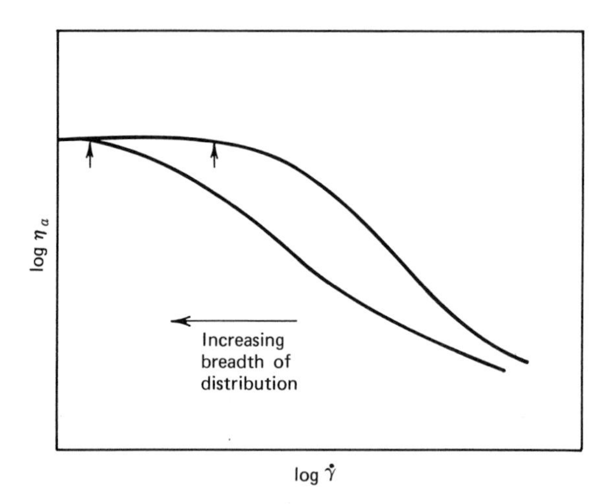

Fig. 10-18. The effect of varying the molecular-weight distribution breadth on the melt-flow curves of polymers of the same chemical type and weight-average molecular weight. Arrows indicate the point of deviation from Newtonian behavior.

GENERAL REFERENCES

Eirich 1956-1969; Severs 1962; Van Wazer 1963; Meares 1965, Chap. 13; Reiner 1966; Berry 1968; Lenk 1968; Mendelson 1968; Brydson 1970; Ferry 1970; Williams 1971, Chap. 12.

B. *Dynamic Mechanical Behavior*

The viscoelastic nature of polymers is well known and is unique in the field of material properties. The term is used to describe the time-dependent mechanical properties of polymers, which in limiting cases can behave either as elastic solids or as viscous fluids. A knowledge of the viscoelastic behavior of polymers and its relation to molecular structure is essential to an understanding of both processing and end-use properties.

The time-dependent changes in polymers subjected to constant stress (creep) or constant strain (stress relaxation) give insight into viscoelastic behavior (*Textbook*, Chap. 6C), but more information can be obtained, often on a more convenient time scale, by the study of dynamic mechanical properties. Here the response (stress) in a material subjected to a periodic stimulus (strain) is measured. Generally, the applied deformation and the resulting stress both vary sinusoidally with time.

The modulus curve

Dynamic experiments yield both the elastic modulus of the material and its mechanical damping, or energy dissipation, characteristics. These properties can easily be determined as functions of frequency (time) and temperature. Application of the time-temperature equivalence principle, as described for flow curves in Sec. A, yields master curves of modulus and damping factor like those shown in Fig. 10-20. The five regions outlined there are typical of polymer viscoelastic behavior.

In the glassy region, the polymer is below T_g and has a modulus, typically, of about 10^{10} dynes/cm^2. The transition region includes T_g, taken here as the point of inflection of the modulus curve or the maximum in the damping curve. In this region the modulus drops by about a factor of 1000. The third region is termed the rubbery plateau, where the modulus remains roughly constant. This is followed by a region of elastic or rubbery flow, and finally liquid flow with very little elastic recovery. Here the modulus may drop below 10^5 dynes/cm^2.

Effect of structure Variation in the molecular structure of a polymer has a pronounced effect on its modulus curve. Increasing molecular weight extends the region of rubbery flow to higher temperatures, while broadening the molecular-weight distribution results in a lower sensitivity of the modulus to temperature in the rubbery- and liquid-flow regions. These changes are illustrated schematically in Fig. 10-21. It should also be noted that at temperatures below T_g, molecular weight has essentially no effect on the modulus as long as the molecular weight is high enough (Fujita 1957*a*,*b*; Ninomiya 1957, 1959; Tobolsky 1958*a*). Crosslinking affects the modulus curve above T_g in a profound way. Even slight crosslinking inhibits polymer flow and

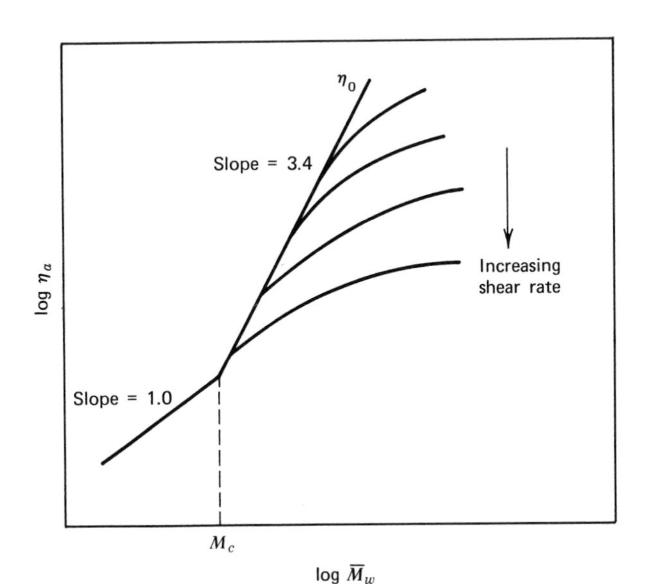

Fig. 10-19. The effect of varying the shear rate on the melt flow curve of a typical polymer.

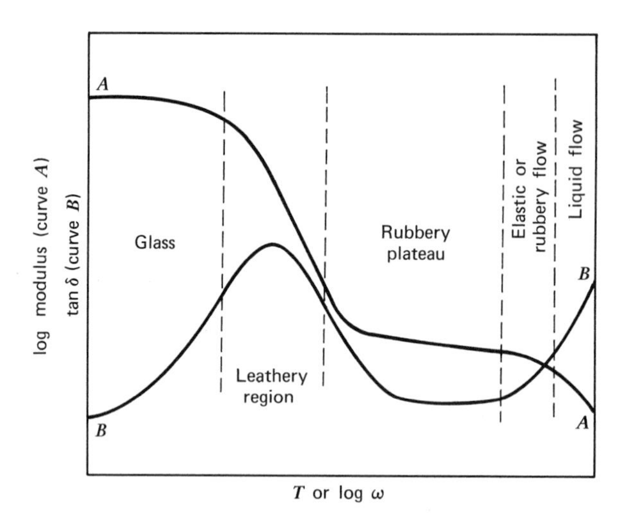

Fig. 10-20. Master curves of the modulus and damping factor of a typical polymer, showing the various regions in which characteristic physical properties are exhibited.

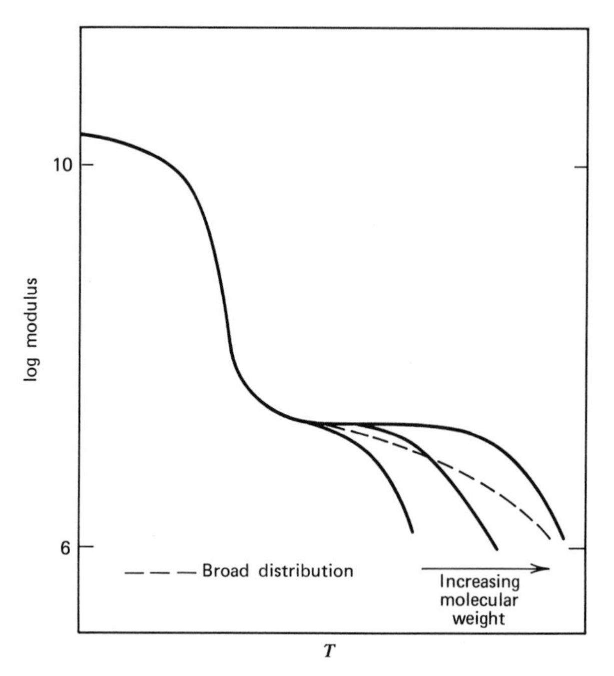

Fig. 10-21. The effect of varying molecular weight and molecular-weight distribution on the master modulus curve.

extends the rubbery plateau. High degrees of crosslinking increase the modulus above T_g as illustrated in Fig. 10-22 (Drumm 1956; Tobolsky 1964). Crystallinity also has a significant effect on the modulus-temperature curve, increasing the modulus somewhat like crosslinking except that at the melting point the modulus drops off rapidly as shown in Fig. 10-23 (Nielsen 1962, 1963; Tobolsky 1963; Krigbaum 1964). The presence of crystalline regions produces a temperature range between T_g and T_m where the modulus is high and the impact resistance is good (Takayanagi 1963). Crystalline polymers (such as linear polyethylene with T_g = -80°C and T_m = 138°C) have this combination of high modulus and good impact strength in a temperature range of great practical importance.

For amorphous polymers, the loss tangent or damping maximum reflects the glass transition, and changes in molecular structure variables affect this maximum in much the same manner as they affect the glass-transition temperature. The secondary loss peak due to crystallinity usually occurs above the glass transition. Crystallinity also increases the magnitude of the damping (Nielsen 1962), as does the addition of some compounding additives such as fillers. Secondary transitions which are often below the glass transition are due to the specific mobility of short segments of the molecular chain, or to side-chain mobility, and have been related to the strength of the material (Boyer 1968). These transitions below T_g cannot be detected by thermal methods. Useful information can also be obtained from the damping curve for polyblends or graft and block polymers, where a single broad peak is generally observed for a compatible blend or block copolymer.

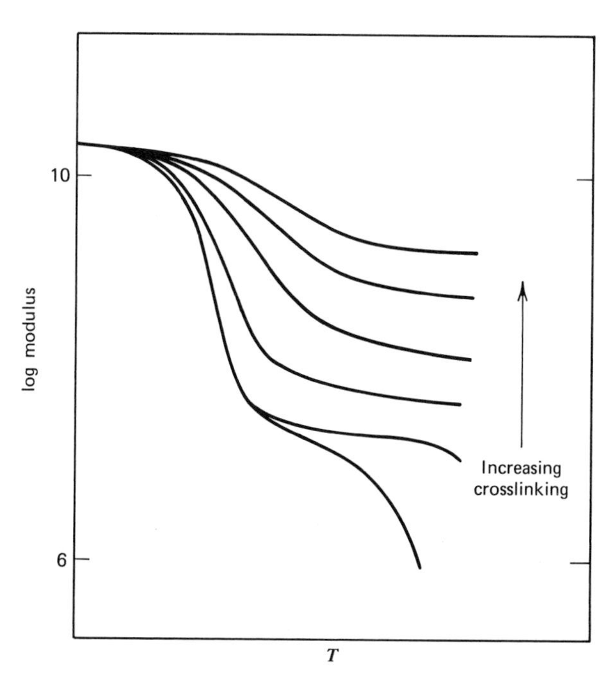

Fig. 10-22. The effect of varying degree of crosslinking on the master modulus curve.

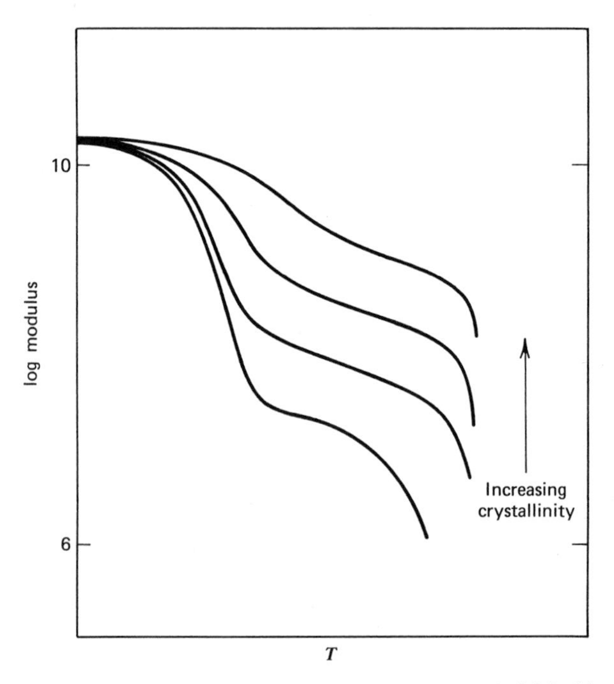

Fig. 10-23. The effect of varying degree of crystallinity on the master modulus curve.

Theory

If a sinusoidal elongation strain is applied to a specimen at a frequency ω and with a small amplitude, the stress also varies sinusoidally (for linear viscoelastic behavior) but is out of phase with the strain by a phase angle δ. Thus,

$$\gamma = \gamma_0 \sin \omega t \tag{10-14}$$

and

$$s = s_0 \sin(\omega t + \delta) \tag{10-15}$$

where γ_0 and s_0 are the peak strain and peak stress, respectively (Fig. 10-24). Expanding Eq. 10-15

$$s = s_0 \sin \omega t \cos \delta + s_0 \cos \omega t \sin \delta \tag{10-16}$$

The peak stress can be resolved into a component $s_0 \cos \delta$ which is in

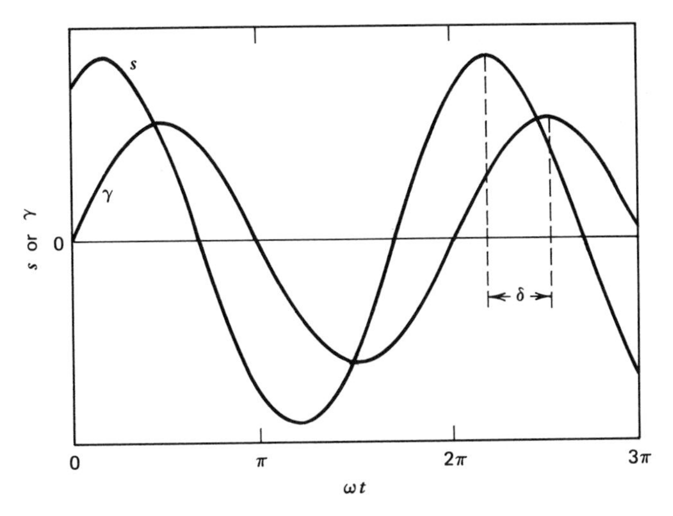

Fig. 10-24. Stress and strain as a function of time in the application of a sinusoidal strain to a viscoelastic specimen.

phase with the strain, related to the stored elastic energy, and a component $s_0 \sin \delta$ which is 90° out of phase with the strain, related to viscous loss of energy. The ratio of the two, $\tan \delta$, is the *dissipation factor*, an indicator of the relative importance of the viscous as compared to the elastic aspects of the material's behavior.

Two moduli* can be defined:

$$G' = (s_0/\gamma_0)\cos \delta \tag{10-17}$$

and

$$G'' = (s_0/\gamma_0)\sin \delta \tag{10-18}$$

whose ratio is also the dissipation factor. The modulus can also be expressed in complex form:

$$G^* = G' + i\, G'' \tag{10-19}$$

whose absolute value is

$$G = (G'^2 + G''^2)^{1/2} \tag{10-20}$$

*While these are Young's moduli, conventionally given the symbol E, we use G for consistency with the *Textbook*, Chap. 6 (wherein see the footnote on p. 196).

The complex modulus is a function describing the dynamic mechanical behavior of the material completely for the case of small tensile strain.

If one adopts as a model the Maxwell element (*Textbook*, p. 203) of a spring and dashpot in series, G is the modulus of the spring. The dashpot is characterized by a viscosity η, and the ratio η/G is a *relaxation time* τ. The behavior of G' and G'' as a function of ω is shown in Fig. 10-25: the inflection in G' and the maximum in G'' locate the point where $\omega\tau = 1$ and allow evaluation of the relaxation time. Real polymers, however, require a distribution of relaxation times to describe their behavior, as discussed in the *Textbook*, p. 205.

Measurement

Of many methods for measuring modulus and dissipation factor in a dynamic experiment, we describe only the use of the Rheovibron dynamic viscoelastometer (Imass, Fig. 10-26 and Exp. 27). In this instrument, the sample is clamped between strain gauges and subjected to a small sinusoidal tensile strain at fixed frequency. The value of $\tan \delta$ is read directly, and the storage and loss moduli can easily be calculated from instrument readings and sample dimensions, as described in Exp. 27.

Measurements are generally made as a function of temperature at one of four fixed frequencies: 3.5, 11, 35, or 110 Hz. A variable low-frequency accessory (0.01 - 1 Hz) is available. Typical results, obtained with an epichlorohydrin rubber, are shown in Fig. 10-27.

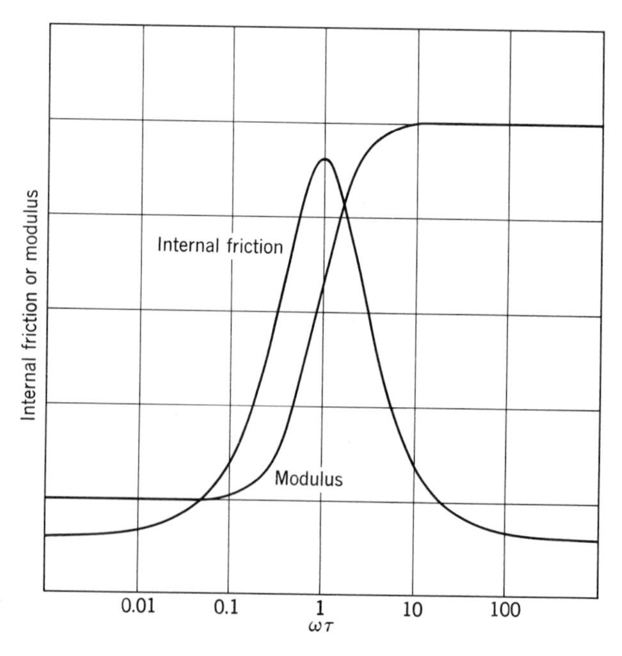

Fig. 10-25. Variation of the dynamic modulus G' and the internal friction G'' with frequency for a viscoelastic system (*Textbook*, Fig. 6-10).

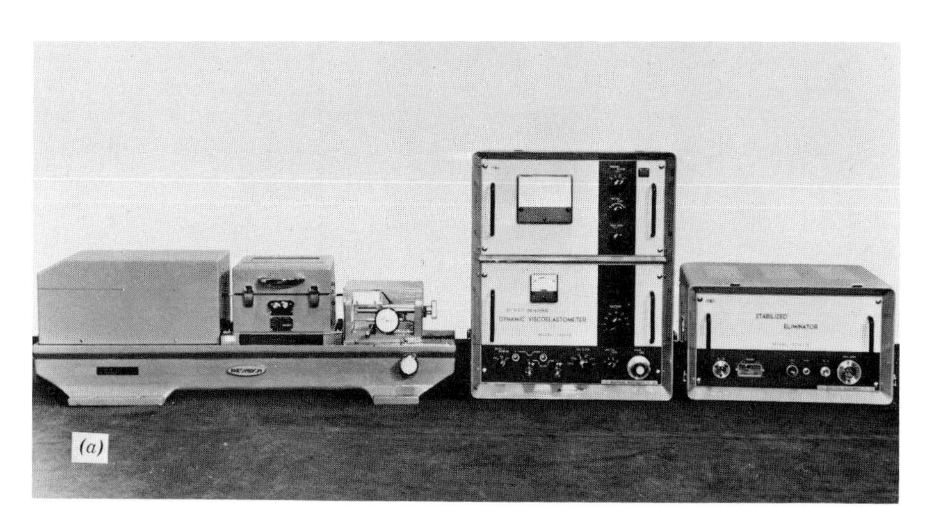

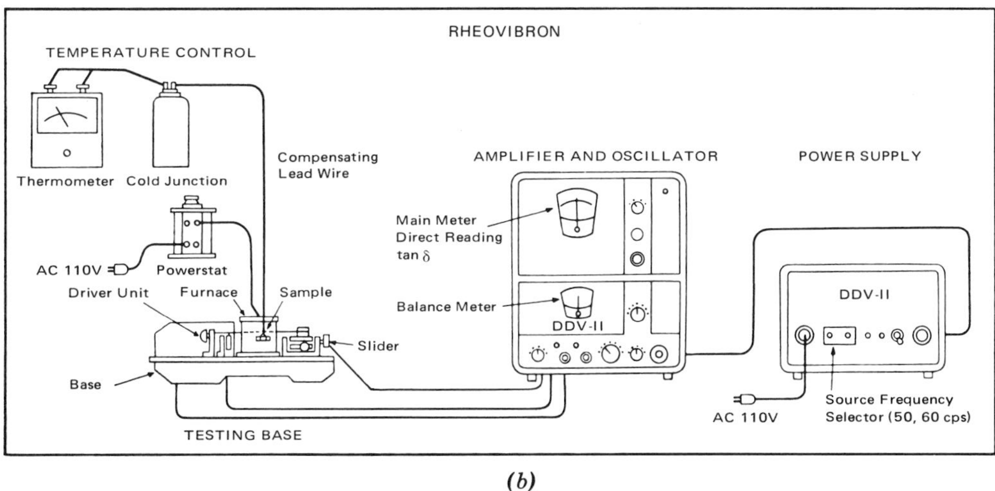

(b)

Fig. 10-26. Photograph (*a*) and block diagram (*b*) of the Rheovibron dynamic viscoelastometer (courtesy Imass).

GENERAL REFERENCES

Textbook, Chap. 6C; Leaderman 1958; Tobolsky 1958*b*, 1960; Nielsen 1962; Miller 1966; Alfrey 1967; Boyer 1968; Ward 1971.

C. *Dynamic Electrical Behavior*

There is a close analogy between measurements of the dynamic mechanical and electrical properties of polymers. The electrical measurements not only allow determination of the dielectric properties

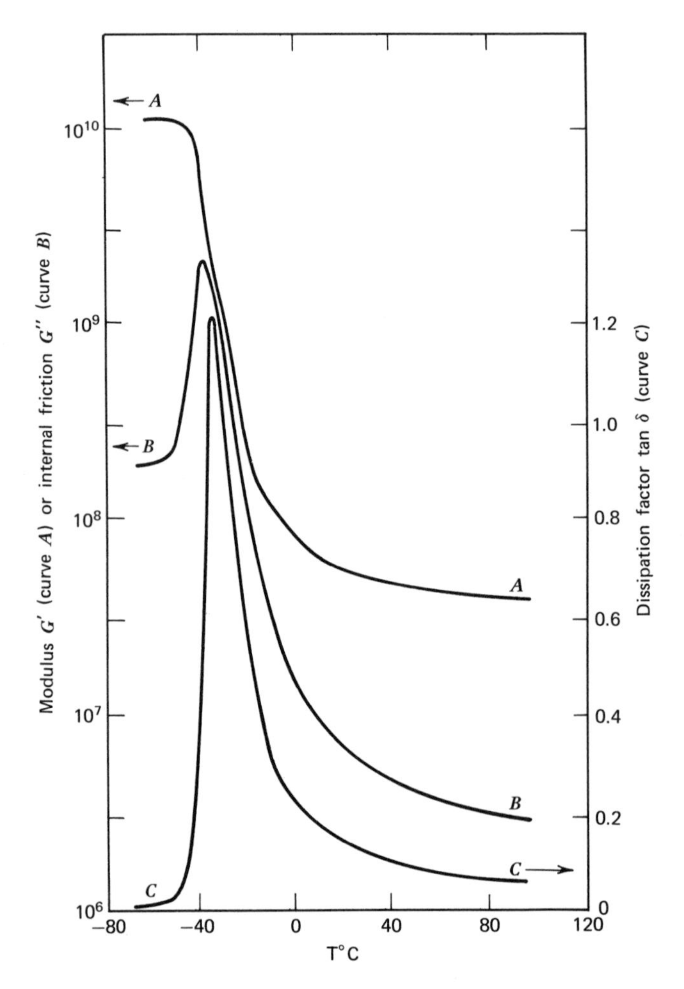

Fig. 10-27. Modulus G', internal friction G'', and dissipation factor tan δ of an epichlorohydrin rubber, measured with the Rheovibron as a function of temperature.

of the material, such as its dielectric constant, dielectric loss or power factor, and conductivity or resistivity, but also provide information on internal friction (analogous to the mechanical damping factor discussed in Sec. B), the distribution of relaxation times, the location of the glass transition, and related properties. As with the mechanical measurements, the electrical tests must be carried out over wide ranges of temperature and frequency to obtain this type of information.

For complex materials such as polymers, there is no adequate theory predicting dielectric behavior. The data from experiments must be correlated with physical, mechanical, and molecular properties in empirical ways. For example, dielectric constant has been correlated with degree of crystallinity, but it is not possible to determine the extent of crystallinity directly from dielectric data without first establishing the correlation. Similarly, dielectric measurements may be used to relate compounding variables to performance, after establishing the correlation. Although it is possible to differentiate among polymers quantitatively through dielectric measurements, especially those containing highly polar groups, it is usually difficult to account for the difference in terms of structure.

Dielectric dispersion arises from the mobility of dipoles, but there is no easy way to identify the nature of the dipoles. Thus, one cannot differentiate between a linear and a branched molecule of the same species. However, in cases where the chemical nature of the branch or graft differs from that of the backbone chain, differentiation may be possible; an example is the correlation of the so-called alpha and beta dispersions in the methacrylate series with mobility of the dipoles in the backbone and carboxyl side chains. Since dielectric measurements are usually made with small amplitudes and low dissipation of energy, as are dynamic mechanical measurements, they are carried out in the linear viscoelastic region and cannot be used to predict ultimate properties such as stress rupture or failure.

Electrical origin of dielectric properties

The capacitance C of a condenser is defined as the ratio of the charge on its plates to the potential difference between them. For a parallel-plate condenser, C can be calculated from the geometry of the system:

$$C = \varepsilon A / 4\pi \ell \tag{10-21}$$

where ε is the *dielectric constant*, a dimensionless property of the material between the plates, which has the value 1.0000 for a vacuum. Here A is the area of one of the plates in cm^2, ℓ is the thickness of the material in cm, and C has the dimensions of cm in the electrostatic system of units. The dielectric constant is thus conveniently the ratio of the capacitance of the condenser with the material in place to its capacitance with vacuum (or air, for which $\varepsilon = 1.0005$ at 20°C) between the plates.

If one considers a loss-free dielectric in an alternating electric field,

$$E = E_0 \sin \omega t \tag{10-22}$$

there would be no dissipation of energy, and the charging voltage and current in a condenser would be exactly 90° out of phase. In practice, there is always some dielectric loss, and in analogy with the case of dynamic mechanical properties, this is accounted for by introducing a complex dielectric constant

$$\varepsilon^* = \varepsilon' - i\varepsilon'' \tag{10-23}$$

It is ε' which is calculated from Eq. 10-21, and ε'' is the loss factor. The power P absorbed per unit volume of a dielectric, measured in erg/sec, is

$$P = E_0^2\, \omega\varepsilon''/8\pi \text{ esu} \tag{10-24}$$

where ω is in $\sec^{-1}$ and E_0 is in esu (1 esu = 300 volt/cm). Since the loss tangent or dissipation factor of the dielectric is

$$\tan\delta = \varepsilon''/\varepsilon' \tag{10-25}$$

(again in analogy with the mechanical case), and using Eq. 10-21, Eq. 10-24 can be written for the power P_c lost in a condenser of capacitance C,

$$P_c = 1/2\, V_0^2\, \omega C \tan\delta \tag{10-26}$$

where V_0 is the peak voltage at the capacitor terminals.

Molecular origin of dielectric properties

When a polar polymer is placed in an electric field, the material reacts in such a way as to reduce the field inside it. Several mechanisms contribute to this process (Kauzman 1942).

Microscopic shifts of electrons with respect to their nuclei, and minor distortions of bond angles, take place essentially instantaneously (about 10^{-13} sec) and lead to what is called optical polarization. That part of the total equilibrium dielectric constant ε_0 contributed by these mechanisms is called ε_∞.

Meanwhile, at a much lower rate (10^{-12} to 1 or more sec), viscoelastic processes take place resulting in orientation of the dipoles in the polymer in such a way as to reduce the internal field still more. This orientation polarization contributes the remainder of the dielectric constant, $\Delta\varepsilon = \varepsilon_0 - \varepsilon_\infty$. The rates of these processes are governed by relaxation times, and for a single relaxation time τ,

$$\varepsilon(t) = \varepsilon_\infty + \Delta\varepsilon(1 - e^{-t/\tau}) \tag{10-27}$$

It is, of course, necessary to consider a distribution of relaxation times $J(\tau)$ for real materials, and more generally

$$\varepsilon(t) = \varepsilon_\infty + \Delta\varepsilon \int_0^\infty J(\tau)(1 - e^{-t/\tau})\,d\tau \tag{10-28}$$

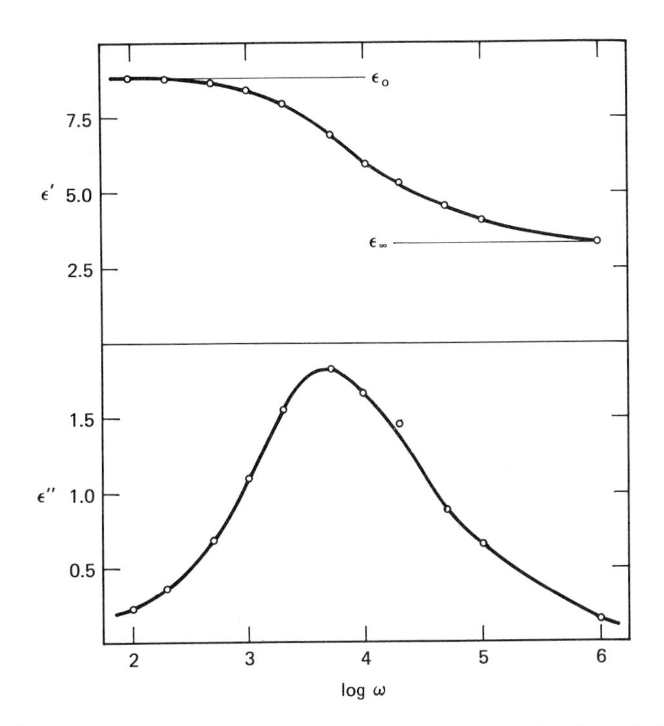

Fig. 10-28. Dependence of the dielectric constant ε' and dielectric loss ε'' for poly(vinyl acetate) on frequency (Bareš 1970).

where $\int_0^\infty J(\tau)d\tau = 1$ to normalize the distribution.

The components of the complex dielectric constant can also be written in terms of the distribution of relaxation times:

$$\varepsilon'(\omega) = \varepsilon_\infty + \Delta\varepsilon \int_0^\infty [J(\tau)/(1 + \omega^2\tau^2)]d\tau \tag{10-29}$$

$$\varepsilon''(\omega) = \Delta\varepsilon \int_0^\infty [J(\tau)\omega\tau/(1 + \omega^2\tau^2)]d\tau \tag{10-30}$$

The frequency dependence of ε' and ε'' for a sample of poly(vinyl acetate) is shown in Fig. 10-28. Note that the peak in ε'' is much broader than that calculated for a single relaxation time in Fig. 10-25. The distribution function $J(\tau)$ can be obtained from experimental data for $\varepsilon(t)$ or $\varepsilon'(\omega)$, $\varepsilon''(\omega)$ by solving the integral equations 10-28 to 10-30 by standard graphical or numerical methods (Klein 1968, Ferry 1970, Bareš 1972).

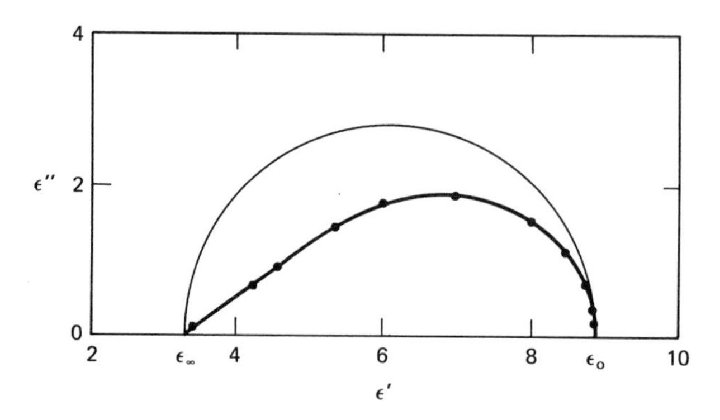

Fig. 10-29. Cole-Cole plot of the data of Fig. 10-28. The asymmetry is characteristic of behavior in the glass-transition region. The semicircle represents the curve for a material with a single relaxation time and the same values of ε_0 and ε_∞ (Bareš 1970).

A useful representation describing the spectrum of relaxation times has been proposed by Cole and Cole (Cole 1941, Böttcher 1952). In a Cole-Cole plot (Fig. 10-29), ε'' is plotted against ε' for all frequencies. If a single relaxation time describes the material, the plot is a perfect semicircle with the center on the axis of ε'. If there is a distribution, the resulting curve is flattened and the center point shifted to below the axis (Bareš 1965, McCrum 1967).

The temperature dependence of the relaxation time can be described by an Arrhenius-type equation:

$$\tau = \tau_0 \exp(\Delta E/kT) \tag{10-31}$$

where τ_0 is about 10^{-13} sec. One can observe the entire relaxation peak by changing the temperature and thus τ, rather than frequency, and in fact the temperature dependence of τ is so strong that a small shift in T accomplishes the same effect as a large shift in ω.

The activation energy ΔE is of the order of 10 kcal/mole for relaxations in the glassy state (Bareš 1965, 1967; McCrum 1967). The more bulky or more sterically hindered the polar group responsible for the relaxation, the higher is ΔE and the longer is τ. In the glass-transition region, cooperative motion of dipoles becomes important, and very high apparent activation energies ($\simeq$ 100 kcal/mole) are observed. These should not be interpreted literally in this region where molecular mechanisms are changing rapidly, however, and in fact the WLF equation describes the temperature dependence of τ better than the Arrhenius representation near and just above T_g.

Experimental methods

Most experiments for the measurement of dielectric constant and

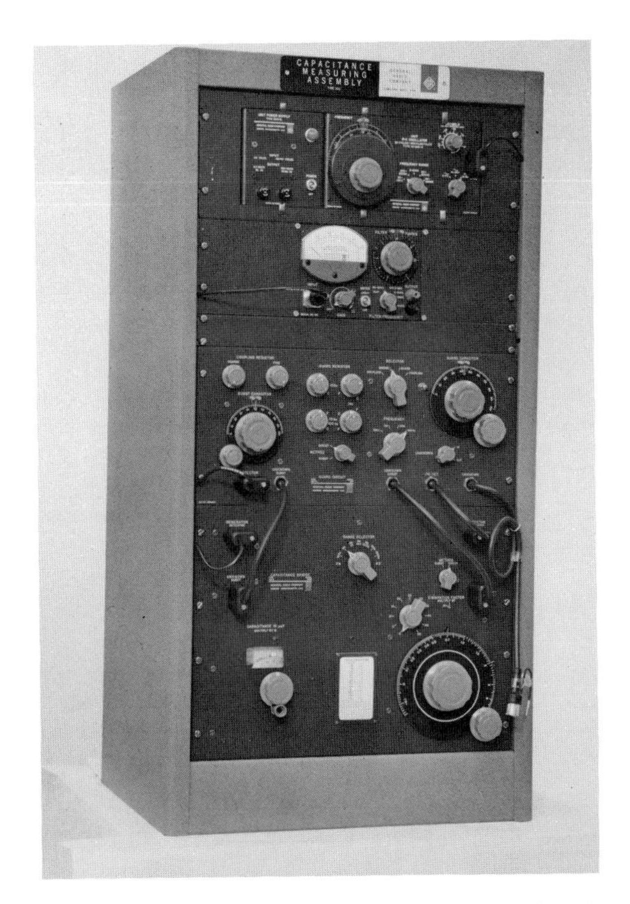

Fig. 10-30. The General Radio Capacitance Measuring Assembly, Type 1610B (courtesy General Radio).

loss are carried out using equipment (Figs. 10-30, 10-31) based on the Schering bridge design (similar in operation to the Wheatstone bridge) (Exp. 28). Since the bridge must operate on alternating current, its elements are impedances instead of resistances, and the balance is achieved by adjusting separately the capacitance and resistance of the variable impedance. Measurement of the capacitance yields ε' via Eq. 10-21, and that of resistance provides tan δ, from which ε'' can be obtained via Eq. 10-25.

The sample is usually a thin disc prepared by molding, calendering, or solvent casting. To insure good contact with the electrodes of the condenser, the sample must have thin electrodes adhered directly to its surfaces. These electrodes can be vacuum deposited, painted on with special conducting paints containing colloidal silver or graphite, or made of aluminum foil carefully oiled in place.

The sample holder (Fig. 10-32) contains the condenser plates. Typically, they are 5 cm in diameter and a sample thickness of the

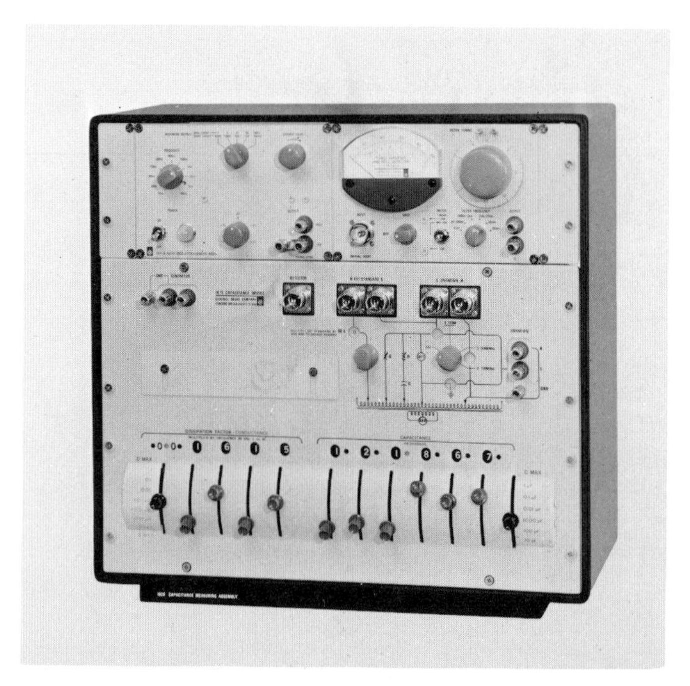

Fig. 10-31. The General Radio Capacitance Measuring Assembly, Type 1620A (courtesy General Radio).

order of 0.5 mm is satisfactory. The ratio of these dimensions must be kept large to minimize edge and cable errors, which can be eliminated with this geometry by a calibration in which C and $\tan \delta$ for the empty condenser are measured.

Typical data Data for the dielectric constant and dissipation factor of poly(vinyl acetate) at 73°C are given in Table 10-3. It is these data which were plotted in Figs. 10-28 and 10-29. Columns 2 and 4 in the table give the observed values of C (in picofarads, 1 pF = 0.899 esu) and $\tan \delta$ for the sample plus holder and leads. Separate measurement with the empty condenser gave C_0 = 25.5 pF; this was subtracted from the data in column 2 to give the values of capacitance of the sample C_s in column 3. For condenser plates 5 cm in diameter, ℓ in mm, and C or C_s in pF, we use Eqs. 10-21 and 10-25, and the fact that $\tan \delta_s$ is given by $(C/C_s)\tan \delta$, to write

$$\varepsilon' = 0.0576\ C_s \ell \tag{10-32}$$

$$\varepsilon'' = 0.0576\ C\ell \tan \delta \tag{10-33}$$

For the data in Table 10-3, ℓ = 0.39 mm.

Fig. 10-32. The General Radio Dielectric Sample Holder, Type 1690A (courtesy General Radio).

GENERAL REFERENCES

Böttcher 1952; Smyth 1955; Fröhlich 1958; Mathes 1966; Daniel 1967; Hill 1969; Ishida 1969; Vaughan 1972.

TABLE 10-3. Dielectric constant and dissipation factor of poly(vinyl acetate) at 73°C (Bareš 1970)

Frequency, Hz	C, pF	C_s, pF	tan δ	ε'	ε"
100	419.1	393.6	0.0236	8.84	0.22
200	417.7	392.2	0.0373	8.81	0.35
500	412.8	387.3	0.0739	8.70	0.69
1,000	401.7	376.2	0.1208	8.45	1.09
2,000	382.6	357.1	0.1816	8.02	1.56
5,000	335.4	309.9	0.2416	6.96	1.82
10,000	294.0	268.5	0.2536	6.03	1.68
20,000	263.2	237.7	0.2461	5.34	1.46
50,000	230.3	204.8	0.1711	4.60	0.89
100,000	208.9	183.5	0.1402	4.12	0.66
1,000,000	176.9	151.4	0.0360	3.40	0.14

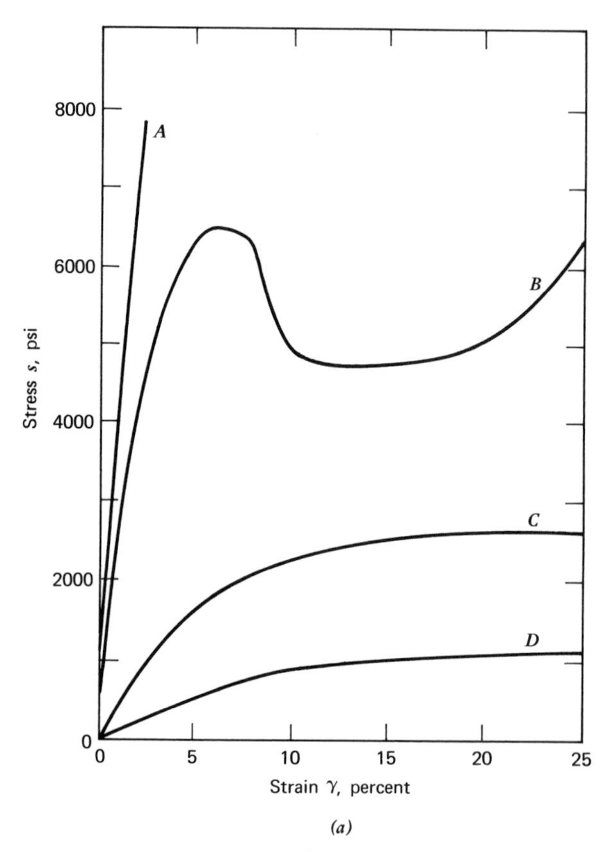

Fig. 10-33. Typical polymer stress-strain curves at various temperatures (*a*), and the corresponding modulus-temperature curve (*b*).

D. *Stress-Strain Properties*

In Chap. 6D the general features of stress-strain curves of various types of polymers were described, with reference to a minimal scheme for characterization. Here our objectives are to consider how stress-strain measurement aids in achieving an understanding of the mechanical behavior of polymers, and to examine briefly the influence of molecular structure on the ultimate properties of polymers as determined by stress-strain testing.

Stress-strain phenomena

While the stress-strain test is usually described as simple, there are many variables which must be controlled if meaningful results are to be obtained, and to carry this out can at times be quite difficult. In addition, the stress-strain experiment is not adequately described theoretically, since at the large deformations involved the relation

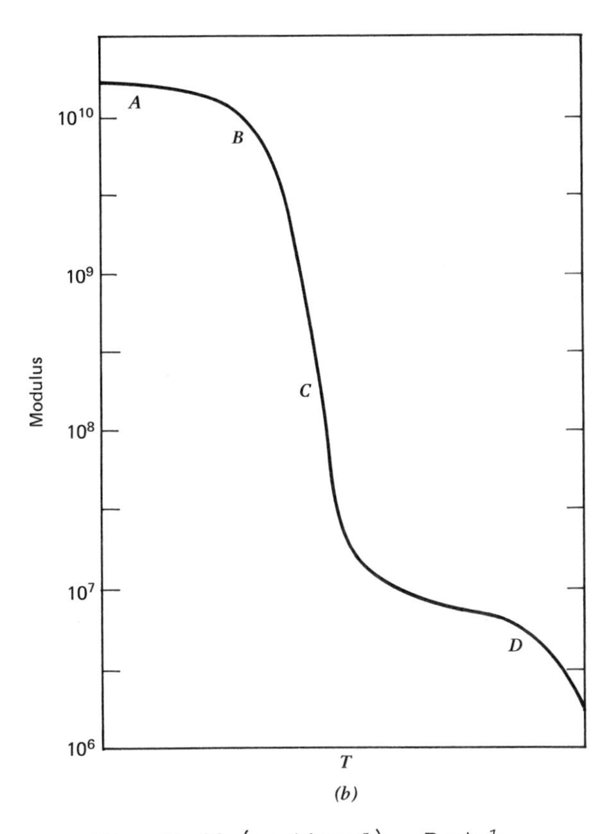

Fig. 10-33 (continued). Part *b*.

between the stimulus (strain or strain rate) and response (stress) is nonlinear. In consequence, it is difficult to extract a universal material function providing a complete description of the phenomena, as we saw could be done for dynamic mechanical and electrical properties, which are determined from experiments in the linear region.

Since stress-strain behavior depends greatly upon temperature, thermal history of the sample, strain rate, magnitude of the strain, and other test conditions, it is necessary to carry out experiments at a variety of conditions to obtain a full spectrum of information.

Types of stress-strain curves In amplification of the qualitative sketches of typical stress-strain curves in Fig. 6-5, the changes in stress-strain behavior within a single amorphous polymer as a function of temperature are shown in Fig. 10-33. A modulus-temperature curve is included to indicate the regions to which the stress-strain curves apply. Below T_g, all polymers exhibit a brittle stress-strain curve, which gradually changes through a ductile to an elastomeric type. As the speed of testing (the strain rate) is increased for an elastomeric material, the trend is reversed, as ultimately the response time of the polymer exceeds the time scale of the experiment, and the material behaves like a brittle solid.

Stress-strain measurements on amorphous polymers can be treated by time-temperature equivalence to produce a master curve which is useful in predicting long-term data from short-term experiments in the rubbery region (Smith 1962, 1969).

The failure envelope Practical information on the ultimate properties of a polymer can be obtained from consideration of its failure envelope (Fig. 10-34), obtained as a plot of stress versus strain at failure over a range of strain rates (Smith 1962; Rodriguez 1970, Chap. 9). As the temperature is increased or the strain rate decreased, the point of failure moves clockwise, the stress at break decreasing continuously while the strain at failure first increases during the transition from brittle to ductile behavior, then decreases again as the material enters the rubbery region.

Effect of molecular parameters on stress-strain behavior

At temperatures far below T_g, the molecular weight of the polymer and such gross structural features as branching or crosslinking have little influence on stress-strain behavior. Near T_g, however, and even somewhat below T_g for the more ductile materials that can cold draw, the elongation at break is greater for samples of high molecular

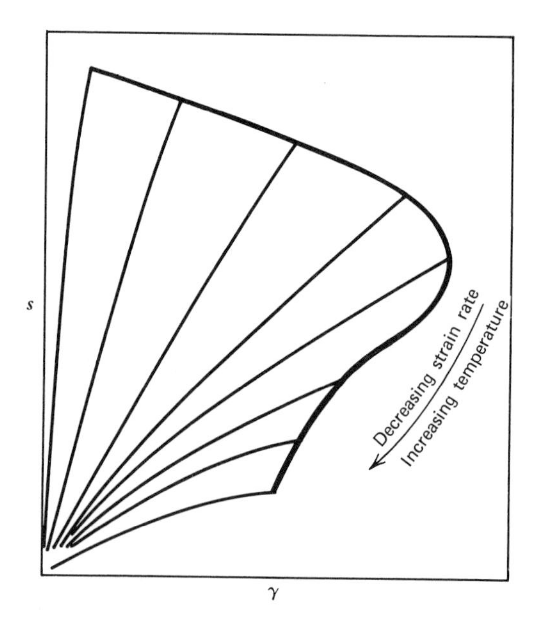

Fig. 10-34. The failure envelope for typical polymer behavior (Smith 1962).

weight compared to the behavior at low molecular weight. In the rubbery state, elongation at break generally increases with molecular weight. As the molecular-weight distribution is broadened, the tensile strength decreases while the elongation increases. Crosslinking tends to decrease the elongation at break. Tensile strength increases with crosslinking initially, but goes through a maximum and decreases at high degrees of crosslinking.

Crystallinity is detrimental to the strength of polymers below T_g, because the crystallites can, like filler particles, act as stress concentrators. However, relatively high degrees of crystallinity (30-85%) can convert a rubbery material into a tough rigid polymer as is the case in polyethylene and polypropylene. The type of morphology has a strong influence on the stress-strain relationship (Clark 1971). An increase in crystallinity generally raises the modulus but decreases the elongation at break. The interpretation of stress-strain curves of crystalline or semi-crystalline polymers is difficult if not impossible without a knowledge of bulk rheological and morphological properties.

Under proper conditions, most polymers will cold draw. The resulting orientation leads to significant changes in material properties. Cold drawing is very sensitive to temperature and strain rate.

In addition to thermoplastics, polymeric materials of practical interest include polyblends, fiber (and filler) reinforced plastics, elastomers and thermoset materials. The stress-strain properties of these materials are governed not only by the nature of the polymer or polymers but also by the adhesive strength of the interface. Consideration of these materials and the effects of compounding additives, while important, is beyond the scope of this book.

Stress-strain measurement

By far the most popular instrument used in stress-strain measurement is the Instron Tensile Tester (Fig. 10-35), although there are others (Baldwin, Tinius Olsen, Scott (Precision)) in wide use. These instruments are essentially devices in which a sample is clamped between grips or jaws which are pulled apart at constant strain rates varying from 0.5 to 500 mm/min. The stress on the sample is followed with load cells whose capacity ranges from 2 g to 500 kg or even more. The elongation of the sample can be measured separate from the motion of the jaws to improve precision and avoid errors arising from slippage of the sample in the jaws. Jaws are available to hold a wide variety of samples, from filaments or fibers to large bars. Jaw design and specimen shape and preparation are selected so as to minimize the introduction of extraneous stresses or strains. To grip soft specimens without influencing the result is a formidable task, as is avoiding the effects of frozen-in strain or orientation caused by the sample molding process. The Instron can be preset to strain a sample at a given rate to a given extension; to return the crosshead automatically to the initial point after a given maximum extension of the sample; to cycle the crosshead between two adjustable points of extension (for hysteresis measurement); to stop the extension at a preselected point for stress-relaxation studies; or to stop the return

Fig. 10-35. The Instron tensile testing machine (courtesy Instron).

travel of the crossheads at a preset gauge length to assure reproducibility in the length of the test sample.

The strain of the sample is simply measured in terms of its initial length L_0 and length L at time t:

$$\gamma = (L - L_0)/L_0 = kt/L_0 \tag{10-34}$$

where k is rate of extension. The stress s is simply the force F measured by the load cell divided by the cross-sectional area of the sample. If the initial area A_0 is used, the resulting figure is called the engineering stress; calculation of the true stress requires the use of the actual cross-sectional area

$$A = A_0 L_0 / L \qquad (10\text{-}35)$$

reduced because of the elongation of the sample. The tensile modulus of elasticity, E, is given by the ratio of stress to strain and is determined from the initial slope of the stress-strain curve:

$$E = s/\gamma = FL_0/A(L - L_0) \qquad (10\text{-}36)$$

Note that E and the shear modulus G (Sec. B) are not the same quantity, but are related through elasticity theory.

GENERAL REFERENCES

Alfrey 1948, 1967; Nielsen 1962, Chap. 5; Thorkildsen 1964; Marin 1965; Rodriguez 1970, Chap. 9; Bikales 1971; Ward 1971; Williams 1971, Chap. 11.

E. *Swelling of Network Polymers*

The production of crosslinked networks in polymers is important in the gelation of polyfunctional condensation polymerization (*Textbook*, Chap. 8D) and in the post-polymerization steps of crosslinking or vulcanization (Chap. 1G; *Textbook*, Chaps. 12C, 19B). Here we comment briefly on one means of characterizing network polymers, namely measurement of their swelling properties in contact with solvent, in anticipation of Exp. 30. The entire subject of network topology and structure, and their effect on the chemical, physical, and mechanical properties of polymers, is both extremely complex and beyond the scope of this book. It is well covered in the literature, to which reference is made at the end of this section.

The first step in the solution of a polymer is swelling, as described in Chap. 6B, and in the case of a crosslinked material it is also the last step since the chains cannot be further separated because of the presence of the crosslinks. Swelling is accompanied by an increase in the volume of the polymer as it imbibes the solvent, and it is usual to assume, consistent with the Flory-Huggins theory of the thermodynamics of polymer solutions (*Textbook*, Chap. 2C), that this mixing occurs without change in the total volume of the system. As a result we can write an expression for the volume V_s of the swollen polymer in terms of the original weight w_0 and swollen weight w_s and the densities ρ_2 and ρ_1 of the polymer and solvent, respectively:

$$V_s = w_0/\rho_2 + (w_s - w_0)/\rho_1 \qquad (10\text{-}37)$$

We now consider the thermodynamics of the system at equilibrium. The chemical potential of the system is lowered by the mixing of the solvent with the polymer; this is the driving force for solution in the ordinary case. With crosslinked polymer, the network chains are forced out of their equilibrium conformations as a consequence of the swelling, and the elastic reaction of the network gives an increase in the chemical potential. At equilibrium, the resultant of these two is zero, and the situation is defined by the concentration of polymer in the swollen polymer-solvent mixture at equilibrium. Calling this concentration c, we write following Flory (1953)

$$-\ [\ln(1-c) + c + \chi_1 c^2] = (V_1 \rho_2 / M_x)(c^{1/3} - c/2) \qquad (10\text{-}38)$$

where the left-hand side represents the lowering of the chemical potential due to mixing and the right-hand side its increase from the elastic reaction of the network. The equation as written here applies to a perfect network (no dangling chain ends) for which the molecular weight per crosslinked unit is M_x and the crosslinks are tetrafunctional. The quantity V_1 is the molar volume of the solvent, and χ_1 is the polymer-solvent interaction parameter of the Flory-Huggins theory. This equation can readily be solved for M_x. The concentration c at equilibrium is given by

$$c = V_s w_0 / \rho_2 \qquad (10\text{-}39)$$

The value of M_x determined from swelling measurements will not in general be in good agreement with the value determined from equilibrium stress-strain measurements because a different value of χ_1 is required for each elongation (Flory 1953). From another point of view, the two results are based on different time scales which reflect the dynamics of the network.

GENERAL REFERENCES

Flory 1953; Treloar 1958; Alfrey 1963; Lloyd 1962; Miller 1966, Chap. 7; Ferry 1970.

F. Permeability of Polymers to Gases and Liquids

The permeability of polymers to gases, vapors, and liquids is important for many applications, in which either high barrier properties are required for protective purposes, or very selective permeation is required for use of the polymer as a membrane for efficient separation of permeating mixtures. In either case, the transport of penetrant molecules through a polymer material (in the form of a film or membrane which is free from cracks, pinholes, or other gross flaws) normally occurs by an activated-diffusion process. The condensed penetrant is considered to dissolve in the polymer surface layer, migrate through the bulk material by a cooperative ("activated") polymer segmental motion-penetrant molecule jump process under the influence of a concentration-gradient driving force, and evaporate from the other surface of the membrane. After a relatively short transient-state buildup, steady-state flow is attained with a constant rate of transmission of penetrant provided constant pressure or concentration difference is maintained across the film.

The activated-transport process is very sensitive to the nature of the penetrant molecule and the polymer composition and structure. The permeation rate depends on both the solubility of penetrant and the diffusivity of its molecules in the polymer medium. The highly selective transport process is typically characterized by a large positive temperature dependence. The corresponding transport of gaseous penetrants, by capillary or connective flow mechanisms through polymers with substantial pore, void, or related defect structures, shows relatively little difference in transport rates for various gases and a small negative temperature dependence mainly due to changes in gas viscosity. For some polymer systems the transmission of penetrant may occur by a combination of activated diffusion and capillary flow, and the overall permeation characteristics must be assessed with caution to achieve a realistic interpretation (Rogers 1964, 1965).

Steady-state theory

The rate of permeation (transmission rate, diffusion flow, or flux), J, is defined as the amount Q of penetrant passing through a surface of unit area A perpendicular to the direction of flow in unit time:

$$J = Q/At \tag{10-40}$$

For the activated-flow process at the steady state, J may be expressed by Fick's equation,

$$J = -D(\partial c/\partial x) \tag{10-41}$$

where D is the diffusion coefficient in cm^2/sec and $\partial c/\partial x$ is the concentration gradient in the direction of flow.

When D is independent of c and x, the equation may be integrated to

$$J = D(c_1 - c_2)/\ell \tag{10-42}$$

where c_1 and c_2 are the steady-state concentrations of penetrant in the incoming and outgoing film surfaces, and ℓ is the film thickness. The equilibrium concentrations in the surface layers can be related to the partial pressures of gas or vapor penetrants in ambient gaseous phases by the Henry's Law expression,

$$c = Sp \tag{10-43}$$

where S is the solubility coefficient of the penetrant in the polymer and p the gas pressure. For ambient liquid phases, the corresponding distribution equation

$$c = Kc' \tag{10-44}$$

may be used to relate the concentration c of penetrant in the membrane to the concentration c' of penetrant in the adjacent liquid phase. When either of these distribution relationships is obeyed, Fick's equation can be expressed as

$$J = DS(p_1 - p_2)/\ell \tag{10-45}$$

or

$$J = DK(c_1' - c_2')/\ell \tag{10-46}$$

The permeability constant is then defined as the product

$$P = DS \tag{10-47}$$

for transport of gas or vapor penetrants and

$$P = DK \tag{10-48}$$

for transport of dissolved penetrants from one solution to another through the polymer membrane.

The permeability constant for gas or vapor flow is calculated as

$$P = (\Delta Q/\Delta t)\ell/A(p_1 - p_2) \tag{10-49}$$

where ΔQ is the quantity of gas or vapor which has permeated through the film of effective area A and average thickness ℓ in the time interval Δt from a pressure reservoir of pressure p_1 to a reservoir of pressure p_2. If the experimental method is such that $p_1 >> p_2 \simeq 0$, the quantity $\Delta Q/\Delta t$ can be calculated from a plot of the pressure increase with time, $\Delta p_2/\Delta t$, in a constant-volume reservoir of volume V. The amount of gas transmitted at any temperature should be reduced to standard temperature and pressure by the gas-law equations.

Unfortunately, values of P, J, D, and S are reported in the literature in a wide variety of units. Reasonable units for P are ml gas at standard temperature and pressure (STP) passing per second through 1 cm^2 of polymer film 1 cm thick under a pressure gradient of 1 at. The diffusion coefficient is then in units of cm^2/sec and the solubility coefficient is in units of ml (STP)/ml polymer × at. Conversion factors for some of the more common units have been tabulated (Rogers 1964).

Non-steady-state treatment

Steady-state measurements of P do not allow the evaluation of the diffusion constant D, but only its product with the solubility coefficient S or partition coefficient K, as may be seen from Eqs. 10-47 and 10-48. Use of the non-steady-state portion of the curve of Q versus

t (Fig. 10-36) allows separation of these parameters: If the linear portion of this curve is extrapolated to the axis of abscissas, the resulting time t_L, called the lag time, can be used to determine D by the relation

$$D = \ell^2/6\, t_L \tag{10-50}$$

Effect of molecular structure

Extensive reviews, cited in the general references for this section, describe the effects on solubility, diffusion, and permeation of such factors as temperature, concentration, penetrant size and shape, penetrant-polymer interactions, and the chemical and physical characteristics of the penetrant and polymer. These references also contain descriptions of many different experimental procedures for the determination and interpretation of the transport and solution parameters.

In general, diffusion coefficients are smaller for glassy or stiff chain polymers as compared to rubbery or flexible polymers of similar chemical constitution. With increasing crystallinity, permeability decreases rapidly because of the lower solubility of gases, vapors, and liquids in the crystalline phase. However, the morphology of a crystalline polymer depends strongly on the mechanical and thermal history of the sample, and by controlling these factors the permeability of a given crystalline polymer can be varied over quite wide limits. For example, linear polyethylene quenched rapidly from the

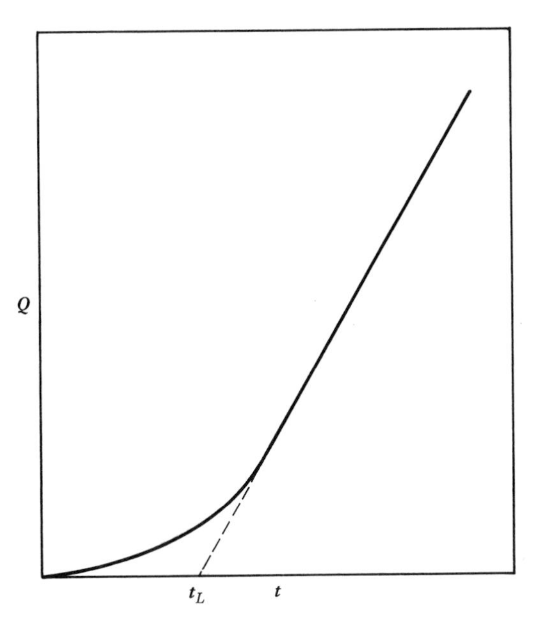

Fig. 10-36. Permeation of vapor through a polymer film as a function of time in the non-steady-state region.

melt has significantly higher gas permeability than the same polymer cooled slowly from the melt.

Measurement of permeability

While permeability apparatus is inherently simple and easy to fabricate, Exp. 31 is written for the inexpensive Custom Scientific CS-135 Permeability Cell (Fig. 10-37) (Stern 1964). The cell consists of two stainless steel discs about 15 cm in diameter. Machined depressions in each disc form a cavity which contains the membrane to be tested. The two discs are clamped to form a pressure-tight fit. A glass capillary tube with a short S-shaped section is connected to the top disc, and gas inlet and vent lines are provided on both sides of the cell. Gas is supplied at constant pressure to the bottom of the cell; after permeating through the membrane, the gas expands into the capillary and displaces a short plug of mercury upward. The movement of this plug of mercury is measured as a function of time and provides a direct measure of the rate of permeation of the gas through the membrane. The S-shape bend at the base of the capillary keeps the mercury plug from falling inside the cell. The permeability is calculated from Eq. 10-49 after correcting to STP.

A typical set of data was obtained for polyethylene at 25°C and 745 torr pressure of CO_2 for a film thickness of 0.0762 mm, a cell diameter of 92 mm, and a capillary diameter of 1 mm. The slope of the resulting straight-line plot of height versus time (Fig. 10-38) was determined to be 0.397 cm/min. This corresponds to $\Delta Q/\Delta t = 5.2 \times 10^{-5}$ ml/sec for the 1-mm capillary used, before correction to STP, or 4.67×10^{-5} ml/sec at STP. Substitution into Eq. 10-49 yields

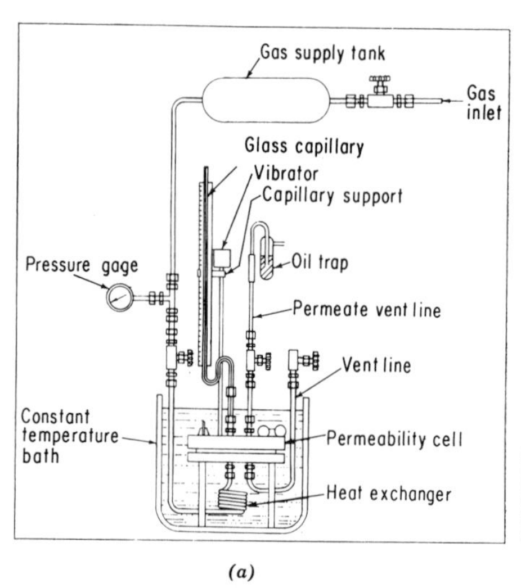

(*a*)

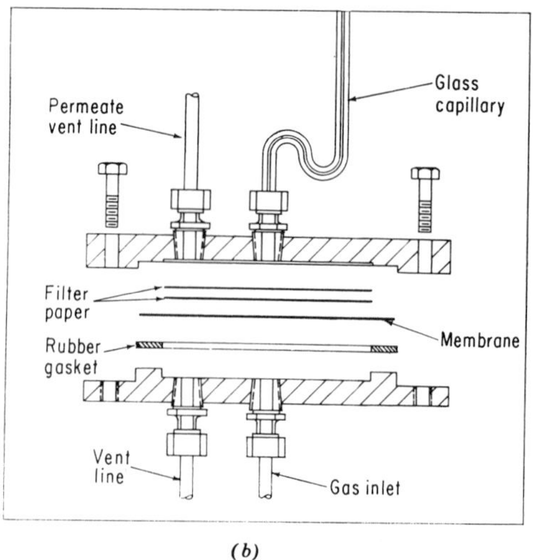

(*b*)

Fig. 10-37. The Custom Scientific CS-135 permeability apparatus (*a*) and cell (*b*) (Stern 1964).

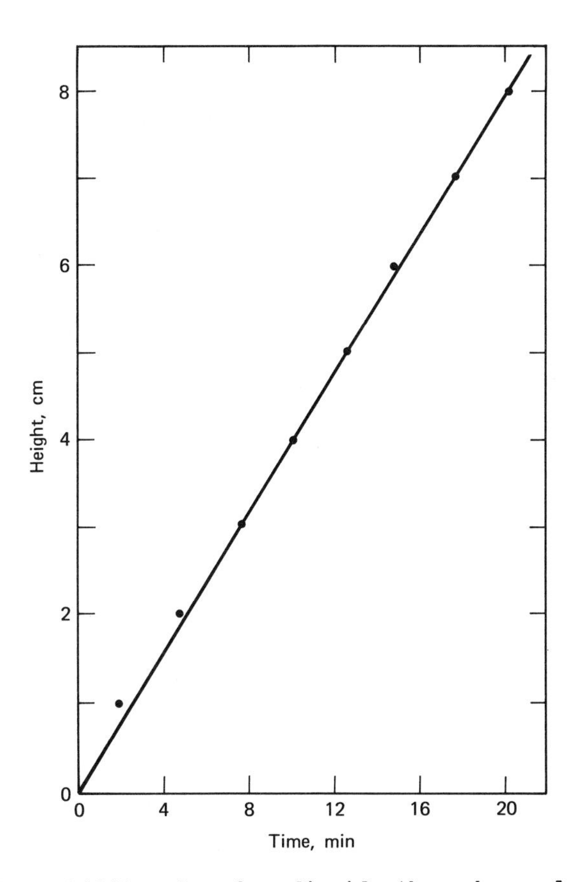

Fig. 10-38. Permeability of carbon dioxide through a polyethylene film; see the text for details.

$$P = (4.67 \times 10^{-5}\ \text{cm}^3/\text{sec} \times 0.00762\ \text{cm})/(66.4\ \text{cm}^2 \times 1\ \text{at.})$$

$$= 5.35 \times 10^{-5}\ \text{cm}^2/\text{sec-at.}$$

GENERAL REFERENCES

Rogers 1964, 1965; Van Amerongen 1964; Paul 1965; Stannett 1965; Crank 1968.

G. *Summary*

While it is nearly fruitless to attempt to summarize the structure-property relationships in high polymers by a few words, it can be said that there is merit in considering general trends in physical, chemical, and mechanical properties as structural variables are changed systematically. Quite often, the recognition of a trend is all that is required to define a problem or provide insight into the design of a new material.

In Fig. 10-39 some of these trends have been listed to show how selected performance properties change as molecular-structure parameters are varied. While the trends indicated are general, many exceptions can be found, so these data cannot be considered infallible.

	Tensile strength	Elongation	Yield strength	Toughness	Brittleness	Hardness	Abrasion res.	Softening temp.	Melt viscosity	Adhesion	Chem. resistance	Solubility
Increase mol. wt. *a*	+	+	+	+	+	+	+	+	+	−	+	−
Narrow the mol. wt. distribution	+	−	−	+	−	−	+	+	+	−	+	0
Increase branching or crosslinking	*M*	−	+	−	*M*	+	+	+	+	−	−	*M*
Add polar chain units	+	+	+	+	+	+	+	*b*	+	+	−	+
Add polar side chains	+	+	+	+	+	+	+	+	+	+	−	+
Stiffen main chain	+	−	+	−	+	+	−	+	+	−	+	−
Increase crystallinity	−	−	+	−	−	+	+	+	+	−	+	−
Add crystallizable branches	+	+	+	+	+	+	+	0	*c*	0	−	+

Fig. 10-39. Profile of performance property dependence on molecular-structure parameters for typical polymers. Key: +, property goes up; -, property goes down; 0, little change; *M*, property passes through a maximum and then decreases; *a*, for amorphous polymers; *b*, result depends on the melting points involved; *c*, result depends on the temperature.

Consider, by way of example, the need to produce a material with improved tensile strength by modification of an existing polymer. The figure shows that tensile strength can be enhanced by narrowing the molecular-weight distribution, introducing a small amount of branching, crosslinking, or crystallinity, or adding polar side-chain groups. Further, any change in structure which tends to stiffen the chain or reduce the number of chain ends (free volume) should lead to higher tensile strength. It is also important to remember that most polymer properties deteriorate progressively as the temperature is increased (if one is careful to define deterioration appropriately).

Of course it cannot be expected that a single property such as tensile strength could be varied while all others remain unchanged. Thus the figure must be examined to assess any side effects that may result, and compromises or trade-offs must be formulated in order to achieve a balance among desirable properties.

More drastic alterations of properties can be achieved by going to another type of polymer structure. Guidance here is provided by Figs. 10-40 and 10-41, which provide profiles of the important properties of some general-purpose and specialty polymers and of some engineering plastics, respectively.

	Poly(methyl methacrylate)	Modacrylic	Cellulose acetate	Cellulose acetate-butyrate	Cellulose propionate	Ethyl cellulose	Chlorinate polyether	Epoxy (cast)	FEP fluorocarbon	Ionomer	Melamine-formaldehyde (cellulose filled)	Phenol-formaldehyde (cellulose filled)	Polybutylene	Polyethylene (low density)	Poly(phenylene oxide) (modified)	Polystyrene	Poly(vinyl chloride)	Poly(vinyl chloride) (copolymer)	Poly(vinylidene chloride)	Poly(vinylidene fluoride)	SAN	Silicone	Urea-formaldehyde (cellulose filled)
Price	0	0	0	0	0	0	0	0	–	0	0	+	0	+	0	+	+	+	0	–	+	–	0
Processability	+	+	+	+	+	+	+	0	+	+	0	0	0	+	+	+	0	0	+	+	0	–	+
Tensile strength	+	0	0	0	0	0	0	+	–	–	0	0	–	–	0	+	0	–	0	0	+	–	+
Stiffness	+	+	+	0	0	0	–	0	–	–	+	+	–	–	0	+	+	–	–	–	+	–	+
Impact strength	–	0	0	0	+	0	–	–	+	+	–	–	+	+	–	–	+	+	–	0	–	–	–
Hardness	0	+	+	+	+	+	0	+	–	–	+	+	–	–	+	+	–	–	+	–	+	–	+
Useful temperature range	0	–	0	0	0	–	0	+	0	–	0	0	0	0	0	–	–	–	0	0	0	0	–
Resistance to chemicals	0	0	–	–	–	–	–	0	+	0	0	–	+	+	+	–	+	+	+	+	–	0	–
Resistance to weather	+	+	0	0	0	0	0	0	+	0	0	0	–	–	–	–	0	0	0	+	–	+	0
Resistance to water	0	0	–	–	–	–	+	–	+	0	0	0	+	+	0	0	0	0	0	0	0	0	0
Flammability	–	–	0	–	–	–	0	–	+	–	0	–	–	–	0	–	0	0	0	0	–	0	0

Fig. 10-40. Profile of the properties of some general-purpose and specialty plastics (*Textbook*, Fig. 7-10, from Billmeyer 1968). Key: +, outstanding in the property indicated, among the best performers available; 0, acceptable performance in this property, still suitable in most cases; -, not recommended if this property is important to the intended use.

The engineering or structural plastics deserve special mention, since they are the strong, tough, stiff, abrasion-resistant materials capable of offering high performance over wide ranges of temperature and environmental conditions. The outstanding properties of these materials depend upon the judiciously balanced introduction of such structural features as high crystallinity, high polarity or rigidity of the chains, and crosslinking. Any of these leads to enhancement of the "engineering" properties of materials, with synergistic effects when more than one can be utilized.

Finally, it is never too early for the student to learn or the industrial scientist to be reminded that price is often a polymer's most important property. The key parameters here are the trade-offs among price, performance, and production volume. Examination of the

	ABS	Acetal	Polytetra-fluoroethylene	Polychlorotri-fluoroethylene	Nylon	Phenoxy	Polycarbonate	Polyimide	Poly(phenylene oxide)	Polyethylene (high density)	Polypropylene	Polysulfone
Price	0	0	−	−	−	−	0	−	−	+	+	−
Processability	0	+	−	+	+	0	0	−	0	+	+	+
Tensile strength	0		−	0	0	0	0	+	+	−	0	+
Stiffness	0		−	0	0	0	0	+	0	−	0	0
Impact strength	0	−	0	0	−	+	+	−	−	+	−	0
Hardness	0	+	−	0	0	+	+	+	+	−	0	+
Useful temperature range	−	0	+	0	0	−	0	+	0	0	0	0
Resistance to chemicals	0	0	+	+	0	0	0	+	0	+	+	+
Resistance to weather	0	0	+	+	−	0	0	+	0	−	−	0
Resistance to water	0	0	+	+	−	0	0	0	+	+	+	0
Flammability	−	−	+	+	0	0	0	+	0	+	+	0

Fig. 10-41. Profile of the properties of some engineering plastics (*Textbook*, Fig. 7-12, from Billmeyer 1968). Key same as for Fig. 10-40.

inverse relation between price and production volume illustrated in Fig. 10-42 will show that it is no coincidence that some of the highest performance materials have achieved high production volume and low cost (Platzer 1971). In other cases, of course, it is low price that has led to high volume despite less than exciting properties. The movement of a polymer to the right on the price-volume curve is, however, profoundly accelerated by high performance.

BIBLIOGRAPHY

Alfrey 1948. Turner Alfrey, Jr., *Mechanical Behavior of High Polymers,* Interscience Publishers, New York, 1948.

Alfrey 1962. T. Alfrey, Jr. and W. G. Lloyd, "Network Polymers. I. Theoretical Remarks," *J. Polymer Sci. 62,* 159-165 (1962).

Alfrey 1967. Turner Alfrey and Edward F. Gurnee, *Organic Polymers,* Prentice-Hall, Englewood Cliffs, New Jersey, 1967.

ASTM D 1238. Standard Method of Measuring Flow Rates of Thermoplastics by Extrusion Plastometer. ASTM Designation: D 1238. American Society for Testing and Materials, Philadelphia, Pennsylvania.

Bauer 1967. Walter H. Bauer and Edward A. Collins, "Thixotropy and Dilatancy," Chap. 8 in Frederick R. Eirich, ed., *Rheology--Theory and Applications,* Vol. 4, Academic Press, New York, 1967.

Bagley 1957. E. B. Bagley, "End Corrections in the Capillary Flow of Polyethylene," *J. Applied Phys. 28,* 624-627 (1957).

Bagley 1969. E. B. Bagley and H. P. Schreiber, "Elasticity Effects in Polymer Extrusion," Chap. 3 in Frederick R. Eirich, ed., *Rheology--Theory and Applications,* Vol. 5, Academic Press, New York, 1969.

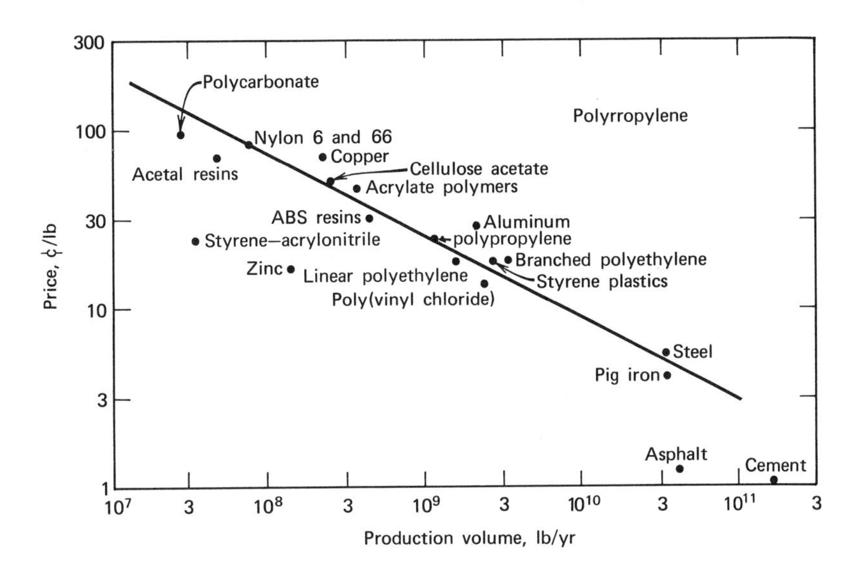

Fig. 10-42. Relationship between prices and production volumes of some important polymers and other materials (Platzer 1971).

Baldwin. Wiedemann Division, The Warner and Swasey Co., Pittsburgh, Pennsylvania 15220.

Bareš 1965. J. Bareš and J. Janáček, "Structure and Properties of Hydrophilic Polymers and Their Gels. IV. Low Temperature Dielectric Relaxation of Polyethyleneglycolmethacrylate, Polydiethyleneglycolmethacrylate and Polytriethyleneglycolmethacrylate," *Coll. Czech. Chem. Comm. 30*, 1604-1610 (1965).

Bareš 1967. J. Bareš and J. Janáček, "Low Temperature Relaxation in Some Polyamides," *J. Polymer Sci. C16,* 293-304 (1967).

Bareš 1970. J. Bareš, M. Pegoraro, J. Kolařík, and J. Hasa, "Measuring Condensers for Dielectric Studies of Polymers" (in Italian), *Mat. Plast. e. Elast. 36*, 513-516 (1970).

Bareš 1972. J. Bareš, "Use of Computer in Polymer Characterization, Viscoelastic Applications: WLF equation, Moduli and Spectra, Dilatometry" (in Italian), *Chim. Ind. Suppl., Ing. Chim. Ital. 8,* 27-34 (1972).

Berry 1968. G. C. Berry and T. G Fox, "The Viscosity of Polymers and their Concentrated Solutions," *Fortschr. Hochpolym. Forsch. (Advances in Polymer Sci.) 5,* 261-357 (1968).

Bikales 1971. Norbert M. Bikales, ed., *Mechanical Properties of Polymers (Encyclopedia Reprints)*, Wiley-Interscience, New York, 1971.

Billmeyer 1968. Fred W. Billmeyer, Jr., and Reneé Ford, "The Anatomy of Plastics," *Science and Technol. No. 73,* 22-37, 81-82 (1968).

Billmeyer 1969. Fred W. Billmeyer, Jr., "Molecular Structure and Polymer Properties," *J. Paint Technol. 41,* 3-16 (1969).

Böttcher 1952. Carl Johan Friedrich Böttcher, *Theory of Dielectric Polarization,* Elsevier Publishing Co., New York, 1952.

Boyer 1968. R. F. Boyer, "Dependence of Mechanical Properties on Molecular Motions in Polymers," *Polymer Eng. Sci. 8*, 161-185 (1968).
Brodnyan 1957. J. G. Brodnyan, F. H. Gaskins, and W. Philippoff, "On Normal Stresses, Flow Curves, Flow Birefringences, and Normal Stress of Polyisobutylene Solutions, Part II, Experimental," *Trans. Soc. Rheol. 1*, 109-118 (1957).
Brookfield. Brookfield Engineering Laboratories, Inc., Stoughton, Massachusetts 02072.
Brydson 1970. J. A. Brydson, *Flow Properties of Polymer Melts,* Van Nostrand-Reinhold, New York, 1970.
Bueche 1952. F. Bueche, "Viscosity, Self-Diffusion, and Allied Effects in Solid Polymers," *J. Chem. Phys. 20,* 1959-1964 (1952).
Bueche 1953. F. Bueche, "Bulk Viscosity of the System Polystyrene-Diethyl Benzene," *J. Applied Phys. 24,* 423-427 (1953).
Bueche 1956. F. Bueche, "Viscosity of Polymers in Concentrated Solution," *J. Chem. Phys. 25,* 599-600 (1956).
Bueche 1958. F. Bueche and S. W. Harding, "A New Absolute Molecular Weight Method for Linear Polymers," *J. Polymer Sci. 32,* 177-186 (1958).
Bueche 1959. F. Bueche, "Mobility of Molecules in Liquids Near the Glass Temperature," *J. Chem. Phys. 30,* 748-752 (1959).
Bueche 1960. F. Bueche and Frank N. Kelley, "Shear Viscosity-Molecular Weight Relationships for Polymers," *J. Polymer Sci. 45,* 267-269 (1960).
Bueche 1962. F. Bueche, *Physical Properties of Polymers,* Interscience Div., John Wiley and Sons, New York, 1962.
Chinai 1964. Suresh N. Chinai and William C. Schneider, "Flow Curves of Concentrated Polymer Solutions, an Empirical Relationship with the Polydispersity of the Solute," *Rheologica Acta 3,* 148-155 (1964).
Chinai 1965. Suresh N. Chinai and William C. Schneider, "Shear Dependence of Viscosity-Molecular Weight Transitions. A Study of Entanglement Effects," *J. Polymer Sci. A3*, 1359-1371 (1965).
Clark 1971. E. S. Clark and C. A. Garber, "Effect of Industrial Processing on the Morphology of Crystalline Polymers," *Intl. J. Polymeric Mat. 1,* 31-46 (1971).
Cole 1941. R. H. Cole and K. S. Cole, "Dispersion and Absorption in Dielectrics, I. Alternating Current Characteristics," *J. Chem. Phys. 9,* 341-351 (1941).
Collins 1965. E. A. Collins and W. H. Bauer, "Analysis of Flow Properties in Relation to Molecular Parameters for Polymer Melts," *Trans. Soc. Rheol. 9 (2),* 1-16 (1965).
Collins 1970. E. A. Collins and A. P. Metzger, "Polyvinyl Chloride Melt Rheology II. The Influence of Molecular Weight on Flow Activation Energy," *Polymer Eng. Sci. 10,* 57-65 (1970).
Crank 1968. J. Crank and G. S. Park, eds., *Diffusion in Polymers,* Academic Press, New York, 1968.
Custom. Custom Scientific Instruments, Inc., Whippany, New Jersey 07981.
Daniel 1967. Vera V. Daniel, *Dielectric Relaxation,* Academic Press, New York, 1967.
Doolittle 1951. A. K. Doolittle, "Studies in Newtonian Flow II. The Dependence of the Viscosity of Liquids on Free-Space," *J. Applied Phys. 22,* 1471-1475 (1951).

Doolittle 1952. A. K. Doolittle, "Studies in Newtonian Flow III. The Dependence of the Viscosity of Liquids on Molecular Weight and Free-Space (in Homologous Series)," *J. Applied Phys. 23*, 236-239 (1952).

Drumm 1956. M. F. Drumm, C. W. H. Dodge, and L. E. Nielsen, "Cross Linking of a Phenol Formaldehyde Novolac," *Ind. Eng. Chem. 48*, 76-81 (1956).

Edelmann 1954. K. Edelmann, "Viscosity and Molecular Weight of Macromolecules," pp. 107-115 in V. G. W. Harrison, ed., *Proceedings of the Second International Congress on Rheology*, Academic Press, New York, 1954.

Edelmann 1960. Kurt Edelmann, "The Flow Behavior of Branched Macromolecules" (in German), *Faserforch. und Textiltech. 11*, 107-113 (1960).

Eirich 1956-1969. Frederich R. Eirich, ed., *Rheology--Theory and Applications*, Academic Press, New York, Vol. 1, 1956; Vol. 2, 1958; Vol. 3, 1960; Vol. 4, 1967; Vol. 5, 1969.

Ferguson 1964. J. Ferguson, B. Wright, and R. N. Howard, "The Flow Properties of Polyethylene Whole Polymers and Fractions," *J. Applied Chem. 14*, 53-63 (1964).

Ferranti. Ferranti Electric, Inc., Plainview, New York 11803.

Ferry 1953. J. D. Ferry, L. Grondine, Jr., and D. Udy, "Viscosities of Concentrated Polymer Solutions, III. Polystyrene and Styrene-Maleic Acid Copolymer," *J. Colloid Sci. 8*, 529-539 (1953).

Ferry 1961. John D. Ferry, *Viscoelastic Properties of Polymers*, John Wiley and Sons, New York, 1961.

Ferry 1970. John D. Ferry, *Viscoelastic Properties of Polymers*, 2nd ed., John Wiley and Sons, New York, 1970.

Flory 1953. Paul J. Flory, *Principles of Polymer Chemistry*, Cornell University Press, Ithaca, New York, 1953.

Fox 1955. T. G Fox and S. Loshaek, "Isothermal Viscosity-Molecular Weight Dependence for Long Polymer Chains," *J. Applied Phys. 26*, 1080-1082 (1955).

Fox 1956. T. G Fox, Serge Gratch, and S. Loshaek, "Viscosity Relationship for Polymers in Bulk and in Concentrated Solution," Chap. 12 in Frederick R. Eirich, ed., *Rheology--Theory and Applications*, Vol. 1, Academic Press, New York, 1956.

Fox 1965. T. G Fox, "Polymer Flow in Concentrated Solutions and Melts," *J. Polymer Sci. C9*, 35-44 (1965).

Fröhlich 1958. Herbert Fröhlich, *Theory of Dielectrics: Dielectric Constant and Dielectric Loss*, 2nd ed., Oxford University Press, Oxford, England, 1958.

Fujita 1957a. Hiroshi Fujita and Kazuhiko Ninomiya, "Dependence of Mechanical Relaxation Spectra of Linear Amorphous Polymers on the Distribution of Molecular Weights," *J. Polymer Sci. 24*, 233-260 (1957).

Fujita 1957b. Hiroshi Fujita and Kazuhiko Ninomiya, "Note on the Calculation of the Molecular Weight Distribution of a Linear Amorphous Polymer from its Relaxation Distribution in the Rubbery Region," *J. Phys. Chem. 61*, 814-817 (1957).

Fujita 1961. H. Fujita and A. Kishimoto, "Interpretation of Viscosity Data for Concentrated Polymer Solutions," *J. Chem. Phys. 34*, 393-398 (1961).

Fujita 1962. H. Fujita and E. Maekawa, "Viscosity Behavior of the System Polymethyl Methacrylate and Diethyl Phthalate Over the Complete Range of Composition," *J. Phys. Chem. 66*, 1053-1058 (1962).
General Radio. General Radio Company, Concord, Massachusetts 01742.
Graessley 1965. W. W. Graessley, "Molecular Entanglement Theory of Flow Behavior in Amorphous Polymers," *J. Chem. Phys. 43*, 2696-2703 (1965).
Graessley 1967. W. W. Graessley, "Viscosity of Entangling Polydisperse Polymers," *J. Chem. Phys. 47*, 1942-1948 (1967).
Graessley 1968. William W. Graessley and James S. Prentice, "Viscosity and Normal Stresses in Branched Polydisperse Polymers," *J. Polymer Sci. A-2 6*, 1887-1902 (1968).
Gruver 1964. J. T. Gruver and G. Kraus, "Rheological Properties of Polybutadienes Prepared by *n*-Butyllithium Initiation," *J. Polymer Sci. A2*, 797-810 (1964).
Haake. Haake Instruments, Inc., Saddle Brook, New Jersey 07662.
Hayahara 1967. Takuro Hayahara and Seiji Takao, "Relationship Between Polymer Concentration and Molecular Weight in the Viscosity Behavior of Concentrated Solutions," *Kolloid Z.--Z. Polymere 225*, 106-111 (1967).
Hayahara 1968. Takuro Hayahara and Seiji Takao, "Elastic Behavior of Concentrated Solutions of Acrylonitrile Copolymers. II. The Shear Stress Dependence of the Shear Modulus of a Pseudo Network in Concentrated Solutions," *J. Applied Polymer Sci. 12*, 1755-1763 (1968).
Hill 1969. Nora E. Hill, Worth E. Vaughan, A. H. Price, and Mansel Davies, *Dielectric Properties and Molecular Behavior*, Van Nostrand-Reinhold, New York, 1969.
Imass. Imass, Accord, Massachusetts 02018.
Instron. Instron Corporation, Canton, Massachusetts 02021.
Ishida 1969. Yoichi Ishida, "Dielectric Relaxation of High Polymers in the Solid State," *J. Polymer Sci. A-2 7*, 1835-1861 (1969).
Johnson 1961. Gordon C. Johnson, "Flow Characteristics of Linear, End-Blocked Dimethyl Polysiloxane Fluids," *J. Chem. Eng. Data 6*, 275-278 (1961).
Kaelble 1969. D. H. Kaelble, "Free Volume and Polymer Rheology," Chap. 5 in Frederick R. Eirich, ed., *Rheology--Theory and Applications*, Vol. 5, Academic Press, New York, 1969.
Kauzmann 1942. W. Kauzmann, "Dielectric Relaxation as a Chemical Rate Process," *Revs. Modern Phys. 14*, 12-44 (1942).
Kirschke 1962. K. Kirschke, "On the Viscometry of Non-Newtonian Liquids at High Velocity Gradients" (in German), *Rheologica Acta 2*, 147-159 (1962).
Kirschke 1969. K. Kirschke and H. Merves, "Characteristics of Different Methods of Viscometry at High Rates of Shear," pp. 517-528 in Shigeharu Onogi, ed., *Proceedings of the Fifth International Congress on Rheology*, Vol. 1, University Park Press, Baltimore, Maryland, 1969.
Klein 1968. I. Klein and D. A. Marshall, eds., *Computer Programs for Plastic Engineers*, Reinhold Publishing Corp., New York, 1968.
Kraus 1965. Gerard Kraus and J. T. Gruver, "Rheological Properties of Multichain Polybutadienes," *J. Polymer Sci. A3*, 105-122 (1965).
Krigbaum 1964. W. R. Krigbaum, R.-J. Roe, and K. J. Smith, Jr., "A Theoretical Treatment of the Modulus of Semi-Crystalline Polymers," *Polymer 5*, 533-542 (1964).

Leaderman 1958. Herbert Leaderman, "Viscoelastic Phenomena in Amorphous High Polymeric Systems," Chap. 1 in Frederick R. Eirich, ed., *Rheology--Theory and Applications,* Vol. 2, Academic Press, New York, 1958.

Lenk 1968. R. S. Lenk, *Plastics Rheology,* Interscience Div., John Wiley and Sons, New York, 1968.

Lloyd 1962. W. G. Lloyd and T. Alfrey, Jr., "Network Polymers. II. Experimental Study of Swelling," *J. Polymer Sci. 62,* 301-316 (1962).

Long 1964. V. C. Long, G. C. Berry, and L. M. Hobbs, "Solution and Bulk Properties of Branched Polyvinyl Acetates. IV. Melt Viscosity," *Polymer 5*, 517-524 (1964).

Lyons 1970. P. F. Lyons and A. V. Tobolsky, "Viscosity of Polypropylene Oxide Solutions Over the Entire Concentration Range," *Polymer Eng. Sci. 10*, 1-3 (1970).

Marin 1965. Joseph Marin, "Mechanical Relationships in Testing for Mechanical Properties of Polymers," Chap. 3 in John V. Schmitz, ed., *Testing of Polymers*, Vol. 1, Interscience Div., John Wiley and Sons, New York, 1965.

Mathes 1966. K. N. Mathes, "Electrical Properties," pp. 528-628 in Herman F. Mark, Norman G. Gaylord, and Norbert M. Bikales, eds., *Encyclopedia of Polymer Science and Technology*, Vol. 5, Interscience Div., John Wiley and Sons, New York, 1966.

McCrum 1967. N. G. McCrum, B. Read, and G. Williams, *Anelastic and Dielectric Effects in Polymeric Solids,* John Wiley and Sons, New York, 1967.

Meares 1965. Patrick Meares, *Polymers: Structure and Bulk Properties,* D. Van Nostrand Co., Princeton, New Jersey, 1965.

Mendelson 1968. Robert A. Mendelson, "Melt Viscosity," pp. 587-620 in Herman F. Mark, Norman G. Gaylord, and Norbert M. Bikales, eds., *Encyclopedia of Polymer Science and Technology,* Vol. 8, Interscience Div., John Wiley and Sons, New York, 1968.

Mendelson 1970. R. A. Mendelson, W. A. Bowles and F. L. Finger, "The Effect of Molecular Structure on Polyethylene Melt Rheology. 1. Low Shear Behavior," *J. Polymer Sci. A-2 8*, 105-126 (1970).

Metzger 1965. A. P. Metzger and J. R. Knox, "The Effect of Pressure Losses in the Barrel on Capillary Flow Measurements," *Trans. Soc. Rheol. 9,* 13-25 (1965).

Miller 1966. M. L. Miller, *The Structure of Polymers,* Reinhold Publishing Corp., New York, 1966.

Mills 1961. D. R. Mills, G. E. Moore and D. W. Pugh, "The Effect of Molecular Weight Distribution on the Flow Properties of Polyethylene," *SPE Trans. 1*, 40-46 (1961).

Moore 1959. Louis D. Moore, Jr., "Relations Among Melt Viscosity, Solution Viscosity, Molecular Weight and Long Chain Branching in Polyethylene," *J. Polymer Sci. 36*, 155-172 (1959).

Mullowney. F. J. Mullowney Company, Trenton, New Jersey 08600.

Nagasawa 1972. Mitsuru Nagasawa and Teruo Fujimoto, "Preparation, Characterization, and Viscoelastic Properties of Branched Polymers," pp. 263-314 in Seizo Okamura and Motowo Takayanagi, eds., *Progress in Polymer Science, Japan,* Vol. 3, Halsted Press, Div. John Wiley and Sons, New York, 1972.

Nielsen 1962. Lawrence E. Nielsen, *Mechanical Properties of Polymers,* Reinhold Publishing Corp., New York, 1962.

Nielsen 1963. L. E. Nielsen and F. D. Stockton, "Theory of the Modulus of Crystalline Polymers," *J. Polymer Sci. A1*, 1995-2002 (1963).

Ninomiya 1957. Kazuhiko Ninomiya and Hiroshi Fujita, "Stress Relaxation Behavior of Polyvinyl Acetate Films," *J. Colloid Sci. 12*, 204-229 (1957).

Ninomiya 1959. Kazuhiko Ninomiya, "Effect of Blending on the Stress-Relaxation Behavior of Polyvinyl Acetate in the Rubbery Region," *J. Colloid Sci. 14*, 49-58 (1959).

Onogi 1963. Shigeharu Onogi, Tadashi Kobayashi, Yasuhiro Kojima, and Koshishige Taniguchi, "Non-Newtonian Flow of Concentrated Solutions of High Polymers," *J. Applied Polymer Sci. 7*, 847-859 (1963).

Onogi 1966. S. Onogi, H. Kato, S. Ueki, and T. Ibaragi, "Rheological Properties of Polystyrene Melts," *J. Polymer Sci. C15*, 481-494 (1966).

Onogi 1968. S. Onogi, T. Masuda, and T. Ibaragi, "Rheological Properties of Poly(Methyl Methacrylate) and Poly(Vinyl Acetate) in the Molten State," *Kolloid Z. 222*, 110-124 (1968).

Osaki 1971. Kunihiro Osaki and Yoshiyuki Einaga, "Viscoelastic Properties of Concentrated Polymer Solutions," pp. 321-375 in Minoro Imoto and Shigeharu Onogi, eds., *Progress in Polymer Science, Japan*, Vol. 1, Halsted Press, Div. John Wiley and Sons, New York, 1971.

Paul 1965. D. R. Paul and A. T. DiBenedetto, "Diffusion in Amorphous Polymers," *J. Polymer Sci. C10*, 17-44 (1965).

Pezzin 1966. G. Pezzin and N. Gligo, "Viscosity of Concentrated Polymer Solutions. 1. Polyvinyl Chloride in Cyclohexanone," *J. Applied Polymer Sci. 10*, 1-19 (1966).

Philippoff 1936. W. Philippoff and K. Hess, "The Viscosity Problems in Organic Colloids" (in German), *Z. Phys. Chem. B31*, 237-255 (1936).

Philippoff 1958. Wladimer Philippoff, "Viscosity Measurements on Polymer Modified Oils," *Am. Soc. Lubrication Eng. Trans. 1*, 82-86 (1958).

Platzer 1971. Norbert Platzer, "Performance Characteristics of Thermoplastics," *Chem. Tech. 1*, 165-175 (1971).

Porter 1960. Roger S. Porter and Julian F. Johnson, "Viscosity of Polyethylenes: Dependence on Molecular Weight at High Shear," *J. Applied Polymer Sci. 3*, 200-205 (1960).

Porter 1964. Roger S. Porter and Julian F. Johnson, "Viscosity/-Temperature Dependence for Polyisobutene Systems. The Effect of Molecular Weight Distribution," *Polymer 5*, 201-206 (1964).

Porter 1966a. Roger S. Porter and Julian F. Johnson, "The Entanglement Concept in Polymer Systems," *Chem. Revs. 66*, 1-27 (1966).

Porter 1966b. Roger S. Porter and Julian F. Johnson, "Temperature Dependence of Polymer Viscosity. The Influence of Shear Rate and Stress," *J. Polymer Sci. C15*, 365-371 (1966).

Precision. Precision Scientific Company, Chicago, Illinois 60647.

Rabinowitsch 1929. B. Rabinowitsch, "The Viscosity and Elasticity of Sols" (in German), *Z. physik. Chem. A145*, 1-26 (1929).

Rebinder 1954. P. A. Rebinder, "Coagulation and Thixotropic Structures," *Disc. Faraday Soc. 18*, 151-160 (1954).

Reiner 1966. M. Reiner, "Deformation," pp. 620-647 in Herman F. Mark, Norman G. Gaylord, and Norbert M. Bikales, eds., *Encyclopedia of Polymer Science and Technology*, Vol. 4, Interscience Div., John Wiley and Sons, New York, 1966.

Rodriguez 1970. Ferdinand Rodriguez, *Principles of Polymer Systems*, McGraw-Hill Book Co., New York, 1970.

Rogers 1964. C. E. Rogers, "Permeability and Chemical Resistance," pp. 609-688 in Eric Baer, ed., *Engineering Design for Plastics*, Reinhold Publishing Corp., New York, 1964.

Rogers 1965. C. E. Rogers, "Solubility and diffusivity," Chap. 6 in David Fox, Mortimer M. Labes, and A. Weissberger, eds., *Physics and Chemistry of the Organic Solid State*, Vol. II, Interscience Div., John Wiley and Sons, New York, 1965.

Rudd 1960. John F. Rudd, "The Effect of Molecular Weight Distribution on the Rheological Properties of Polystyrene," *J. Polymer Sci. 44*, 459-474 (1960).

Rudin 1970. A. Rudin, "Effect of Processing History on Melt Flow Defects," *Polymer Eng. Sci. 10*, 94-101 (1970).

Schaefgen 1948. John R. Schaefgen and Paul J. Flory, "Synthesis of Multi Chain Polymers and Investigation of Their Viscosities," *J. Am. Chem. Soc. 70*, 2709-2718 (1948).

Schreiber 1962. H. P. Schreiber and E. B. Bagley, "The Newtonian Melt Viscosity of Polyethylene: An Index of Long Chain Branching," *J. Polymer Sci. 58*, 29-48 (1962).

Schurz 1959. J. Schurz, "Interpretation of Flow Curves in Terms of Molecular Properties," *J. Colloid Sci. 14*, 492-500 (1959).

Schurz 1970. Josef Schurz, "Rheological Structure Investigation with Concentrated Polymer Solutions," pp. 215-230 in Shigeharu Onogi ed., *Proceedings of the Fifth International Congress on Rheology*, Vol. 4, University Park Press, Baltimore, Maryland, 1970.

Severs 1962. Edward T. Severs, *Rheology of Polymers*, Reinhold Publishing Corp., New York, 1962.

Smith 1962. Thor L. Smith, "Stress-Strain-Time-Temperature Relationships for Polymers," pp. 60-89 in *ASTM Special Technical Publication No. 325*, American Society for Testing and Materials, Philadelphia, Pennsylvania, 1962.

Smith 1969. Thor L. Smith, "Strength and Extensibility of Elastomers," Chap. 4 in Frederick R. Eirich, ed., *Rheology--Theory and Applications*, Vol. 5, Academic Press, New York, 1969.

Smyth 1955. Charles Phelps Smyth, *Dielectric Behavior and Structure*, McGraw-Hill Book Co., New York, 1955.

Stabler 1967. H. G. Stabler, R. N. Howard, and B. Wright, "The Flow Properties of High-Density Polyethylene in Capillary Rheometry and Extrusion," pp. 327-347 in *SCI Monograph No. 26, Advances in Polymer Science and Technology*, Gordon and Breach Science Publishers, New York, 1967.

Stannett 1965. V. Stannett and H. Yasuda, "The Measurement of Gas and Vapor Permeation and Diffusion in Polymers," Chap. 13 in John V. Schmitz, ed., *Testing of Polymers*, Vol. I, Interscience Div., John Wiley and Sons, New York, 1965.

Stern 1964. S. A. Stern, T. F. Sinclair, and P. J. Gareis, "An improved permeability apparatus of the variable-volume type," *Modern Plastics 42* (2), 154-206 (4 pages) (October, 1964).

Stratton 1966. Robert A. Stratton, "The Dependence of Non-Newtonian Viscosity on Molecular Weight for 'Monodisperse' Polystyrene," *J Colloid Interface Sci. 22*, 517-530 (1966).

Takayanagi 1963. Motowo Takayanagi, "Viscoelastic Behavior of Crystalline Polymers," pp. 161-187 in E. H. Lee and Alfred L. Copley, eds., *Proceedings of the Fourth International Congress on Rheology*, Part 1, Interscience Div., John Wiley and Sons, New York, 1965.

Thorkildsen 1964. R. L. Thorkildsen, "Mechanical Behavior," Chap. 5 in Eric Baer, ed., *Engineering Design for Plastics*, Reinhold Publishing Corp., New York, 1964.

Tinius Olsen. Tinius Olsen Testing Machine Company, Willow Grove, Pennsylvania 19090.

Tobolsky 1958a. A. V. Tobolsky, A. Mercurio, and K. Murakami, "Molecular Weight Distribution by Stress Relaxation," *J. Colloid Sci. 13*, 196-197 (1958).

Tobolsky 1958b. Arthur V. Tobolsky, "Stress Relaxation Studies of the Viscoelastic Properties of Polymers," Chap. 2 in Frederick R. Eirich, ed., *Rheology--Theory and Applications*, Vol. 2, Academic Press, New York, 1958.

Tobolsky 1960. Arthur V. Tobolsky, *Properties and Structures of Polymers*, John Wiley and Sons, New York, 1960.

Tobolsky 1963. A. V. Tobolsky and V. D. Gupta, "The Modulus of Polyethylene," *Textile Res. J. 33*, 761 (1963).

Tobolsky 1964. A. V. Tobolsky, D. Kotz, M. Takahashi, and R. Schaffhauser, "Rubber Elasticity in Highly Crosslinked Systems: Crosslinked Styrene, Methyl Methacrylate, Ethyl Acrylate and Octyl Acrylate," *J. Polymer Sci. A2*, 2749-2758 (1964).

Tordella 1969. John P. Tordella, "Unstable Flow of Molten Polymers," Chap. 2 in Frederick R. Eirich, ed., *Rheology--Theory and Applications*, Vol. 5, Academic Press, New York, 1969.

Treloar 1958. L. R. G. Treloar, *The Physics of Rubber Elasticity*, 2nd ed., Clarendon Press, Oxford, England, 1958.

Van Amerongen 1964. G. J. Van Amerongen, "Diffusion in Elastomers," *Rubber Chem. Tech. 37*, 1065-1152 (1964).

Van Wazer 1963. J. R. Van Wazer, J. W. Lyons, K. Y. Kim, and R. E. Colwell, *Viscosity and Flow Measurement. A Laboratory Handbook of Rheology*, Interscience Div., John Wiley and Sons, New York, 1963.

Vaughan 1972. Worth E. Vaughan, Charles P. Smyth, and Jack Gordon Powles, "Determination of Dielectric Constant and Loss," Chap. 5 in Arnold Weissberger and Bryant W. Rossiter, eds., *Physical Methods of Chemistry*, Vol. 1 of A. Weissberger, ed., *Techniques of Chemistry*, Part 4, Interscience-Wiley, New York, 1972.

Ward 1971. I. M. Ward, *Mechanical Properties of Solid Polymers*, Interscience Div., John Wiley and Sons, New York, 1971.

Weber 1956. N. Weber and Walter Bauer, "Flow Properties of Aluminum Dilaurate-Toluene Gels," *J. Phys. Chem. 60*, 270-273 (1956).

West 1967. G. H. West, R. N. Howard, and B. Wright, "Sensitivity of Viscosity of Thermoplastics to Temperature and Diluent Content, A Test of Free Volume Theory," pp. 348-369 in *SCI Monograph No. 26, Advances in Polymer Science and Technology*, Gordon and Breach Science Publishers, New York, 1967.

Williams 1955. M. L. Williams, R. F. Landel, and J. D. Ferry, "The Temperature Dependence of Relaxation Mechanisms in Amorphous Polymers and Other Glass Forming Liquids," *J. Am. Chem. Soc. 77*, 3701-3707 (1955).

Williams 1971. David J. Williams, *Polymer Science and Engineering*, Prentice-Hall, Englewood Cliffs, New Jersey, 1971.

Zabusky 1964. H. H. Zabusky and R. F. Heitmiller, "Properties of High Density Polyethylene with Bimodal Molecular Weight Distribution," *SPE Trans*. *4*, 17-21 (1964).

III

EXPERIMENTS

A

Synthesis Experiments

EXPERIMENT 1. BULK POLYCONDENSATION OF ω-AMINOUNDECANOIC ACID

A. Introduction

Poly(ω-aminoundecanoic acid) (11 nylon) is prepared by bulk polycondensation. Six samples are prepared with varying polymerization times, allowing the kinetics of polymerization to be followed if degrees of polymerization are obtained by end-group analysis in Exp. 10. The calculations and report of this experiment, if kinetics is followed, cannot be completed until Exp. 10 has been performed. The polymerization of Exp. 1 requires approximately three hours.

B. Principle

11 nylon is prepared as the result of the reaction of amino and acid groups to amide groups, with the elimination of water. The stoichiometric equivalence required to obtain high molecular weights is insured by the structure and purity of the monomer.

Pinner (1961) describes the bulk polymerization of 66 nylon in his Exp. B.7.3, and Braun (1971) includes Exp. 4.1.2.1 on the bulk polymerization of 6 nylon.

C. Applicability

This procedure can be used for the polycondensation of any ω-amino acid except those containing 4 or 5 carbon atoms, which form stable rings.

D. Precision and Accuracy

E. *Safety Considerations*

ω-AMINOUNDECANOIC ACID IS A POTENTIALLY HARMFUL CHEMICAL FOR WHICH THE POSSIBLE DANGERS OF EXPOSURE HAVE NOT BEEN FULLY STUDIED. NORMAL PRECAUTIONS SHOULD BE OBSERVED TO AVOID CONTACT OF THIS CHEMICAL WITH THE SKIN. USE APPROPRIATE PRECAUTIONS IN HANDLING HOT APPARATUS. WRAP TEST TUBES IN CLOTH OR PAPER TOWELING BEFORE BREAKING THEM AT THE END OF THE EXPERIMENT. SAFETY GOGGLES MUST BE WORN DURING ALL LABORATORY OPERATIONS.

F. *Apparatus*

1. Constant-temperature bath (Fisher Hi-Temp or equivalent) with silicone fluid, capable of operation at 220°C, and fitted with a rack for test tubes.
2. 100-ml test tubes with serum stoppers.
3. Long and short syringe needles.
4. Source of dry nitrogen at controllable low pressure.

G. *Reagents and Materials*

1. ω-Aminoundecanoic acid, reagent grade.

H. *Preparation*

I. *Procedure*

1. Place approximately 3 g monomer in each of 6 test tubes, and cap with serum stoppers. Place the test tubes in the rack in the bath at 220°C.
2. Insert a long syringe needle, extending to the bottom, in each test tube as a nitrogen inlet, and a short needle in each as an outlet.
3. Apply *slight* nitrogen flow (CAUTION: The monomer is a dry powder which can blow about) in each tube for 1-2 min to remove air.
4. When the monomer melts, record the time as the start of polymerization, and increase the nitrogen flow until a steady stream of bubbles rises in each tube.
5. At each of the following times (10, 20, 30, 45, 60, and 90 min), remove one test tube from the bath, remove the syringe needles, and cool the tube quickly under tap water. Rotate the test tube while cooling it, so that the polymer forms a thin film on the sides of the tube. This facilitates breaking up the polymer in step 6.
6. Wrap the tube in cloth or paper toweling and break it with a heavy tool (OBSERVE CAUTION). Isolate the polymer from glass fragments, crush it into small pieces, and dry in a vacuum oven for 24 hr at 100°C. Retain for use in Exp. 10.

J. *Fundamental Equations*

1. *Chemical*

$$x\ H_2N(CH_2)_{11}COOH \rightarrow H[HN(CH_2)_{11}CO]_xOH + (x-1)\ H_2O$$

2. *Kinetic*

$$\bar{x}_n = kc_0t + 1 \qquad (1\text{-}4)$$

$$\bar{x}_n = 1/(1-p) \qquad (1\text{-}1)$$

K. *Calculations*

(*Note*: Completion of Secs. K and L requires that Exp. 10 has been performed to determine p for the polymers prepared in Sec. I.)

1. Calculate c_0 from the monomer molecular weight, assuming that its density is 1.0.
2. From Exp. 10, determine $\bar{x}_n$ for each of the six polymer samples.
3. Following Eq. 1-4, plot corresponding values of $\bar{x}_n$-1 and t, and determine k and its standard deviation s_k by a least-squares analysis (Appendix IV).

L. *Report*

1. Describe the apparatus and experiment in your own words.
2. Prepare a table listing t, p, $\bar{x}_n$, and a comment on the properties of the polymer (for example, waxy, brittle, tough, etc.).
3. Include a graph of $\bar{x}_n$-1 versus t with the least-squares line included.
4. Tabulate intermediate quantities in the least-squares calculation, and values of c_0, k, and s_k.
5. Comment on the magnitude of s_k compared to k, and discuss possible errors contributing to s_k, including the effects of moisture in the monomer, uncertainty in c_0, inaccuracies in determining t, temperature fluctuations, etc.
6. How would the polycondensation proceed if the test tubes were flushed with dry nitrogen and sealed?

M. *Comments*

Because of the long reaction time and high temperature, and high monomer purity required, this method is not used for the commerical production of 11 nylon (which constitutes about 2% of the polyamides made commercially). Instead, catalyzed polycondensation or ring-opening polymerization is used.

N. *General References*

Chap. 1B; *Textbook*, Chaps. 8B, 8C, 15A; Aélion 1948; Floyd 1958, pp. 59-61; Sokolov 1968; Sorenson 1968, pp. 81-82; Sweeney 1969.

O. *Bibliography*

Aélion 1948. M. R. Aélion, "Preparation and Structure of Some New Types of Polyamides" (in French), *Ann. Chim.* Ser. 14 *3*, 5-61 (1948).

Braun 1971. Dietrich Braun, Harald Cherdon, and Werner Kern, *Techniques of Polymer Syntheses and Characterization*, Wiley-Interscience, New York, 1971.

Floyd 1958. Don E. Floyd, *Polyamide Resins*, Reinhold Publishing Corp., New York, 1958.

Pinner 1961. S. H. Pinner, *A Practical Course in Polymer Chemistry*, Pergamon Press, New York, 1961.

Sokolov 1968. B. Sokolov, *Synthesis of Polymers by Polycondensation*, Israel Scientific Translations, Jerusalem, Israel, 1968.

Sorenson 1968. Wayne R. Sorenson and Tod W. Campbell, *Preparative Methods of Polymer Chemistry*, 2nd ed., Interscience Div., John Wiley and Sons, New York, 1968.

Sweeney 1969. W. Sweeney and J. Zimmerman, "Polyamides," pp. 483-597 in Herman F. Mark, Norman G. Gaylord, and Norbert M. Bikales, eds., *Encyclopedia of Polymer Science and Technology*, Vol. 10, Interscience Div., John Wiley and Sons, New York, 1969.

EXPERIMENT 2. INTERFACIAL POLYCONDENSATION OF HEXAMETHYLENE DIAMINE AND SEBACOYL CHLORIDE

A. *Introduction*

In this experiment, 610 nylon, poly(hexamethylene sebacamide), is formed by the rapid, room-temperature method of interfacial polycondensation. The experiment requires about 1.5 hours.

B. *Principle*

In interfacial polymerization, two complementing monomers (for example, a diamine and a diacid chloride as in this experiment) are dissolved in two immiscible solvents, neither of which is a solvent for the resulting polymer. When the solutions are mixed, a film of high polymer is immediately formed at the interface. This can be pulled out as a continuous fiber (the "nylon rope trick"); or, if the solutions are mixed with violent agitation to produce a large surface area, polymer can be formed in high yield very rapidly. This nonequilibrium method eliminates three of the disadvantages of bulk polycondensation: the need for high temperatures, long reaction times, and exact stoichiometric equivalence.

See McCaffery (1970), Exp. 8, and Braun (1971), Exp. 4.1.2.3b, for similar experiments.

C. *Applicability*

This method is applicable to virtually all complementary pairs of monomers for polycondensation, and has been used to produce polyamides, polyurethanes, polyureas, polysulfonamides, and poly(phenyl esters) among others. It is particularly useful for preparing polymers which are unstable at the elevated temperatures required in bulk polycondensation.

D. *Precision and Accuracy*

E. *Safety Considerations*

EACH OF THE REAGENTS IN THIS EXPERIMENT IS AN IRRITANT CAPABLE OF CAUSING SKIN BURNS. CARE MUST BE TAKEN TO AVOID SKIN CONTACT WITH REAGENTS OR SOLUTIONS. THE SOLVENTS USED ARE LIKEWISE HAZARDOUS BECAUSE OF TOXICITY OR FLAMMABILITY. APPROPRIATE PRECAUTIONS ARE REQUIRED. THE BLENDER USED MUST BE EQUIPPED WITH EITHER AN EXPLOSION-PROOF MOTOR OR PROVISION TO PURGE THE MOTOR COMPARTMENT WITH NITROGEN. SAFETY GLASSES MUST BE WORN IN THE LABORATORY AT ALL TIMES.

F. *Apparatus*

1. "Waring" blender or equivalent, explosion-proof or equipped for purging with nitrogen. Cover drilled to admit stem of a separatory funnel.
2. 300-ml separatory funnel.
3. Large fritted-glass filter, 4-5 inch diameter, or equivalent.

G. *Reagents and Materials*

1. Hexamethylene diamine, reagent grade.
2. Sebacoyl chloride, reagent grade.
3. Sodium hydroxide, reagent grade.
4. Tetrachloroethylene, reagent grade.
5. Acetone, reagent grade.
6. Methanol, reagent grade.

H. *Preparation*

I. *Procedure*

1. Dissolve 0.02 mole (2.32 g) hexamethylene diamine and 0.04 mole

(1.6 g) sodium hydroxide in 330 ml distilled water, and place in blender.

2. Dissolve 0.02 mole (4.78 g) sebacoyl chloride in 250 ml tetrachloroethylene and place in separatory funnel. Support the separatory funnel with its stem passing through the cover of the blender.

3. Turn the blender on to maximum speed, and add the contents of the separatory funnel rapidly (15 sec). Continue to stir vigorously for about 2 min.

4. Stop the blender, and collect the polymer on the fritted-glass filter.

5. Return the polymer to the blender, and wash with 500 ml of 1:1 water:methanol for 2-5 min. Then repeat step 4.

6. Repeat step 5 using a water:acetone mixture.

7. Dry the polymer in a vacuum oven for 24 hr at about 80°C. Retain for use in later experiments.

J. *Fundamental Equations*

$$x\ H_2N(CH_2)_6NH_2 + x\ ClOC(CH_2)_8COCl \rightarrow$$

$$H[HN(CH_2)_6NHCO(CH_2)_8CO]_xCl + 2x-1\ HCl$$

K. *Calculations*

L. *Report*

1. Describe the experiment and apparatus in your own words.
2. Calculate the yield of polymer recovered.
3. Answer the following questions:
   a. What purpose is served by the sodium hydroxide?
   b. Why are the particular solvent mixtures used to wash the polymer?

M. *Comments*

1. Carbon tetrachloride can be used in place of tetrachloroethylene.

2. 610 nylon is made commercially by bulk rather than interfacial polycondensation for economic reasons associated with the solvents and their recovery. The polymer is widely used in monofilament form for brushes and bristles and in sports equipment.

N. *General References*

Chaps. 1B, 2C; *Textbook*, Chaps. 8B, 8C, 15A; Morgan 1959, 1962, 1963, 1965, pp. 488-490; Sweeney 1969; Braun 1971, pp. 218-219; Williams 1971, pp. 74-77.

O. Bibliography

Braun 1971. Dietrich Braun, Harald Cherdon, and Werner Kern, *Techniques of Polymer Syntheses and Characterization,* Wiley-Interscience, New York, 1971.

McCaffery 1970. Edward M. McCaffery, *Laboratory Preparation for Macromolecular Chemistry,* McGraw-Hill Book Co., New York, 1970.

Morgan 1959. Paul W. Morgan and Stephanie L. Kwolek, "The Nylon Rope Trick," *J. Chem. Educ. 36*, 182-184 (1959).

Morgan 1962. Paul W. Morgan and Stephanie L. Kwolek, "Interfacial Polycondensation. XII. Variables Affecting Stirred Polycondensation Reactions," *J. Polymer Sci. 62,* 33-58 (1962).

Morgan 1963. P. W. Morgan and S. L. Kwolek, "Poly(hexamethylenesebacamide) by Interfacial Polycondensation," pp. 13-16 in C. G. Overberger, ed., *Macromolecular Syntheses,* Vol. 1, John Wiley and Sons, New York, 1963.

Morgan 1965. Paul W. Morgan, *Condensation Polymers: By Interfacial and Solution Methods,* Interscience Div., John Wiley and Sons, New York, 1965.

Sweeney 1969. W. Sweeney and J. Zimmerman, "Polyamides," pp. 483-597 in Herman F. Mark, Norman G. Gaylord, and Norbert M. Bikales, eds., *Encyclopedia of Polymer Science and Technology,* Vol. 10, Interscience Div., John Wiley and Sons, New York, 1969.

Williams 1971. David J. Williams, *Polymer Science and Engineering,* Prentice-Hall, Englewood Cliffs, New Jersey, 1971.

EXPERIMENT 3. BULK POLYMERIZATION OF METHYL METHACRYLATE

A. Introduction

Methyl methacrylate is polymerized in bulk in the presence and absence of a chain-transfer agent. The time dependence of conversion is observed for the polymerization without added chain-transfer agent. The dependence of molecular weight upon concentration of the transfer agent is determined if dilute-solution viscosities are measured as in Exp. 15. If this is done, the calculations and report sections of this experiment can be completed only after the viscometric results are obtained. The polymerization of Exp. 3 requires 3 hours.

B. Principle

The free-radical polymerization of methyl methacrylate is initiated by azobisisobutyronitrile (AIBN) as described in Chap. 1C. The degree of conversion is determined by refractometry using the method developed in Chap. 5D. *n*-Butyl mercaptan is used as a chain-transfer agent, and its effect on the molecular weight of the polymer is determined by measuring the dilute-solution viscosity of the poly(methyl methacrylates) and calculating molecular weights by means of the Mark-Houwink equation.

Both Pinner (1961) in Exp. B.1.2 and McCaffery (1970) in Exp. 5

describe the bulk polymerization of methyl methacrylate in experiments somewhat different from this one, while Braun (1971) has a similar procedure in Exp. 3-14*b*.

C. Applicability

This experiment is applicable to a large number of monomers polymerizable by radical mechanisms. Table 5-2 gives monomer and polymer refractive indices for several monomers, allowing the course of polymerization to be followed by refractometry. Information on suitable chain-transfer agents is widely available; see Sec. N.

D. Precision and Accuracy

E. Safety Considerations

DANGEROUS CHEMICALS ARE USED IN THIS EXPERIMENT. AIBN IS HIGHLY TOXIC; THE MONOMER, CHAIN-TRANSFER AGENT, AND CHLOROFORM ARE TOXIC, IRRITANT, AND NOXIOUS; AND HEXANE OR PETROLEUM ETHER IS HIGHLY FLAMMABLE. ADEQUATE VENTILATION, INCLUDING USE OF A FUME HOOD, IS ESSENTIAL. THE POLYMERIZATION IS HIGHLY EXOTHERMIC. TO AFFORD PROTECTION IN CASE IT GETS OUT OF CONTROL, IT IS NECESSARY TO SHIELD THE LARGE TEST TUBES CONTAINING MONOMER WITH WIRE SCREENS (Fig. 4-23). SAFETY GLASSES MUST BE WORN IN THE LABORATORY AT ALL TIMES.

F. Apparatus

1. Constant-temperature bath regulated at 70°C.
2. Vacuum bench and supply of low-pressure dry nitrogen (Fig. 4-19).
3. Refractometer, Abbe type, with circulating water bath to keep prisms at 25°C.
4. Large fritted-glass filter for collecting polymer.
5. 20-ml syringe.
6. Microliter syringe (Fig. 4-8*c*).

G. Reagents and Materials

1. Methyl methacrylate, redistilled as described in step H-1.
2. 2,2'-Azobisisobutyronitrile, recrystallized as described in step H-2.
3. *n*-Butyl mercaptan, reagent grade.
4. Hexane or petroleum ether, practical grade.

H. Preparation

1. Prepare methyl methacrylate in advance as follows:
   a. Wash to remove inhibitor: The inhibitor, usually an

aromatic such as hydroquinone or *t*-butylpyrocatechol, is removed by washing the monomer with 10% aqueous NaOH. Roughly equal parts of the basic solution and the monomer are placed in a separatory funnel and mixed by tumbling. The heavier aqueous phase is drained off. The procedure is repeated once or twice until the liquids remain clear. The monomer is then washed with distilled water until litmus paper shows that all the base has been removed.

b. Dry: A drying agent such as anhydrous Na_2SO_4 (Table 3-1) is added to the monomer (100 g/ℓ). With occasional tumbling, drying is complete in about 1/2 hr.

c. Distill: Add about 1 g/ℓ CuCl stabilizer to the monomer and distill under dry N_2 at 60 torr and 33-35°C. The distillation setup of Fig. 4-20 is suitable.

d. Check before use: Test for the presence of polymer in redistilled monomer just before use by adding a drop of methanol. Turbidity indicates the presence of polymer.

2. Recrystallize AIBN as required by preparing a saturated solution in methanol and cooling in a refrigerator. Collect the crystals on a fritted-glass filter when about 60% of the solvent has evaporated.

I. Procedure

1. Place 110 mg AIBN in a dry 250-ml flask. Cap the flask with a serum stopper and insert syringe needles (Fig. 4-7). Flush the flask with dry N_2 and introduce 110 ml monomer with a syringe (see steps M-1 and M-2). Swirl occasionally to dissolve the AIBN.
2. Prepare 10 small and 5 large dry test tubes by capping, and evacuating and filling with N_2 twice.
3. With the microliter syringe, place 0.5, 1.0. 1.5, and 2.0 mole percent (0.09, 0.18, 0.27, and 0.36 ml) *n*-butyl mercaptan in four large test tubes.
4. With the syringe, place 20 ml AIBN-monomer mixture in each large test tube and 1 ml in each small test tube.
5. Place all test tubes (the large ones protected by screening) in the 70°C constant-temperature bath at the same time.
6. Remove the small test tubes from the bath at about 5-min intervals, cool quickly (note the time), place a drop of the contents on the refractometer prism, and measure the refractive index. (CAUTION: Wash the prisms thoroughly *at once* with chloroform to remove all traces of polymer.)
7. Remove the five large test tubes simultaneously from the bath after about 1/2 hr. Cool them quickly, dilute their contents with a little chloroform if necessary to obtain fluidity, and pour the contents quantitatively into 200-300 ml hexane or petroleum ether. Collect the precipitated polymer quantitatively on fritted-glass filters (see step M-3), and dry in a vacuum oven at 80°C for later use in Exp. 15. Weigh the polymer when dry.

J. Fundamental Equations

1. Chemical

$$x\ CH_3{=}\underset{\underset{COOCH_3}{|}}{C}(CH_3) \xrightarrow{AIBN} -(CH_2{-}\underset{\underset{COOCH_3}{|}}{C}(CH_3){-})_x$$

$$M_n^{\cdot} + C_4H_9SH \rightarrow M_nH + C_4H_9S\cdot$$

$$C_4H_9S\cdot + M \rightarrow C_4H_9SM\cdot \quad , \text{ etc.}$$

2. *Refractometry*

$$p = 1.0684\ (n_p - 1.4140)/(0.2137\ n_p - 0.2359) \qquad (5\text{-}9)$$

3. *Kinetic*

$$1/\bar{x}_n = (1/\bar{x}_n)_0 + C_S[S]\ /\ [M] \qquad (\text{from } 1\text{-}14)$$

K. *Calculations*

1. Calculate the extent of polymerization p from Eq. 5-9, and plot it as a function of polymerization time.
2. Determine C_S, the chain-transfer constant, from the slope of the straight-line plot of Eq. 1-14. See step M-5, and calculate $\bar{x}_n$ from the dilute-solution viscosities determined in Exp. 15.

L. *Report*

1. Describe the apparatus and experiment in your own words.
2. Prepare a table of refractive index and conversion versus time.
3. Include the graph of step K-1.
4. Prepare a table of polymer yield (see step M-4), [S], [S]/[M], appropriate viscosity functions from Exp. 15, $\bar{M}_v$, and $\bar{x}_n$.
5. Include the graph of step K-2.
6. Give your value for C_S, and compare it with the literature value of 0.66. Comment on the cause of any differences.

M. *Comments*

1. Caution is required in using syringes in vessels which are sealed or under positive or negative pressure. Positive pressure will force the plunger out, whereas withdrawing liquid from a sealed vessel will create a negative pressure, allowing air to leak in and contaminate the contents. It is advisable to keep the vessel on the nitrogen

line, with inlet and outlet tubes, when adding or withdrawing any liquid.

2. Do not pierce rubber serum stoppers more than once at a given spot; leaks may develop, and pieces of rubber may be torn off to contaminate the reagent. Also, do not use the same needle for different reagents.

3. Be sure that the fritted-glass filter is clean, and remove the polymer carefully when dry to avoid any contamination from particles of glass or any other source. Such contamination can cause great inaccuracies in Exp. 15.

4. Note that Eq. 1-14 holds for very low conversion only. Be sure that your samples are made to less than 10% conversion.

5. The procedure of Exp. 15 includes the estimation of the intrinsic viscosity from measurement at a single concentration, for poly(methyl methacrylate). Subsequently, use the Mark-Houwink constants in Table 7-7 to calculate $\overline{M}_v$; and for this polymer made to low conversion at 70°C, take $\overline{M}_n = \overline{M}_v/1.85$.

6. Poly(methyl methacrylate) is polymerized in bulk for commercial use as sheets, rods, and tubes (see Pinner 1961, Exp. B.1.2). For use as a molding plastic, it is polymerized either in bulk or in a suspension system (*Textbook*, Chap. 14B).

N. *General References*

Chaps. 1C, 2A, 5D, 7D; *Textbook,* Chaps. 9B, 9C, 14B; Schildknecht 1952, Chap. IV: Ringsdorf 1965; Braun 1971, Chap. 3.

O. *Bibliography*

Braun 1971. Dietrich Braun, Harald Cherdon, and Werner Kern, *Techniques of Polymer Syntheses and Characterization,* Wiley-Interscience, New York, 1971.

McCaffery 1970. Edward M. McCaffery, *Laboratory Preparation for Polymer Chemistry,* McGraw-Hill Book Co., New York, 1970.

Pinner 1961. S. H. Pinner, *A Practical Course in Polymer Chemistry,* Pergamon Press, New York, 1961.

Ringsdorf 1965. H. Ringsdorf, "Bulk Polymerization," pp. 642-666 in Herman F. Mark, Norman G. Gaylord, and Norbert M. Bikales, eds., *Encyclopedia of Polymer Science and Technology,* Vol. 2, Interscience Div., John Wiley and Sons, New York, 1965.

Schildknecht 1952. Calvin E. Schildknecht, *Vinyl and Related Polymers,* John Wiley and Sons, New York, 1952.

EXPERIMENT 4. EMULSION POLYMERIZATION OF STYRENE

A. *Introduction*

Polystyrene is prepared by emulsion polymerization. The dependence of the polymerization rate on the concentration of emulsifier (soap) is studied, and the polymer is reserved for use in later experiments. The polymerization requires 3 hours.

B. Principle

At the beginning of an emulsion polymerization, three phases are present: a continuous aqueous phase containing the initiator; suspensed droplets of monomer, kept from coagulating to a continuous organic phase by agitation; and soap micelles containing a small proportion of monomer. Initiator fragments enter some of the micelles, where polymerization takes place, supplied with monomer by diffusion from the monomer droplets through the aqueous phase. As these micelles (now more properly called monomer-polymer particles) grow, they are stabilized by more soap at the expense of uninitiated micelles, which eventually disappear. The polymerization rate depends upon the number of micelles, and this in turn on the concentration of the soap.

Pinner (1961) in Exp. B.3.2, McCaffery (1970) in Exp. 7, and Braun (1971) in Exp. 3-03 describe similar experiments.

C. Applicability

The emulsion system can be applied to a wide variety of vinyl, acrylic, and diene monomers with water solubility in the proper range, usually 0.001-1%.

D. Precision and Accuracy

E. Safety Considerations

POTASSIUM PERSULFATE, HYDROQUINONE, STYRENE, AND METHANOL ARE TOXIC CHEMICALS AND MUST BE HANDLED WITH PROPER PRECAUTIONS. ADEQUATE VENTILATION IS IMPORTANT. DESPITE THE SMALL AMOUNTS OF MONOMER USED, THERE IS DANGER OF TEMPERATURE AND PRESSURE BUILDUP AND EXPLOSION SHOULD THE POLYMERIZATION RATE INCREASE PRECIPITOUSLY. IT IS ESSENTIAL THAT THE SHAKING BATH BE SURROUNDED WITH A SAFETY SHIELD. BOTTLES MUST BE WRAPPED IN A TOWEL UPON REMOVAL AND OPENED *IMMEDIATELY*. SAFETY GLASSES MUST BE WORN IN THE LABORATORY AT ALL TIMES.

F. Apparatus

1. Shaker bath maintained at 70°C.
2. Soda bottles with crown caps.
3. Bottle capper and opener.
4. Source of low-pressure dry nitrogen.

G. Reagents and Materials

1. Styrene, reagent grade, redistilled as described in step H-1.

2. Potassium persulfate, reagent grade, recrystallized if necessary as described in step H-2.
3. Sodium hydrogen phosphate, reagent grade.
4. Sodium lauryl sulfate, 30% active ingredients (Emulsifier 104, Procter & Gamble).
5. Aluminum sulfate, reagent grade.
6. Methanol, reagent grade.
7. Distilled water.

H. Preparation

1. Prepare styrene in advance as follows:

a. Wash to remove inhibitor: The inhibitor, usually an aromatic such as hydroquinone or *t*-butylpyrocatechol, is removed by washing the monomer with 10% aqueous NaOH. Roughly equal parts of the basic solution and the monomer are placed in a separatory funnel and mixed by tumbling. The heavier aqueous phase is drained off. The procedure is repeated once or twice until the liquids remain clear. The monomer is then washed with distilled water until litmus paper shows that all the base has been removed.

b. Dry: A drying agent such as anhydrous Na_2SO_4 (Table 3-1) is added to the monomer (100 g/ℓ). With occasional tumbling, drying is complete in about 1/2 hr.

c. Distill: Add about 1 g/ℓ CuCl stabilizer to the monomer and distill under dry N_2 at 20 torr and 40-43°C. The distillation setup of Fig. 4-20 is suitable.

(*Note:* It is not essential that the monomer be dried and distilled after step a if it is to be used only for emulsion polymerizations; however, nitrogen should be bubbled through it to remove oxygen. Monomer treated in this way should be stored in a refrigerator for no more than 24 hr before use.)

2. Recrystallize potassium persulfate if necessary from a saturated solution in distilled water at about 0°C. Collect the crystals on a fritted-glass filter when about 60-70% of the water has evaporated.

I. Procedure

1. Into each of 10 clean soda bottles place 60 ml distilled water, 0.03 g $K_2S_2O_8$, 0.03 g Na_2HPO_4, and 20 ml styrene.
2. Place the following amounts of sodium lauryl sulfate in the bottles: 0.06, 0.08, 0.12, 0.15, 0.20, 0.25, 0.30, 0.40, 0.50, and 0.60 g.
3. Bubble nitrogen through the contents of each bottle to disperse the monomer and remove air. Then quickly cap the bottle, placing aluminum foil between the top of the bottle and the gasket in the cap. Label the bottle.
4. Place the bottles in the shaker bath (Fig. 4-17) at 70°C and shake at a frequency not less than 5 Hz.
5. Prepare a solution of 2.5 g $Al_2(SO_4)_3$ in 100 ml distilled water.
6. After 1-1/2 hr. polymerization, remove the bottles from the

bath, open them *at once* (see Sec. E, Safety Precautions) and add 10 ml of the alum solution (a coagulant for the latex) to each bottle.

7. Transfer the coagulated polymer (still swollen with styrene) to beakers, and wash twice with methanol, then twice with distilled water. Do not stir violently, since the coagulum may re-emulsify. Dry the polymer to constant weight in a vacuum oven at 80°C, and weigh to determine the yield.

J. *Fundamental Equations*

1. *Chemical*

$$S_2O_8^{=} \rightarrow 2\ SO_4^{\overline{\cdot}}$$

$$SO_4^{\overline{\cdot}} + CH_2 = CH(C_6H_5) \rightarrow H_2C(SO_4) - CH^{\cdot}(C_6H_5), \quad \text{etc.}$$

2. *Kinetic*

$$v_p \simeq [E]^{\alpha}$$

K. *Calculations*

1. Calculate the yield for each sample.
2. Calculate the exponent α in the kinetic equation, as the slope of the straight line obtained by plotting log [E] versus log (yield).

(*Note:* To a good approximation, polymerization rate is constant for the time period of this polymerization, hence the rate is proportional to the yield.)

L. *Report*

1. Describe the apparatus and experiment in your own words.
2. Tabulate yields and values of [E].
3. Include the plot of log [E] versus log (yield).
4. State your estimate of α, and discuss it in relation to the literature value of 0.5-0.6.

M. *Comments*

Retain the polymer sample(s) of high yield for use in later experiments.

N. *General References*

Chaps. 1C, 2B; *Textbook*, Chap. 12B; Kolthoff 1951; Bartholomé 1956; Van der Hoff 1956, 1962, 1969; Gerrens 1959; Duck 1966; Sorenson 1968, p. 220; Braun 1971, Chap. 3.

O. *Bibliography*

Bartholomé 1956. E. Bartholomé, H. Gerrens, R. Herbeck, and H. M. Weite, "The Kinetics of Emulsion Polymerization of Styrene" (in German), *Z. Elektrochem. 60*, 334-359 (1956).

Braun 1971. Dietrich Braun, Harald Cherdon, and Werner Kern, *Techniques of Polymer Syntheses and Characterization*, Wiley-Interscience, New York, 1971.

Duck 1966. Edward W. Duck, "Emulsion Polymerization," pp. 801-859 in Herman F. Mark, Norman G. Gaylord, and Norbert M. Bikales, eds., *Encyclopedia of Polymer Science and Technology*, Vol. 5, Interscience Div., John Wiley and Sons, New York, 1966.

Gerrens 1959. H. Gerrens, "Kinetics of Emulsion Polymerization" (in German), *Fortschr. Hochpolym. Forsch. (Advances in Polymer Sci.) 1*, 234-328 (1959).

Kolthoff 1951. I. M. Kolthoff, E. J. Meehan, and C. W. Carr, "Studies on Rate of Emulsion Polymerization of Butadiene-Styrene (75:25) as a Function of Amount and Kind of Emulsifier Used. III. Effect of Amount of Soap on Rate of Polymerization," *J. Polymer Sci. 6*, 73-81 (1951).

McCaffery 1970. Edward M. McCaffery, *Laboratory Preparation for Polymer Chemistry*, McGraw-Hill Book Co., New York, 1970.

Pinner 1961. S. H. Pinner, *A Practical Course in Polymer Chemistry*, Pergamon Press, New York, 1961.

Procter & Gamble. Procter & Gamble Co., Cincinnati, Ohio 45200.

Sorenson 1968. Wayne R. Sorenson and Tod W. Campbell, *Preparative Methods of Polymer Chemistry*, 2nd ed., Interscience Div., John Wiley and Sons, New York, 1968.

Van der Hoff 1956. B. M. E. Van der Hoff, "On the Mechanism of Emulsion Polymerization of Styrene," *J. Phys. Chem. 60*, 1250-1254 (1956).

Van der Hoff 1962. B. M. E. Van der Hoff, "Kinetics of Emulsion Polymerization," pp. 6-31 in Norbert A. J. Platzer, ed., "Polymerization and Polycondensation Processes," *Advances in Chemistry Series*, No. 34, American Chemical Society, Washington, D. C., 1962.

Van der Hoff 1969. John W. Van der Hoff, "Mechanism of Emulsion Polymerization," Chap. 1 in George E. Ham, ed., *Vinyl Polymerization*, Vol. 1, Part II, Marcel Dekker, New York, 1969.

EXPERIMENT 5. RADICAL COPOLYMERIZATION OF STYRENE AND METHYL METHACRYLATE

A. *Introduction*

The homopolymer of methyl methacrylate and four copolymers of styrene and methyl methacrylate are prepared by bulk radical polymerization with azobisisobutyronitrile (AIBN). The reactivity ratios for this copolymerization can be determined by analyzing the copolymers for monomer content by infrared-absorption spectroscopy in Exp. 32. If this is done, the calculations and report of this experiment cannot be completed until Exp. 32 has been carried out. The polymerization of Exp. 5 requires 3 hours.

B. *Principle*

Styrene and methyl methacrylate form a copolymer not too far from random, with each type of growing chain end preferring to add the other monomer by about 2:1. The reactivity ratios expressing these facts are determined from the composition of polymer formed at low conversion from known mixtures of the monomers, according to the methods outlined in Chap. 1F and the *Textbook*, Chap. 11A. Pinner (1961) describes a similar copolymerization in his Exp. B.8.3.

C. *Applicability*

Similar copolymerizations can be carried out with many pairs of vinyl, acrylic, and diene monomers. The analysis by infrared spectroscopy is specific, however, and it will not be possible to find a similar analysis technique for all pairs of monomers.

D. *Precision and Accuracy*

E. *Safety Considerations*

DANGEROUS CHEMICALS ARE USED IN THIS EXPERIMENT. AIBN IS HIGHLY TOXIC, AND THE MONOMERS AND SOLVENTS ARE TOXIC AND IRRITANT AS WELL AS HIGHLY FLAMMABLE. ADEQUATE VENTILATION IS ESSENTIAL. THE POLYMERIZATION IS HIGHLY EXOTHERMIC. TO AFFORD PROTECTION IN CASE IT GETS OUT OF CONTROL, THE POLYMERIZATION TUBES MUST BE SHIELDED WITH WIRE SCREENS (Fig. 4-23). SAFETY GLASSES MUST BE WORN IN THE LABORATORY AT ALL TIMES.

F. *Apparatus*

1. Constant-temperature bath regulated at 70°C.
2. 20-ml syringe.
3. Vacuum manifold and low-pressure nitrogen supply (Fig. 4-19).

G. Reagents and Materials

1. Styrene, reagent grade, redistilled (see step H-1).
2. Methyl methacrylate, reagent grade, redistilled (see step H-2).
3. 2,2'-Azobisisobutyronitrile, recrystallized (see step H-3).
4. Benzene, reagent grade.
5. Chloroform, reagent grade.
6. Petroleum ether, hexane, or methanol, reagent grade.

H. Preparation

1. Prepare styrene as described in Exp. 4, step H-1.
2. Prepare methyl methacrylate as described in Exp. 3, step H-1.
3. Recrystallize AIBN, if necessary, as described in Exp. 3, step H-2.

I. Procedure

1. Weigh 20 mg AIBN into each of four large test tubes, and weigh 10 mg AIBN into a fifth tube. Cap the tubes with serum stoppers, evacuate them, and flush with dry nitrogen (Fig. 4-7). See steps M-1 and M-2.

2. Inject with the syringe the following amounts of the monomers into the tubes: In the first four tubes, 4 ml methyl methacrylate (MMA) and 16 ml styrene (S); 8 ml MMA and 12 ml S; 12 ml MMA and 8 ml S; 14 ml MMA and 6 ml S; and in the last tube, 10 ml MMA.

3. Place the tubes, protected with wire screens, in the constant-temperature bath at 70°C, and heat for the following times (in the same order as that of composition in step 2): 10, 10, 20, 20, and 20 min.

4. At the proper time, remove each tube from the bath and cool it quickly under the water tap. Open the tube, and dilute the mixture with benzene or chloroform, if necessary, to obtain fluidity.

5. Precipitate each copolymer quantitatively by pouring the polymerizate into about 300 ml petroleum ether, hexane, or methanol. Collect the precipitate on a fritted-glass filter, and dry to constant weight in a vacuum oven at 80°C. Determine the yield and see that it does not exceed 10% for the copolymers.

J. Fundamental Equations

1. *Chemical*

$$CH_2 = \underset{\displaystyle COOCH_3}{C(CH_3)} \quad + \quad CH_2{=}CH\text{–}C_6H_5 \quad \xrightarrow{\text{AIBN}} \quad \text{Copolymer}$$

2. *Kinetic*

$$r_2 = r_1 H^2/h \quad + \quad H(1-h)/h \qquad (1\text{-}18)$$

where r_1 and r_2 are the reactivity ratios, h is the mole ratio of monomer 1 to monomer 2 in the copolymer, and H the same ratio in the monomer feed. The equation is correct only for low conversion.

K. *Calculations*

1. Obtain values of h from Exp. 32, and calculate the quantities appearing in Eq. 1-18.
2. Plot H^2/h versus $H(1-h)/h$ and determine r_1 and r_2 from the slope and intercept of the resulting straight line.

L. *Report*

1. Describe the apparatus and experiment in your own words.
2. Prepare a table listing H, h from Exp. 32, the calculated quantities in Eq. 1-18, and the yield, for each polymerization.
3. Include the graph of step K-2.
4. Give your values of r_1 and r_2, and discuss them in comparison with those given in Chap. 1F.

M. *Comments*

1. Caution is required in using syringes in vessels which are sealed or under positive or negative pressure. Positive pressure will force the plunger out, whereas withdrawing liquid from a sealed vessel will create a negative pressure, allowing air to leak in and contaminate the contents. It is advisable to keep the vessel on the nitrogen line, with inlet and outlet tubes, when adding or withdrawing any liquid.
2. Do not pierce rubber serum stoppers more than once at a given spot; leaks may develop, and pieces of rubber may be torn off to contaminate the reagent. Also, do not use the same needle for different reagents.

N. *General References*

Chap. 1F, *Textbook*, Chap. 11A; Mayo 1944; Ham 1964, 1966.

O. *Bibliography*

Ham 1964. George E. Ham, "Theory of Copolymerization," Chap. 1 in George E. Ham, ed., *Copolymerization*, Interscience Div., John Wiley and Sons, New York, 1964.

Ham 1966. George E. Ham, "Copolymerization," pp. 165-244 in Herman F. Mark, Norman G. Gaylord, and Norbert M. Bikales, eds., *Encyclopedia of Polymer Science and Technology*, Vol. 4, Interscience Div., John Wiley and Sons, New York, 1966.

Mayo 1944. Frank R. Mayo and Frederick M. Lewis, "Copolymerization. I. A Basis for Comparing the Behavior of Monomers in Copolymerization, The Copolymerization of Styrene and Methyl Methacrylate," *J. Am. Chem. Soc. 66*, 1594-1601 (1944).
Pinner 1961. S. H. Pinner, *A Practical Course in Polymer Chemistry*, Pergamon Press, New York, 1961.

EXPERIMENT 6. SOLID STATE POLYMERIZATION OF TRIOXANE

A. *Introduction*

Trioxane is polymerized in the solid state by a cationic ring-opening reaction, using boron trifluoride etherate as the catalyst. About one hour is required to set up the polymerization, and a few days later, another half hour to wash the polymer and prepare it for drying.

B. *Principle*

Ring-opening polymerization (Chap. 1E; *Textbook*, Chap. 10E) proceeds by a variety of mechanisms depending on the choice of monomer and catalyst. In this instance, the chain carrier is the cation formed by the addition of a catalyst fragment to the monomer. The particular features of interest in this experiment are its simplicity and the fact that the polymerization takes place in the solid state.

C. *Applicability*

This method can be used to polymerize a variety of aldehydes and other cyclic compounds.

D. *Precision and Accuracy*

E. *Safety Considerations*

BORON TRIFLUORIDE ETHERATE IS TOXIC AND IRRITANT. GOOD VENTILATION OR THE USE OF A FUME HOOD IS ESSENTIAL IN HANDLING THIS CATALYST. USE OF AN EXPLOSION-PROOF BLENDER OR PURGING THE BLENDER MOTOR WITH NITROGEN IS NECESSARY. SAFETY GLASSES MUST BE WORN IN THE LABORATORY AT ALL TIMES.

F. *Apparatus*

1. Blender, Waring or equivalent, explosion-proof or equipped for purging motor with nitrogen.
2. Microliter syringe (Fig. 4-8*c*).

3. Vacuum manifold with supply of low-pressure nitrogen (Fig. 4-19).

G. *Reagents and Materials*

1. Trioxane, reagent grade.
2. Boron trifluoride etherate, reagent grade.
3. Sodium hydroxide, reagent grade.
4. Methanol, reagent grade.
5. Distilled water.

H. *Preparation*

I. *Procedure*

1. Weigh accurately about 10 g trioxane in a test tube and cap it with a serum stopper.
2. Purge the tube with dry nitrogen for about 15 min (Fig. 4-7), or evacuate the tube and fill it with dry nitrogen twice.
3. Inject about 0.1 ml boron trifluoride etherate into the tube with the microliter syringe. (*Note*: 20 ml boron trifluoride or silicon tetrafluoride gas may be used as catalyst instead of the etherate.) Leave the tube at room temperature for several days.
4. Open the tube and transfer the polymer to the blender. Wash it, with agitation, with 5% aqueous NaOH, then with distilled water, and finally with methanol.
5. Collect the polymer on a fritted-glass filter, and dry it to constant weight in a vacuum oven at about 80°C.

J. *Fundamental Equations*

$$(CH_2O)_3 \xrightarrow{BF_3\cdot(C_2H_5)_2O} -CH_2OCH_2OCH_2O- \text{ , etc.}$$

K. *Calculations*

1. Calculate the yield of polymer.

L. *Report*

1. Describe the experiment in your own words.
2. Report the yield of polymer.
3. Answer the following questions:
   a. Why is it not necessary to dry the monomer?
   b. Why is the polymer washed with aqueous NaOH?

M. Comments

1. Retain the polymer for use in later experiments.

N. General References

Chap. 1E; *Textbook*, Chap. 10E; Sorenson 1968, p. 405; Furukawa 1969; Braun 1971.

O. Bibliography

Braun 1971. Dietrich Braun, Harald Cherdon, and Werner Kern, *Techniques of Polymer Syntheses and Characterization*, Wiley-Interscience, New York, 1971.

Furukawa 1969. Junji Furukawa and Koichi Tada, "Mechanism and Kinetics of Trioxane Polymerization," pp. 173-188 in Kurt C. Frisch and Sidney L. Reegen, eds., *Ring-Opening Polymerization*, Marcel Dekker, New York, 1969.

Sorenson 1968. Wayne R. Sorenson and Tod W. Campbell, *Preparative Methods of Polymer Chemistry*, 2nd ed., Interscience Div., John Wiley and Sons, New York, 1968.

EXPERIMENT 7. SOLUTION AND BULK PREPARATION OF A POLYETHER-BASED POLYURETHANE

A. Introduction

A polyurethane elastomer is prepared by either bulk or solution polymerization. Either preparation can be completed in 3 hours.

B. Principle

A polyether glycol (Polymeg 1000) is reacted with a diisocyanate to produce a prepolymer terminated with isocyanate (NCO) groups. This is chain-extended or crosslinked by subsequent reaction with a polyol or a polyamine. The structure of the polymer (random or block) can be varied by altering the nature and amount of chain extender and the reaction sequence; see step M-1.

C. Applicability

The methods of this experiment can be varied in many ways, leading to polymers with a variety of physical properties. Some suggested alterations are: Use of a higher molecular-weight polyol such as Polymeg 2000; use of a hydroxy-terminated polyester (Multrathane R-14, Mobay), a polyfunctional polyester polyol (Multron R-12A, Mobay), or other polyols (Rucoflex S series, Ruco), or such extenders as 1,4-ethoxy(βhydroxybenzene); or use of other diisocyanates such as hexamethylene or tolylene diisocyanate.

D. *Precision and Accuracy*

E. *Safety Considerations*

ISOCYANATES ARE STRONG SKIN IRRITANTS AND MUST BE HANDLED WITH PROPER PRECAUTIONS. SAFETY GLASSES MUST BE WORN IN THE LABORATORY AT ALL TIMES. GLOVES MUST BE WORN AND A WELL-VENTILATED FUME HOOD USED WHEN MOLTEN ISOCYANATES ARE HANDLED. ANY SKIN CONTACTING ISOCYANATES MUST BE WASHED IMMEDIATELY WITH SOAP AND WATER.

F. *Apparatus*

Procedure I. Bulk Polymerization

1. Hot plate or sand bath.
2. Disposable clean 1-pint paint cans (alternatively, 250-ml beakers).
3. Disposable aluminum pan (*ca.* 8 × 10 in. by 1 in. deep).
4. Variable-speed air stirrer or other high-torque stirrer.
5. Metal (Weston) thermometer, range 0-250°C.

Procedure II. Solution Polymerization

1. 1-Liter resin kettle.
2. Reflux condenser.
3. Air stirrer.
4. Heating mantle.

(*Note*: The following items are desirable but not essential.)

5. Brookfield viscometer.
6. Thermowatch control.

G. *Reagents and Materials*

Procedure I. Bulk Polymerization

1. Poly(tetramethylene glycol) (Polymeg 1000, Quaker).
2. 1,4-Butanediol (GAF).
3. *p,p'*-Diphenylmethane diisocyanate (MDI) (Multrathane M, Mobay).
4. No. 34 emulsion release agent (Dow Corning).

Procedure II. Solution Polymerization

1. Items 1-3 from Procedure I.
2. Dimethyl formamide, reagent grade.
3. *n*-Propanol, reagent grade.

H. Preparation

1. Dry the Polymeg in vacuum (<1 torr) for 15 min at 110-120°C. (*Note*: See step M-2.)
2. If the Polymeg is not used immediately after drying, heat it to 70°C to melt it, then keep it at 40°C until used.
3. For Procedure I, coat the disposable aluminum pan with a mixture of equal parts of water and No. 34 emulsion release agent, and dry in an air oven at 110°C.

I. Procedure

Procedure I. Bulk Polymerization

1. Weigh 102.2 g Polymeg and 9.0 g 1,4-butanediol into a pint paint can.
2. Weigh 50.0 g MDI (CAUTION: Observe safety precautions) into a second paint can.
3. Heat both cans to 100°C.
4. Set up the variable-speed air stirrer (with metal blade and shaft) in the MDI can.
5. With continuous stirring, add the Polymeg-butanediol mixture to the MDI. (*Note*: Be prepared to increase the power to the stirrer as the viscosity of the mixture rises sharply.)
6. Measure and record the reaction temperature every 15 sec.
7. (*Note*: Be prepared to perform this step rapidly, as the mixture soon becomes too viscous to pour.) Stop the agitation after 4 min or when two consecutive temperature readings are the same. Remove the stirrer and pour the contents of the can into the precoated aluminum pan. Spread the polymer to a thin layer and quickly cool the pan in ice water. (CAUTION: Do not splash water on the hot polymer!)
8. Dry the polymer in an air oven at 80°C for 24 hr.
9. Repeat the polymerization using (a) 52.5 g and (b) 55.0 g MDI.

Procedure II. Solution Polymerization

1. Weigh carefully (±0.01 g) 102.20 g Polymeg, 9.00 g 1,4-butanediol, and 539 g dimethyl formamide into the resin kettle.
2. Set up the resin kettle with stirrer, reflux condenser, thermometer, and heating mantle. Heat the contents to 55-60°C, with stirring.
3. Weigh out 50.00 g MDI, and add to the kettle. The temperature should rise, indicating reaction.
4. Heat the kettle to 100°C and maintain that temperature. If available, a Thermowatch control can be used.
5. If a Brookfield viscometer is available (see Exp. 26), follow the course of polymerization by withdrawing, at 15-min intervals, 2-3 oz samples into 4-oz jars, cooling to 25°C, and determining their viscosity in centipoise. A viscosity of about 25,000 cp is desirable, and corresponds to $\overline{M}_n$ = 35,000-40,000.
6. When the above viscosity is reached, or after 1.5 hr, add 5-10 ml *n*-propanol to the kettle.
7. Pour the reaction mixture slowly into water, with high-speed

agitation. Collect the polymer on a fritted-glass filter and dry in a vacuum oven at 80°C for 24 hr. (*Note*: An alternative isolation procedure is described in step M-3.)

J. *Fundamental Equations*

Polyol plus MDI

$$HO[(CH_2)_4O]_xH \;+\; 2\; OCN\text{-}C_6H_{10}\text{-}CH_2\text{-}C_6H_{10}\text{-}NCO \;\rightarrow$$

$$OCN\text{-}C_6H_{10}\text{-}CH_2\text{-}C_6H_{10}\text{-}NHCOO[(CH_2)_4O]_xCONH\text{-}C_6H_{10}\text{-}CH_2\text{-}C_6H_{10}\text{-}NCO$$

Chain Extension

$$HO(CH_2)_4OH \;+\; \text{(product above)} \;\rightarrow\; \text{(random or block copolymer)}$$

K. *Calculations*

Calculate the percent urethane groups in the polymer as follows:

$$\%\ \text{MDI} = 100(\text{wt MDI})/(\text{wt MDI} + \text{Polymeg} + \text{butanediol})$$

$$\%\ \text{urethane} = (\%\ \text{MDI})(\text{mole wt 2NHCOO})/(\text{mole wt MDI})$$

$$= (\%\ \text{MDI}) \times 118/250$$

L. *Report*

1. Describe the apparatus and experiment in your own words.
2. Calculate the yield of polymer and the percent urethane linkages.
3. a. (Procedure I only) Include a graph of reaction temperature versus time.
   b. (Procedure II only) Include a graph of Brookfield viscosity versus time if this was obtained.
4. Comment on the reactions possible among the ingredients in the recipe, emphasizing differences to be expected between the solution and bulk polymerizations.
5. Calculate the recipe required for the preparation of 0.1 mole of a polyether (M = 1000) based polyurethane having 10% urethane groups.
6. Write equations for chemical reactions leading to branch formation and crosslinking in the urethane polymerization of this experiment.
7. Answer the following questions:
   a. What is the purpose of holding the polymer for 24 hr at 80°C?

b. What would be the effect on the polymerization and the polymer formed of unrecognized moisture in the polyether?

M. Comments

1. In the bulk polymerization, the mode of mixing the ingredients can change the structure of the polymer drastically. Instead of mixing the polyol and the chain extender, then adding this to the diisocyanate, possible variations are: (a) Premix the polyol and diisocyanate and add to the extender; and (b) premix the extender and diisocyanate and add to the polyol. While the overall urethane content will remain the same, there will be significant differences in one or more properties of the polymer. A challenging experiment would be to prepare and characterize these materials.

Mixing the ingredients all together at once leads to a random structure, whereas the abovementioned methods yield materials more like block copolymers. The nature of the block sequence ("hard," i.e. aromatic, or "soft," i.e. aliphatic) has a profound effect on the physical and mechanical properties of the polymer.

2. Since the isocyanate group reacts with water, the moisture content of the reactants must be considered in calculating the amount of MDI required; the calculated amount in step 2 of Procedure I and step 3 of Procedure II assumes exact stoichiometry and perfectly dry reagents.

The moisture content can best be determined by use of the Karl Fischer reagent (Chap. 3D). An example of the correction to be made based on the result of this measurement is the following: If the Polymeg were found to contain 0.2% water, there would be $(0.002 \times 102.2)/18 = 0.0114$ moles water in the system; this would use up $0.114 \times 250 = 2.85$ g MDI; this is the additional amount required.

3. An alternative means of isolating the polymer from Procedure II as a fine-particle crumb is as follows: Remove the DMF by adding the reaction mixture to a 70:30 mixture of pentane and ether. A sticky coacervate (two-phase liquid-liquid system) will form if the polymer molecular weight is not too high. Adjust the amount of the pentane-ether mixture so that a coacervate free from discrete aggregates is formed. Decant the fluid layer, which contains the pentane-ether mixture and most of the DMF. Dilute the polymer-rich phase with 1-2 volumes of isopropyl acetate to redisperse the coacervate and reduce its viscosity. Now add this dispersion to 5 volumes of 85:15 pentane: ether mixture. Polymer should precipitate as a fine powder. If it still contains DMF as evidenced by stickiness or odor, further extraction with pentane may be required.

4. The hydroxyl number or end-group analysis of the polyol can be carried out by Exp. 10.

5. The polymer should be retained for use in several later experiments.

N. General References

Chap. 1B; *Textbook*, Chaps. 8B, 11D, 15D; Saunders 1962, 1964; Dombrow

1965; Lenz 1967, Chap. 7; Blokland 1968; Buist 1968; Blackfan 1969; Bruins 1969; Carvey 1969; David 1969; Pigott 1969; Wright 1969; Braun 1971.

O. Bibliography

Blackfan 1969. C. L. Blackfan, "Thermoplastic Polyurethane Film and Sheet," pp. 364-368 in Sidney Gross, ed., *Modern Plastics Encyclopedia 1969-1970* (McGraw-Hill Book Co., New York) *46* (10A), October, 1969.

Blokland 1968. R. Blokland, *Elasticity and Structure of Polyurethane Networks*, Gordon and Breach Science Publishers, New York, 1968.

Braun 1971. Dietrich Braun, Harald Cherdon, and Werner Kern, *Techniques of Polymer Syntheses and Characterization*, Wiley-Interscience, New York, 1971.

Bruins 1969. Paul F. Bruins, ed., *Polyurethane Technology*, Interscience Div., John Wiley and Sons, New York, 1969.

Buist 1968. J. M. Buist and H. Gudgeon, eds., *Advances in Polyurethane Technology*, MacLaren and Sons, London, 1968.

Carvey 1969. R. M. Carvey, "Urethane Elastomers," p. 215 in Sidney Gross, ed., *Modern Plastics Encyclopedia 1969-1970* (McGraw-Hill Book Co., New York) *46* (*10A*), October 1969.

David 1969. D. J. David and H. B. Staley, *Analytical Chemistry of the Polyurethanes*, Interscience Div., John Wiley and Sons, New York, 1969.

Dombrow 1965. Bernard A. Dombrow, *Polyurethanes*, 2nd ed., Reinhold Publishing Corp., New York, 1965.

Dow Corning. Dow Corning Corp., S. Saginaw Road, Midland, Michigan 48640.

GAF. GAF Corporation, 140 West 51st Street, New York, New York 10020.

Lenz 1967. Robert W. Lenz, *Organic Chemistry of Synthetic High Polymers*, Interscience Div., John Wiley and Sons, New York, 1967.

Mobay. Mobay Chemical Co., Pittsburgh, Pennsylvania 15205.

Pigott 1969. K. A. Pigott, "Polyurethanes," pp. 506-563 in Herman F. Mark, Norman G. Gaylord, and Norbert M. Bikales, eds., *Encyclopedia of Polymer Science and Technology*, Vol. 11, Interscience Div., John Wiley and Sons, New York, 1969.

Quaker. Quaker Oats Co., Chemicals Division, Merchandise Mart Plaza, Chicago, Illinois 60654.

Ruco. Ruco Division, Hooker Chemical Co., Hicksville, New York 11802.

Saunders 1962. J. H. Saunders and K. C. Frisch, *Polyurethanes: Chemistry and Technology. Part I: Chemistry*, Interscience Div., John Wiley and Sons, New York, 1962.

Saunders 1964. J. H. Saunders and K. C. Frisch, *Polyurethanes: Chemistry and Technology. Part 2: Technology*, Interscience Div., John Wiley and Sons, New York, 1964.

Wright 1969. P. Wright and A. P. C. Cunning, *Solid Polyurethane Elastomers*, Gordon and Breach Science Publishers, New York, 1969.

EXPERIMENT 8. PREPARATION OF CROSSLINKED POLYMERS
I. POLY(STYRENE-CO-DIVINYL BENZENE)
II. POLY(ETHYL ACRYLATE-CO-ETHYLENE GLYCOL DIMETHACRYLATE)

A. *Introduction*

Two procedures are given for the preparation of polymers with controlled degree of crosslinking. Emulsion polymerization is used. The polymers are studied informally for solubility properties in this experiment, and are retained for use in later experiments. Either procedure requires approximately 3 hours.

B. *Principle*

In radical polymerization, crosslinked polymers are formed when small amounts of difunctional or polyfunctional monomers are added to the vinyl monomer used.

C. *Applicability*

The procedures described in this experiment can be applied to most vinyl, acrylic, or diene monomers with solubility in water in a range making them suitable for emulsion polymerization.

D. *Precision and Accuracy*

E. *Safety Considerations*

INHALATION OF MONOMER OR SOLVENT VAPORS, OR CONTACT OF THESE CHEMICALS WITH THE SKIN, MUST BE AVOIDED. EXTREME CARE MUST BE USED IN HANDLING INITIATORS. THE POLYMERIZATION SHOULD BE CARRIED OUT IN A FUME HOOD, WITH THE APPARATUS BEHIND A SAFETY SHIELD. SAFETY GLASSES MUST BE WORN IN THE LABORATORY AT ALL TIMES.

F. *Apparatus*

1. 1-liter 3-neck round-bottom flask.
2. Stirrer and motor, variable speed.
3. Condenser.
4. Thermometer, 0-100°C.
5. Graduated addition funnel, 250-ml.
6. Water bath with provision for heating and cooling.
7. "Waring" blender or other high-speed stirrer.
8. Source of dry nitrogen at low pressure.

G. *Reagents and Materials*

Procedure I

1. Styrene, redistilled (see step H-1).
2. Divinyl benzene, redistilled (see step H-1).
3. Sodium lauryl sulfate, 30% active ingredients. (Emulsifier 104, Procter & Gamble Co., Cincinnati, Ohio.)
4. Sequestrene Na Fe.
5. Potassium persulfate, reagent grade.
6. Sodium metabisulfite, reagent grade.
7. Aluminum sulfate, reagent grade.

Procedure II

1. Ethyl acrylate, inhibitor free (see step H-2).
2. Ethylene glycol dimethacrylate, inhibitor free (see step H-2).
3. Sodium lauryl sulfate, 30% active ingredients.
4. Ammonium persulfate, reagent grade.
5. Aluminum sulfate, reagent grade.

H. *Preparation*

1. Distill styrene and divinyl benzene as specified in Exp. 4, step H-1.
2. Low-inhibitor or inhibitor-free grades of acrylic monomers are suitable for use without further purification.

I. *Procedure*

Procedure I

1. Weigh out 40 g distilled water, 23.4 g sodium lauryl sulfate, 200 g styrene, and 1.0 g divinyl benzene. Mix together in a blender or with a high-speed stirrer until a stable emulsion is formed. Place this mixture in the addition funnel.
2. Purge the reaction flask with nitrogen, and add 150 ml distilled water.
3. Weigh out 0.020 g Sequestrene Na Fe and add to the reaction flask.
4. Add 10% of the monomer premix (26.5 g) to the reaction flask. Heat flask and contents to 60°C, still purging with nitrogen.
5. Weigh out 1.0 g potassium persulfate and dissolve in 50 ml distilled water.
6. Weigh out 0.4 g sodium metabisulfite, dissolve it quickly in the solution of potassium persulfate, and add this mixture to the reaction flask at once.
7. When the temperature has stabilized and "seed" has formed, as indicated by a bluish color in the reaction flask, begin metering in the remainder of the monomer mixture. Do this at a rate (not exceeding 5 ml/min) at which temperature can be controlled. Total addition of premix should take place over about 1 hr.

8. After all materials have been added, increase the temperature to 80°C for 20-30 min. Then cool the reaction mixture to room temperature.

9. Withdraw 5-10 g of the mixture, and determine total solids by evaporating a weighed amount to dryness in a tared aluminum weighing dish. (CAUTION: Evaporate off water and any residual monomers in a fume hood.)

10. Retain 50 ml of the reaction mixture for use in Exp. 14.

11. Weigh out 5 g aluminum sulfate and dissolve in 1500 ml distilled water. Heat to 95°C.

12. Pour the reaction mixture slowly into the hot aluminum sulfate solution, with good agitation. (CAUTION: Perform this operation in a fume hood.)

13. Filter the mixture through a Buchner funnel, and wash several times with hot water, until no more emulsifier (soap) is removed. Check a small sample of the first filtrate for complete coagulation.

14. Dry the polymer in a vacuum oven for 24 hr at 80°C.

Procedure II

1. Purge the reaction flask with nitrogen, and add 235 ml distilled water, 17.0 g sodium lauryl sulfate, and 0.25 g ammonium persulfate. Heat the reactor to 65°C.

2. Add to the addition funnel 100 g ethyl acrylate and 3.0 g ethylene glycol dimethacrylate.

3. Add the monomer mixture to the reaction flask slowly and steadily, at about 2 ml/min. If the temperature rises appreciably above 65°C, decrease the rate of addition. The total addition of monomers should take place over about 1 hr.

4. Complete the experiment by carrying out steps 8-14 of Procedure I.

J. *Fundamental Equations*

Procedure I

$$C_6H_5CH{=}CH_2 + CH_2{=}CH{-}C_6H_4{-}CH{=}CH_2 \rightarrow$$

$$\text{---}\ CH_2\ CH(C_6H_5)\ CH_2\ CH\ CH_2\ CH(C_6H_5)\ \text{---}$$

$$|$$

$$C_6H_4$$

$$|$$

$$\text{---}\ CH_2\ CH(C_6H_5)\ CH_2\ CH\ CH_2\ CH(C_6H_5)\ \text{---}$$

Procedure II

$$CH_2{=}CHCOOC_2H_5 \;+\; CH_2{=}C(CH_3)COOCH_2CH_2OCOC(CH_3){=}CH_2 \;\rightarrow$$

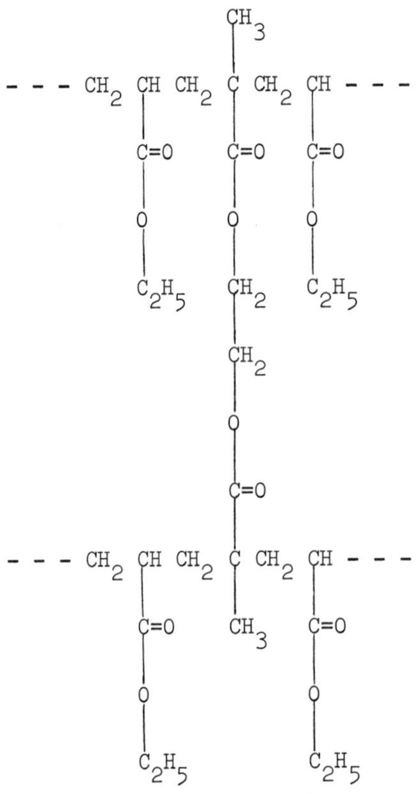

```
                        CH3
                        |
--- CH2  CH  CH2  C  CH2  CH ---
         |        |       |
         C=O      C=O     C=O
         |        |       |
         O        O       O
         |        |       |
         C2H5     CH2     C2H5
                  |
                  CH2
                  |
                  O
                  |
                  C=O
                  |
--- CH2  CH  CH2  C  CH2  CH ---
         |        |       |
         C=O      CH3     C=O
         |                |
         O                O
         |                |
         C2H5             C2H5
```

K. *Calculations*

L. *Report*

1. Describe the experiment and apparatus in your own words.
2. Determine the yield of polymer and percent conversion.
3. From Table 4-1 or Table 6-2, select a suitable solvent for the polymer prepared and test its solubility. Does it dissolve? Swell? Explain.
4. Answer the following questions:
   a. What is the mechanism of the polymerization? Describe it briefly.
   b. Why was the temperature raised after the reactants were added?
   c. Why was the coagulation carried out at elevated temperature?

M. *Comments*

1. Polymers prepared in this experiment can be studied further in later experiments, in particular Exp. 31.
2. Pinner (1961) describes the suspension preparation of poly-(styrene-*co*-divinyl benzene) in his Exp. B.2.2.
3. A variation to this experiment can be carried out using Procedure II where a mixture of the monomers ethyl acrylate and methyl methacrylate is used in the ratio 70/30, together with the 3.0 parts ethylene glycol dimethacrylate. The resulting polymer can be evaluated in the same manner as already indicated. However, in addition it can be examined by infrared spectroscopy.

N. *General References*

Chaps. 1E, 3A, 4B; *Textbook*, pp. 7, 366; Schildknecht 1952, pp. 68-72; Flory 1953, pp. 54-55, 357-361, 391-392; Duck 1966; Smith 1968; Odian 1970.

O. *Bibliography*

Duck 1966. Edward W. Duck, "Emulsion Polymerization," pp. 801-859 in Herman F. Mark, Norman G. Gaylord, and Norbert M. Bikales, eds., *Encyclopedia of Polymer Science and Technology*, Vol. 5, Interscience Div., John Wiley and Sons, New York, 1966.

Flory 1953. Paul J. Flory, *Principles of Polymer Chemistry*, Cornell University Press, Ithaca, New York, 1953.

Odian 1970. George Odian, *Principles of Polymerization*, McGraw-Hill Book Co., New York, 1970.

Pinner 1961. S. H. Pinner, *A Practical Course in Polymer Chemistry*, Pergamon Press, New York, 1961.

Schildknecht 1952. Calvin E. Schildknecht, *Vinyl and Related Polymers*, John Wiley and Sons, New York, 1952.

Smith 1968. Derek A. Smith, "Polymer Structures and Polymerization Techniques," Chap. 1 in Derek A. Smith, ed., *Addition Polymers: Formation and Characterization*, Plenum Press, New York, 1961.

EXPERIMENT 9. ANIONIC POLYMERIZATION OF STYRENE

A. *Introduction*

In this experiment, styrene is polymerized by an anionic mechanism, using sodium naphthalide catalyst and tetrahydrofuran as the solvent. This polymerization requires 3 hours.

B. *Principle*

Sodium naphthalide reacts rapidly with styrene to give radical ions which quickly dimerize, destroying the radical activity and leaving a dianion capable of growth at both ends. This initiation process is quite rapid compared to propagation; hence the growth of all the polymer chains is started almost at one time. Polymer with an unusually narrow molecular-weight distribution results; $\overline{M}_w/\overline{M}_n$ is 1.1 or less in the usual case.

Both the catalyst and the growing chains are easily destroyed by oxygen, water, carbon dioxide, alcohols, and many other impurities. If these substances are rigorously eliminated, the chains are capable of further growth even if the supply of monomer is depleted; they constitute a "living" polymer system, defined as one in which there is no termination (Chap. 1D). Polymerization can continue when more monomer of the same or another type is added; in the latter case, a block copolymer is formed, and the experiment can easily be modified to do this.

C. *Applicability*

This polymerization can be carried out with a variety of monomers, including isobutylene and butadiene as well as styrene. Other strongly solvating media, for example ethylene glycol or dimethyl ether, can be used instead of tetrahydrofuran as the solvent.

D. *Precision and Accuracy*

E. *Safety Considerations*

STYRENE, METHANOL, AND TETRAHYDROFURAN ARE ALL TOXIC AND FLAMMABLE. CONTACT WITH THE SKIN SHOULD BE AVOIDED, AND SPECIAL CARE SHOULD BE TAKEN IN THE DISTILLATION OF THESE LIQUIDS. SODIUM AND POTASSIUM REACT VIOLENTLY WITH WATER TO PRODUCE STRONG BASES, AND MUST BE HANDLED WITH CARE. SAFETY GLASSES MUST BE WORN IN THE LABORATORY AT ALL TIMES.

F. *Apparatus*

1. 100-ml flask (one-neck or three-neck) fitted with stirring magnet and rubber serum stoppers.
2. 2-ml and 20-ml syringes.
3. Syringe needles, long and short.
4. Vacuum bench.
5. Nitrogen, dry, at low pressure.

G. *Reagents and Materials*

1. Tetrahydrofuran, dry, distilled from sodium-potassium alloy (see step H-1).
2. Styrene, redistilled (see step H-2).
3. Sodium, reagent grade.
4. Naphthalene, resublimed.

5. Methanol, reagent grade.

H. Preparation (see step M-1)

1. Dry tetrahydrofuran with $CaCl_2$ for several hours, reflux with sodium-potassium alloy, and distill from the alloy at atmospheric pressure under dry nitrogen.

2. Distill styrene as specified in Exp. 4, step H-1.

3. Prepare solution of sodium naphthalide in tetrahydrofuran as follows: Place 50 ml purified tetrahydrofuran, 1.5 g resublimed dry naphthalene, and 1.5 g metallic sodium (cut into small pieces) in a flask equipped with a stirring magnet. Purge the flask with dry nitrogen, and maintain nitrogen pressure slightly above atmospheric. Stir for two hours. Dark green sodium naphthalide forms.

I. Procedure (see step M-1)

1. Flame a clean, dry 100-ml 3-neck flask containing a stirring magnet, and close the necks with rubber serum stoppers while the flask is hot. Purge it with dry nitrogen for at least 15 min, using syringe needles for inlet and outlet, and protecting the outlet with a drying tube.

2. Maintaining the nitrogen flow, introduce 50 ml tetrahydrofuran and then 5 ml styrene, using syringes (see steps M-2 and M-3).

3. Place the flask in a 1000-ml beaker so that the flask touches the bottom of the beaker. Fill the beaker to above the liquid level in the flask with an acetone-dry ice mixture (CAUTION: do not allow the serum stoppers to be wet by acetone). Place flask and beaker on the magnetic stirrer.

4. After stirring for about 10 min to allow the monomer-solvent mixture to reach a temperature below -70°C, inject 1.0 ml sodium naphthalide solution (see step M-4). The mixture should rapidly change color from dark green to red-orange as the styrene anion forms (see step M-5). Polymer should begin to form at once.

5. After a few minutes, the polymerization should be complete. Terminate the chains by injecting 2 ml methanol. Disconnect the nitrogen line, and allow the flask to warm to room temperature.

6. Wash the polymer with methanol, and dry it in a vacuum oven for 24 hr at about 80°C. Retain the polymer for use in later experiments.

J. Fundamental Equations

1. Chemical

$$C_{10}H_8^{-}Na^{+} + CH_2{=}CH(C_6H_5) \rightarrow CH_2{=}CH(C_6H_5)\cdot^{-}Na^{+} + C_{10}H_8$$

$$2\ CH_2{=}CH(C_6H_5)\cdot^{-}Na^{+} \rightarrow Na^{+-}CH(C_6H_5)CH_2CH_2CH(C_6H_5)^{-}Na^{+}$$

2. *Kinetic*

$$\bar{x}_n = 2\,[M] / [C]$$

where [M] and [C] are the initial concentrations of monomer and catalyst, respectively.

K. *Calculations*

1. Calculate the degree of polymerization, provided the purity of the reagents was sufficiently high (see step M-5).

L. *Report*

1. Describe the apparatus and experiment in your own words.
2. Report yield and calculated degree of polymerization.
3. If the molecular weight of the polymer is measured in Exp. 11 or 15, the result may differ from that calculated from kinetic considerations. Explain, and predict the direction of the difference.
4. Why is the polymerization carried out at low temperature?

M. *Comments*

1. This experiment is unusually sensitive to traces of air and moisture. All vessels must be thoroughly purged with, and maintained under positive pressure of, dry nitrogen.
2. Caution is required in using syringes in vessels which are sealed or under positive or negative pressure. Positive pressure will force the plunger out, whereas withdrawing liquid from a sealed vessel will create a negative pressure, allowing air to leak in and contaminate the contents. It is advisable to keep the vessel on the nitrogen line, with inlet and outlet tubes, when adding or withdrawing any liquid.
3. Do not pierce rubber serum stoppers more than once at a given spot; leaks may develop, and pieces of rubber may be torn off to contaminate the reagent. Also, do not use the same needle for different reagents.
4. When adding catalyst solution to the cold monomer-solvent mixture, do not immerse the needle in the liquid, as the catalyst solution will freeze in the needle.
5. If the bright red-orange color quickly disappears, the mixture is "poisoned" with air, water, or another substance. It may still be possible to carry out the polymerization by adding more catalyst, but the molecular weight will differ from that calculated from stoichiometry, and the molecular-weight distribution will probably be broadened.
6. While best results are obtained with freshly-prepared reagents, the catalyst solution, monomer and solvent can be kept, refrigerated and properly sealed, for a few days. New serum stoppers should be used, and a stopcock between flask and serum stopper provides additional protection.

7. For class use, it is advisable to provide separate portions of the reagents, especially monomer and solvent, for each experiment in separate vessels. Then, if accidental contamination occurs, the entire supply of reagent is not lost.

8. To standardize the catalyst, quench an aliquot with methanol and titrate with standardized HCl. As prepared above, the catalyst contains about 1.6 meq/ml.

N. General References

Chaps. 1D, 4A; *Textbook*, Chap. 10C; Szwarc 1956; Waack 1957; Margerison 1967, p. 252; Sorenson 1968, p. 281; Braun 1971, pp. 145-151.

O. Bibliography

Braun 1971. Dietrich Braun, Harald Cherdon, and Werner Kern, *Techniques of Polymer Syntheses and Characterization*, Wiley-Interscience, New York, 1971.

Margerison 1967. D. Margerison and G. C. East, *An Introduction to Polymer Chemistry*, Pergamon Press, New York, 1967.

Sorenson 1968. Wayne R. Sorenson and Tod W. Campbell, *Preparative Methods of Polymer Chemistry*, 2nd ed., Interscience Div., John Wiley and Sons, New York, 1968.

Szwarc 1957. M. Szwarc, M. Levy, and R. Milkovich, "Polymerization Initiated by Electron Transfer to Monomer. A New Method of Formation of Block Polymers," *J. Am. Chem. Soc. 78*, 2656-2657 (1956).

Waack 1957. R. Waack, A. Rembaum, J. D. Coombes, and M. Szwarc, "Molecular Weights of 'Living' Polymers," *J. Am. Chem. Soc. 79*, 2026-2027 (1957).

B

Experiments Measuring Molecular Weight and Distribution

EXPERIMENT 10. END-GROUP ANALYSIS

A. Introduction

This experiment provides two procedures for the determination of number-average molecular weight by end-group analysis. In one, the titration of amine groups with standard acid is used to determine the extent of reaction and number-average molecular weight of polyamides prepared from ω-amino acids. In the second procedure, hydroxyl end groups are acetylated by treatment with acetic anhydride in pyridine, and determined by difference through the amount of anhydride used. Either procedure requires about 3 hours.

B. Principle

The determination of $\overline{M}_n$ by end-group analysis is based on the principle that, by definition, the number-average molecular weight is inversely related to the number of molecules per unit weight of sample. If the number of chemically determinable end groups per molecule is known, the application of the method is straightforward.

C. Applicability

End-group analysis in general requires that the molecules in question be linear, with determinable end groups, and that it be known whether such groups are on one or both ends of each molecule. Procedure I, for the determination of amine groups in polyamides, is applicable only when the fraction of amine end groups is known; in the case of polyamides made from ω-amino acids, as in Exp. 1, this is known exactly provided that side reactions were not present during polymerization. Procedure II for hydroxyl end groups is applicable, among other cases, to the polyethers used in polyurethane preparation in Exp. 7.

D. *Precision and Accuracy*

The method is precise to about ±2% of the end-group concentration. The corresponding uncertainty in $\overline{M}_n$ will vary from case to case because of the reciprocal relation between the two.

E. *Safety Considerations*

DANGEROUS CHEMICALS ARE USED IN THIS EXPERIMENT. METHANOL IS BOTH TOXIC AND FLAMMABLE, WHEREAS PYRIDINE IS ESPECIALLY NOXIOUS. IN PARTICULAR, PHENOL IS HIGHLY TOXIC, BEING ABSORBED RAPIDLY THROUGH UNDAMAGED SKIN WITH THE INCIDENCE OF SEVERE BURNS. SPECIAL PRECAUTIONS ARE REQUIRED TO AVOID CONTACT WITH PHENOL, AND ANY SPILLED ON THE SKIN MUST BE WASHED OFF IMMEDIATELY. SINCE PHENOL HAS ANAESTHETIC PROPERTIES, ITS PRESENCE MAY NOT BE NOTICED UNLESS THE USER IS AWARE OF THE HAZARD. PHENOL IS PREFERABLY USED AS A SOLID TO LIMIT ITS VAPOR PRESSURE AND REDUCE THE POSSIBILITY OF SPILLAGE. ADEQUATE VENTILATION IS NECESSARY. SAFETY GLASSES MUST BE WORN IN THE LABORATORY AT ALL TIMES.

F. *Apparatus*

Procedure I. Amine End Groups

1. 100-ml 3-neck flasks with heating mantles and controllers.
2. Condensers.
3. 5-ml microburets with 0.01-ml graduations.
4. Stirring motors or magnetic stirrers.
5. (Optional) Conductance bridge and conductometric cell with platinum-black electrodes.

Procedure II. Hydroxyl End Groups

1. 250-ml iodine flasks.
2. 50-ml burette.
3. 10-ml pipettes.
4. Hot plate with magnetic stirrer.

G. *Reagents and Materials*

Procedure I. Amine End Groups

1. Poly(ω-aminoundecanoic acid)(11 nylon) from Exp. 1 or another polyamide prepared from an ω-amino acid (see step H-1).
2. Phenol, crystal, reagent grade, free flowing (see step H-2).
3. Methanol, reagent grade (see step H-3).
4. Hydrochloric acid, standardized, in one or more of the following concentrations: 0.05, 0.1, 0.2, 0.5, 1.0 *N* (see step H-4).
5. Thymol blue indicator (thymosulfonphthalein), 0.1% in distilled water.

Procedure II. Hydroxyl End Groups

1. Poly(tetramethylene glycol), Polymeg 1000, or Polymeg 2000 from Exp. 7; or another hydroxyl-terminated polymer.
2. Sodium hydroxide, reagent grade (see step H-5).
3. Methanol, reagent grade (see step H-5).
4. Potassium acid phthalate, primary standard grade (see step H-5).
5. Acetic anhydride, reagent grade (see step H-6).
6. Pyridine, reagent grade (see step H-6).
7. Phenolphthalein, cresol red, and thymol blue indicators (see step H-7).
8. *n*-Butanol, reagent grade.

H. *Preparation*

1. For Procedure I, the polyamide *must* be broken or ground into pieces no larger than 1 mm in greatest dimension, or dissolution is far too slow for the time allotted. It is recommended that the sample be ground ahead of time. Since the polyamides are tough, they are best cooled to liquid-nitrogen temperature and crushed or ground in a small laboratory mill.
2. If reagent-grade phenol from a previously unopened container is not available for use in Procedure I, distill phenol from 1 g/ℓ BaO, collecting the constant-boiling portion of the distillate.
3. If reagent-grade methanol from a previously unopened container is not available, distill methanol for use in Procedure I from 1 g/ℓ KOH (pellets), discarding the first 10% of the distillate.
4. Standardized HCl for use in Procedure I can be purchased from laboratory supply houses or prepared by usual analytical techniques. For the polyamides prepared in Exp. 1, convenient titers are obtained by using 1*N* acid for the sample polymerized 10 min; 0.5*N*, 20 min; 0.2*N*, 30 min; 0.1*N*, 45 min and 60 min; and 0.05*N*, 90 min. For conductometric titration, double these concentrations.
5. Prepare ahead of time for use in Procedure II 0.40*N* methanolic NaOH by dissolving 16 g NaOH in a minimum amount of water. Dilute to 1 liter with methanol, allow to stand overnight, and standardize against primary standard-grade potassium acid phthalate using phenolphthalein indicator.
6. Prepare acetylating reagent ahead of time for use in Procedure II by dissolving the amount of acetic anhydride calculated below in pyridine to a total volume of 250 ml. Calculate volume of acetic anhydride as $68.85 \times N$ where N is the normality of the NaOH from step H-5. This is calculated to make a reagent giving a blank titration of slightly less than 50 ml standard NaOH.
7. Prepare the mixed indicator used in Procedure II by adding 1 part 0.1% aqueous cresol red to 3 parts 0.1% aqueous thymol blue, both indicators having been neutralized with NaOH.

I. *Procedure*

Procedure I. Amine End Groups

1. Place 35 g phenol and 15 g (19 ml) methanol in each of 3 100-ml 3-neck flasks.
2. Add 1.5-2 g (accurately weighed) finely ground polyamide to each flask.
3. Fit each flask with two stoppers and condenser, and heat to reflux until the sample is dissolved completely. Cool to room temperature.
4. Replace condenser with stirrer, remove one stopper, and add 0.2 ml thymol blue solution.
5. Titrate to pink end point. Use 0.2-ml increments of standardized HCl.
6. (Optional) For conductometric titration, do not add thymol blue. Add 5 ml distilled water and titrate using 0.05-ml increments of standardized HCl. Take as the end point the titer at which a curve of conductivity versus ml HCl changes slope.

Procedure II. Hydroxyl End Groups

II A. Hydroxyl Equivalent

1. Weigh accurately the amount specified below of the polymer to be analyzed into each of two 250-ml iodine flasks. If the approximate $\overline{M}_n$ of the sample is about 400, weigh in 1.2 g; 800, 1.8 g; 1200, 2.0 g; 1600, 2.3 g; 2000, 3.0 g.
2. Prepare two empty flasks as blanks.

(*Note*: Follow steps 3-5 for each of the four flasks prepared in steps 1 and 2.)

3. Pipette accurately 10.0 ml acetylating reagent (step H-6) into the flask, and stopper immediately.
4. Place a magnetic stirring bar in the flask, and place it on the stirring hot plate. Add 10 ml distilled water and 10 ml pyridine, and heat to ca. 100°C for 5 min. Remove and cool to room temperature.
5. Add 10 ml *n*-butanol and 6 drops mixed indicator (step H-7), and titrate with 0.4*N* NaOH (step H-6) to a neutral end point.

II B. Acid Equivalent

1. Weigh accurately 2 to 3 g polymer into each of two 250-ml iodine flasks.
2. Prepare two empty flasks as blanks.

(*Note*: Follow steps 3 and 4 for each of the four flasks prepared in steps 1 and 2.)

3. Add 25 ml pyridine and a magnetic stirring bar, and heat on a stirring hot plate at 105-110°C until the sample is dissolved. Add 10 ml distilled water and heat 3 min, then cool to room temperature.
4. Add 10 ml *n*-butanol and 6 drops of mixed indicator (step H-7), and titrate with 0.4*N* NaOH (step H-6) to the first blue end point. (*Note*: This end point fades rapidly.)

J. Fundamental Equations

Procedure I

$$\overline{M}_n = (\text{sample weight} \times 1000)/(\text{titer, ml,} \times \text{normality})$$

$$\overline{x}_n = \overline{M}_n/M_0$$

where M_0 is the molecular weight of the repeat unit (monomer less water).

Procedure II

Hydroxyl equivalent = [(titer of blank - titer of sample) × normality] / sample weight

where the sample and blank are those of Sec. I, Procedure II A.

Acid equivalent = [(titer of sample - titer of blank) × normality] / sample weight

where the sample and blank are those of Sec. I, Procedure II B.

$$\overline{M}_n = 2000/(\text{hydroxyl equivalent} + 2 \times \text{acid equivalent})$$

K. *Calculations*

Procedure I

Calculate $\overline{M}_n$ and $\overline{x}_n$ for each sample.

Procedure II.

Calculate hydroxyl equivalent, acid equivalent, and $\overline{M}_n$ from average titers for duplicate samples and blanks.

L. *Report*

Procedure I

1. Describe the apparatus and experiment in your own words.
2. Tabulate, for each sample, values of sample weight, normality of the acid used, titer, $\overline{M}_n$, and $\overline{x}_n$. Add comments if any.
3. Why is it necessary to avoid large titers in this experiment?
4. (Optional) Explain the change in slope of the conductivity-titer curve at the end point in the conductometric titration.

Procedure II

1. Describe the apparatus and experiment in your own words.
2. Tabulate the individual titers and their average values for each step, the acid and hydroxyl equivalents, and $\overline{M}_n$.

3. Calculate the error that would result if the correction for the acid number were neglected.

M. Comments

N. General References

Chaps. 1B, 5B; *Textbook*, Chap. 3A; Ogg 1945; Waltz 1947; Schaefgen 1950; Price 1959; Hellman 1962; Sorenson 1968, p. 155.

O. Bibliography

Hellman 1962. Max Hellman and Leo A. Wall, "End-Group Analysis," Chap. V in Gordon M. Kline, ed., *Analytical Chemistry of Polymers*, Part III, Interscience Div., John Wiley and Sons, New York, 1962.

Ogg 1945. C. L. Ogg, W. L. Porter, and C. O. Willits, "Determining the Hydroxyl Content of Certain Organic Compounds; Macro and Semi-micro Methods," *Ind. Eng. Chem., Anal. Ed. 17*, 394-397 (1945).

Schaefgen 1950. John R. Schaefgen and Paul J. Flory, "Multilinked Polyamides," *J. Am. Chem. Soc. 72*, 689-701 (1950).

Sorenson 1968. Wayne R. Sorenson and Tod W. Campbell, *Preparative Methods of Polymer Chemistry*, 2nd ed., Interscience Div., John Wiley and Sons, New York, 1968.

Waltz 1947. J. E. Waltz and Guy B. Taylor, "Determination of the Molecular Weight of Nylon," *Ind. Eng. Chem., Anal. Ed. 19*, 448-450 (1947).

EXPERIMENT 11. MEMBRANE OSMOMETRY

A. Introduction

In this experiment, the number-average molecular weight of polystyrene is determined using a high-speed automatic membrane osmometer. Approximately 3 hours is needed for this experiment. Procedures are given for 3 widely used high-speed automatic osmometers.

B. Principle

In membrane osmometry (Chap. 7A, *Textbook*, Chap. 3B), a polymer solution and the pure solvent are placed in two compartments of an osmometer (*Textbook*, Figs. 3-2 to 3-4), separated by a membrane permeable to solvent molecules but not to polymer molecules. The chemical potential of the solvent in the two compartments is not equal, if both are at the same temperature and pressure. In response to the thermodynamic drive towards equilibrium, solvent flows through the membrane into the solution, increasing its pressure until the pressurized solvent in the solution has the same activity as the pure solvent at ambient pressure. In a conventional osmometer, the pressure is generated by a column of liquid in a capillary rising from the solution compartment; in an automatic osmometer, it is measured as the pressure necessary to maintain one sealed compartment (solution or solvent) at constant volume.

Thermodynamic analysis relates the change in free energy, required to restore the chemical potential of the solvent in the solution to that of the pure solvent, to the mole fraction of solute. If the weight concentration of solute is known, its number-average molecular weight $\bar{M}_n$ can be calculated.

Pinner (1961) (Sec. C.2), McCaffery (1970) (Exp. 24), and Bettelheim (1971) (Exp. 43) include experiments on membrane osmometry.

C. *Applicability*

This method is applicable only when there is no diffusion of low-molecular-weight polymer through the osmotic membrane, as evidenced by an osmotic pressure which is completely stable with time. Unless the molecular-weight distribution of the sample is known in advance, the level of $\bar{M}_n$ above which no diffusion occurs is difficult to estimate. As a rough rule of thumb, the method as ordinarily practiced is applicable to narrow-distribution polymers with $\bar{M}_n$ > 10,000-20,000, depending upon the retentiveness of the membrane; or to normal-distribution polymers with $\bar{M}_n$ > 50,000.

D. *Precision and Accuracy*

With care, a precision of about ±5% for repeat experiments with the same membrane can be obtained for normal values of $\bar{M}_n$, but the second virial coefficient A_2 may be expected to vary considerably more than this. The precision falls off for values of $\bar{M}_n$ in excess of a few hundred thousand, as the osmotic height becomes small. The accuracy of both $\bar{M}_n$ and A_2 depend strongly upon details of membrane behavior in ways not well understood at the present time, and it is not safe to assume that accurate results are being obtained unless the performance of the membrane being used has been tested by measuring a polymer of known $\bar{M}_n$.

E. *Safety Considerations*

CARE MUST ALWAYS BE TAKEN TO USE ORGANIC SOLVENTS ONLY IN WELL-VENTILATED AREAS AND IN SMALL QUANTITIES. AVOID BREATHING FUMES OR EXCESSIVE CONTACT WITH SKIN AND PARTICULARLY EYES. AVOID THE USE OF OPEN FLAMES OR SOURCES OF ELECTRIC SPARKS. SAFETY GLASSES MUST BE WORN IN THE LABORATORY AT ALL TIMES.

F. *Apparatus*

1. Automatic membrane osmometer: Shell design (Hallikainen,

Stabin), Mechrolab design (Hewlett-Packard), or Reiff design (Melabs, Wescan), fitted with gel cellophane or equivalent membrane for operation in toluene at or slightly above ambient temperature, and tested for accuracy with a narrow-distribution polystyrene of known $\bar{M}_n$.

2. 10-ml stoppered volumetric flasks.

G. *Reagents and Materials*

1. Polystyrene. Source may be polymer prepared in Exp. 4 or 9, commercial material, or a narrow-distribution polymer of known $\bar{M}_n$.

2. Toluene, reagent grade.

H. *Preparation*

Prepare ahead of time in 10-ml volumetric flasks four solutions of polystyrene in toluene at accurately known concentrations approximating 2, 4, 6, and 8 g/ℓ, as follows:

1. Tare four 10-ml volumetric flasks.
2. Accurately weigh approximately 20, 40, 60, and 80 mg finely divided polystyrene into the four flasks.
3. Fill the flasks 2/3 with toluene. Stopper and agitate gently; do not shake, as particles of polymer may adhere to the walls above the liquid level and fail to dissolve.
4. Repeat agitation from time to time until polymer is completely dissolved. Allow at least 24 hours.
5. Place the flasks in a constant-temperature bath regulated to the temperature of operation of the osmometer.
6. After temperature equilibrium has been reached (allow 15 min), fill the flasks to the mark with toluene maintained at the bath temperature. Mix well and return to the bath.
7. If the solutions contain obvious dust or dirt, filter them through a medium or fine fritted-glass filter or a 0.2-μm "Millipore" filter.

I. *Procedure*

Procedure I. Stabin-Shell Osmometer (Holleran 1967)

Instrument Check

(*Note*: It is assumed that the instructor has prepared the osmometer with a satisfactory membrane, with solvent at the proper level in both compartments, and free from leaks.)

1. See that the POWER switch is ON, the SERVO switch is at MANUAL, the CHART switch is OFF, the SAMPLE INLET valve is OPEN, and the SAMPLE OUTLET, SOLVENT INLET, and SOLVENT OUTLET valves are closed. The cell should be at temperature equilibrium, with the thermometer reading the actual measurement temperature T. Note the value of T. The heater meter should show a steady deflection, and the oscillator meter should be almost at full scale.

Solvent Measurement

2. Close the SAMPLE INLET valve. (CAUTION: Keep this valve closed only long enough to make the measurements called for.)

3. Open, then immediately close the SAMPLE OUTLET valve to insure that the sample compartment is at atmospheric pressure.

4. Turn the SERVO switch to AUTOMATIC and the CHART switch ON.

5. Observe the recorder chart for about 10 min. The pen must reach a constant reading near the bottom of the chart, with random noise motion of less than 0.1 mm. At some time during this measurement, open and immediately close the SAMPLE OUTLET valve. The pen should deflect, but rapidly return to its previous position.

If the above conditions are not met, obtain help from the laboratory assistant. When a satisfactory base line is obtained, record the counter reading on the chart.

6. Open the SAMPLE INLET valve, turn the SERVO switch to MANUAL and the CHART switch OFF, and use the MANUAL switch to return the bob to its lowest position.

Sample Measurement

7. Fill the sample stack with the least concentrated polymer solution to be measured. (CAUTION: Use only 2.5-3 ml. This step uses up most of the polymer solution prepared, so there is no margin for error.) Open the SAMPLE OUTLET valve, allowing the solution to flow into the cell. Close the valve just before the liquid level reaches the capillary. Repeat with two more portions, retaining the third for measurement.

8. After temperature equilibrium has been reached, close the SAMPLE INLET valve, open and close the SAMPLE OUTLET valve to equilibrate pressure, turn the SERVO switch to AUTOMATIC and the CHART switch ON.

9. Observe the recorder chart for 2-3 min or until a steady reading is conclusively reached. Record the counter reading and the exact solution concentration on the chart.

10. Open the SAMPLE INLET valve, turn the SERVO switch to MANUAL and the CHART switch OFF, and use the MANUAL switch to return the bob to its lowest position.

11. Repeat steps 7-10 for the remaining polymer solutions in order of increasing concentration.

12. Immediately after the last solution has been run, repeat Step 7 with pure solvent. Leave the osmometer as described in Step 1.

Procedure II. Hewlett-Packard (Mechrolab) Osmometer (Paglini 1968)

Instrument Check

(*Note*: It is assumed that the instructor will see that the osmometer is properly filled with solvent, a satisfactory membrane is in place, the bubble is of the proper size and location, the gain is correctly set, and the instrument is free from leaks.)

1. See that the following switches on the control unit are set as indicated: SERVO and THERMO switches ON, RECORDER switch OFF, CONTROL switch to NORMAL, and SELECTOR switch to RUN.

Solvent Measurement

2. Fill the sample stack (top of instrument) with fresh solvent, preheated to the temperature of operation of the osmometer to avoid prolonging the time to achieve thermal equilibrium.

3. Observing the solvent level through the viewing tube (near top of instrument), open the SIPHON valve (top of instrument) until the liquid level in the stack reaches the scribe mark. Close the SIPHON valve.

4. Turn the RECORDER switch ON (if the recorder is being used). Allow the instrument to run until thermal equilibrium has been established as indicated by a steady reading on the counter or recorder. Record the reading. Turn the RECORDER switch OFF.

5. Repeat steps 2 and 3 several times, until the solvent reading is repeatable to 0.02 cm.

Sample Measurement

6. Fill the sample stack with the least concentrated polymer solution to be measured. Open the SIPHON valve, and draw the liquid level down to the mark seen through the viewing tube. Close the valve.

7. Repeat step 5 twice, retaining the third portion of solution for measurement.

8. Turn the RECORDER switch ON, and allow the instrument to run until a steady reading is conclusively reached. Record the reading, and turn the RECORDER switch OFF.

9. Repeat steps 5-7 for the remaining polymer solutions in order of increasing concentration.

10. Immediately after the last solution has been run, repeat steps 5 and 6 with pure solvent. Leave the osmometer as described in step 1.

Procedure III. Melabs or Wescan (Reiff) Osmometer

Instrument Check

(*Note*: It is assumed that the instructor will have prepared the osmometer by installing a suitable membrane, insuring that the instrument is free from leaks, and calibrating the strain gauge.)

1. See that there is solvent in both compartments, that the POWER switch is ON, and that the temperature of the instrument has stabilized.

Solvent Measurement

2. Fill the sample stack with fresh solvent, and open the SOLUTION DRAIN valve (center, on top of instrument) until the liquid level meter reads 60. Wait until equilibrium is reached, as indicated by a steady recorder reading.

3. Repeat step 2 with a second portion of fresh solvent. When a steady reading is reached, set the recorder to zero using the RECORDER ZERO potentiometer on the osmometer.

Sample Measurement

4. Fill the sample stack with the least concentrated polymer solution to be measured. Open the SOLUTION DRAIN valve, and allow the liquid to drain into the cell. Then close the valve.

5. Refill the stack, and again open the valve to drain the liquid into the cell. SLOWLY close the SOLUTION DRAIN valve (in order that the pressure may have time to equalize), setting the level so that the meter reads 60. Allow the system to attain equilibrium (about 15-20 min).

6. Readjust the level to a meter reading of 60 by slowly opening and closing the SOLUTION DRAIN valve. After equilibrium has been reached, read the osmotic height on the recorder.

7. Repeat steps 5 and 6 to insure that the solution was not diluted through incomplete rinsing in step 4. Accept the result if the heights in steps 6 and 7 differ by not more than 2-5%; otherwise, repeat again.

8. Repeat steps 4-7 for the remaining polymer solutions in order of increasing concentration.

9. Immediately after the last solution has been run, repeat step 4 with solvent. Leave the osmometer as described in step 1.

J. *Fundamental Equations*

$$\pi = RT(A_1c + A_2c^2 + A_3c^3 + \cdots) \tag{7-5}$$

$$\pi/RTc = 1/\overline{M}_n + A_2c \tag{7-6}$$

$$\overline{M}_n = RT/(\pi/c)_0 \tag{7-8}$$

$$A_2 = [(\pi/c)_{c_2} - (\pi/c)_{c_1}]/RT(c_2-c_1) \tag{7-10}$$

K. *Calculations*

1. Calculate π/c for each polymer concentration, and plot π/c versus c. Draw the best straight line through the data points, and read $(\pi/c)_0$ as its intercept at $c = 0$. Alternatively, carry out a least squares analysis, as described in App. IV, to obtain $(\pi/c)_0$.

2. Calculate $\overline{M}_n$ from Eq. 7-8 and A_2 from Eq. 7-10. In calculating R, use the density of toluene at room temperature, 0.865.

L. *Report*

1. Describe the apparatus and the experiment in your own words.
2. Include the graph prepared in step K-1.
3. Report the values of $\overline{M}_n$ and A_2 calculated in step K-2. Include the units for both quantities.

4. Comment on the results of the experiment.
5. Answer the following questions:

a. Sketch a curve showing the osmotic pressure as a function of time for the normal case and for the case of diffusion of low-molecular-weight material through the osmotic membrane. Explain the latter case, including its effect on $\overline{M}_n$ and the possibilities of correcting for diffusion. Could you explain a curve deviating from the normal case in the opposite direction from the deviation caused by diffusion? How? What would be the effect on the osmotic pressure-time curve of a leak in the membrane, and how could this be distinguished, if at all, from diffusion?

b. Using Eq. 3-7 in the *Textbook* (p. 67), test your results using the square-root plot. Demonstrate whether or not your data require this treatment within experimental error. Calculate A_2 from this treatment.

c. Discuss the significance of the second virial coefficient A_2. What is the meaning of negative values of A_2, and of $A_2 = 0$?

d. In calculating $\tilde{R}$, why is it proper to use the density of toluene at room temperature even though T is different? (The data in Table 7-3 were obtained in a different type osmometer totally immersed in a constant-temperature bath at 105°C.)

M. Comments

N. General References

Chap. 7A; *Textbook*, Chap. 3B; Armstrong 1968, Coll 1968; Elias 1968.

O. Bibliography

Armstrong 1968. Jerold L. Armstrong, "Critical Evaluation of Commercially Available Hi-Speed Membrane Osmometers," pp. 51-55 in D. McIntyre, ed., *Characterization of Macromolecular Structure*, National Academy of Sciences Publication No. 1573, Washington, D.C., 1968.

Bettelheim 1971. Frederick C. Bettelheim, *Experimental Physical Chemistry*, W. B. Saunders Co., Philadelphia, Pennsylvania, 1971.

Coll 1968. Hans Coll and F. H. Stross, "Determination of Molecular Weights by Equilibrium Osmotic-Pressure Measurements," pp. 10-27 in *Characterization of Macromolecular Structure*, D. McIntyre, ed., National Academy of Sciences Publication No. 1573, Washington, D.C., 1968.

Elias 1968. Hans-Georg Elias, "Dynamic Osmometry," pp. 28-50 in *Characterization of Macromolecular Structure*, D. McIntyre, ed., National Academy of Sciences Publication No. 1573, Washington, D.C., 1968.

Holleran 1967. Peter M. Holleran, "Rapid Osmometry with Diffusable Polymers," M. Sc. Thesis, Rensselaer Polytechnic Institute, Department of Chemistry, 1967.

McCaffery 1970. Edward L. McCaffery, *Laboratory Preparation for Macromolecular Chemistry*, McGraw-Hill Book Co., New York, 1970.

Paglini 1968. Severo Paglini, "Experimental Aspects of the High-Speed Membrane Osmometer," *Anal. Biochem. 23,* 247-251 (1968).
Pinner 1961. S. H. Pinner, *A Practical Course in Polymer Chemistry,* Pergamon Press, New York, 1961.

EXPERIMENT 12. VAPOR-PHASE OSMOMETRY

A. *Introduction*

In this experiment, the number-average molecular weight of a polyester or poly(methyl methacrylate) is determined using a "vapor-pressure osmometer." About 3 hours is needed for this experiment.

B. *Principle*

In vapor-phase osmometry (Chap. 7A, *Textbook*, Chap. 3B), droplets of a polymer solution and the pure solvent are placed on two thermistors suspended in the thermostated chamber of a "vapor pressure osmometer" (Figs. 7-6 and 7-7), saturated with the solvent vapor. The chemical potential of the solvent in the two droplets is not equal. Because the vapor pressure of the solvent in the solution is less than that of the pure solvent, condensation takes place on the solution droplet. It thus becomes both diluted and heated by the latent heat of vaporization of the solvent. The temperature of the solution droplet rises until the vapor pressure of the solvent in it, at the new higher temperature, reaches that of the pure solvent at the original temperature. The difference in temperature of the two droplets is detected by the two thermistors, which are arranged in a Wheatstone bridge circuit.

The temperature difference is predicted from thermodynamics by combining Raoult's law with the Clapeyron equation as

$$\Delta T = RT^2 n_2 / \Delta H_v \tag{12-1}$$

where n_2 is the mole fraction of the solute and ΔH_v the heat of vaporization of the solvent. If the full temperature difference required by thermodynamics were achieved, the number-average molecular weight $\overline{M}_n$ of the solute could be calculated, knowing its weight concentration. Heat losses, primarily by conduction along the leads of the thermistors, limit the temperature difference achieved, however, and it is customary to calibrate the instrument with a substance of known molecular weight rather than to make the thermodynamic calculation.

McCaffery (1970) includes an experiment (Exp. 26) on vapor-phase osmometry.

C. *Applicability*

This method is applicable to any soluble polymer whose molecular weight does not exceed the limit set by the sensitivity of the instrument, approximately 20,000 in the usual case. Not all solvents are suitable; some criteria for suitable solvents are given, and several solvents are listed, in Sec. M-1. It should be noted that the method is sensitive to the presence of small amounts of low-molecular-weight impurities, since all non-volatile molecules other than the solvent species contribute to the effect measured.

D. *Precision and Accuracy*

With care, a precision of ±5% can be obtained for repeat measurements of the difference in thermistor resistance, equivalent to the difference in temperature between the two droplets. The corresponding precision of $\bar{M}_n$ varies, of course, because of the inverse relation between the two, and falls off for values of $\bar{M}_n$ in excess of 20,000 or so.

E. *Safety Considerations*

CARE MUST ALWAYS BE TAKEN TO USE ORGANIC SOLVENTS ONLY IN SMALL QUANTITIES AND IN WELL-VENTILATED AREAS. AVOID BREATHING FUMES OR EXCESSIVE CONTACT WITH SKIN AND PARTICULARLY EYES. AVOID THE USE OF OPEN FLAMES OR SOURCES OF ELECTRIC SPARKS. WEAR SAFETY GLASSES AT ALL TIMES IN THE LABORATORY.

F. *Apparatus*

1. "Vapor pressure osmometer": Mechrolab design (Hewlett-Packard) or Hitachi-Perkin Elmer design (Coleman). Preassembled for use with acetone as solvent at 37°C.
2. 10-ml stoppered volumetric flasks.

G. *Reagents and Materials*

1. Poly(tetramethylene glycol) (precursor of polyurethane prepared in Exp. 7); poly(methyl methacrylate) prepared in Exp. 3, sample of lowest molecular weight; or other acetone-soluble polymer with $\bar{M}_n$ expected to be less than 20,000.
2. Acetone, reagent grade.

H. *Preparation*

Prepare ahead of time in 10-ml volumetric flasks four solutions of polymer in acetone at accurately known concentrations approximating 20, 40, 60, and 80 g/ℓ, as follows:

1. Tare four 10-ml volumetric flasks.

2. Accurately weigh approximately 0.2, 0.4, 0.6, and 0.8 g finely divided polymer into the flasks.

3. Fill the flasks 2/3 with acetone. Stopper and agitate gently; do not shake, as particles of polymer may adhere to the walls above the liquid level and fail to dissolve.

4. Repeat agitation from time to time until polymer is completely dissolved. Allow at least 24 hr.

5. Place the flasks in a constant-temperature bath regulated to 37°C.

6. After temperature equilibrium has been reached (allow 15 min), fill the flasks to the mark with acetone maintained at the bath temperature. Mix well and return to the bath.

7. If the solutions contain obvious dust or dirt, filter them through a medium or fine fritted-glass filter or a 0.2-μm "Millipore" filter.

I. Procedure

Procedure I. Mechrolab-Design "Vapor-Pressure Osmometer"

Instrument Set-Up

1. Turn on the THERMOSTAT switch at least 4 hrs in advance, and turn on the NULL DETECTOR switch 30 min in advance of measurements.

2. Remove the instrument syringes from their holders, and rinse them with acetone. Fill syringes 5 and 6 with pure solvent (1-2 ml is sufficient) and insert them carefully in the "up" or counter-clockwise position. Use the viewing mirror to check that they are not in the "down" position where they could damage the thermistor bead.

3. Rinse and fill syringes 1-4 with the four polymer solutions, placing the least concentrated solution in syringe 1, etc.

4. Using the viewing mirror to observe the operation, turn syringe 6 clockwise until the tip of the needle is near the sample (front) thermistor bead (CAUTION: Do not allow the needle to touch the bead). Turn the syringe feed-screw knob clockwise to wash the bead with at least 3 drops of solvent. Finally, deposit a drop on the bead which approximately doubles its diameter.

5. Repeat step 4 with syringe 5 to place solvent on the reference (rear) thermistor bead.

Solvent Measurement

6. Turn the BRIDGE ON-OFF switch ON, and turn the MEASURE-TEST switch to TEST.

7. Depress the ZERO button and center the meter needle with the ZERO control; then release the button.

8. Rotate the TEST potentiometer knob until the needle is near the center of the scale: If it drifts to the right, allow more time for thermal equilibration.

9. Turn the MEASURE-TEST switch to MEASURE, and set the ΔR resistance dials to zero.

10. (Repeat steps 4, 5, and 7 if they were not just completed.) Center the meter needle with the BALANCE control. If balance cannot

be achieved, use the knob on top of the instrument to provide a coarse setting, and the BALANCE control for fine adjustment.

(*Note*: Some instruments have only one BALANCE control instead of two.)

11. Rebalance every minute for 3-5 min, then repeat step 4 to place fresh solvent on the sample thermistor, and rebalance twice more. A steady balance position should be achieved.

Sample Measurement

12. Lower syringe 1, as in step 4, rinse the sample bead with about 6 drops of solution, and finally deposit a drop for measurement. Start a stopwatch or timer.

13. Check the zero by repeating step 7.

14. After 2 min, center the meter needle by adjusting the ΔR dials. Record the value of ΔR. Repeat every minute until a steady value of ΔR is achieved. This should take 8-10 min.

15. Repeat steps 12-14 with syringes 2-4.

(*Note*: The highest value of ΔR obtained should be no greater than 8-10 ohms. If necessary, the solution concentrations should be adjusted to meet this requirement.)

16. Repeat steps 4, 5, and 7, and check the balance position with the ΔR dials set to zero. The meter needle should be no more than 0.5 mm from center. Obtain assistance from the instructor if this cannot be achieved.

Calibration

17. Remove syringes 1-4 and rinse and fill them with solutions of 0.1, 0.2, 0.4, and 0.8 g/ℓ benzil in acetone. Determine the values of ΔR for these solutions by repeating steps 12-16.

Final Steps

18. Turn the BRIDGE, THERMOSTAT, and NULL DETECTOR switches to OFF, and the MEASURE-TEST switch to TEST.

19. Remove all syringes, rinse them thoroughly with acetone, and replace them in the counter-clockwise (up) position.

Procedure II. Hitachi-Perkin Elmer Design "Vapor-Pressure Osmometer"

Instrument Set-Up

1. Clean and assemble the instrument and turn it on at least 3 hrs before making measurements. At that time, set the temperature controller to the desired temperature (37°C), and inject 10 ml solvent into the outer injection port and 5 ml solvent into the inner injection port. Draw excess solvent from the drain vat with a syringe immediately after solvent injection and at intervals thereafter.

Solvent Measurement

2. Assemble two instrument syringes, wash them three times with

acetone, and insert them into the osmometer. Inject about 1/3 of the contents of each syringe into the instrument to clean the lines.

3. Inject two drops from each syringe and wait until equilibrium is achieved as indicated by the galvanometer needle becoming steady (about 4 min is required). Set the Δr dial to 45 units, and adjust the BALANCE CONTROL potentiometer to bring the galvanometer to zero.

4. Inject two drops from each syringe, wait 4 min, and bring the galvanometer to zero with the Δr dial. Record the dial reading to the nearest half unit. Repeat this injection and rebalancing six times. After each injection, attempt to draw solvent from the drain vat.

5. Continue taking readings as in step 4 until six successive readings are obtained which lie within a range of ±1 unit of Δr. Average these six readings, taking the average as the solvent reading.

(*Note*: Injection of 1/3 the contents of the syringe will temporarily change the temperature of the sample cell, requiring a few minutes for equilibration. Consequently, the first one or two readings may differ slightly from subsequent ones.)

Sample Measurement

6. Wash and fill a syringe with the most dilute sample solution. Insert it in the instrument and inject about 1/3 of the contents to flush the system.

7. Injecting two drops from the sample syringe and two drops from the reference solvent syringe each time, repeat steps 4 and 5 until six successive readings lying within a range of ±1 in Δr are obtained, or until 12 readings have been taken. Average the six readings, or all 12 if the specified reproducibility was not obtained.

8. Immediately repeat steps 3-5 with the solvent.

9. Repeat steps 6-8 with the remaining sample solutions in order of increasing concentration.

(*Note*: For samples with $\overline{M}_n$ above 10,000, four to six solutions should be run, with concentrations adjusted to give values of Δr for (sample less solvent) spaced evenly between 25 and 150 units; for $\overline{M}_n$ between 3,000 and 10,000, three solutions giving readings between 50 and 150; for $\overline{M}_n$ below 3000, one solution reading about 75 is usually adequate.)

Calibration

10. Repeat steps 6-8 for five solutions of benzil (M = 210) with concentrations evenly spaced between 0.0005 and 0.01 molal.

11. Be sure that the last material in the sample cell is solvent, flush it thoroughly, withdraw solvent from the drain vat, and leave the instrument as described in step 1.

J. *Fundamental Equations*

$$\Delta r/kc = 1/\overline{M}_n + A_2c \qquad (7\text{-}13)$$

K. *Calculations*

Procedure I

1. Plot Δr versus time for each solution measured in steps I-14, I-15, and I-17. From these, determine the "steady" values of Δr for use in steps 2 and 3.

2. Calculate the calibration constant k from the steady-value data of step I-17, using Eq. 7-13. Use the formula weight of benzil (210) as $\overline{M}_n$.

3. Calculate $\Delta r/kc$ from the steady-value data of steps I-14 and I-15 and the result of step 2, and plot this quantity versus c. Extrapolate to zero concentration and determine $\overline{M}_n$ and A_2 using Eq. 7-13.

Procedure II

1. For each calibration and sample solution measured in steps I-7 and I-10, calculate Δr(solution, ave.) - Δr(solvent, ave.). For the solvent reading, take the mean of the two average readings obtained immediately before and after the solution was measured.

2. Repeat step 2 for Procedure I using the calibration data calculated in step 1.

3. Repeat step 3 for Procedure I using the sample data calculated in step 1.

L. *Report*

1. Describe the apparatus and experiment in your own words.
2. Include the graphs prepared in Sec. K.
3. Report the values of k calculated in step K-2 and of $\overline{M}_n$ and A_2 calculated in step K-3. Include the units for each.
4. Comment on the results of the experiment.
5. Answer the following questions:

   a. Why is it necessary to standardize such variables as drop size and time of measurement in this experiment?

   b. Would you expect drop size to change during the course of a measurement? If so, how and why?

   c. Write the equivalent of Eq. 3-7 in the *Textbook* (p. 67) for the case of vapor-phase osmometry, and test your data using it. Does the use of this square-root plot alter your values of $\overline{M}_n$ and A_2 significantly?

M. *Comments*

1. Unusual purity is required in solvents for use in vapor-phase osmometry. Especial care should be taken that the solvent is dry; this is particularly important with acetone. From Eq. 12-1 it is predicted that solvents with lower heats of vaporization give a larger response; this is confirmed at least qualitatively. The vapor pressure of the solvent should be in the range 100-400 torr at the temperature of measurement. A few solvents suitable for use at 37°C are acetone, benzene, carbon tetrachloride, chloroform, cyclohexane, ethyl acetate,

and ethanol.

2. Cleanliness of equipment and purity of the solvent affect the time of equilibration greatly. Pure solvents in a clean instrument equilibrate rapidly; wet or impure solvents may require 20-30 min to equilibrate. If the instrument is not clean, particularly if it contains traces of polymers from previous experiments, prolonged drifting may take place. Severe problems of this sort may require washing the thermistor beads with a series of different solvents over a period of several days to remove the offending material.

N. *General References*

Chap. 7A; *Textbook*, Chap. 3B; Pasternak 1962; Wegmann 1962; Tomlinson 1963; Simon 1966; Van Dam 1968; Wachter 1969.

O. *Bibliography*

McCaffery 1970. Edward L. McCaffery, *Laboratory Preparation for Macromolecular Chemistry*, McGraw-Hill Book Co., New York, 1970.

Pasternak 1962. R. A. Pasternak, P. Brady, and H. C. Ehrmantraut, "Apparatus for the Rapid Determination of Molecular Weight," *Dechema Monograph 44*, 205-207 (1962).

Simon 1966. W. Simon, J. T. Clerc, and R. E. Dohner, "Thermoelectric (Vaporometric) Determination of Molecular Weight in 0.001 Molar Solutions: A New Detector for Liquid Chromatography," *Microchem. J. 10*, 495-508 (1966).

Tomlinson 1963. C. Tomlinson, Ch. Chylewski, and W. Simon, "The Thermoelectric Determination of Molecular Weight - III," *Tetrahedron 19*, 949-960 (1963).

Van Dam 1968. J. Van Dam, "Vapor-Phase Osmometry," pp. 336-342 in D. McIntyre, ed., *Characterization of Macromolecular Structure*, National Academy of Sciences Publication 1573, Washington, D.C., 1968.

Wachter 1969. Alfred H. Wachter and Wilhelm Simon, "Molecular Weight Determination of Polystyrene Standards by Vapor Pressure Osmometry," *Anal. Chem. 41*, 90-94 (1969).

Wegmann 1962. Dorotheé Wegmann, C. Tomlinson, and W. Simon, "Thermoelectric Microdetermination of Molecular Weights, Part II: Routine Determination," *Microchem J.* (Symp. Ser.) *2*, 1069-1085 (1962).

EXPERIMENT 13. LIGHT-SCATTERING MOLECULAR WEIGHT

A. *Introduction*

In this experiment, the weight-average molecular weight, radius of gyration, and second virial coefficient are determined for polystyrene dissolved in toluene. The determination of the particle size of a latex by light scattering is performed in Exp. 14. This experiment requires approximately 3 hours. Procedures are included for use with several commercially available light-scattering photometers.

B. *Principle*

In light scattering (Chap. 7B, *Textbook*, Chap. 3C) the intensity of light scattered from a polymer solution is measured in a photoelectric photometer as a function of angle of observation and concentration. The origin of the scattering was shown by Debye (1944, 1947) to result from fluctuations, on a microscopic scale, of the concentration of the solution. This leads to transient regions with refractive index different from the surrounding medium, and this is the requirement for the scattering of light. The intensity of the scattered light is proportional to the square of the difference in refractive index between the polymer solution and the pure solvent, to the inverse fourth power of the wavelength, and to the molecular weight of the solute. It is shown in the *Textbook* (p. 78) that the experiment yields the weight-average molecular weight $\bar{M}_w$, defined by Eq. 7-14.

The angular dependence of light scattering arises from the interference of light rays scattered from different parts of the same polymer molecule, provided that it is large enough (at least 1/10 to 1/20 of the wavelength). In the usual case, this corresponds to a rather low molecular weight, say 50,000. The interference effect gives information on the molecular size, expressed as the z-average radius of gyration, independent of the molecular weight. The effect disappears as the angle of observation θ approaches zero; after extrapolation to $\theta = 0$ the correct value of $\bar{M}_w$ is obtained regardless of molecular size. Finally the second virial coefficient, a measure of polymer-solvent interactions, is determined from the slope of a plot of scattering data versus concentration after extrapolation to $\theta = 0$. McCaffery (1970) in Exp. 25 and Bettelheim (1971) in Exp. 45 include experiments on the determination of $\bar{M}_w$ by light scattering.

C. *Applicability*

The light-scattering method is applicable to any polymer which can be dissolved in a solvent with refractive index sensibly different from that of the polymer itself. In most cases, the intensity of light scattering from solutions of polymers with $\bar{M}_w < 10{,}000$ differs so little from that of the solvent that the determination is not precise. At the other end of the scale, few accurate determinations have been made at $\bar{M}_w > 10{,}000{,}000$ because of the need to measure at very small values of θ, outside the range of most light-scattering photometers, to obtain accurate extrapolation to $\theta = 0$. Special treatment is required for copolymers and polyelectrolytes, and in the case of polymers dissolved in mixtures of two or more solvents. One of the most serious problems in light scattering is the need to free the solution from any impurities which scatter light significantly. The method is inapplicable if this cannot be accomplished.

D. *Precision and Accuracy*

With care, a precision of about ±5% can be obtained for repeat light-scattering measurements. Since scattered intensity and $\overline{M}_w$ are proportional, precision in $\overline{M}_w$ is about the same. Absolute accuracy, however, depends upon the care with which the calibration of the photometer and the determination of dn/dc, the specific refractive increment, are carried out; in most cases it is not better than ±10%.

E. *Safety Considerations*

CARE MUST ALWAYS BE TAKEN TO USE ORGANIC SOLVENTS IN WELL-VENTILATED AREAS AND IN SMALL QUANTITIES. AVOID BREATHING FUMES OR EXCESSIVE CONTACT OF SOLVENT WITH SKIN OR, PARTICULARLY, EYES. AVOID THE USE OF FLAMES OR SOURCES OF ELECTRIC SPARKS. SAFETY GLASSES MUST BE WORN IN THE LABORATORY AT ALL TIMES. THE PHOTOMULTIPLIER TUBE OF MOST LIGHT-SCATTERING PHOTOMETERS IS POWERED BY DANGEROUSLY HIGH VOLTAGES (*ca.* 1000 V). THE CIRCUITRY IS USUALLY WELL PROTECTED, BUT CAUTION SHOULD BE EXERCISED. CARE SHOULD ALSO BE TAKEN NEVER TO EXPOSE THE PHOTOTUBE TO HIGH LIGHT INTENSITY WHEN VOLTAGE IS APPLIED TO IT. SUCH EXPOSURE CAN RESULT IN PERMANENT DAMAGE TO THE PHOTOMULTIPLIER TUBE.

F. *Apparatus*

1. Light-scattering photometer, with absolute calibration carried out in advance. Procedures follow for the Brice-Phoenix design (Wood) and the SOFICA design (Bausch and Lomb).
2. 1.0-μm "Millipore" filter or equivalent, in appropriate holder, for clarifying solutions.
3. 50-ml stoppered volumetric flask.

G. *Reagents and Materials*

1. Polystyrene, prepared in Exp. 4 or 9, commercial material, or narrow-distribution polymer of $\overline{M}_w > 50{,}000$.
2. Toluene, reagent grade.

H. *Preparation*

Prepare ahead of time in a 50-ml volumetric flask a solution of polystyrene in toluene at an accurately known concentration of approximately 8 g/ℓ, as follows:

1. Tare a clean, dry 50-ml volumetric flask.
2. Accurately weigh approximately 0.4 g finely divided polymer into the flask.
3. Fill the flask 2/3 with toluene. Stopper and agitate gently; do not shake, as particles of polymer may adhere to the walls above the liquid level and fail to dissolve.
4. Repeat agitation from time to time until polymer is completely

dissolved. Allow at least 24 hours.

5. Place the flask and a supply of solvent in or near the cell compartment of the light-scattering photometer.

6. After temperature equilibrium has been reached (allow 30 min), fill the flask to the mark with solvent. Mix well.

(*Note*: This solution must be prepared far enough in advance that it can be clarified during the laboratory period *before* the light-scattering experiment proper.)

I. *Procedure*

Clarification of Solvent and Solution

(*Note*: Carry out this part of the Procedure one week before the light-scattering determination is performed, as a full three-hour period should be available for the latter.)

1. Rinse two scrupulously cleaned 100-ml stoppered Erlenmeyer flasks or equivalent with several portions of dust-free toluene from a freshly opened bottle. Stopper the flasks.

2. Filter about 75 ml toluene through an ultrafine fritted-glass or a 1.0-μm "Millipore" filter in a suitable holder. (CAUTION: DO NOT attempt to speed up the filtration by applying even a small pressure of inert gas; this almost always results in some fine contaminants being forced through the filter. This step MUST be done slowly at atmospheric pressure only.)

3. Catch the filtrate in step 2 in one of the flasks prepared in step 1. Place aluminum foil or equivalent over the top of the flask to prevent dust from entering, leaving just enough space to insert the exit tube of the filter holder. Rinse the flask with several small portions of the filtrate. Collect 50 ml of the filtered solvent. Stopper the flask (rinse the stopper with filtrate also!) and protect it from contamination by placing it under an inverted beaker or equivalent.

4. Similarly filter the polymer solution prepared in Sec. H by repeating steps 2 and 3.

5. Using a scrupulously clean, dry pipette, remove 5 ml of the filtered solution to a tared aluminum weighing dish. Evaporate to dryness (CAUTION: Do not use open flame and evaporate slowly to avoid spattering) to determine the concentration of the polymer.

Procedure I. Brice Photometer (Brice 1950)

Instrument Check

6. See that the high-voltage supply to the photomultiplier tube is ON, and that the mercury arc lamp is lighted. (CAUTION: Never turn the lamp off, since it cannot be restarted at once, and takes many hours to reach constant intensity.) (*Note*: It is assumed that the instructor has prepared and checked out the instrument, and that a working standard of known intensity of scattering is available.)

Solvent Preparation

7. Open the top of the photometer case and carefully remove the lid to the cell.

8. Using scrupulously clean, dry, dust-free pipettes rinsed first with solvent from a freshly-opened bottle and then with filtered solvent, add 25 ml filtered solvent to the cell. Do not allow the tip of the pipette to touch the walls of the cell.

9. Stir the contents of the cell with a clean copper wire carefully rinsed with solvent as were the pipettes in step 8. Keep the wire in a suitable dust-free container.

10. Cap the cell and close the photometer case.

Amplifier Gain

11. Select the wavelength filter to be used, turn the phototube to 150°, open the shutter, and set the PHOTOMULTIPLIER GAIN switches to produce a reading of about 25 divisions on the galvanometer or recorder. Close the shutter, and adjust the galvanometer or recorder zero (dark current). Repeat if necessary. DO NOT READJUST THE PHOTOMULTIPLIER GAIN AFTER THIS STEP.

(*Note*: Alternatively, measure the highest polymer concentration first, and set the gain to produce a reading of about 95 divisions.)

12. If at any time in subsequent steps the galvanometer or recorder goes off scale, insert one of the calibrated neutral-density filters. Note which filter is used, and to which readings its use applies.

Solvent Measurement

13. Read and record the scattered intensity at angles of 30°, 60°, 90°, ..., 150°. Check the dark current for each reading and adjust or correct for it as required.

Reference

14. Return the phototube to 0°, and insert the reference standard. Read the reference intensity, correcting for dark current if necessary. Record any neutral filters used.

Solution Measurement

15. Calculate the volumes of filtered polymer stock solution to be added to the solvent in the cell (25 ml) to give solution concentrations of approximately 0.1, 0.2, 0.3, and 0.4 g/100 ml in successive additions.

16. Using scrupulously clean, dry, dust-free pipettes, add the first of the calculated amounts of stock solution to the cell. Follow the procedure and precautions of steps 7-10.

17. Repeat steps 13 and 14 to measure the solution and reference.

18. Repeat steps 16 and 17 for the remaining three additions.

Cleanup

19. After all measurements are completed, repeat step 7, and remove the cell from the instrument. Empty the cell and rinse the inside of

it several times with solvent. Replace the cell and leave the instrument as described in step 1.

Procedure II. Sofica Photometer (Wippler 1954)

Instrument Check

6. See that the high-voltage supply to the photomultiplier tube is ON, the cooling-water bath for the lamp is circulating properly, and the lamp is lit. If it is not, consult the instructor, who will assist in starting it. (*Note:* It is assumed that the instructor has prepared and checked out the instrument, and that a working standard of known intensity of scattering is available.)

Solvent Preparation

7. Repeat steps 7-10 of Procedure I.

Amplifier Gain

8. Set the phototube to 90°, select the wavelength filter to be used, open the shutter, and set the slits and lamp voltage to obtain a reading of about 100. Close the shutter and adjust the dark current. Repeat if necessary. DO NOT READJUST LAMP VOLTAGE OR SLITS AFTER THIS STEP.

9. If at any time in subsequent steps, too high or too low readings are obtained, readjust the intensity gain switch to correct. Note the changes in switch position, and correct all readings by multiplication by the proper gain factors.

Solvent Measurement

10. Read and record the scattered intensity at preset angles of 30°, 37.5°, 45°, 75°, 90°, 105°, 135°, 142.5°, and 150°. Check the dark current at each reading and adjust or correct for it as required.

Reference

11. Return the phototube to 90°, read and record the scattered intensity again, and then replace the cell with the calibrated glass reference standard. Read and record the intensity of scattering from the standard.

Solution Measurement

12. Repeat steps 15 and 16 of Procedure I to prepare the first solution for measurement.

13. Repeat steps 10 and 11 of this procedure to measure the solution.

14. Repeat steps 12 and 13 for the remaining three additions.

Cleanup

15. Repeat step 19 of Procedure I.

J. *Fundamental Equations*

$$R_\theta = k_w(n_s/n_w)^2 \times \text{(intensity ratio)} \qquad (7\text{-}18)$$

$$K = 2\pi^2 n^2 (dn/dc)^2 / N_0 \lambda^4 \qquad (7\text{-}16)$$

$$Kc/\Delta R_\theta = 1/\overline{M}_w P(\theta) + 2A_2 c \qquad (7\text{-}15)$$

$$1/P(\theta) = 1 + (16\pi^2/3\lambda_s^2)\overline{s_z^2} \sin^2(\theta/2) \qquad (7\text{-}17)$$

K. *Calculations*

1. Tabulate values of scattered intensity, corrected for the dark current, for each scan and $\theta = 30, \ldots, 150°$. Include values for the reference. Correct all readings for differences, if any, in amplifier gain or filters used. Tabulate as in Part A of Table 7-5.
2. Apply corrections for scattering volume and reference intensity, and subtract solvent readings, to prepare the counterparts of Parts B-D of Table 7-5.
(*Note*: These calculations assume that vertically polarized light is used. If unpolarized light is used, each reading must be divided by $1 + \cos^2\theta$. See also step M-1.)
3. Calculate ΔR_θ from suitable modification of Eq. 7-18, using $k_w = 0.419\ \text{cm}^{-1}$, $n_s = 1.513$ for toluene and $n_w = 1.340$. Complete Parts E and F of the table.
4. Compare the highest concentration used to the range of $\sin^2(\theta/2)$ covered, and select a value of k. Compute $\sin^2(\theta/2) + kc$ to complete Part G of the table.
5. Prepare the Zimm plot and read the intercept and the slopes of the $c = 0$ and $\theta = 0$ lines.
6. Calculate K from Eq. 7-16. Take $dn/dc = 0.111$ ml/g, and $\lambda = 436$ nm (see step M-2).
7. Calculate $\overline{M}_w$ from the Zimm-plot intercept using Eq. 7-15; A_2 from the slope of the $\theta = 0$ line using the same equation; and $(\overline{s_z^2})^{1/2}$ from the slope of the $c = 0$ line using Eq. 7-17.
8. As an alternative calculation procedure, carry out the multiple linear regression treatment of App. IV.

L. *Report*

1. Describe the experiment and apparatus in your own words.
2. Include the Zimm plot, the table of data, and the calculated values of $\overline{M}_w$, A_2, and $(\overline{s_z^2})^{1/2}$ with the appropriate units for each.
3. Answer the following questions:
   a. Is your Zimm plot rectilinear? If not, what conclusions can you draw from the nature of the distortions?
   b. If your polymer sample were branched instead of linear,

what changes would be required in procedure and results? What if it were a copolymer, or a polyelectrolyte?

M. Comments

1. McCaffery (1970) includes measurement of dn/dc in his Exp. 25. See also Lorimer 1972 *a,b*.

2. This procedure excludes measurement of and correction for depolarization. If it is desired to include this minor correction, see the instruction manual for the photometer used.

N. General References

Chap. 7B; *Textbook*, Chap. 3C; Billmeyer 1964; McIntyre 1964; Huglin 1972.

O. Bibliography

Bettelheim 1971. Frederick A. Bettelheim, *Experimental Physical Chemistry*, W. B. Saunders Co., Philadelphia, Pennsylvania, 1971.

Billmeyer 1964. Fred W. Billmeyer, Jr., "Principles of Light Scattering," Chap. 56 in I. M. Kolthoff and Philip J. Elving, eds., with the assistance of Ernest B. Sandell, *Treatise on Analytical Chemistry*, Part I, Vol. 5, Interscience Div., John Wiley and Sons, New York, 1964.

Brice 1950. B. A. Brice, M. Halwer, and R. Speiser, "Photoelectric Light-Scattering Photometer for Determining High Molecular Weights," *J. Opt. Soc. Am. 40*, 768-778 (1950).

Debye 1944. P. Debye, "Light Scattering in Solutions," *J. Applied Phys. 15*, 338-342 (1944).

Debye 1947. P. Debye, "Molecular-Weight Determination by Light Scattering," *J. Phys. & Coll. Chem. 51*, 18-32 (1947).

Huglin 1972. M. B. Huglin, ed., *Light Scattering from Polymer Solutions*, Academic Press, New York, 1972.

Lorimer 1972a. J. W. Lorimer, "Refractive Index Increments of Polymers in Solution: I. General Theory," *Polymer 13*, 46-51 (1972).

Lorimer 1972b. J. W. Lorimer and D. E. G. Jones, "Refractive Index Increments of Polymers in Solution: 2. Refractive Index Increments and Light Scattering in Polydisperse Systems of Low Molecular Weight," *Polymer 13*, 52-56 (1972).

McCaffery 1970. Edward M. McCaffery, *Laboratory Preparation for Macromolecular Chemistry*, McGraw-Hill Book Co., New York, 1970.

McIntyre 1964. D. McIntyre and F. Gornick, eds., *Light Scattering from Dilute Polymer Solutions*, Gordon and Breach Science Publishers, New York, 1964.

Wippler 1954. C. Wippler and G. Scheibling, "Description of an Apparatus for the Study of Light Scattering" (in French), *J. Chim. Phys. 51*, 201-205 (1954).

EXPERIMENT 14. PARTICLE SIZE BY LIGHT SCATTERING

A. *Introduction*

The angular dependence of light scattering is used in this experiment to determine the diameters of particles larger than the random-coil molecules considered in Exp. 13. The data are analyzed by the Mie theory, which is exact for isotropic spherical particles. The method is particularly suitable for determining the average particle sizes of latexes, such as those prepared by emulsion polymerization in Exps. 4 and 8.

In this experiment, unlike most of those presented, only broad outlines are suggested for apparatus and procedure, since these are described in more detail in Exp. 13. Such details as are missing can be supplied by the instructor or devised by the able student. The experiment should be capable of completion within 3 hours.

B. *Principle*

As discussed by Van de Hulst (1957) and Billmeyer (1964), the appropriate theory describing light scattering by particles can be selected from knowledge of the two variables $m = n/n_0$, the relative refractive index of the particle compared to the surrounding medium, and $\alpha = \pi d/\lambda_s$, which compares the diameter d of the particle to the wavelength of light in the medium, $\lambda_s = \lambda/n_0$. Restricting the discussion to the scattering of vertically polarized light for simplicity, the following cases are important.

If the particles are small, $\alpha < 0.1$, regardless of the value of m, the intensity of scattered light is independent of angle of observation and is described by Rayleigh's theory. This is the type of scattering exhibited by gases, and accounts for the blue color of the sky. If α lies between about 0.1 and about 1.0, and m is near unity, scattered-light intensity decreases monotonically with increasing angle. This is the Rayleigh-Gans-Debye scattering accounting for the dissymmetry measured in Exp. 13 and used, following the analysis of Chap. 7B, to evaluate the radius of gyration of random-coil polymers. Both of these theories, plus the laws of geometric optics applying to very large particles, are limiting cases of the Mie (1908) theory, which describes the intermediate case.

The Mie theory describes the scattering behavior of isotropic spheres of any size and relative refractive index. This behavior can be quite complex, showing many maxima and minima as a function of angle of observation. The Mie equations are correspondingly complex, and there are many ways to analyze Mie scattering (Kerker 1969). In this experiment we use the approximate analysis of Maron (1963, 1964) which is based on the Rayleigh-Gans-Debye relation for spheres of diameter d (Debye 1947)

$$P(\theta) = (3/u^3)(\sin u - u \cos u)^2 \qquad (14\text{-}1)$$

where $u = 2\alpha \sin(\theta/2) = (2\pi d/\lambda_s) \sin(\theta/2)$. The principle of Maron's method is to locate and identify successive maxima and minima in the scattered intensity as a function of angle. He shows that the ith minimum occurs at an angle θ_i given by

$$(d/\lambda_s)\sin(\theta_i/2) = k_i \tag{14-2a}$$

and the ith maximum at

$$(d/\lambda_s)\sin(\theta_i/2) = K_i \tag{14-2b}$$

where the k_i and K_i are related to m (for m less than 1.55) by

$$k_1 = 1.062 - 0.347\ m \tag{14-3}$$

and the ratios given in Table 14-1.

Since most light-scattering photometers do not measure continuously from $\theta = 0$, any minima and maxima occurring at angles below the lowest at which measurements can be made (often 25-30°) will not be observed. It is necessary to identify the proper value of i for minima and maxima observed at higher angles; this can be done by taking ratios of $\sin(\theta_i/2)$ for successive minima, successive maxima, or a consecutive minimum and maximum, and using the data in columns 4 to 6 of Table 14-1 to make the identification.

Since this method is not discussed in Chap. 7, the treatment of typical data is included in Sec. M.

TABLE 14-1. Ratios of parameters in Eq. 14-2

i	k_i/k_1	K_i/k_1	K_i/k_i	k_{i+1}/k_i	K_{i+1}/K_i
1	1.000	1.283	1.283	1.720	1.578
2	1.720	2.024	1.176	1.410	1.354
3	2.421	2.742	1.130	1.292	1.259
4	3.134	3.452	1.102	1.223	1.202
5	3.833	4.151	1.083	1.182	1.168
6	4.531	4.850	1.070	1.154	1.144
7	5.230	5.548	1.061	1.134	1.126

C. *Applicability*

This method is applicable to dilute suspensions or emulsions of essentially spherical particles, all having essentially the same size, suspended in a medium of similar refractive index. The particle size must be such that at least one minimum occurs in the observable range of angles. Depending on the value of m, this is roughly the range 150 nm to 1.5 μm. If the suspension is polydisperse, the average particle sizes calculated from different maxima and minima may not agree well. It is not certain how deviations from spherical shape affect the results, but this is not a serious problem for the experiment as written. The suspension must be sufficiently dilute that once-scattered light is observed before it can be scattered again, and the measurements must be made with vertically polarized light (electric vector perpendicular to the plane containing the source, sample and detector).

D. *Precision and Accuracy*

For a nearly monodisperse sample, the diameters calculated for different maxima and minima, and for observations at different wavelengths, can be expected to agree to well within 5%. This can be taken as a representative value for the precision of the method. If agreement with diameters observed by electron microscopy is taken as a measure of accuracy, this may be much poorer if the particles change size on drying for observation in the microscope, or similar to the precision if they do not.

E. *Safety Considerations*

OBSERVE THE USUAL PRECAUTIONS IN HANDLING ORGANIC SOLVENTS. OBSERVE CAUTION WITH RESPECT TO THE HIGH VOLTAGE POWERING THE PHOTOMULTIPLIER TUBE. NEVER ALLOW THE PHOTOTUBE TO BE EXPOSED TO BRIGHT LIGHT WHEN THE HIGH VOLTAGE IS ON. SAFETY GLASSES MUST BE WORN IN THE LABORATORY AT ALL TIMES.

F. *Apparatus*

1. Light-scattering photometer capable of measuring scattered intensity versus angle, using vertically polarized light. Absolute calibration through the use of a working standard, as in Exp. 13, is not required.
2. Equipment for filtration of solutions, as specified in Exp. 13.

G. *Reagents and Materials*

1. Polystyrene latex prepared in Exp. 4 or 8, or other suitable emulsion or suspension.
2. Distilled water, or other solvent, for dilution.

H. *Preparation*

I. *Procedure*

1. Refer to Exp. 13 for details of the instrument check and operating procedure for the light-scattering photometer used.

2. Dilute the sample with distilled water to a concentration in the range 0.005 - 0.0005 g/ℓ, depending on anticipated particle size. The larger particles require the more dilute solutions.

3. Filter the solution as described in Exp. 13 to remove dust, agglomerates, and other extraneous material.

4. Using vertically polarized incident light, measure relative scattered intensity at 2° intervals over the useful range of the instrument. Take measurements for both blue (λ = 435.6 nm) and green (λ = 546.1 nm) light. Be sure to note and correct for the use of any range switches or neutral filters required to keep the scattered-light readings on scale.

J. *Fundamental Equations*

$$(d/\lambda_s)\sin(\theta_i/2) = k_i \quad \text{for the } i\text{th minimum} \qquad (14\text{-}2a)$$

$$(d/\lambda_s)\sin(\theta_i/2) = K_i \quad \text{for the } i\text{th maximum} \qquad (14\text{-}2b)$$

$$k_1 = 1.062 - 0.347\, m \qquad (14\text{-}3)$$

K. *Calculations*

1. Tabulate the data of step I-4, and correct each value for range or filter (if required), and for observed volume by multiplying the scattered intensity by sin θ. (See Table 7-5, part B, for the corresponding correction to the data of Exp. 13.)

2. Prepare a plot of corrected scattered intensity versus angle for each set of data obtained in step I-4. (*Note*: It may be convenient to plot these results on semilogarithmic paper as the scattered intensity can vary by several orders of magnitude for large particles.)

3. Read and tabulate the locations θ_i of all observable minima and maxima. If the highest accuracy is desirable, repeat measurements at smaller increments of angle in the neighborhood of minima and maxima.

4. If there is doubt as to whether any minima or maxima occurred in the range of small angles not observed, calculate ratios of $\sin(\theta_i/2)$ for successive minima, successive maxima, and consecutive minima and maxima, and use the procedure of Sec. B and Table 14-1 to identify the values of i for all observed minima and maxima.

5. Calculate k_1, and the k_i and K_i for all observed minima and maxima. For a polystyrene latex in water, take m = 1.20.

6. Calculate d for each minimum and maximum observed with each incident-light wavelength, using Eq. 14-2. Recall that $\lambda_s = \lambda/n_0$, and for water take n_0 = 1.340 for λ = 435.6 nm and n_0 = 1.334 for λ = 546.1 nm. Average the values of d so obtained. (See step M-2.)

L. *Report*

1. Describe the apparatus and experiment in your own words.
2. Include the graphs obtained in step K-1, and a table containing the data of steps K-2 to K-4.
3. Justify your assignment of values of i.
4. Comment on the range of values of d obtained. Can you assign any differences to experimental error or to polydispersity of the sample?
5. If a sample was measured for which average values of d from electron microscopy are available, compare these with your value of d and comment on any differences.

M. *Comments*

1. *Sample Calculation.* Figure 14-1 shows the light-scattering data for a poly(vinyl chloride) latex, determined at λ = 436 nm and plotted according to Sec. I and step K-2. The latex particles are relatively large in diameter, and it is not certain that the first observed minimum is indeed that with i = 1. The data are analyzed by Maron's method in Table 14-2.

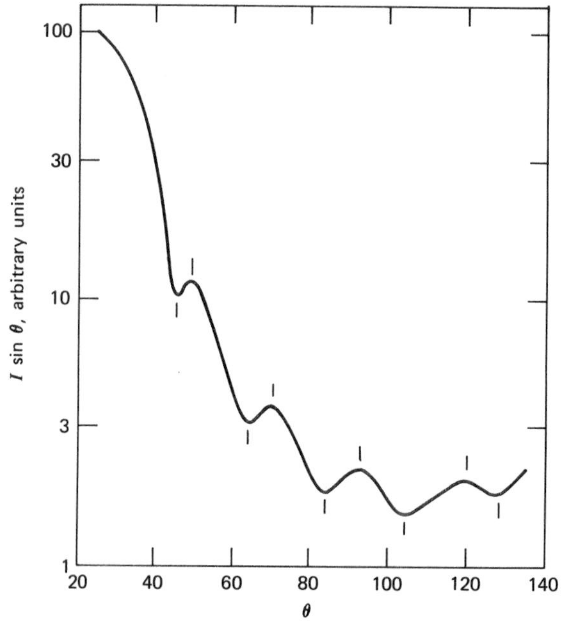

Fig. 14-1. Angular light-scattering data at 436 nm for a poly(vinyl chloride) latex, showing Mie-theory maxima and minima.

TABLE 14-2. Analysis of angular light-scattering data for a poly(vinyl chloride) latex to identify the order i of maxima and minima

Min or Max	θ	sin(θ/2)	k_{i+1}/k_i obs.	k_{i+1}/k_i calc.	K_{i+1}/K_i obs.	K_{i+1}/K_i calc.	K_i/k_i obs.	K_i/k_i calc.	i
A	45	0.383					1.105	1.176	2
A'	50	0.423	1.355	1.410					2
B	64	0.520			*1.355*	*1.354*	1.102	1.130	3
B'	70	0.574	1.272	1.292					3
C	83	0.663			*1.255*	*1.259*	1.185	1.102	4
C'	92	0.720	1.223	1.188					4
D	104	0.788			*1.202*	*1.202*	1.098	1.083	5
D'	120	0.866	1.140	1.182					5
E	128	0.900							6

Inspection of ratios of the k's and K's in comparison with Table 14-1 shows clear-cut agreement for K_{i+1}/K_i, as indicated in italics in Table 14.2. It is found generally that this ratio is most reliable for assigning values of i. After the assignment of i was made on this basis, the calculated values of other ratios were filled in, and no major discrepancies appeared. The identification was considered satisfactory, and values of d were calculated by Eq. 14-2, as follows.

For the poly(vinyl chloride) latex, m = 1.159, whence by Eq. 14-3,

$$k_1 = 1.062 - (0.347 \times 1.159) = 0.660$$

Consider the first maximum, at 50°, for which i = 2. From Table 14-1, K_2/k_1 = 2.024, whence K_2 = 1.34. Then by Eq. 14-2*b*

$$d = K_2\lambda_s/\sin(\theta_2/2) = 1.34 \times (436/1.340)/\sin 25°$$

$$d = 1030 \text{ nm.}$$

N. General References

Chap. 7B; *Textbook*, Chap. 3C; Maron 1963, 1964; Billmeyer 1964; Kerker 1969.

O. Bibliography

Billmeyer 1964. Fred W. Billmeyer, Jr., "Principles of Light Scattering," Chap. 56 in I. M. Kolthoff and P. J. Elving, eds., with the assistance of Ernest B. Sandell, *Treatise on Analytical Chemistry,* Part I, Vol. 5, Interscience Div., John Wiley and Sons, New York, 1964.

Debye 1947. P. Debye, "Molecular-Weight Determination by Light Scattering," *J. Phys. & Coll. Chem. 51*, 18-32 (1947).

Kerker 1969. Milton Kerker, *The Scattering of Light and Other Electromagnetic Radiation,* Academic Press, New York, 1969.

Maron 1963. S. H. Maron and M. E. Elder, "Determination of Latex Particle Size by Light Scattering. I. Minimum Intensity Method," *J. Colloid Sci. 18,* 107-118 (1963).

Maron 1964. S. H. Maron, P. E. Pierce, and M. E. Elder, "Determination of Latex Particle Size by Light Scattering. VI. Minima and Maxima in the Angular Dependence of the Polarization Ratio," *J. Colloid Sci. 19,* 591-601 (1964).

Mie 1908. G. Mie, "Optics of Turbid Media" (in German), *Ann. Physik 25,* 377-445 (1908).

Van de Hulst 1957. H. C. van de Hulst, *Light Scattering by Small Particles,* John Wiley and Sons, New York, 1957.

EXPERIMENT 15. DILUTE-SOLUTION VISCOSITY

A. Introduction

This experiment covers the determination of the dilute-solution viscosity of polymers. Two procedures are included: In Procedure I, the reduced and inherent viscosities are determined at a single solution concentration using an Ostwald-Fenske viscometer; and in Procedure II, the intrinsic viscosity is determined from measurements at several solution concentrations using a Ubbelohde dilution viscometer. Results obtained by Procedure I are used in Exp. 3. Procedure I requires approximately 1 hour, and Procedure II 2 to 3 hours.

B. Principle

As discussed in Chap. 7C and the *Textbook*, Chap. 3D, the viscous drag created by the presence of random-coil polymers in a flowing solvent is a measure of the size, not the mass, of the polymer molecules. Measurement of the dilute-solution viscosity of polymer solutions provides one of the most easily obtained and widely used items of information about the molecular structure of the samples. The various quantities measured and calculated are defined in Table 7-6; in this experiment we use common names and g/100 ml (g/dl) as the unit of concentration. Since the dilute-solution viscosity of a given sample depends upon the type of solvent used and the temperature of the measurement, it is necessary that both these be specified (and the concentration also unless the intrinsic viscosity is calculated) for all measurements.

Provided that it is known that the polymer is linear rather than branched, empirical correlations can be developed between the intrinsic viscosity and the molecular weight of the sample. Such correlations are widely used, for example in the treatment of data obtained in Exp. 3. They are discussed in Chap. 7C, but are not treated further in this experiment.

The use of two different viscometers is covered here. One, the Ostwald-Fenske viscometer used in Procedure I, is a constant-volume device, while the other, the Ubbelohde viscometer utilized in Procedure II, operates independent of the total volume of solution over a considerable range. It is therefore useful as a dilution viscometer, in which solutions having several different concentrations can be prepared and measured *in situ*. This facility results from its construction in which the solution emerging from the lower end of the capillary flows down the walls of bulb A (Fig. 7-15) in a manner which is independent of the liquid level in the main reservoir B. This mode of action is referred to as a suspended level.

Both Pinner (1961) in Exp. C.1.1 and McCaffery (1970) in Exp. 2 describe the measurement of dilute-solution viscosity.

C. *Applicability*

The dilute-solution viscosity measurement is applicable to all polymers which dissolve to give stable solutions at a temperature between ambient and about 150°C. Special techniques, beyond the scope of this experiment, are required in the measurement and data treatment for polyelectrolytes, or for polymers of sufficiently high molecular weight that their solution viscosity depends upon the rate of shear in the viscometer.

D. *Precision and Accuracy*

For most polymer-solvent systems, the reduced, inherent, and intrinsic viscosities can be determined to well within ±0.01 dl/g. Since these quantities are not obtainable by any other means, accuracy of their values has no meaning.

E. *Safety Considerations*

CARE MUST BE TAKEN IN THE USE OF ORGANIC SOLVENTS SUCH AS CHLOROFORM AND TOLUENE, BOTH OF WHICH ARE HIGHLY TOXIC; TOLUENE IS ALSO FLAMMABLE. USE THESE AND OTHER SOLVENTS ONLY IN SMALL QUANTITIES IN WELL VENTILATED AREAS. AVOID THE USE OF FLAMES OR SOURCES OF ELECTRIC SPARKS. SAFETY GLASSES MUST BE WORN IN THE LABORATORY AT ALL TIMES.

F. *Apparatus*

1. Constant-temperature bath, capable of maintaining ±0.01°C at

25.0°C.

2. Viscometers, Ostwald-Fenske for Procedure I or Ubbelohde for Procedure II, with efflux times greater than 100 sec for the solvent.
3. Timer, graduated in 0.1 sec or less.
4. 25-ml volumetric flasks, stoppered.
5. Pipettes as required.
6. Source of filtered dry nitrogen at low pressure.
7. Fine fritted glass or 1.0 μm "Millipore" filter in suitable holder.

G. *Reagents and Materials*

Procedure I

1. Poly(methyl methacrylate) samples prepared in Exp. 3, or other suitable samples.
2. Chloroform, reagent grade.

Procedure II

1. Polystyrene, prepared in Exp. 4 or 9, or other suitable samples.
2. Toluene, reagent grade.

H. *Preparation*

1. Prepare ahead of time solutions of poly(methyl methacrylate) in chloroform at a concentration of 0.5 g/dl, or of polystyrene in toluene at a concentration of 1.0 g/dl. Use 25-ml volumetric flasks, and follow the directions given in Exps. 11-13.

I. *Procedure*

Procedure I. Ostwald-Fenske Viscometer

1. Rinse the viscometer with solvent and let it drain. Then place the viscometer in the constant-temperature bath, securely fastened and with the upright tubes exactly vertical. See that the bath is regulating properly at the desired temperature.
2. With a pipette or syringe, transfer exactly 10.0 ml filtered solvent to the viscometer. (*Note*: If a "Millipore" filter holder attaching to a syringe (Fig. 7-8) is available, it may be used conveniently for this step.)
3. After temperature equilibration has been achieved (a minimum of 10 min), bring the liquid level in the viscometer above the upper graduation mark using a stream of low-pressure filtered nitrogen applied to the arm opposite the capillary. Allow the liquid to drain down the capillary. Start the timer exactly as the meniscus passes the upper graduation mark, and stop it exactly as the meniscus passes the lower mark.
4. Determine the efflux time at least three times. The readings should agree within 0.1 sec or 1.0% of their mean, whichever is larger.

If they do not, repeat and apply the above test until three satisfactory readings or a total of six readings are obtained. Accept the mean of three readings agreeing as indicated above; if they cannot be obtained, it is likely that the variation results from foreign material in the capillary or inadequate temperature control. Locate and remedy before proceeding.

5. Remove the viscometer from the bath, empty it and allow it to drain, then blow it dry with the stream of filtered nitrogen.

6. Repeat steps 2-4 with each polymer solution to be measured.

7. After each solution has been measured, remove the viscometer from the bath, empty it, rinse it several times with filtered solvent, including the capillary and upper bulbs, and blow it dry with the stream of filtered nitrogen.

Procedure II. Ubbelohde Viscometer (Fig. 7-15b)

1. Rinse the viscometer with solvent and let it drain. Then place the viscometer in the constant-temperature bath, securely fastened and with the upright tubes exactly vertical. See that the bath is regulating properly at the desired temperature.

2. With a pipette or syringe, transfer exactly 10.0 ml filtered solvent to the viscometer. *Note*: If a "Millipore" filter holder attaching to a syringe (Fig. 7-8) is available, it may be used conveniently for this step.

3. After temperature equilibration has been achieved (a minimum of 10 min), bring the liquid level in the viscometer above the upper graduation mark of bulb C as follows: Close tube 3 with one finger, and apply a low pressure of filtered nitrogen to tube 1. When the desired liquid level (in bulb D) is reached, open both tube 1 and tube 3. Bulb A will now drain, establishing the "suspended level" at the bottom of the capillary. Allow the liquid to drain down the capillary. Start the timer exactly as the meniscus passes the upper graduation mark, and stop it exactly as the meniscus passes the lower mark.

4. Determine the efflux time at least three times. The readings should agree within 0.1 sec or 0.1% of their mean, whichever is larger. If they do not, repeat and apply the above test until three satisfactory readings or a total of six readings are obtained. Accept the mean of three readings agreeing as indicated above; if they cannot be obtained, it is likely that the variation results from foreign material in the capillary or inadequate temperature control. Locate and remedy before proceeding.

5. Add exactly 5.0 ml filtered polymer solution to bulb B through tube 1. Mix the solution well by closing tube B and applying nitrogen pressure alternately through tubes 1 and 3.

6. Repeat steps 2-4. Add another aliquot of solution as in step 5, and repeat until at least four solutions have been measured. (See step M-1.)

7. Remove the viscometer from the bath, empty it, rinse all parts of it (including capillary and bulbs C and D) with filtered solvent. Drain the viscometer and blow it dry with the stream of filtered nitrogen.

J. *Fundamental Equations*

$$\eta_r = t/t_0 \qquad (15\text{-}1)$$

$$\eta_{sp} = (t - t_0)/t_0 \qquad (15\text{-}2)$$

$$\eta_{inh} = (\ln \eta_r)/c \qquad (15\text{-}3)$$

$$\eta_{sp}/c = [\eta] + k' \, [\eta]^2 \, c \qquad (7\text{-}23)$$

$$(\ln \eta_r)/c = [\eta] + k'' \, [\eta]^2 \, c \qquad (7\text{-}24)$$

K. *Calculations*

Procedure I

1. Calculate η_r by Eq. 15-1 and η_{inh} by Eq. 15-3.

Procedure II

1. Calculate η_r by Eq. 15-1, η_{sp} by Eq. 15-2, and η_{inh} by Eq. 15-3.
2. Plot, on the same piece of graph paper, η_{sp}/c versus c and η_{inh} versus c. Read $[\eta]$ as the common intercept at $c = 0$ of the best straight lines through the two sets of points.
3. Calculate the Huggins constant k' from Eq. 7-23 and the Kraemer constant k'' from Eq. 7-24. As an additional check on the accuracy of your work, check to see that $k' - k'' \simeq 0.5$.

L. *Report*

1. Describe the apparatus and experiment in your own words.
2. If Procedure I was used, report η_{inh} and the solution concentration, solvent, and temperature of measurement, for each sample.
3. If Procedure II was used, include a table of values of c, η_r, η_{sp}, η_{sp}/c, and η_{inh}, and the graph prepared in step K-2. Report $[\eta]$, k', k'', the solvent, and the temperature of measurement.

M. *Comments*

1. Several variations are possible in the schedule of concentrations to be measured in steps 5 and 6 of Procedure II. For example, instead of adding four aliquots of 5 ml each to 10 ml solvent, the following schedules are suggested:

a. Add 10 ml stock solution to 10 ml solvent, then successively dilute with further additions of solvent.

b. Start with stock solution, and successively dilute with solvent. Determine the efflux time of the solvent separately.

N. General References

Chap. 7C; *Textbook*, Chap. 3D; Lyons 1967; Moore 1967; ASTM D 2857.

O. Bibliography

ASTM D 2857. *Standard Method of Test for Dilute-Solution Viscosity of Polymers*. ASTM Designation: D 2857. American Society for Testing and Materials, Philadelphia, Pennsylvania 19103.

Lyons 1967. John W. Lyons, "Measurement of Viscosity," Chap. 83 in I. M. Kolthoff and Philip J. Elving, eds., with the assistance of Ernest B. Sandell, *Treatise on Analytical Chemistry*, Part I, Vol. 7, Interscience Div., John Wiley and Sons, New York, 1967.

McCaffery 1970. Edward M. McCaffery, *Laboratory Preparation for Macromolecular Chemistry*, McGraw-Hill Book Co., New York, 1970.

Moore 1967. W. R. Moore, "Viscosities of Dilute Polymer Solutions," Chap. 1 in A. D. Jenkins, ed., *Progress in Polymer Science*, Vol. 1, Pergamon Press, New York, 1967.

Pinner 1961. S. H. Pinner, *A Practical Course in Polymer Chemistry*, Pergamon Press, New York, 1961.

EXPERIMENT 16. GEL PERMEATION CHROMATOGRAPHY

A. Introduction

Gel permeation chromatography (GPC) is used in this experiment to determine the molecular-weight distribution of a polystyrene, using toluene as the solvent. Although GPC is a separation technique based on differences in molecular size, use is made of the one-to-one relation between size and mass for linear polymers of a single chemical type in making this determination. This experiment requires a full 3-hour laboratory period.

B. Principle

Gel permeation chromatography is a liquid-liquid chromatographic separation in which columns are packed with porous "gel" particles, the pore sizes being of the same order of magnitude as the sizes of dissolved polymer molecules. The gels commonly used in GPC consist of poly(styrene-*co*-divinyl benzene) or glass; in this experiment the former is used. The mobile phase, toluene, is pumped through the columns and subsequently through a differential refractometer detector.

The stationary phase is the solvent inside the porous gel particles. When a solution of polystyrene in toluene is injected as a sample, the polymer molecules tend to diffuse from the mobile phase into the stationary phase, to the extent that the molecules are small enough to be accommodated within the gel, reducing the concentration gradient of polystyrene between the two phases. The larger the solute molecule, the less of the stationary phase is accessible to it, and the less time it remains in the stationary phase as it traverses the column. Thus the largest polymer molecules have the smallest retention volumes, and conversely.

It is customary to measure the retention volumes of a series of narrow-distribution polymers, conveniently anionic polystyrenes, to provide a calibration curve of retention volume versus molecular weight. If polystyrene is used as the sample, as in this experiment, the calibration curve so obtained applies directly; otherwise, account must be taken of the fact that the separation is based upon molecular size rather than mass, as is discussed in Chap. 7D.

McCaffery (1970) includes a GPC experiment as his Exp. 28.

C. *Applicability*

GPC is applicable to a wide variety of species, both low-molecular-weight and polymeric, dissolved in many different solvents of varying polarity. The selection of gel type, gel pore size, solvent, and temperature (ambient up to 150°C) must be made appropriately for each solute, and certain limitations are imposed by the need to avoid adsorption and to match polarities reasonably well. Changes from one solvent to another are not easily made with existing apparatus.

The direct interpretation of GPC results in terms of the molecular-weight distribution is applicable only to linear homopolymers where the calibrating materials and test samples are of the same chemical type. In other situations, the primary interpretation must be made in terms of the molecular-size distribution, and this may or may not be directly relatable to the molecular-weight distribution, as discussed in the text and references.

Use of the differential refractometer detector is applicable to all polymers having refractive indices different from that of the solvent. A correction must be made if the polymer refractive index depends upon molecular size, as for example at very low molecular weights. Other detectors, including infrared, ultraviolet, and flame ionization types, can be used.

D. *Precision and Accuracy*

Repeat chromatograms in GPC usually agree within 1-2%, but the chromatogram is sensitive to such experimental details as the resolution of the columns used, flow rate, range of porosities of the column packings, etc. The extent of absolute agreement from one laboratory to another is uncertain and variable at this writing. Under favorable circumstances, however, GPC results agree with those from other methods to well within experimental error (see, for example, the *Textbook*, Figs. 2-16 and 2-20).

E. Safety Considerations

CARE MUST ALWAYS BE TAKEN TO USE ORGANIC SOLVENTS IN WELL-VENTILATED AREAS AND IN SMALL QUANTITIES. AVOID BREATHING FUMES OR EXCESSIVE CONTACT WITH SKIN OR EYES. AVOID THE USE OF FLAMES OR SOURCES OF ELECTRIC SPARKS. THE GPC INSTRUMENT SHOULD BE VENTED TO A FUME HOOD OR OTHER SAFE EXHAUST SYSTEM. SAFETY GLASSES MUST BE WORN IN THE LABORATORY AT ALL TIMES.

F. Apparatus

1. Gel permeation chromatograph, with 3 "Styragel" columns suitable for analysis of high-molecular-weight polymers, and set up for operation with toluene as the solvent.

(*Note*: The procedure is written for the Waters Model 100 or 200 chromatograph, probably the most widely used instrument at this time. Use of other instruments will require slight modifications in the operating procedure.)

2. 25-ml and 10-ml stoppered volumetric flasks.
3. Fine fritted glass or 1.0 μm "Millipore" filter in suitable holder.

G. Reagents and Materials

1. Polystyrene, polydisperse, prepared in Exp. 4 or commercial material, or narrow distribution, prepared in Exp. 9.
2. Polystyrenes, anionic, narrow distribution, several with $\bar{M}_w$ covering the range of 10,000 to 1,000,000.
3. Toluene, drawn from the chromatograph at time of use.
4. *o*-Dichlorobenzene, reagent grade.

H. Preparation

Prepare ahead of time in volumetric flasks solutions of polystyrene and *o*-dichlorobenzene in toluene at accurately known concentrations approximating 1.2 g/ℓ for polydisperse samples and 0.8 g/ℓ for narrow-distribution samples, as follows:

1. Tare the clean, dry flasks.
2. Accurately weight approximately 30 mg of *o*-dichlorobenzene and of finely divided polydisperse polystyrene into 25-ml volumetric flasks, or 8 mg of the narrow-distribution polymers into 10-ml flasks.
3. Fill the flasks 2/3 with toluene freshly drawn from the SOLVENT DRAWOFF valve of the chromatograph. Stopper the flasks and agitate gently; do not shake, as particles of polymer may adhere to the walls above the liquid level and fail to dissolve.
4. Repeat agitation from time to time until polymer is completely dissolved. Allow 24 hours.
5. Maintain the flasks at 25°C and fill to the mark with solvent freshly drawn from the chromatograph. Mix well.
6. Filter the solutions before use through a fine fritted-glass filter or a 1-μm "Millipore" filter.

I. *Procedure*

Instrument Setup

1. Check to see that the MAIN POWER, PUMP, RECORDER AMPLIFIER, EXHAUST FAN, DEGASSER HEATER, REFRACTOMETER HEATER, AND REFRACTOMETER POWER switches are ON. The refractometer jacket temperature should be approximately 35°C, and the solvent degasser temperature, approximately 90°C. The HEATER switches for the SYPHON, SAMPLE INLET, and OVEN should be OFF. The SAMPLE COLUMN, REFERENCE COLUMN, and PUMP SUCTION valves should be open, and all other valves closed. The INJECTION valve should be turned to RUN, and the recorder SPAN SELECTOR set to 2X.
2. Check the flow rate by timing between successive "count" marks on the recorder chart. These marks coincide with the dumping of the syphon at nominal 5-ml intervals of retention volume. The flow rate should be approximately 1 ml/min.
3. Observe that the base line is steady and free from drift.

Sample Injection

4. Inject the solution of polydisperse polystyrene, or of the sample prepared in Exp. 9, by the procedure of step 5. Number successive counts on the recorder chart, starting with 0 at the time of this injection.
5. a. Draw 40-50 ml solvent from the chromatograph into a beaker for rinsing purposes.
   b. Remove the syringe from the injection port, affix a clean needle to it, and rinse it once with solvent, discarding the solvent.
   c. Draw in 10 ml of the polymer solution. Remove the needle, carefully remove all air from the syringe and tip, and reinsert the syringe into the injection port.
   d. Inject all of the solution into the injection loop (CAUTION: Be sure that the injection valve is in the RUN position, *not* the INJECT position). Allow excess solution to run through the loop into a waste beaker. Leave the syringe in place.
   e. Just as the specified count is marked, turn the injection valve to INJECT. Leave it in this position for well over 2 min, to allow the full 2 ml of the loop to enter the sample stream, but less than 5 min. Then turn the valve back to RUN.
6. Fill the injection loop with the solution of *o*-dichlorobenzene ahead of time, and inject this solution at count no. 2. Use the procedure of step 5.
7. Separate the solutions of narrow-distribution calibration polystyrene into pairs such that values of $\overline{M}_w$ differ by factors of at least 10 between the members of a pair, e.g., nominal $\overline{M}_w$ of 10,000 paired with 100,000, etc. (*Note*: If a single solution remains, pair it with solvent.) Prior to injection, mix the members of a pair and inject the mixture. Inject the first such mixture, following the procedure of step 5, at count 11, the second at count 19, and others at 8-count intervals. (*Note*: Use of more than two pairs may extend the experiment beyond a 3-hr period.)
8. Observe the chromatogram of the polydisperse polystyrene as it appears, beginning around count 10. If the trace appears to be going

off scale, reset the recorder span to 1X. The sharp peak of *o*-dichlorobenzene will follow the polystyrene trace closely.

Shutdown

9. After all samples have eluted and the trace has returned to base line, flush the sample loop with solvent (but do not inject). Leave the syringe in place.

10. Turn the RECORDER AMPLIFIER, EXHAUST FAN, DEGASSER HEATER, REFRACTOMETER HEATER, and REFRACTOMETER POWER switches OFF, but *always* leave the MAIN POWER and PUMP switches ON.

J. *Fundamental Equations*

$$\overline{M}_n = \sum N_i M_i / \sum N_i \tag{7-1}$$

$$\overline{M}_w = \sum N_i M_i^2 / \sum N_i M_i \tag{7-14}$$

K. *Calculations*

1. Read the retention volume V_r in counts at the peak of the chromatograms for each of the narrow-distribution polystyrenes. Estimate these to the nearest tenth count.
2. Plot V_r versus log $\overline{M}_w$ for the above samples, and draw a smooth curve through the points. This is the calibration curve.
3. Read the recorder deflection above base line for the polydisperse polystyrene for at least 1-count and preferably 1/2-count intervals of V_r. Tabulate these data as in Table 7-9.
4. Fill in the molecular-weight column of your table by reference to the calibration curve.
5. Complete your table by calculating values of N_i and $N_i M_i^2$, and calculate $\overline{M}_n$ by Eq. 7-1, $\overline{M}_w$ by Eq. 7-14, and the polydispersity $\overline{M}_w/\overline{M}_n$.
6. Treat your data as in Table 7-10, and calculate and plot the cumulative and differential distribution curves, like those of Figs. 7-25 and 7-26.
7. Read from the chromatogram the retention volume at the peak and the width at the base of the curve for *o*-dichlorobenzene. Calculate the total plate count and the plates per foot of the system as indicated in Fig. 7-21. Three 4-ft columns are used. Compare the result to the value quoted in the text.

L. *Report*

1. Describe the experiment and apparatus in your own words.
2. Include all plots and tables, the chromatograms, and the calculated values of $\overline{M}_w$, $\overline{M}_n$, $\overline{M}_w/\overline{M}_n$, and the plates per foot.
3. Answer the following questions:
   a. Outline two different procedures for determining the

molecular-weight distribution of a chemically different polymer using a polystyrene calibration. Comment on their relative merits.

b. How would you prepare a calibration curve if you were operating in a solvent in which polystyrene is not soluble, and had no narrow-distribution samples available which were soluble in that solvent?

c. Comment on the resolution of the chromatograph (1) in relation to plate count and (2) in relation to separation between the members of a pair of narrow-distribution samples.

d. What changes in procedure or interpretation would be required in order to study in this experiment (1) poly(styrene-*co*-methyl methacrylate), and (2) branched polystyrene? What kind of information could you obtain?

M. Comments

1. Plate count with poly(styrene-*co*-divinyl benzene) gels appears to be determined mainly by mobile-phase effects. Among these, the more important are a velocity-profile effect determined by the ratio of gel particle size to column diameter, and an eddy-diffusion effect. Molecular diffusion is not as important in liquid as in gas chromatography in determining resolution. See Kelley 1970. The plate count is found to improve with decreasing flow rate (at the expense of longer analysis time) and with decreasing gel particle size (at the expense of higher pressure drops across the columns).
2. If the viscosity of the polymer solution in the column is too high compared to that of the solvent, a shift in retention volume and distortion of the chromatogram can result. Thus it is necessary to work at quite low solution concentrations, and for very-high-molecular-weight samples, to measure at several concentrations and extrapolate to $c = 0$.
3. While almost any solvent will work in GPC provided that it wets but does not swell the packing, some precautions are necessary. Solvent selection is influenced by such factors as viscosity (preferred low to reduce pressure drop), stability, refractive index, and volatility.
4. Any impurity in one but not both of the sample and reference streams will grossly affect the refractometer reading. For this reason, solvent for sample preparation and all related purposes is taken directly from the chromatograph.
5. Air in the solvent streams can irreversibly damage the column packing, severely reducing plate counts. All operations should be carried out so as to prevent any air from being introduced into the system, for example by injection, sudden temperature change causing expansion or contraction, or stopping the pump.
6. Precise temperature control, especially in the refractometer, is essential for a steady base line.

N. General References

Chap. 7D; *Textbook*, Chap. 2F; Cazes 1966, 1970; Johnson 1970; McCaffery 1970, Exp. 28.

O. *Bibliography*

Cazes 1966. Jack Cazes, "Topics in Chemical Instrumentation. XXIX. Gel Permeation Chromatography," *J. Chem. Educ. 43*, "Part One" A567-A582, and "Part Two" A625-A642 (1966).

Cazes 1970. Jack Cazes, "Current Trends in Gel Permeation Chromatography 1970," *J. Chem. Educ. 47*, "Part One: Theory and Experiment" A461-A471 and "Part Two: Methodology" A505-A514 (1970).

Johnson 1970. J. F. Johnson and R. S. Porter, "Gel Permeation Chromatography," Chap. 4 in A. D. Jenkins, ed., *Progress in Polymer Science*, Vol. 2, Pergamon Press, New York, 1970.

Kelley 1970. Richard N. Kelley and Fred W. Billmeyer, Jr., "A Review of Peak Broadening in Gel Chromatography," *Sep. Sci. 5*, 291-316 (1970).

McCaffery 1970. Edward M. McCaffery, *Laboratory Preparation for Macromolecular Chemistry*, McGraw-Hill Book Co., New York, 1970.

EXPERIMENT 17. TURBIDIMETRIC TITRATION

A. *Introduction*

This experiment is designed to allow the determination of relative molecular-weight distribution by turbidimetric titration, in which the solubility of a polymer is continuously altered by changing solvent composition (constant-temperature procedure) or temperature (constant-volume procedure). Turbidity (scattered intensity or loss in transmittance) is used as a measure of the amount of precipitated polymer, and the word titration refers to the original technique of continuous addition of nonsolvent to a polymer solution.

In this experiment, unlike most of those presented, only broad outlines of the apparatus and procedure requirements are given. Details can be provided by the instructor, depending on the apparatus available, or can be devised by advanced students. Suggestions are given for both the constant-temperature and constant-volume procedures. With details of either procedure worked out, the experiment should be capable of completion within a three-hour laboratory period.

B. *Principle*

Turbidimetric titration is an analytical method of polymer fractionation based on two assumptions: 1. As solvent power is reduced, polymer of a specified molecular weight always precipitates at the same solvent composition and temperature; and 2. The turbidity of the mixture is a direct measure of the total amount of polymer precipitated. These remarks are stated for linear homopolymers; if a branched polymer, a copolymer, or a mixture of polymers is used, solubility is primarily determined by chemical composition rather than molecular weight, as discussed in the *Textbook*, Chap. 2A.

Neither assumption holds accurately. It is predicted by polymer solution thermodynamics and observed experimentally (as demonstrated in Fig. 7-29) that the point of precipitation of a given species varies depending upon the rest of the polymer distribution. The requirement of proportionality between weight of precipitated polymer and turbidity imposes the severe condition that the polymer precipitates in particles of constant size and refractive index, and that these variables remain fixed throughout the experiment. It is usually possible to find a polymer-solvent-nonsolvent system for which this is approximately true, but considerable searching may be required. Fortunately, Guzmán (1961) and Giesekus (1967) list such systems for several common polymers. In summary, turbidimetric titration is an approximate rather than an exact technique; its use is best restricted to the qualitative comparison of molecular-weight or compositional differences among rather similar polymers.

The turbidity τ of the solution can be measured as scattered-light intensity, or related to the transmittance of the solution I/I_0 by the usual Lambert's-law expression $I/I_0 = \exp(-\tau \ell)$ where ℓ is the path length of the cell. In the constant-temperature method, the turbidity must be corrected for the change in volume of the solution: If V_s is the original volume of solvent, and V_n the volume of nonsolvent added in the titration, and it is assumed that the two mix without change in volume, the corrected turbidity is inversely proportional to the volume fraction of solvent:

$$\tau_{corr} = \tau_{obs}(V_s + V_n)/V_s \tag{17-1}$$

The titration curve is obtained by plotting τ_{corr} as a function of the volume fraction of nonsolvent,

$$\gamma_n = V_n/(V_s + V_n) \tag{17-2}$$

Bettelheim (1971) includes as his Exp. 46 the determination of the cohesive-energy density of polymers by turbidimetric titration.

C. Applicability

Turbidimetric titration is applicable to any polymer for which a suitable solvent-nonsolvent pair can be found. The search for and selection of such a pair is beyond the scope of this book, so it is recommended that the experiments be limited to polymer-solvent-nonsolvent systems listed as suitable by Guzmán (1961) and Giesekus (1967).

In both the constant-temperature and constant-volume procedures, care must be taken that the polymer precipitates as a swollen amorphous rather than a crystalline phase, if information about the molecular-weight distribution is desired. Thus, the method is not universally applicable to crystalline polymers.

D. *Precision and Accuracy*

With proper experimental procedures and equipment, turbidity titration curves can be reproduced to within a few percent, or better. Accuracy, however, has no meaning since the curves are not obtained from other experiments.

E. *Safety Considerations*

THE SOLVENTS USED IN THIS EXPERIMENT ARE TOXIC AND FLAMMABLE. PROPER PRECAUTIONS MUST BE TAKEN INCLUDING ADEQUATE VENTILATION AND AVOIDANCE OF FLAMES OR SOURCES OF ELECTRICAL SPARKS. IF THE CONSTANT-VOLUME METHOD IS USED, CARE MUST BE TAKEN TO AVOID BURNS FROM HOT SOLVENT AND APPARATUS. SAFETY GLASSES MUST BE WORN IN THE LABORATORY AT ALL TIMES.

F. *Apparatus*

1. Photometer. Almost any light-scattering photometer or spectrophotometer can be used. If the titration cannot be carried out in the instrument, the constant-temperature procedure can still be applied by titrating outside and transferring the solutions to the instrument for measurement. For application of the constant-volume method, it is preferable to provide for controlled heating or cooling of the cell compartment (see step F-3). The photometer should fulfill the usual requirements for stability. It is advantageous to be able to measure scattering at several angles (45°, 90°, 135°) and transmittance at 0°, and to provide for automatic recording of the turbidity.

2. Titrator, constant-temperature procedure. A source of nonsolvent delivered at a constant but adjustable rate, for example an infusion pump. The titration cell should be equipped for efficient agitation at a constant but adjustable rate, and should be thermostatted for control of temperature.

3. Titration cell, constant-volume procedure. The solution cell should be equipped for efficient agitation at a constant but adjustable rate. Either the cell or the cell compartment of the photometer should be thermostatted to allow heating or cooling at a programmed rate.

G. *Reagents and Materials*

(*Note*: This section lists reagents and materials required for the analysis of polystyrene. Other polymers can be used, but the solvents required will differ.)

1. Polystyrene, prepared in Exp. 4 or 9, standard samples such as NBS 705 and NBS 706 (App. I), or commercial material.
2. Toluene, reagent grade (solvent for Procedure I).
3. Methanol, reagent grade (nonsolvent for Procedure I).
4. Methylcyclohexane, reagent grade (solvent for Procedure II).

H. Preparation

Prepare ahead of time solutions of 0.1% polystyrene in toluene (for Procedure I) or methylcyclohexane (for Procedure II). (*Note*: Methylcyclohexane is a solvent for polystyrene only at elevated temperature (Θ = 70.5°C). Prepare the solution at 90°C and maintain it above 80°C to prevent precipitation until the start of the experiment.)

I. Procedure

Procedure I. Constant-Temperature Titration

1. Adjust the cell thermostat to control temperature at 30 ± 0.5°C.
2. Place a measured amount of the solution of polystyrene in toluene in the cell. Use just enough so that the liquid level is above the light beam while the agitator is operating.
3. Start the agitator and adjust for efficient mixing without turbulence or bubbles.
4. Fill the infusion pump with methanol and clear the lines of air.
5. Set the photometer controls so that a low scattered intensity (or absorbance, if the transmittance is measured) is indicated. Correct for dark current if necessary.
6. When temperature equilibrium has been reached, start the infusion pump and record turbidity versus volume of methanol added.

Procedure II. Constant-Volume Cooling

1. Adjust the cell thermostat to hold temperature at approximately 100°C.
2. Fill the cell to above the level of the light beam with hot solution of polystyrene in methylcyclohexane.
3. Start the agitator and adjust for efficient mixing without turbulence or bubbles.
4. Set the photometer controls so that a low scattered intensity (or absorbance, if transmittance is measured) is indicated. Correct for dark current if necessary.
5. Start programming the temperature downward, and record turbidity as a function of temperature. (*Note*: It is also possible to run the experiment in reverse by cooling the solution until the maximum turbidity is reached, and then programming the temperature upward, recording the decrease in turbidity with increasing temperature.)

J. Fundamental Equations

Dilution Correction

$$\tau_{corr} = \tau_{obs}(V_s + V_n)/V_s \qquad (17\text{-}1)$$

Volume Fraction Nonsolvent

$$\gamma_n = V_n/(V_s + V_n) \tag{17-2}$$

K. *Calculations*

Procedure I

1. Calculate τ_{corr} from τ_{obs} by Eq. 17-1.
2. Calculate γ_n by Eq. 17-2.

L. *Report*

1. Describe the apparatus and experiment in your own words.
2. If Procedure I was used, plot τ_{corr} versus γ_n. If Procedure II was used, no calculations are necessary; plot τ_{obs} versus temperature.
3. Comment on the differences observed in the curves of step 2 for any different samples measured, in the light of what you know about the molecular weights and distributions of those samples from other experiments.
4. Include other observations pertinent to the experiment, such as the temperature at which precipitation began or the value of γ_n at the onset of precipitation.
5. If you devised your own procedure, include this as part of the report.

M. *Comments*

N. *General References*

Chap. 7D; *Textbook*, Chaps. 2A and 2D; Guzmán 1961; Giesekus 1967; Urwin 1972.

O. *Bibliography*

Bettelheim 1971. Frederick A. Bettelheim, *Experimental Physical Chemistry*, W. B. Saunders Co., Philadelphia, Pennsylvania, 1971.

Giesekus 1967. Hanswalter Giesekus, "Turbidimetric Titration," Chap. C.1 in Manfred J. R. Cantow, ed., *Polymer Fractionation*, Academic Press, New York, 1967.

Guzmán 1961. G. M. Guzmán, "Fractionation of High Polymers," pp. 113-183 in J. C. Robb and F. W. Peaker, eds., *Progress in High Polymers*, Vol. I, Academic Press, New York, 1961.

Urwin 1972. J. R. Urwin, "Molecular Weight Distributions by Turbidimetric Titration," Chap. 18 in M. B. Huglin, ed., *Light Scattering from Polymer Solutions*, Academic Press, New York, 1972.

C

Morphology and Thermal Property Experiments

EXPERIMENT 18. X-RAY DIFFRACTION

A. *Introduction*

In this experiment, wide-angle x-ray diffraction patterns are recorded photographically from a variety of materials, including a crystalline inorganic compound for calibration purposes, an amorphous polymer, an unoriented (partially) crystalline polymer, and the same polymer drawn into an oriented fiber. If a suitable camera is available, a small-angle pattern is recorded photographically for a similar fiber sample. This experiment requires 3 hours, excluding the time required to develop conventional photographic films.

B. *Principle*

A collimated beam of essentially monochromatic x-rays passes through the sample and is absorbed in a lead cup. Diffracted beams are recorded on photographic film (conventional or Polaroid). From an amorphous polymer, the diffraction pattern usually consists of one broad diffuse ring (the amorphous halo). The diameter of the ring yields, from Bragg's law, an estimate of the average nearest-neighbor spacing in the sample.

An unoriented crystalline polymer usually exhibits two or more sharp diffraction rings superimposed on the amorphous halo (Fig. 8-1); if the polymer is oriented as in a drawn fiber, only segments of the rings or spots, mostly on the equator (perpendicular to the fiber axis) are seen (Fig. 8-13). Application of Bragg's law yields the interplanar distances corresponding to these diffraction maxima, but the further interpretation of the data in terms of a crystal structure is difficult. The method of so doing is illustrated in this experiment for the well-known structure of polyethylene, as described in Secs. K and M-1.

With the equipment for which this experiment is written, the optimum exposure time for a small-angle diffraction pattern exceeds the 3-hour laboratory period. Within this time, however, it is possible to record at least an asymmetrical reflection pattern extending along the equator. The interpretation of these "equatorial tails", without support of other experiments, is not unequivocal. Longer exposures should yield meridional spots interpreted in terms of lamellar thickness or intercrystallite spacings.

Bettelheim (1971) includes an experiment (Exp. 48) on x-ray diffraction.

C. *Applicability*

The x-ray diffraction experiment is applicable to all crystalline and amorphous solids, and all liquids. However, x-ray scattering power depends on the number of electrons in the atom. For polymers, usually containing only light atoms, the information obtainable is limited by this fact; no information is obtained, for example, about the locations of hydrogen atoms.

D. *Precision and Accuracy*

The precision of interplanar spacings, as determined in this experiment, is about ±1%; use of standard but more time-consuming crystallographic techniques can give precision as good as ±0.01%. The accuracy of a structure deduced from such spacings depends on the number of reflections observed. For a semicrystalline polymer showing only 3 or 4 reflections, only a very rough estimate of the structure can be obtained, whereas the structure of a metal or inorganic salt, giving up to a thousand reflections, can be determined with very high reliability.

E. *Safety Considerations*

X-RADIATION IS HARMFUL TO LIVING TISSUE. LOW AND PROLONGED DOSES MAY HAVE CHRONIC CONSEQUENCES, WHILE SHORT EXPOSURE TO THE X-RAY BEAM CAUSES UNPLEASANT, SLOW-HEALING BURNS. *BE SURE* THAT THE BEAM SHUTTERS ARE ALWAYS CLOSED EXCEPT DURING MEASUREMENT, THAT THE CAMERA AREA IS WELL SHIELDED BY LEAD WHEN THE EXPOSURE IS IN PROGRESS, AND THAT ALL PERSONNEL STAY CLEAR WHEN THE BEAM IS ON. SAFETY GLASSES MUST BE WORN IN THE LABORATORY AT ALL TIMES.

Personnel badges to check cumulative exposure, required for all persons using the equipment for extended periods, are not needed by those present only for the duration of this experiment. The equipment must be checked periodically for stray radiation and overall level following prescribed safety regulations.

OBSERVE THE USUAL PRECAUTIONS REGARDING THE HIGH VOLTAGE APPLIED TO THE X-RAY TUBE, AND IN THE USE OF THE HOT BATH FOR FIBER PREPARATION. SODIUM FLUORIDE IS A DEADLY POISON IF INGESTED; OBSERVE FULL PRECAUTIONS.

F. Apparatus

1. General Electric XRD-3 Diffractometer (Fig. 8-12) or equivalent, equipped with water-cooled Coolidge x-ray tube with copper target, beryllium window, and nickel filter (see step M-2).
2. For wide-angle scattering, flat film camera with "Polaroid" back and intensifying screen, placed 50 mm from sample (Fig. 18-1).
3. For small-angle scattering, Statton camera (Fig. 8-12, Warhus).
4. For drawing fiber samples, oil bath regulated at 160°C.
5. Photographic developing facilities for film used with Statton camera.

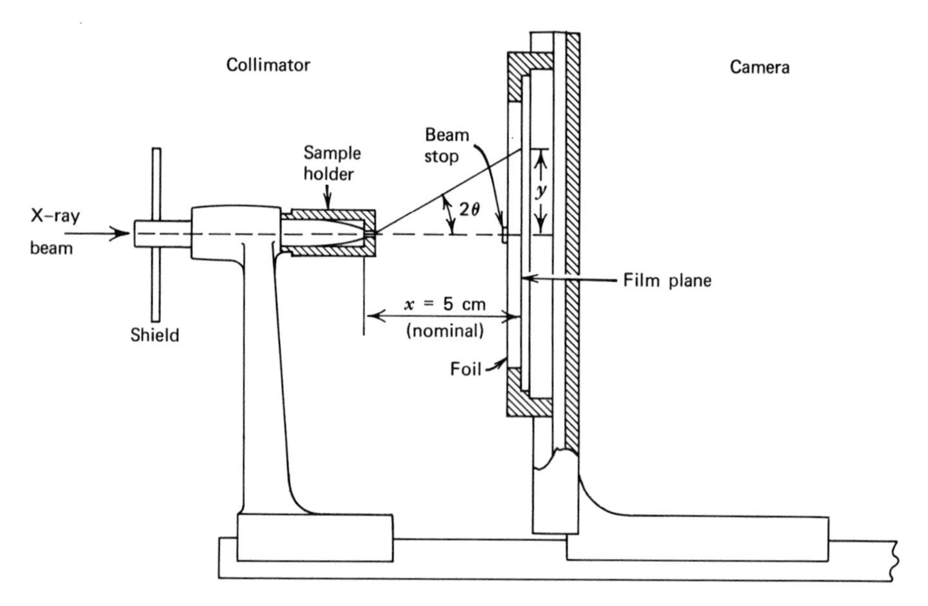

Fig. 18-1. Sketch of the specimen and camera arrangement for wide-angle x-ray photography with the General Electric XRD-3 diffractometer.

G. Reagents and Materials

1. Sodium fluoride, reagent grade (calibration standard).
2. Amorphous polymer, for example polystyrene or poly(methyl methacrylate).
3. Linear polyethylene (Marlex 50 or equivalent).
4. Polaroid Type 57 film, 4" × 5".
5. X-ray film, 3" × 4", Kodak Industrial type K or Ilford G.

H. Preparation

1. *Sample Mounting*

a. Pack powder samples into a hole 1-2 mm in diameter in a piece of metal or plastic about 1.5 mm thick. Mount this piece on the specimen holder with pressure-sensitive tape so that the x-ray beam passes through the hole.

b. Mount pieces of polymer, and fibers if stiff enough to stand alone, on the specimen holder (Fig. 18-1) with pressure-sensitive tape. Make samples at least 1.5 mm thick.

c. Mount thin fibers on a suitable frame (a paper clip is often satisfactory), then mount this on the specimen holder. Several fibers should be made into a bundle, taking care that they are parallel. Make the bundle at least 1.5 mm thick.

(*Note*: The Statton camera is provided with a specimen holder suitable for fibers as well as bulk samples.)

2. *Preparation of fibers*

(*Note*: Prepare fibers ahead of time if they are to be used for the small-angle experiment, or during the first part of the laboratory period if for the wide-angle experiment only.)

a. Place granules of linear polyethylene in a test tube and melt by immersing in the oil bath at 160°C for 30 min.

b. Draw fibers with tweezers as shown in Fig. 6-3.

I. Procedure

Startup

1. See that all shutters on the x-ray tube are closed and that the cameras are prepared for the experiments to be performed.
2. Energize the low-voltage circuits as follows:

a. Turn the TIMER switch OFF and the timer DAY-NIGHT switch to DAY.

b. Set the current and voltage controls to give 20 ma (C setting) and 50 kV.

c. Depress the LINE ON button and hold down for several sec to allow water pressure to build up. Allow at least 2 min before energizing the high-voltage circuit in step 7.

Wide-Angle Experiments

(*Note*: It is suggested that experiments be run in the following order with the indicated exposure times: sodium fluoride calibration, 30 min; amorphous polymer, 30 min; bulk polyethylene, 30 min; polyethylene fibers, 60-90 min. If the small-angle experiment is run simultaneously, skip to step 11.)

3. Set the desired exposure time on the timer and turn the TIMER ON-OFF switch to ON.
4. Put the specimen holder with sample in place on the collimator (Fig. 18-1).

5. Place the camera lever in the LOAD position and insert the Polaroid film fully into the camera. Lift the envelope back, and pull the EXPOSURE knob up.

6. Open the shutter on the x-ray tube.

7. Depress the X-RAY ON button and hold down for a few sec. This starts the high voltage and thus the x-ray beam, and the timer, if used. At the end of the exposure time set on the timer, the high voltage is turned off.

8. Close the x-ray tube shutter.

9. Remove and develop the Polaroid film as follows: Push the EXPOSURE knob down, push the envelope back down, and turn the lever to DEVELOP. Pull the film envelope out of the camera with a smooth uniform motion. After 15 sec open the envelope and coat the surface of the print.

10. Repeat steps 3-9 for the remaining samples.

Small-Angle Experiment

(*Note*: If the wide-angle and small-angle experiments are run simultaneously, the timer is not used. For the wide-angle experiment, follow steps 4-6 and 8-10, timing the exposure by clock.)

11. After steps 1 and 2 have been performed, load the Statton camera as follows:

a. Open the camera and insert the specimen holder.

b. Load the film cassette in a photographic darkroom, and insert it in the camera.

c. Replace the cover to the camera and turn on its vacuum pump.

12. Open the shutter on the x-ray tube.

13. Start the exposure by depressing the X-RAY ON button for a few sec.

14. At the end of the exposure period, turn off the x-ray beam by depressing the X-RAY OFF button.

15. Stop the vacuum pump, bleed air into the camera, and open the camera and remove the film cassette for development in a photographic darkroom. Develop Kodak Industrial film, type K, for 10 min. at 20°C in Kodak Rapid developer, or Ilford G film for 8 min. at 20°C in Ilford ID 42 developer. Fix in the usual manner, wash, and dry.

15. After all experiments are completed, shut the apparatus down by depressing the LINE OFF button.

J. *Fundamental Equations*

$$n \lambda = 2 d \sin \theta \tag{8-1}$$

(For this experiment, $n = 1$)

$$\log d_{hk0} = \log a - 1/2 \log[h^2 + k^2(a/b)^2] \tag{18-1}$$

where h and k are the first two of the Miller indices hkl.

K. *Calculations*

1. Calculate the exact sample-film distance for the wide-angle camera from the calibration sample, NaF. For example, note that y/x (Fig. 18-1) = tan 2θ, and search for the (200) reflections of NaF, for which d = 2.319 A, assuming $x \simeq 50$ mm. Take λ = 1.541 A, use Eq. 8-1 to find sin 2θ, look up tan 2θ, and note that the diameter of the (200) ring is $2y$.
2. Calculate the average nearest-neighbor distance for the amorphous polymer sample.
3. Calculate values of d for all observed equatorial reflections for the bulk polyethylene sample and the polyethylene fiber. (*Note*: If tables of interplanar distances are available as a standard accessory to the instrument, only 2θ need be calculated; then d may be looked up in the tables.)
4. Using the Hull-Davey chart, Fig. 18-2, as described in Sec. M-1, determine the Miller indices of the reflections, the ratio b/a, and the corresponding ratio a/b. (As an aid to finding the fit, one of these ratios is marked on the chart.)
5. Using Eq. 18-1 and the values of a/b or b/a and of log d from step 4, determine a and b. Assign the proper numerical values for a and b (not the reverse pair), and compare your answers to accepted values, by reference to Fig. 8-4.

L. *Report*

1. Describe the apparatus and experiment in your own words.
2. Include all Polaroid prints and the small-angle scattering pattern, if obtained.
3. Give the sample-film distance obtained from the calibration experiment, and the nearest-neighbor distance for the amorphous polymer.
4. Tabulate Miller indices, 2θ, and d for all identified equatorial reflections from polyethylene.
5. Give your values of a and b, and discuss them in relation to literature values.

M. *Comments*

1. A great problem in determining atomic arrangements from x-ray diffraction results lies in the identification of the reflections. In general, these can result from sets of planes with any combination of Miller indices hkl. For polymers, reflections with any index greater than 4 or 5 are seldom observed, however, and the use of oriented samples allows further simplifications, since reflections with related indices are grouped regularly on the film.

The procedure for identification can be illustrated using as an example the well-known crystal structure of polyethylene (Bunn 1939). We assume that it is known that the unit cell is orthorhombic. With

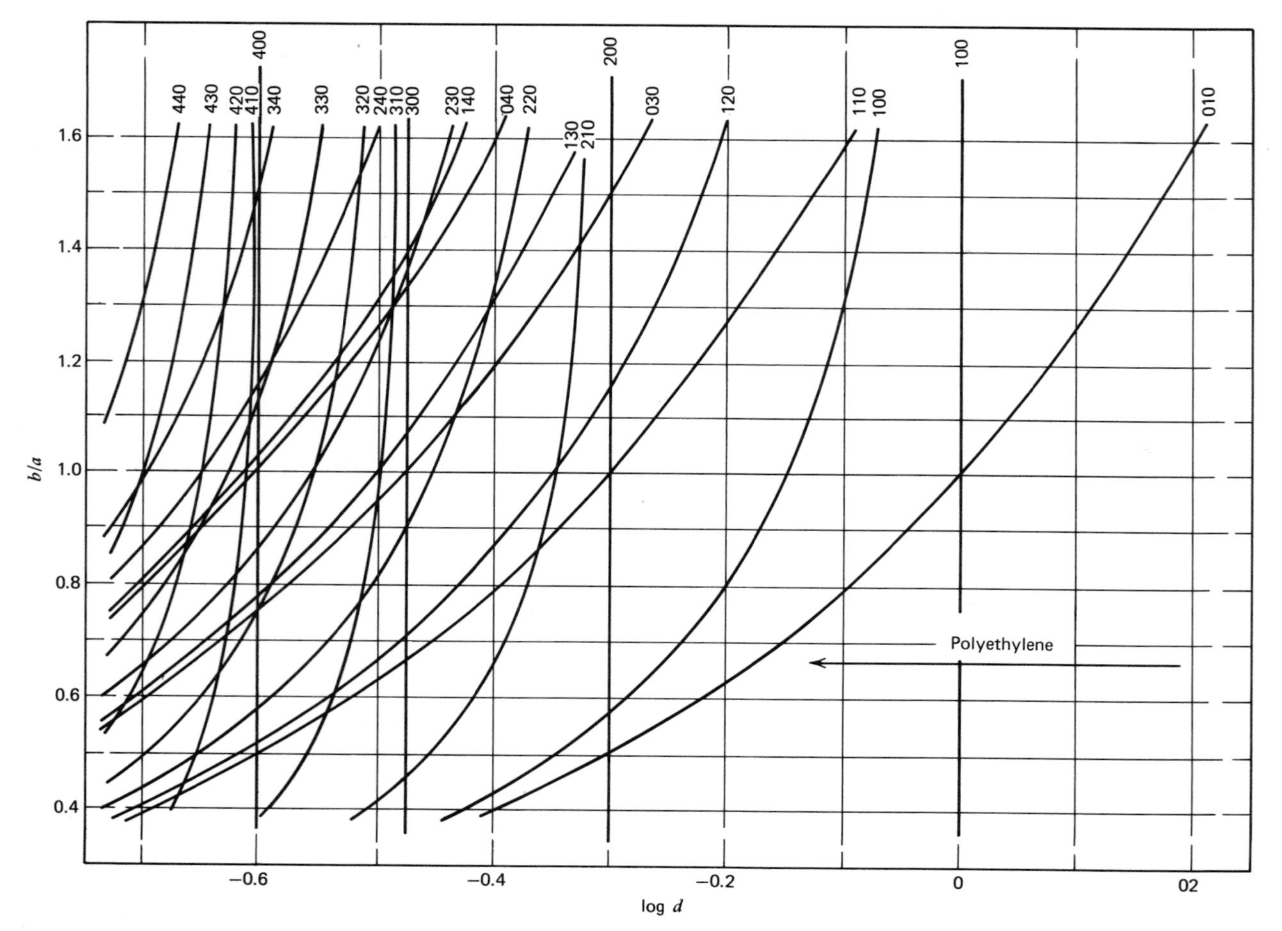

Fig. 18-2. Hull-Davey chart for determining the unit-cell dimensions of an orthorhombic lattice.

a well oriented sample, reflections along the equator arise only from planes with indices $hk0$. These two indices can be identified from a modified Hull-Davey chart (Hull 1921, Davey 1934) (Fig. 18-2) constructed from Eq. 18-1, which is a transformation of the basic equation for interplanar distances from planes $hk0$ in an orthorhombic lattice:

$$(1/d)^2 = (h/a)^2 + (k/b)^2 \qquad (18\text{-}2)$$

where a and b are the unit-cell dimensions in the directions perpendicular to the chain (in this case also the fiber) axis c.

To use Fig. 18-2, one plots values of log d_{hk0} on a strip of paper, to the same scale as the log d axis in the figure. Then the strip of paper is shifted vertically and horizontally along the figure (but kept parallel to the log d axis) until an exact correspondence is found between chart lines and observed reflections. The ratio b/a and the exact value of log d corresponding to any reflection can then be read. (A similar fit should also be found at the inverse ratio a/b). The subsequent indexing and determination of a and b (as yet unidentified as to which is which) is straightforward, particularly if data at the fits for both b/a and a/b are used.

2. The x-ray emission spectrum of copper contains as its most prominent lines the widely separated doublets K_α and K_β, with wavelengths 1.541 and 1.389 A, respectively (the doublets are not resolved). The two doublets are separated by a filter of nickel foil, which has an absorption edge at 1.49 A, exhibiting low absorbance at slightly higher wavelengths and much higher absorbance at slightly lower wavelengths. With a Ni foil about 20 µm thick, the ratio of the intensities of the K_α and K_β lines is reduced from about 1/2 to 1/100.

N. *General References*

Chap. 8B; *Textbook*, Chap. 4C; Alexander 1969; Jeffery 1971.

O. *Bibliography*

Alexander 1969. Leroy E. Alexander, *X-Ray Diffraction Methods in Polymer Science*, Interscience Div., John Wiley and Sons, New York, 1969.

Bettelheim 1971. Frederick A. Bettelheim, *Experimental Physical Chemistry*, W. B. Saunders Co., Philadelphia, Pennsylvania, 1971.

Bunn 1939. C. W. Bunn, "The Crystal Structure of Long Chain Normal Paraffin Hydrocarbons. The 'Shape' of the $>CH_2$ Group," *Trans. Faraday Soc. 35*, 482-491 (1939).

Davey 1934. W. P. Davey, *A Study of Crystal Structure and Its Applications*, McGraw-Hill Book Co., New York, 1939.

Hull 1921. Albert P. Hull and Wheeler P. Davey, "Graphical Determination of Hexagonal and Tetragonal Crystal Structures From X-Ray Data," *Phys. Rev. 17*, 549-570 (1921).

Jeffery 1971. J. W. Jeffery, *Methods in X-Ray Crystallography*, Academic Press, New York, 1971.
Warhus. William H. Warhus, Wilmington, Delaware 19803.

EXPERIMENT 19. ELECTRON MICROSCOPY

A. *Introduction*

In this experiment, the manipulation of a small, relatively low-magnification electron microscope for the examination of polymer samples is described. It is recommended that sample preparation be carried out in advance by the instructor or staff familiar with these techniques. The observations described in this experiment require less than 3 hours.

B. *Principle*

The JEM 30 electron microscope (JEOL, Chap. 8C, *Textbook*, Chap. 4F, Figs. 8-14 and 8-15) consists of a heated filament and high-voltage accelerating system to produce an electron beam; magnetic condenser, objective and projector lenses; a sample stage; a fluorescent viewing screen; and (optionally) provision for photographing the image, all contained in a column which is evacuated by a two-stage (mechanical and diffusion) pumping system. The principle of operation, except for the need for evacuation, is similar to that of the conventional light microscope. The wavelength of the electron beam is of the order of 0.07 A and the resolution of the microscope is about 100-150 A.

C. *Applicability*

Observation in the electron microscope can be carried out for virtually all materials which can be prepared in a sufficiently thin section and in a form completely nonvolatile at a pressure of 10^{-5} torr or less. This represents a severe limitation to the applicability of the method, and preparation of a replica of the surface of the specimen is often resorted to in lieu of direct observation.

D. *Precision and Accuracy*

E. *Safety Considerations*

ALL GENERAL LABORATORY SAFETY PRECAUTIONS SHOULD BE OBSERVED IN THIS EXPERIMENT. ALTHOUGH THE ELECTRON MICROSCOPE IS WELL PROTECTED, IT CONTAINS POTENTIALLY DANGEROUS HIGH VACUUM AND HIGH VOLTAGE COMPONENTS, AND MUST NOT BE TAMPERED WITH. FOLLOW OPERATING INSTRUCTIONS AT ALL TIMES. SAFETY GLASSES MUST BE WORN IN THE LABORATORY AT ALL TIMES.

F. *Apparatus*

JEM 30 or JEM 50 "Superscope" electron microscope (JEOL, Fig. 8-15), or equivalent.

G. *Reagents and Materials*

A variety of samples, consisting of polymers, replicas of polymer surfaces, latex particles, etc., prepared on electron microscope grids and ready for observation.

H. *Preparation*

I. *Procedure*

Startup

1. About 30-45 min before observations are to begin, turn the instrument power on by depressing the POWER button on the main panel. The pilot light over this pushbutton should come on.
2. Push the ROTARY PUMP knob in and turn it clockwise. This closes the leak and starts the mechanical pump.
3. Turn the VALVING lever clockwise from position C to position P. This opens the valve between the mechanical pumping system and the column.
4. Start the diffusion pump heater, by pushing the DIFFUSION PUMP button on the vacuum panel (see step M-1). The pilot light should come on.
5. After 5-10 min, the VACUUM 1 light should come on. This indicates that the column pressure has been reduced to the point where the diffusion pump can be connected to the column. Do this by turning the VALVING control clockwise to position L (see step M-2). After about one min continue turning the control to position E.

Sample Insertion

6. Remove the specimen holder rod by pulling it out from the instrument until it stops. Then turn it counterclockwise 90° and pull it completely out of the instrument.
7. Using tweezers, remove the cylindrical cap from the side of the rod, and remove the sample grid, if any, below the cap. Put the new sample grid in place and replace the cap.
8. Insert the specimen rod into the instrument, see that the guide pin is positioned in its slot, and push the rod all the way in. Now turn it 90° clockwise and push it into the instrument to its final position. (*Note*: At this point a small amount of air is introduced into the column and high vacuum is temporarily lost. The VACUUM 2 light, if previously lit, will go out.)

Observation

9. When the pressure level for operation of the microscope has been reached, the VACUUM 2 light will come on. Dim the room lights for better observation of the viewing screen.

10. Depress the HIGH VOLTAGE pushbutton; the pilot light above this button should light, and the BEAM CURRENT meter should read approximately 30 μA.

11. Turn the FILAMENT control slowly clockwise to the position of its stop, at which point the BEAM CURRENT meter should read 70-80 μA, in the red zone of the meter. (Do not exceed this reading.) The viewing screen should now be lighted, a bright green if there is no sample in the beam, or dimmer if a sample is in place. If the screen is dark, it may be because a wire of the sample grid is intercepting the beam. If so, a slight adjustment of the SPECIMEN CONTROLS should move it to one side.

12. Adjust the SPECIMEN CONTROLS until a suitable area is seen, and adjust the FOCUS knob for best focus.

13. Set the MAGNIFICATION control to 2000 ×, 3000 ×, or 4000 × as desired, readjusting the FOCUS if necessary, and scan the sample with the SPECIMEN CONTROL knobs. Observe and record morphological features and other information about the sample.

Change Sample

14. Turn the FILAMENT knob counterclockwise to zero, and turn off the high voltage circuits by depressing the HIGH VOLTAGE pushbutton; the pilot light will go off.

15. Repeat steps 6-8 to change the sample, then steps 9-13 to observe the new sample.

Shutdown

16. Repeat step 14, then turn the VALVING lever to position L and depress the DIFFUSION PUMP pushbutton to turn off the diffusion pump heater. WAIT 30-45 MIN for the pump to cool (see step M-1), then turn the VALVING control counterclockwise to position C.

17. Turn the ROTARY PUMP control counterclockwise to OFF. Turn the power off by depressing the POWER pushbutton.

J. *Fundamental Equations*

K. *Calculations*

L. *Report*

1. Describe the apparatus and experiment in your own words.

2. Describe the morphological and structural features of the samples examined, as specified by the instructor.

3. If a latex sample was observed, measure the diameters of several particles during the observations and, assuming the indicated

magnification to be exact, calculate an average particle size of the latex for comparison, for example, with the results of Exp. 14. Alternatively, use the true size of the latex particles to calculate the correct magnification of the instrument.

M. Comments

1. The diffusion pump can be used only after the system has been evacuated to about 0.1 torr, and must never be exposed to atmospheric pressure while hot. Therefore the rotary pump must be placed in operation before the diffusion pump is turned on, and must be kept in operation until after it has cooled down at the end of the experiment.

2. Position L of the VALVING control is used when the column of the microscope must be opened, as for changing photographic film or the filament. After this, the control must be turned to P for rough evacuation, then through L to E after the VACUUM 1 light has come on.

N. General References

Chap. 8C; *Textbook*, Chap. 4F; Geil 1966; Hall 1966; Sjöstrand 1967; Hayat 1970; Swift 1970.

O. Bibliography

Geil 1966. Phillip H. Geil, "Electron Microscopy," pp. 662-669 in Herman F. Mark, Norman G. Gaylord, and Norbert M. Bikales, eds., *Encyclopedia of Polymer Science and Technology*, Vol. 5, Interscience Div., John Wiley and Sons, New York, 1966.

Hall 1966. Cecil E. Hall, *Introduction to Electron Microscopy*, 2nd ed., McGraw-Hill Book Co., New York, 1966.

Hayat 1970. M. Arif Hayat, *Principles and Techniques of Electron Microscopy, Biological Applications*, Van Nostrand-Reinhold, New York, 1970.

JEOL. Japan Electron Optics Laboratory Co., Ltd., Medford, Massachusetts 02155.

Sjöstrand 1967. Fritiof S. Sjöstrand, *Electron Microscopy of Cells and Tissues*, Vol. I, *Instrumentation and Techniques*, Academic Press, New York, 1967.

Swift 1970. H. F. Swift, *Electron Microscopes*, Barnes and Noble, New York, 1970.

EXPERIMENT 20. LIGHT MICROSCOPY

A. *Introduction*

In this experiment, a polarizing microscope equipped with hot stage and Berek compensator is used to observe various features of the morphology of crystalline polymers. Two major experiments are described: 1. The determination of crystalline melting point and morphology as a function of crystallization temperature; and 2. The determination of birefringence as a function of sample elongation during cold drawing. The experiment requires approximately 3 hours.

B. *Principle*

As described in Chap. 8 and in the *Textbook*, Chap. 5, both the morphology of a crystalline polymer and its melting point depend on the size and perfection of the crystalline regions. These in turn depend upon the rate of crystallization, which is a function of the degree of supercooling of the sample below its equilibrium melting point during the crystallization process. Thus, by carrying out crystallization at different temperatures it is possible to prepare samples with different morphologies and a range of melting points. These operations are readily carried out and observed with a polarizing microscope equipped with a hot stage (Fig. 8-19).

The detection of crystallinity in this experiment depends upon the observation of birefringence, the rotation of the plane of polarized light by the anisotropic crystalline regions (lamellae or spherulites) in the sample. Since the polymer melt is isotropic, no birefringence is present in the absence of crystallinity. Birefringence is observed as light areas in the dark field produced by crossed polarizer and analyzer. The crystalline melting point is taken as the temperature of disappearance of the last traces of birefringence on heating the sample, and the onset of crystallization as the point of its first reappearance on cooling.

Birefringence can be determined quantitatively by means of the Berek compensator (Fig. 8-20). The amount of birefringence observed in any polymer, crystalline or noncrystalline, increases (from zero in the case of an amorphous sample) as the degree of orientation, and thus the anisotropy, in the specimen is increased. This observation forms the basis of the second part of this experiment.

McCaffery (1970) includes the observation of spherulites in his Exp. 3.

C. *Applicability*

The polarizing microscope can be used to observe morphological features of all crystalline polymers which can be prepared as thin films. Birefringence measurement is applicable to all polymers in which orientation or anisotropy can be induced.

D. *Precision and Accuracy*

The precision of the melting-point determination using the hot stage depends upon heating rate: At 10°C/min, T_m can be estimated to no better than ±1°C, whereas at 0.2°C/min, a precision of ±0.1°C can be reached. Accuracy is difficult to specify since the melting point of any specimen depends upon its previous thermal history.

The Berek compensator can be read with a precision equivalent to three significant figures, but other variables in the experiment, such as the ability to set the positions of the analyzer and compensator and to determine the specimen thickness accurately, limit the accuracy of the birefringence determination to ±5% at best, when monochromatic light is used; with white light, the errors are much greater.

E. *Safety Considerations*

OBSERVE ALL NORMAL LABORATORY SAFETY PRECAUTIONS DURING THIS EXPERIMENT. SAFETY GLASSES MUST BE WORN IN THE LABORATORY AT ALL TIMES.

F. *Apparatus*

1. Polarizing microscope, any of several manufacturers, equipped with crosshair eyepiece, calibrated fine focus, and provision for insertion of compensator. A moderate magnification, say 320× (32× objective and 10× ocular) is desirable.
2. Mercury arc with monochromatizing filter for λ = 546 nm as microscope illuminator.
3. Hot stage, Mettler, or equivalent (see Sec. M-1).
4. Berek compensator (Burri 1950).
5. Mechanical stage with vernier scales.
6. Usual microscope accessories including slides and cover glasses.

G. *Reagents and Materials*

1. Linear polyethylene (Marlex 50), polyoxymethylene, or other highly crystalline polymer.
2. Poly(hexafluoropropylene-*co*-tetrafluoroethylene) (Teflon FEP, Du Pont), polyethylene wrapping or bag film, or other polymer which can be cold drawn without "necking down," in thin (0.05-0.2 mm) film form. About 5-10 sq cm is required.
3. Other semicrystalline polymers, such as the nylon prepared in Exp. 1.

H. *Preparation*

Prepare samples for the cold-drawing experiment by cutting 6-8 specimens about 2 mm × 2-3 cm from the film described in step G-2. Exact width is unimportant but the specimens must have constant width and edges free from nicks or irregularities; for example, cut with a sharp razor blade in one continuous motion. Alternatively, cut dumbbell-shaped specimens like the tensile bars of Exp. 29, preferably by using the area between two holes punched close together by a sharp paper punch.

I. Procedure

Microscope Setup

1. Turn on the mercury arc illuminator about 10 min ahead of time to allow development of full intensity. Insert the monochromatizing filter.
2. Remove the microscope eyepiece, set the polarizer and analyzer to 0°, and (looking down the microscope tube) adjust the position of the lamp and mirror to provide optimum even illumination. Replace the ocular.
3. Turn the analyzer to the crossed position and adjust it slightly as required to obtain complete extinction (the darkest field possible).

Crystallization and Melting of Linear Polyethylene

4. Cut a small piece of polyethylene, not more than 0.5 mm on a side, and place it about 1 cm from the end of a clean microscope slide. Put a cover glass over the specimen.
5. Open the hot-stage unit (Fig. 8-19*b*) and place the slide in it with the specimen above the center of the aperture. Fasten the slide with the spring clip, close the unit, and place it on the microscope stage.
6. Carefully lower the microscope objective until it almost touches the window of the hot stage.
7. Focus on the sample by raising the objective (CAUTION: Do not lower it) and adjust the condenser focus and diaphragm for even illumination at a convenient level.
8. Turn the hot-stage control unit (Fig. 8-19*a*) ON and increase the hot-stage temperature rapidly to 165°C (about 30°C above T_m; Table 9-2). (*Note*: At the most rapid heating or cooling rate, the sample temperature lags behind the indicated temperature, reaching it only when the control light glows steadily.)
9. Cool the sample rapidly to 95°C (about 40°C below T_m), and watch for the onset of crystallization (probably even before temperature equilibrium is reached).
10. When crystallization appears to be complete, heat at 10°C/min and determine T_m as follows: When the sample starts to melt, record the temperature as T_A (by depressing button A); when the light intensity is reduced to about 1/2, record T_B; and when the field becomes completely dark, record T_C. Write down these temperatures and clear the display.
11. Repeat steps 8-10, but in step 9 cool in successive experiments to 115, 120, 122, and 124°C. (*Note*: At the higher crystallization temperatures, wait at least 15 min for the completion of crystallization.)
12. Determine T_m for any other crystalline polymers by repeating

steps 4-7, then step 10, starting from about 30°C below the anticipated value of T_m.

13. Finally, turn the hot stage control OFF and remove the hot stage from the microscope.

Birefringence

14. Turn the drum of the Berek compensator (Fig. 8-20) to the specified angle for insertion and insert it in the opened slot in the microscope barrel. (CAUTION: DO NOT ATTEMPT TO INSERT THE COMPENSATOR WITHOUT SETTING IT TO THE PROPER ANGLE.) Mount the mechanical stage on the microscope.

15. Using a fine pen and ink, mark the samples prepared in Sec. H with two lines about 1 mm apart, in order to determine their elongation.

16. Place the sample on a slide mounted on the mechanical stage of the microscope, aligned so that the ink marks are parallel to one axis of the stage, and focus on its center part (between the ink marks), so that both sample and surrounding field can be seen. With the compensator set for no compensation and the polarizer and analyzer set for extinction, rotate the microscope stage, observing the intensity of light transmitted through the sample. A point of minimum intensity should easily be found. (If it is not, the sample must be annealed using the hot stage to reduce its anisotropy.) Set the stage to the position of the minimum, and read its position using the scale and vernier on its circumference.

17. Set the stage exactly 45° from the position found in step 16. Turn the compensator drum until the sample and field show equal brightness. Record the drum reading to the nearest 0.1°, using the vernier. Return the drum to its original position and rotate it in the opposite direction until a second position of equal brightness is found. Record this angle, and record half the difference between the two readings as the angle θ.

18. Shift the mechanical stage until first one, then the other, ink mark coincides with the ocular crosshair. Record both positions by reading the mechanical stage scale and vernier, and subtract the two to find the length L of the sample.

19. Measure the sample thickness by focusing carefully on its upper, and then lower, surface. At each point read the setting of the fine-focus knob. The difference in the readings is the sample thickness ℓ.

20. Remove the sample and stretch it slightly by gripping it between two pair of tweezers or pliers and pulling slowly and evenly. (*Note*: Care and some practice is required to avoid breaking the sample.)

21. Repeat steps 16-20 for several stages of elongation of the sample. Keep the elongations small (less than 20% or so for samples thicker than 20 μm); otherwise, the range of the compensator will be exceeded.

22. Finally, reset the compensator to the specified angle and remove it from the microscope. (SEE CAUTION IN STEP 14.) Remove the sample and mechanical stage, and turn off the microscope lamp.

J. *Fundamental Equations*

Percent elongation:

$$E = 100(L-L_0)/L_0 \tag{20-1}$$

where L_0 and L are the initial and final lengths of the sample at each stage of elongation.

Retardation

$$\Gamma = Cf(\theta) \;;\quad \delta = \Gamma(2\pi/\lambda) \tag{20-2}$$

where C for λ = 546 nm and tables of $f(\theta)$ are supplied with the compensator.

Birefringence:

$$\Delta n = \lambda\delta/2\pi\ell = \Gamma/\ell \qquad \text{(from 8-5)}$$

where ℓ is the sample thickness.

K. *Calculations*

Calculate E, Γ, δ, and Δn from the equations in Sec. J for each elongation of the sample.

L. *Report*

1. Describe the apparatus and experiment in your own words.
2. Describe the morphology of the linear polyethylene sample as observed at each crystallization temperature. Comment on any differences.
3. Tabulate crystallization temperatures and melting temperature range T_A, T_B and T_C for the polyethylene sample. Plot T_A, T_B, and T_C versus crystallization temperature and include the graph.
4. Give values of T_m observed for any other samples measured, and compare them to literature values.
5. Tabulate L (or L_0), E, ℓ, θ, Γ, δ, and Δn. Plot Δn versus E and include graph.

M. *Comments*

1. The Mettler hot stage (Fig. 8-19) consists of (a) a control unit which provides rapid temperature changes in either direction, and heating at three controlled rates (10, 2, and 0.2°C/min); (b) a remote pushbutton unit for holding the stage temperature constant momentarily and for recording three selected temperatures for later digital display;

and (c) the hot stage proper, small enough to be placed on the microscope stage, with insulated heaters and air cooling to protect the microscope.

N. General References

Chaps. 8A, 8D; *Textbook*, Chaps. 4F, 5E, 5F; Mandelkern 1964; Stein 1964, 1969; Sharples 1966; Wilchinski 1968; Hallimond 1970; Hartshorne 1970; Loveland 1970; Richardson 1971.

O. Bibliography

Burri 1950. Conrad Burri, *The Polarizing Microscope* (in German), Birkhauser, Basel, Switzerland, 1950.
Du Pont. E. I. du Pont de Nemours and Co., Plastics Department, Wilmington, Delaware 19898.
Hallimond 1970. Arthur F. Hallimond, *The Polarizing Microscope*, 3rd ed., Vickers Ltd., York, England, 1970.
Hartshorne 1970. N. H. Hartshorne and A. Stuart, *Crystals and the Polarizing Microscope*, 4th ed., American Elsevier, New York, 1970.
Loveland 1970. R. P. Loveland, *Photomicrography*, Interscience Div., John Wiley and Sons, New York, 1970.
Mandelkern 1964. Leo Mandelkern, *Crystallization of Polymers*, McGraw-Hill Book Co., New York, 1964.
McCaffery 1970. Edward M. McCaffery, *Laboratory Preparation for Macromolecular Chemistry*, McGraw-Hill Book Co., New York, 1970.
Mettler. Mettler Instrument Corp., Hightstown, New Jersey 08520.
Richardson 1971. James H. Richardson, *Optical Microscopy for the Materials Sciences*, Marcel Dekker, New York, 1971.
Sharples 1966. Allan Sharples, *Introduction to Polymer Crystallization*, St. Martin's Press, New York, 1966.
Stein 1964. R. S. Stein, "Optical Methods of Characterizing High Polymers," pp. 155-206 in Bacon Ke, ed., *Newer Methods of Polymer Characterization*, Interscience Div., John Wiley and Sons, New York, 1964.
Stein 1969. Richard S. Stein, "Studies of the Deformation of Crystalline Polymers," pp. 297-351 in Frederick R. Eirich, ed., *Rheology--Theory and Applications*, Vol. 5, Academic Press, New York, 1969.
Wilchinski 1968. Zigmond W. Wilchinski, "Orientation," pp. 624-648 in Herman F. Mark, Norman G. Gaylord, and Norbert M. Bikales, eds., *Encyclopedia of Polymer Science and Technology*, Vol. 9, Interscience Div., John Wiley and Sons, New York, 1968.

EXPERIMENT 21. DIFFERENTIAL THERMAL ANALYSIS

A. Introduction

The thermal properties of polymers are studied in this experiment, which includes measurement of the glass transition temperature T_g, the crystallization temperature T_c, and the crystalline melting point, T_m. The time required for measurement of benzoic acid and two polymers is 2 hours; one or more additional polymer samples can be measured within a 3-hour laboratory period. Procedures are included for the Du Pont 900 and Perkin-Elmer DSC 1B instruments.

B. Principle

Differential thermal analysis (DTA) is a technique for detecting thermal changes which occur on heating or cooling materials. Such changes usually accompany physical changes such as phase transitions or alterations in heat capacity, or chemical changes such as degradation or loss of adsorbed or entrapped material. Among the changes detected are crystallization temperature and crystalline melting, the glass transition, thermal degradation, and others. In the DTA experiment, the sample and a thermally inert reference material are heated or cooled at a programmed rate. The temperature difference between them, ΔT, is recorded as a function of temperature, T, as described in Chap. 9C. In the absence of thermal transitions in the sample, ΔT remains approximately constant.

McCaffery (1970) includes the determination of T_m and T_g in his Exp. 22 on DTA.

C. Applicability

DTA is applicable to the study of the thermal behavior of virtually all polymers and many other materials. Thermal changes can be detected over a temperature range of -150 to 600°C, but the method is not suited for quantitative measurement of such quantities as heat of fusion or heat capacity; for this, differential scanning calorimetry (Exp. 22) should be used.

D. Precision and Accuracy

Most DTA instruments use thermocouples for the measurement of T and ΔT; these are accurate to about ±0.25°C, but usually the recorder chart cannot be read with this precision.

The interpretation of DTA data as thermodynamic transition temperatures can introduce further inaccuracies. Since DTA is a dynamic method, the results depend upon rate of heating or cooling, and this must be specified for completeness. To obtain equilibrium transition temperatures, it is necessary to measure as a function of rate and extrapolate to zero heating rate. Additionally, phenomena such as reactions or degradation may be difficult to interpret without prior knowledge since more than one thermal reaction may occur simultaneously, for example exothermic decomposition accompanied by endothermic gas evolution.

E. Safety Considerations

OBSERVE NORMAL LABORATORY SAFETY PRECAUTIONS DURING THIS EXPERIMENT. THE USE OF LIQUID NITROGEN FOR MEASUREMENTS BELOW AMBIENT TEMPERATURE REQUIRES SPECIAL PRECAUTIONS. THIS MATERIAL MUST NEVER BE CONFINED IN A GAS-TIGHT VESSEL. CONTACT OF LIQUID NITROGEN WITH THE SKIN CAN CAUSE SEVERE FROSTBITE. SAFETY GLASSES MUST BE WORN IN THE LABORATORY AT ALL TIMES.

F. Apparatus

1. DTA equipment: Du Pont 900, Perkin-Elmer DSC 1B, or other.
2. Source of dry nitrogen (or other gas if controlled atmosphere is desired) at low pressure.

G. Reagents and Materials

1. Polymer samples, synthesized in Exp. 1-9 or obtained from the sources listed in App. I. Suggested samples include polyethylene, polystyrene (Exp. 4 or 9) and polyoxymethylene (Exp. 6) (for which procedures are written), nylons (Exp. 1 or 2), or poly(ethylene terephthalate), and *cis*-polybutadiene or natural rubber for experiments below ambient temperature (see step M-1).
2. Benzoic acid, reagent grade.
3. Liquid nitrogen, if experiments below ambient temperature are to be carried out.

H. Preparation

1. One hour in advance, turn the apparatus on.
2. See that polymer samples are finely ground and free from monomer, water, solvents, or other impurities (see step M-2).

I. Procedure

Procedure I. Du Pont 900 DTA

1. *Startup.* It is assumed that the apparatus has been prepared for measurement. If electronic cold-junction compensation is not used, see that the cold-junction compensation thermocouples for the controlling and sample thermocouples are properly connected and placed in an ice bath (Dewar flask) prepared with ice made from distilled or demineralized water. See that a low-pressure nitrogen source is connected to the PURGE inlet on the apparatus, and the purge valve is open.

2. *Standby.* Turn the POWER switch to STAND BY, the PROGRAM MODE switch to HEAT OFF, the T and ΔT ZERO SHIFT controls to 0, the BASELINE SLOPE control OFF (see Note), the T SCALE control to 50°C/in. and the ΔT SCALE control to 1°C/in, and raise the pen. (*Note*: The BASELINE SLOPE control is used to correct a sloping baseline, but it is not normally used for the first run with a given sample since the slope cannot be predicted accurately.)

3. *Sample preparation.* Fill a micro sample tube (2 mm diameter) to a depth of 2 mm for benzoic acid or 3 mm for polymers: Push the open end of the tube into the powdered sample, invert, and pack the material to the bottom by tapping the sealed end of the tube on a flat surface. Fill control and reference tubes to a depth of 2 mm with glass beads. Cut the tubes 3 cm from the end by scoring them with the cutter provided and bending them gently. Insert a thermocouple with ceramic sleeve into each tube until the top of the thermocouple is centered and touching the bottom of the tube.

4. *Sample cell assembly.* Remove the glass bell jar from the cell assembly and insert the sample tubes in the proper holes. Connect the thermocouple leads using long tweezers to hold the pin connectors. Refer to Fig. 21-1, and be sure that the leads are connected to the correct pins. (*CAUTION*: Do not touch the power pins when the instrument is in operation, as they carry 110 V.) Check to see that the thermocouples still touch the bottom center of the tubes; bend them or support them with masking tape as necessary to insure this. Place the sample-cell assembly on the instrument, centering the plugs and pins properly. Place the O-ring in the groove on the plate, replace the bell-jar cover, and clamp it into position.

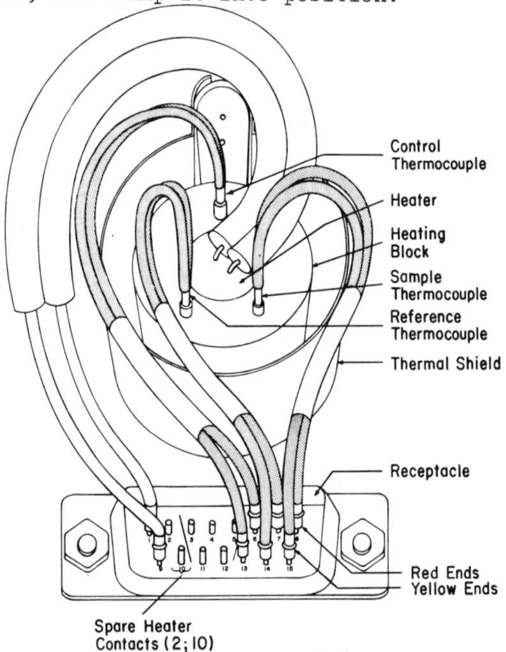

Fig. 21-1. Heating block and thermocouple connections for the Du Pont 900 DTA cell (courtesy Du Pont).

5. *Gas flow.* Open the VACUUM valve by turning it fully counterclockwise. (*Note*: This provides an exhaust for the inert-gas flow. If the sample is to be measured while under vacuum, close the PURGE valve and connect a vacuum source to the VACUUM outlet.) Adjust the inert-gas flow to about 10 ml/min (pressure of 4-5 psi).

6. *Prepare instrument.* Select a recorder chart with a range appropriate to the experiment and place it on the recorder. Turn the POWER switch to RECORD, set the T SCALE to 20°C/inch, and adjust the T ZERO SHIFT to place the pen near the top of the chart.

7. *Set scales.* Set the T ZERO SHIFT (which sets the chart zero) and the T SCALE (which sets the span on the temperature axis) as follows:

Chart paper range	T ZERO SHIFT	T SCALE
-100 to +100	5	20
0 to 200	0	20
0 to 500	0	50

Set the ΔT SCALE to 0.5°C/inch for the benzoic acid experiment, or to 0.2°C/inch for polymer samples. (*Note*: This setting depends upon heating rate, sample size, and phenomenon to be detected as well as type of material. Higher rates and larger samples generally require lower sensitivity (lower settings). Detection of T_g requires higher sensitivity than the observation of melting phenomena. Using larger samples with the macro cell at 10°C/min, typical settings are 0.2°C/in. for the detection of T_g and 1-2°C/in. for large melting peaks. The setting of the ΔT SCALE can be changed during a run, if necessary, to keep the trace on scale.)

8. *Heating.* Push the RESET button to erase previous heating-program instructions. Set the TEMPERATURE RATE to 20°C/min. Turn the STARTING TEMPERATURE control counterclockwise until the heater voltage meter reaches its minimum value, then clockwise until it just begins to move upscale. Lower the pen and set the PROGRAM MODE to HEAT.

9. *Cooling.* When the maximum desired temperature is reached (150°C for benzoic acid and polyethylene, 210°C for polyoxymethylene, 130°C for polystyrene), raise the pen, set the PROGRAM MODE to COOL, and set the ΔT SCALE to 2°C/in. Reposition the pen near the bottom of the chart by means of the ΔT ZERO SHIFT, and lower the pen. When the temperature reaches 100°C (for polyethylene or polyoxymethylene) or 80°C (for benzoic acid or polystyrene), turn the POWER switch to STAND BY.

10. Repeat step 2.

11. *Remove sample.* Remove the cell assembly from the instrument and take off the bell jar. Disconnect the sample thermocouple, remove it from the sample tube, and discard the tube. Clean as much of the sample as possible from the thermocouple and heat the thermocouple to red heat (not hotter) in a bunsen flame to burn off remaining material.

12. Repeat steps 3-11 for subsequent samples.

13. *Shutdown.* Repeat step 2, turn the POWER OFF, and stop the inert gas flow.

Procedure II. Perkin-Elmer DSC 1B

1. *Startup.* It is assumed that the apparatus has been prepared for measurement. See that the sample-holder assembly is in place, the DIFFERENTIAL and AVERAGE TEMPERATURE CALIBRATION and SLOPE controls are set to pre-established calibration values, the HEATER switch is OFF (it is used only for rapid heating from temperatures below ambient), the pen is set in the middle of the chart, and the chart speed is set at 1 in./min. Turn the recorder power ON.

2. *Standby.* Turn the toggle switch to NEUTRAL, the RANGE switch to STANDBY, turn the recorder chart OFF, and set the TEMPERATURE control to 273 K.

3. *Sample preparation.* (*CAUTION*: Never touch the sample pans, holders or associated parts of the apparatus with fingers; use tweezers at all times.) Place about 7 mg benzoic acid, 10 mg polyethylene or polyoxymethylene, or 15 mg polystyrene (powdered sample or molded disc 6 mm in diameter) in a sample pan, cover it with a pan cover, and crimp the cover in place as follows: Depress the crimper table slightly to form a recess into which the sample pan can be centered. Release the table slowly to avoid shifting the pan. Press the crimper handle down with steady pressure until the table touches the base; hold for a few seconds. Release the handle, depress the table, and remove the crimped pan. Rub the bottom of the pan gently over a sheet of clean paper to ensure that it is flat.

4. *Sample mounting.* Remove the cover to the sample-holder assembly, and place the sample from step 3 in the right holder and an empty pan in the left holder. Replace the cover.

5. *Gas flow.* Adjust the flow of inert gas to approximately 20 ml/min (range 10-30); this usually requires about 20 psi pressure.

6. *Set scales.* Set the INCREASE and DECREASE scan speeds and the RANGE (ΔT sensitivity in millicalories/second) according to the following table, which includes suggested upper and lower temperatures:

	Benzoic acid	Poly-ethylene	Poly-oxymethylene	Polystyrene
INCREASE, °C/min	10	10	10	10
DECREASE, °C/min.	5	5	5	5
RANGE, mc/sec	64	32	32	2
Upper T, K	423	423	483	403
Lower T, K	323	373	373	323

(See Note to step 7, Procedure I, for further comments on RANGE settings.)

7. *Heat.* Raise the temperature to the desired starting level by setting the TEMPERATURE control. When the green AVERAGE TEMPERATURE light comes on, the sample is at the indicated temperature. Turn the recorder chart ON, and start the scan by turning the toggle switch to INCREASE.

8. *Cool.* When the desired upper temperature is reached, turn the toggle switch to DECREASE after making any desired changes in settings.

9. Repeat step 2 when the desired lower temperature is reached.

10. *Remove sample.* Wait 10-20 min for the sample holder to cool, then remove the sample, following the reverse procedure of step 4.

11. Repeat steps 3-10 for subsequent samples.
12. *Shutdown.* Turn the RANGE switch OFF and the recorder power off, and shut off the inert gas flow.

J. *Fundamental Equations*

K. *Calculations*

Read the transition temperatures from the charts. Take the inflection of the transition region as T_g, and the maximum peak temperature as T_m. These points are usually better defined and more reproducible than others.

L. *Report*

1. Describe the apparatus and experiment in your own words.
2. Report transition temperatures and ranges for all experiments.
3. Include all charts.
4. Answer the following questions:

a. What effect does heating rate have on the glass-transition temperature?

b. What is the effect of heating rate on T_m?

c. How does cooling rate affect the value of T_m observed in a subsequent experiment?

d. Why is higher ΔT sensitivity needed in measurement of T_g in comparison to T_m?

e. What parameters affect the base line, and why?

f. How do values of T_m obtained by DTA compare with those obtained by microscopy (Exp. 19)? Explain any differences, using your data if available.

M. *Comments*

1. For analyses below room temperature with the Du Pont 900 DTA, the low-temperature sample cell and bell jar equipped with tubing connections should be used. Liquid nitrogen is poured slowly and steadily onto the sample block until the desired low temperature is reached. Avoid using too much nitrogen during cooling (excessive boil-off) but keep the assembly wet with coolant to avoid the condensation of moisture. Close off or replace the bell jar as soon as possible.

2. Better reproducibility and fewer spurious effects are obtained if samples are heated at 10-20°C/min to about 130°C, then cooled, before measurement. This is particularly valuable for the samples synthesized in Exp. 1-9, where residual volatiles may be present initially. This annealing treatment also relieves mechanical stresses which may cause spurious thermal effects.

3. Variations of the experiment include study of the effect of atmosphere (for example, vacuum, air, oxygen, nitrogen, ozone) on thermal degradation, and the effects of heating and cooling rates on the various transition temperatures (see also Wunderlich 1964 *a*,*b*).

N. *General References*

Chap. 9C; *Textbook*, Chap. 4E; Ke 1964; Wendlandt 1964; Barrall 1966, 1970; David 1966; Jaffe 1967; Mackenzie 1970; Wunderlich 1971.

O. *Bibliography*

Barrall 1966. E. M. Barrall, II, and Julian F. Johnson, "Instrumentation, Techniques and Applications of Differential Thermal Analysis," Chap. 1 in Philip E. Slade, Jr. and Lloyd T. Jenkins, eds., *Techniques and Methods of Polymer Evaluation*, Vol. I, *Thermal Analysis*, Marcel Dekker, New York, 1966.

Barrall 1970. E. M. Barrall, II, and Julian F. Johnson, "Differential Scanning Calorimetry--Theory and Applications," Chap. 1 in Philip E. Slade, Jr. and Lloyd T. Jenkins, eds., *Techniques and Methods of Polymer Evaluation*, Vol. 2, *Thermal Characterization Techniques*, Marcel Dekker, New York, 1970.

David 1966. J. J. David, "Transition Temperatures by Differential Thermal Analysis," Chap. 2 in Philip E. Slade, Jr. and Lloyd T. Jenkins, eds., *Techniques and Methods of Polymer Evaluation*, Vol. I, *Thermal Analysis*, Marcel Dekker, New York, 1966.

Du Pont. E. I. du Pont de Nemours and Co., Instrument Products Division, Wilmington, Delaware 19898.

Jaffe 1967. M. Jaffe and B. Wunderlich, "Melting of Polyoxymethylene," *Kolloid Z.--Z. Polymere 216-217*, 203-217 (1967).

Ke 1964. Bacon Ke, "Differential Thermal Analysis," Chap. IX in Bacon Ke, ed., *Newer Methods of Polymer Characterization*, Interscience Div., John Wiley and Sons, New York, 1964.

Mackenzie 1970. R. C. Mackenzie, *Differential Thermal Analysis, Fundamental Aspects*, Academic Press, New York, 1970.

McCaffery 1970. Edward M. McCaffery, *Laboratory Preparation for Macromolecular Chemistry*, McGraw-Hill Book Co., New York, 1970.

Perkin-Elmer. Perkin-Elmer Corporation, Instruments Division, Norwalk, Connecticut 06812.

Wendlandt 1964. Wesley Wm. Wendlandt, *Thermal Methods of Analysis*, Vol. XIX in P. J. Elving and I. M. Kolthoff, eds., *Chemical Analysis*, Interscience Div., John Wiley and Sons, New York, 1964.

Wunderlich 1964a. Bernhard Wunderlich, David M. Bodily, and Mark H. Kaplan, "Theory and Measurement of the Glass-Transformation Interval of Polystyrene," *J. Applied Phys. 35*, 95-102 (1964).

Wunderlich 1964b. Bernhard Wunderlich and David M. Bodily, "Dynamic Differential Thermal Analysis of the Glass Transition Interval," *J. Polymer Sci. C6*, 137-148 (1964).

Wunderlich 1971. Bernhard Wunderlich, "Differential Thermal Analysis," Chap. VIII in Arnold Weissberger and Bryant W. Rossiter, eds., *Physical Methods of Chemistry*, Vol. 1 of A. Weissberger, ed., *Techniques of Chemistry*, Part 5, Interscience Div., John Wiley and Sons, New York, 1971.

EXPERIMENT 22. DIFFERENTIAL SCANNING CALORIMETRY

A. *Introduction*

This experiment includes the measurement of specific heat, heat of fusion, and heat of crystallization of a crystalline polymer, and the change in specific heat at the glass transition for an amorphous polymer. The time required is 3 hours. Procedures are given for two widely-used differential scanning calorimeters, the Du Pont 900 with DSC cell, and the Perkin-Elmer DSC 1B.

B. *Principle*

In contrast to differential thermal analysis (DTA) (Exp. 21), in which the temperatures of a sample and reference are compared, differential scanning calorimetry (DSC) operates by measurement of the amount of heat required to maintain the temperature of the sample at the value given by the temperature program. In the Perkin-Elmer DSC, this measurement is made by determining the power input to the sample container (less that to a similar empty container), and in the Du Pont DSC cell, the heat flow from the sample to an external thermocouple sensor (similarly corrected). In each case, the measurement is compared to that for a reference material having a known specific heat.

The specific heat of a sample is determined by comparing the instrument reading, corrected for the blank, at constant temperature (zero heat required) to that at a constant heating or cooling rate. The heat delivery is proportional to the heating rate, the sample weight, and its heat capacity. The latent heat of a first-order transition, for example the heat of fusion, is determined by integrating the heat delivery over the time interval including the transition, from the area under the fusion curve (Fig. 9-20). The reference for this measurement must be selected with care, and it is usual to establish a calibration curve from measurements of several materials with known heats of fusion.

C. *Applicability*

This determination is applicable to the measurement of transition temperatures, specific heats, and heats of transition or reaction for all nonvolatile materials or materials which do not evolve significant amounts of volatiles by reaction. The temperature range covered is -100 to 600°C, but measurements below ambient temperature require a cooling accessory.

D. *Precision and Accuracy*

With careful calibration and accurate weighing, heats of transition and specific heats can be determined with a precision of about ±2%. Accuracy of this order of magnitude requires additional careful attention to such corrections as those for thermocouple nonlinearity and variation of heating rates from nominal values.

E. *Safety Considerations*

NORMAL SAFETY PRECAUTIONS FOR LABORATORY WORK AND THE USE OF ELECTRONIC INSTRUMENTS MUST BE OBSERVED. IF LIQUID NITROGEN IS USED FOR EXPERIMENTS BELOW AMBIENT TEMPERATURE, SPECIAL CARE MUST BE TAKEN TO AVOID CONTACT WITH THE SKIN OR CONFINING THE LIQUID IN A SEALED CONTAINER. SAFETY GLASSES MUST BE WORN IN THE LABORATORY AT ALL TIMES.

F. *Apparatus*

1. DSC equipment: Du Pont 900 with DSC cell, Perkin-Elmer DSC 1B, or other.
2. Source of dry nitrogen at low pressure.
3. Cooling accessory if experiments are to be carried out below room temperature.

G. *Reagents and Materials*

1. Polymer samples, synthesized in Exp. 1-9 or obtained from the sources listed in App. I. Suggested samples include linear polyethylene, polystyrene (Exp. 4 or 9), and polyoxymethylene (Exp. 6).
2. Liquid nitrogen, if experiments are to be carried out below room temperature.

H. *Preparation*

1. One hour in advance, turn the apparatus on.
2. See that polymer samples are finely ground and free from monomer, water, solvents, or other impurities (see step M-1).

I. *Procedure*

Procedure I. Du Pont 900 DSC

A. Heat of Fusion of Polyoxymethylene or Calibration Material

1. *Startup*. It is assumed that the apparatus has been prepared for measurement. See that the cold junction compensation thermocouples (if used) are placed in an ice bath (Dewar flask) prepared with ice made from distilled or demineralized water. See that a low-pressure nitrogen source is connected to the PURGE inlet on the apparatus, and that the PURGE valve is open.

2. *Standby*. Turn the POWER switch to STAND BY, the PROGRAM MODE switch to HEAT OFF, the T ZERO SHIFT to 0, the T SCALE to 20, the ΔT SCALE to 0.5°C/min, and the PROGRAM MODE and BASELINE SLOPE controls OFF. (*Note*: The setting of the ΔT SCALE depends upon the size and type of sample and the heating rate. Faster rates or larger samples

require lower settings. The BASELINE SLOPE control is not used on a first run, but can be used subsequently to correct a sloping base line.)

3. *Sample preparation*. Obtain the tare weight of a sample pan and lid (the type which can be hermetically sealed is not required), add about 10 mg polyoxymethylene or calibration standard, and weigh again. Weigh accurately and carefully to 4 significant figures. (*Note*: It is essential to select pans with flat undistorted bottoms, so that good contact with the cell platforms is assured.) Similarly weigh about 10 mg glass beads into a second pan as reference material.

4. *Mount Samples*. Remove the glass bell jar and cover from the cell assembly. Place the sample pan on the *rear* platform and the reference pan on the *front* platform, centering them carefully. (*Note*: This inverts the thermogram from usual DTA practice but yields a sample temperature rise linear with time. The small correction of the chart temperature to the true sample temperature can easily be made if desired.) (See step M-2.) Replace the cell cover and bell jar, and fasten the hold-down clamps.

5. *Gas flow*. Open the VACUUM valve and adjust the nitrogen pressure as required (4-5 psi) to obtain a flow of about 10 ml/min.

6. *Prepare instrument*. Select a 0-200°C recorder chart and place it on the instrument. Turn the POWER switch to RECORD, and set the pen on scale on the vertical axis by means of the ΔT ZERO SHIFT control.

7. *Heating*. Set the TEMPERATURE RATE control to 10°C/min, adjust the STARTING TEMPERATURE control so that the HEATER VOLTAGE meter indicates just above zero, and press the RESET button. Start the heating cycle by turning the PROGRAM MODE switch to HEAT.

8. After the sample has heated 10-20°C, adjust the ΔT ZERO SHIFT control to bring the pen about 1/3 of the chart width above the bottom, and lower the pen.

9. At the end of the transition (followed by an adequate section of base line), turn the PROGRAM MODE control to OFF, raise the pen, and repeat step 2.

B. Specific Heat of Polystyrene

10. Repeat steps 3-7 of part A, weighing in polystyrene in step 3; but in step 4 place the sample pan on the *front* platform and the reference pan on the *rear* platform.

11. After the heating program has been initiated (step 8), and at a temperature below that for which the specific heat is to be measured (say, 40°C), turn the PROGRAM MODE control to HOLD and allow the temperature to equilibrate. Lower the pen, turn the PROGRAM MODE control to HEAT, and allow the instrument to heat through the temperature range of interest. Again set to HOLD and allow equilibration. Lift the pen and repeat step 2, allowing the instrument to cool to room temperature.

12. Repeat steps 4-7 and 10-11 with an empty sample pan. Use the same chart paper and settings, and switch from HEAT to HOLD and *vice versa* at exactly the same temperatures as in step 11.

13. *Shutdown*. Repeat step 2, turn the POWER switch OFF, and stop the inert gas flow.

Procedure II. Perkin-Elmer DSC 1B

A. Heat of Fusion and Crystallization of Linear Polyethylene

1. *Startup.* It is assumed that the apparatus has been prepared for measurement. See that the sample-holder assembly is in place, the DIFFERENTIAL and AVERAGE TEMPERATURE CALIBRATION and SLOPE controls are set to pre-established calibration values, the HEATER switch is OFF (it is used only for rapid heating from temperatures below ambient), the pen is set in the middle of the chart, and the chart speed is set at 1 in./min. Turn the recorder power ON.

2. *Standby.* Turn the toggle switch to NEUTRAL, the RANGE switch to STANDBY, turn the recorder chart OFF, and set the TEMPERATURE control to 273 K.

3. *Sample preparation.* (*CAUTION*: Never touch the sample pans, holders or associated parts of the apparatus with fingers; use tweezers at all times.) Tare a sample pan and cover, and add approximately 10 mg linear polyethylene (powdered sample or molded disc 6 mm in diameter). Weigh again, carefully and accurately, to 4 significant figures. Crimp the cover in place as follows: Depress the crimper table slightly to form a recess into which the sample pan can be centered. Release the table slowly to avoid shifting the pan. Press the crimper handle down with steady pressure until the table touches the base; hold for a few seconds. Release the handle, depress the table, and remove the crimped pan. Rub the bottom of the pan gently over a sheet of clean paper to ensure that it is flat.

4. *Sample mounting.* Remove the cover to the sample-holder assembly, place the sample from step 3 in the right holder and an empty pan in the left holder. Replace the cover.

5. *Gas flow.* Adjust the flow of inert gas to approximately 20 ml/min (range 10-30); this usually requires about 20 psi pressure.

6. Set the INCREASE scan speed to 10°C/min, the DECREASE to 5°C/min, and the RANGE to 32 mc/sec. (*Note*: The RANGE setting depends upon the size and type of sample and the heating rate. Faster rates or larger samples require lower settings.)

7. *Heat.* Raise the temperature to 100°C by setting the TEMPERATURE control to 373 K. When the green AVERAGE TEMPERATURE light comes on, turn the recorder chart ON and start the scan by turning the toggle switch to HEAT.

8. *Cool.* When the temperature reaches 150°C (423 K), turn the toggle switch to DECREASE.

9. Repeat step 2 when the temperature reaches 100°C. Allow the apparatus to cool to below 30°C (303 K).

B. Specific Heat of Polyethylene

10. Set the RANGE to 2 mc/sec and the TEMPERATURE control to 303 K. When this temperature has equilibrated, wait about 2 min to obtain a steady base line, then turn the toggle switch to INCREASE and scan to 40°C. Turn the switch to NEUTRAL and wait 2 min to obtain a steady base line. Repeat, raising the temperature in 10° increments up to 120°C (393 K).

11. Repeat step 2, allow the instrument to cool to room temperature, and remove the sample.

12. *Shutdown.* Turn the RANGE switch OFF and recorder power off, and stop the inert gas flow.

J. *Fundamental Equations (Procedure I only)*

$$\Delta H/m = (K/mq) \int_{T_i}^{T_f} \Delta T \, dT \tag{9-9a}$$

$$c_p = K \, \Delta T/qm \tag{9-8b}$$

K. *Calculations*

Enthalpy

1. Read K from a calibration curve, obtained with materials with known heats of fusion (usually metals such as gallium, indium, zinc, and tin), at the temperature of the peak in the melting curve.
2. Evaluate the integral in Eq. 9-9*a* as the area between the peak and base line, in square inches, times the x- and y-axis scale factors in deg. C/in.
3. Evaluate $\Delta H/m$ from Eq. 9-9*a* (see step M-3).

Specific Heat

4. Evaluate K (*Note*: This is not the same, numerically, as that in step 1).
5. Evaluate ΔT as illustrated in Fig. 22-1.
6. Evaluate c_p from Eq. 9-8*b* (see steps M-3 and M-4).

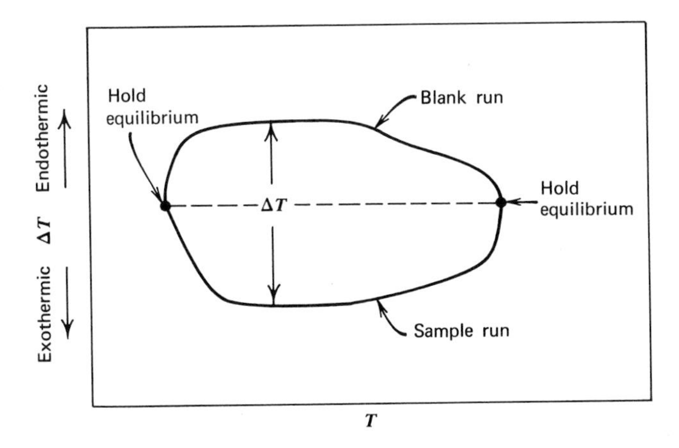

Fig. 22-1. DSC thermogram for the evaluation of specific heat with the Du Pont 900 DSC cell.

L. Report

1. Describe the apparatus and experiment in your own words.
2. Report values for heats of fusion or crystallization, and for heat capacities above and below transitions, for all experiments.
3. Include all graphs from the instruments, properly labeled.
4. Compare heats of fusion and heat capacities with literature values where available (see Boundy 1952, Inoue 1961, and the *Textbook*, Table 5-1, among other sources). Comment on any differences observed.
5. Compare the heats of fusion and crystallization, measured on the same polymer, and discuss their difference, if any.
6. Answer the following questions:
   a. How would you measure c_p at the fusion temperature?
   b. How is the effect of c_p eliminated in the measurement of ΔH?
   c. What parameters affect the base line, and why?

M. Comments

1. Better reproducibility and fewer spurious effects are obtained if samples are heated at 10-20°C/min to about 130°C, then cooled, before measurement. This is particularly valuable for the samples synthesized in Exp. 1-9, where residual volatiles may be present initially. This annealing treatment also relieves mechanical stresses which may cause spurious thermal effects.
2. For measurements below room temperature, place the cooling accessory in position and add liquid nitrogen until the desired sample temperature is reached, then remove the accessory.
3. For most accurate results, measure the heating rate by timing the change in temperature directly instead of relying on the nominal programmed rate.
4. Recrystallization of the sample on heating can cause poor reproducibility in the measurement of c_p.

N. General References

Chap. 9C; *Textbook*, Chap. 4E; Ke 1960; Wunderlich 1965, 1967, 1971; O'Neill 1966; Gray 1968; Baxter 1969; Barrall 1970.

O. Bibliography

Barrall 1970. E. M. Barrall, II, and Julian F. Johnson, "Differential Scanning Calorimetry--Theory and Applications," Chap. 1 in Philip E. Slade, Jr., and Lloyd T. Jenkins, eds., *Techniques and Methods of Polymer Evaluation*, Vol. 2, *Thermal Characterization Techniques*, Marcel Dekker, New York, 1970.

Baxter 1969. R. A. Baxter, "A Scanning Microcalorimetry Cell Based on a Thermoelectric Disc--Theory and Applications," pp. 65-84 in R. F. Schwenker and P. D. Garn, eds., *Thermal Analysis*, Vol. 1, *Instrumentation, Organic Materials and Polymers*, Academic Press, New York, 1969.

Boundy 1952. Ray H. Boundy and Raymond F. Boyer, eds., *Styrene, Its*

Polymers, Copolymers and Derivatives, Reinhold Publishing Corp., New York, 1952.
Du Pont. E. I. du Pont de Nemours and Co., Instrument Products Division, Wilmington, Delaware 19898.
Gray 1968. Allan P. Gray, "A Simple Generalized Theory for the Analysis of Dynamic Thermal Measurements," pp. 209-218 in Roger S. Porter and Julian F. Johnson, eds., *Analytical Calorimetry*, Plenum Press, New York, 1968.
Inoue 1961. M. Inoue, "Heat of Fusion and Interaction of Polyoxymethylene with Diluents," *J. Polymer Sci. 51,* S18-S20 (1961).
Ke 1960. Bacon Ke, "Characterization of Polyolefins by Differential Thermal Analysis," *J. Polymer Sci. 42*, 15-23 (1960).
O'Neill 1966. M. J. O'Neill, "Measurement of Specific Heat Functions by Differential Scanning Calorimetry," *Anal. Chem. 38*, 1331-1336 (1966).
Perkin-Elmer. Perkin-Elmer Corporation, Instruments Division, Norwalk, Connecticut 06812.
Wunderlich 1965. Bernhard Wunderlich, "Specific Heat of Polyethylene Single Crystals," *J. Phys. Chem. 69*, 2708-2081 (1965).
Wunderlich 1967. Bernhard Wunderlich and C. M. Cormier, "Heat of Fusion of Polyethylene," *J. Polymer Sci. A-2 5*, 987-988 (1967).
Wunderlich 1971. Bernhard Wunderlich, "Differential Thermal Analysis," Chap. VIII in Arnold Weissberger and Bryant W. Rossiter, eds., *Physical Methods of Chemistry*, Vol. 1 of A. Weissberger, ed., *Techniques of Chemistry*, Part 5, Interscience Div., John Wiley and Sons, New York, 1971.

EXPERIMENT 23. THERMAL CONDUCTIVITY

A. Introduction

In this experiment, the thermal conductivity of polystyrene is measured using a Colora Thermoconductometer (Dynatech). The experiment requires 2-3 hours time.

B. Principle

A polymer sample held between two silver plates is heated by contact of the vapor of a boiling liquid with the bottom plate. Heat flows through the sample, heating a lower-boiling liquid in contact with the upper plate. The time for a given amount of the lower-boiling liquid to distill is a measure of the heat flow through the sample.

C. Applicability

The Colora Thermoconductometer can be used for all solid materials having thermal conductivities below 0.2 cal/cm-sec°C, and operates over the temperature range from 23°C to 170°C. The measurement of polymers above their glass transition can present some difficulties, however, because they behave as viscoelastic solids (see step M-1).

D. *Precision and Accuracy*

The accuracy claimed for this method is ±3%. However, because polymer materials are relatively poor heat conductors, the accuracy is probably closer to ±6%. For most practical purposes this is, however, quite acceptable.

E. *Safety Considerations*

AT LOW TEMPERATURES HIGHLY VOLATILE, FLAMMABLE, AND SOMETIMES TOXIC SOLVENTS MAY BE USED AND THEY MUST BE HANDLED WITH CAUTION. NO OPEN FLAMES SHOULD BE ALLOWED, AND IT IS RECOMMENDED THAT THE APPARATUS BE OPERATED IN A FUME HOOD. THE COLORA GLASSWARE IS VACUUM JACKETED AND IF CRACKED COULD IMPLODE VIOLENTLY. GREAT CARE MUST BE USED IN HANDLING THE GLASSWARE. *SAFETY GLASSES MUST BE WORN WHEN OPERATING THIS EQUIPMENT;* THEY ARE REQUIRED IN THE LABORATORY AT ALL TIMES.

F. *Apparatus*

1. Colora Thermoconductometer (Dynatech) (Figs. 9-21, 9-22).
2. Stopwatch.
3. Minute timer.
4. Micrometer.

G. *Reagents and Materials*

1. Polystyrene, prepared in Exp. 4 or 9, or commercial material; or other polymers (recommended are poly(methyl methacrylate), polyethylene).
2. Acetone, reagent grade.
3. *n*-Pentane, reagent grade.

H. *Preparation*

1. Samples must be pressed into films or sheets about 1.5 mm thick and 15-17 mm in diameter. (*Note*: Samples must be thick enough that centering rings in the apparatus do not touch.)

I. *Procedure*

(*Note*: It is assumed that the Colora instrument is prepared for use. If a calibration curve is not furnished for the liquid pair acetone-pentane, the glass calibration standards must be measured as well as the polymer samples; see step M-2.)

1. Fill vessel A (Fig. 9-22) approximately half full with acetone. Refer to the instrument instructions for details. Fill the dispensing bottle (Fig. 9-21, left) with pentane up to the level of the metering reservoir. Fill the metering reservoir by tipping the dispensing bottle 45°. Then tip vessel B 45°, insert the dispensing tube, and pump about 12 ml pentane from the metering reservoir into B. Add 5 or 6 boiling stones to vessels A and B. (CAUTION: If boiling stones other than those provided with the Colora are used they should be crushed and a size between 8 and 12 mesh should be used. Larger stones can become jammed in the glass tubing making the removal of old or addition of fresh boiling stones impossible.)

2. Put a centering ring on the lower silver plate S_1, place a drop of oil on the plate, and put the sample in place. Place a drop of oil on the sample, attach a centering ring to the upper plate S_2 (at the bottom of vessel B), and place vessel B on the sample, centering it carefully and holding it vertical. Carefully lower the spring clamp until the rubber end presses against vessel B, and advance it until the indicator reaches the red mark.

3. Attach the measuring burette M to vessel B, start the water flow through the condensers, and initiate heating.

4. When 1 ml pentane has boiled over, conduct successive runs without interruption on a rapid fixed time schedule, in order to maintain temperature equilibrium and make all measurements under exactly the same conditions:

a. Start the minute timer.

b. Remove tube M and release and raise the spring clamp.

c. Lift vessel B off of A, empty it, and place it in the holder (Fig. 9-21, left).

d. Attach the aspirator hose and turn on the aspirator.

e. Remove the sample and both centering rings. Wipe off oil from the rings and silver plates.

f. Wipe off the sample and measure its thickness (this can be done later unless the sample is to be rerun).

g. Just before 2 min on the timer, replace the lower centering ring and place a small drop of oil on plate S_1.

h. At 2 min, place the sample on S_1 and center it carefully. Press it down lightly with a pencil eraser to set it firmly in place, with good contact through the oil over its entire area.

i. Clean plate S_2 and attach a centering ring to it.

j. Just before 3.5 min, empty any accumulated solvent from the metering reservoir.

k. At 3.5 min, remove the aspirator hose and insert the dispensing tube into vessel B.

l. At 3.75 min, fill the metering reservoir.

m. At 4 min, pump pentane into vessel B, using moderate air pressure. Avoid blowing excess air into B.

n. Place a small drop of oil on the center of the sample, carefully center plate S_2 and vessel B on the sample, and press down firmly but do not rotate B. (*Note*: This is an important step and must be done carefully. Plates S_1 and S_2 must be held firmly parallel until the spring pressure is applied.)

o. Bring down the spring clamp and apply the load slowly up to the red mark while holding vessel B firmly to keep the plates parallel.

p. Empty tube M and attach it gently; avoid moving vessel B.

q. Initiate the run and time it with the stopwatch. When the run is complete, record the run time and return to step a.

J. *Fundamental Equations*

$$K = \ell/RA \tag{9-12}$$

where R is the thermal resistance read from a calibration curve (like that of Fig. 9-23) for the run time measured in step I-4q, and ℓ and A are the thickness and area of the sample, respectively.

K. *Calculations*

Calculate K by Eq. 9-12.

L. *Report*

1. Describe the apparatus and experiment in your own words.
2. Compare your results with those in the literature. See Knapp (1969) for the thermal conductivity of polystyrene, Eiermann (1962) and Reese (1966) for poly(methyl methacrylate), and Sheldon (1965) for polyethylene.
3. Answer the following questions:

a. What is the effect of molecular weight on thermal conductivity?

b. In what way do crystallinity and orientation affect thermal conductivity?

c. What advantages and disadvantages does the method of this experiment have over the classical guarded hot plate method (ASTM C 177)?

M. *Comments*

1. For measurements near or above the glass-transition temperature, extra care must be taken to anneal the sample since distortion of the sample can cause large deviations. In any case, sample thickness should be measured after the experiment.

2. A calibration curve is prepared using a selected liquid pair and several standard materials of known thermal resistance. A set of glass samples of different thicknesses is provided with the Colora instrument, together with values for their thermal resistance and thermal conductivity over the temperature range 20 to 200°C. The measured times for distillation of 1 ml of liquid are determined and plotted against the known heat resistance of the calibration samples. Because this plot yields a straight line, it is usually sufficient to measure only two or three calibration samples for each liquid pair of interest. A calibration curve need only be prepared once for a given

apparatus, but should be checked periodically.

3. Samples which are too thick will take longer to come to temperature equilibrium, and the time it takes to distill 1 ml of solvent will also be increased. Ideally, the sample thickness should be selected so that the thermal resistance falls in the midrange of the calibration curve.

4. The selection of the solvent pair is made on the basis of the desired measurement temperature and solvent toxicity, flammability, purity, availability, and other factors.

5. Any unevenness in the sample can have a large effect on the result. Great care should be taken to assure a flat sample and good contact between the sample and the silver plates.

N. General References

Chap. 9D; Schroeder 1959-1960, 1963; Tye 1969; Knappe 1971.

O. Bibliography

ASTM C 177. Standard Method of Test for Thermal Conductivity of Materials by Means of the Guarded Hot Plate. ASTM Designation: C 177. American Society for Testing and Materials, Philadelphia, Pennsylvania 19103.

Dynatech. Dynatech Corporation, Cambridge, Massachusetts 02139.

Eiermann 1962. K. Eiermann and K. H. Hellwege, "Thermal Conductivity of High Polymers from -180°C to 90°C," *J. Polymer Sci. 57*, 99-106 (1962).

Knappe 1971. W. Knappe, "Thermal Conductivity of Polymers" (in German), *Fortschr. Hochpolym. Forsch. (Advances in Polymer Sci.) 7*, 477-535 (1971).

Reese 1966. W. Reese, "Temperature Dependence of the Thermal Conductivity of Amorphous Polymers: Polymethylmethacrylate," *J. Applied Phys. 37*, 3227-3230 (1966).

Schroeder 1959-1960. J. Schroeder, "A Simplified Method for Determining the Thermal Conductivity of Solids," *Philips Technical Review 21*, 364-368 (1959-1960).

Schroeder 1963. J. Schroeder, "Apparatus for Determining the Thermal Conductivity of Solids in the Temperature Range from 20° to 200°C," *Rev. Sci. Inst. 34*, 615-621 (1963).

Sheldon 1965. R. P. Sheldon and Sister K. Lane, "Thermal Conductivity of Polymers. II. Polyethylene," *Polymer 6*, 205-212 (1965).

Tye 1969. R. P. Tye, *Thermal Conductivity*, Academic Press, New York, 1969.

EXPERIMENT 24. THERMOGRAVIMETRIC ANALYSIS

A. *Introduction*

The thermal decomposition behavior of a polymer is examined in this experiment, employing a dynamic method of temperature control. The time required for this experiment is 3 hours.

B. *Principle*

Thermogravimetric analysis (TGA) is an analytical technique for determining the weight change of a sample as a function of temperature. The sample is placed in a furnace while being suspended from one arm of a precision balance. The change in sample weight is recorded as the sample is maintained at constant high temperature or dynamically heated by programmed heating.

A TGA curve may be presented either in the form of the weight loss of a polymer as a function of temperature or in differential form, where the change of weight with time is plotted as a function of temperature (Fig. 9-24).

McCaffery (1970) includes an experiment (Exp. 21) on TGA.

C. *Applicability*

The technique of TGA can be applied to virtually all polymers.

D. *Precision and Accuracy*

A precision of 0.1 microgram and 0.05% accuracy can be achieved with the Cahn Electrobalance. However, most recorders can be read only to 0.1%, which sets a maximum to the observable precision. Reproducibility is usually significantly poorer. Absolute accuracy has no real meaning in this experiment. (See also step M-1.)

E. *Safety Considerations*

ALL THE NORMAL LABORATORY SAFETY PRACTICES SHOULD BE EMPLOYED DURING THIS EXPERIMENT. SPECIAL CARE SHOULD BE TAKEN TO AVOID BURNS FROM THE FURNACE. IF THE OPERATION IS CARRIED OUT UNDER VACUUM OR WITH INERT GAS PURGING, CARE SHOULD BE TAKEN TO PREVENT EXCESSIVE GAS PRESSURE BUILDUP, AND PURGE GASES MUST BE PROPERLY VENTED. SAFETY GLASSES MUST BE WORN IN THE LABORATORY AT ALL TIMES.

F. *Apparatus*

1. Perkin-Elmer TGA apparatus, consisting of Cahn Electrobalance and temperature programmer (Fig. 9-26), or equivalent. (*Note*; The procedure is written for this unit, and other TGA equipment will

require different procedural details.)

2. 50-mg class M weight for calibration.
3. X-Y recorder.
4. Vacuum pump (mechanical).
5. Mercury U-tube manometer or accurate vacuum gauge, range 0-1 at.

G. Reagents and Materials

1. Polymer sample in finely divided form; recommended are polystyrene prepared in Exp. 4 or 9, other polymers prepared in Exp. 1-9, or sample selected from App. I.
2. Helium gas cylinder with regulator.

H. Preparation

I. Procedure

Instrument Setup and Controls

It is assumed that the balance is set up ready for the introduction of a sample. The following controls are used: The MASS RANGE selector sets the weight range of the balance in mg; for example, 1-100 mg. The MASS control is a 10-turn potentiometer reading in percent of the MASS RANGE; for example, if a 50-mg sample is used, set the MASS RANGE to 1-100 mg and the MASS control to 0.5000. The RECORDER RANGE and CALIBRATE RECORDER are, respectively, coarse and fine recorder span controls, and the FACTOR switch controls a 10× attenuator to extend the recorder range. The 0/10 control is a zero weight adjustment used with empty pan, and the SET 5 control is the zero adjustment used with a calibrating weight. The FILTER switch is used to reduce the noise level at the recorder. (*Note*: If the filter is used, all calibration and measurement must be done at the same filter setting. The filter slows the recorder response, but at ordinary heating rates (10°C min.) the speed of response is still adequate.)

Calibration

1. With an alcohol burner, heat the balance wires and pans to remove any residues and moisture. (CAUTION: Do not overheat or burn the wires.)
2. Set the MASS RANGE to 0-100 and the MASS control to 0.5000. Zero the weight axis of the recorder.
3. With forceps, place the 50-mg class M weight on the sample pan, and counterbalance it with student weights. (*Note*: Small pieces of wire can also be used for counterbalancing, since the exact weight of the counterbalance is not required.)
4. Set the RECORDER RANGE to 1 and the FACTOR switch to 1. Adjust the SET 5 control to bring the recorder to zero. Readjust the counterbalance weight if required to bring the pen to zero.
5. If the recorder trace is noisy, increase the FILTER setting,

but do not change it hereafter.

6. Remove the calibrating weight, set the MASS control to 0.0000, and zero the recorder with the 0/10 control. Replace the calibrating weight, reset the MASS control to 0.5000, and readjust the SET 5 control to zero the recorder.

7. Repeat step 6 as required until the recorder zero remains the same for both conditions. DO NOT reset the 0/10 or SET 5 controls hereafter.

Adjust for Sample

8. Place 50-75 mg sample on the pan. Adjust the MASS control until the recorder reads zero. The MASS control dial now reads the exact sample weight. Record this value.

9. Set the MASS control to 0.0000. Bring the pen to a convenient point near the top of the chart by means of the CALIBRATE RECORDER and, if necessary, RECORDER RANGE controls. The balance is now ready for the experiment.

Atmosphere

(*Note*: Use of helium or helium-oxygen atmosphere rather than nitrogen or air is recommended to reduce "noise" resulting from disturbance of the pan by convection currents. See step M-2.)

10. Be sure that all valves and joints are lightly greased, close all openings, *close* the valve between the vacuum pump and balance, and start the pump.

11. VERY CAREFULLY AND SLOWLY open the vacuum valve to the balance while observing the wire and pan. Sudden vacuum may pull the sample off the pan or disrupt the balance. When full vacuum is finally achieved, pump for several minutes, then close the vacuum valve and slowly bleed in helium. AGAIN OBSERVE CARE TO CHANGE PRESSURE VERY SLOWLY. (*CAUTION*: Use an accurate vacuum gauge, preferably a mercury U-tube manometer) to see that 1 at. pressure is NEVER exceeded; even a slight excess pressure can blow off glass fittings and damage the balance.

Heating

12. Adjust the temperature axis on the recorder so that the pen is set to ambient temperature. Set the temperature programmer for a heating rate of 10°C/min, and initiate the heating cycle. If the programmer has an upper temperature limit shutoff, set this appropriately. Otherwise, monitor the temperature and shut off the heater at the proper point. Record the curve of sample weight versus temperature.

13. Allow the apparatus to cool down, and repeat steps 1 and 8-12 for subsequent samples.

J. *Fundamental Equations*

$$\text{Percent weight loss} = 100(w_0 - w)/w_0 \quad (24\text{-}1)$$

$$dw/dt = kw^n \quad (24\text{-}2)$$

$$k = A \exp(-\Delta E/RT) \tag{24-3}$$

K. Calculations

1. Calculate percent weight loss at the temperature of decomposition.

2. Calculate and plot the differential weight loss and dw/dT (see Fig. 9-24).

3. Calculate the decomposition rate constant dw/dt and the order of the reaction n.

L. Report

1. Describe the apparatus and experiment in your own words.

2. Include the thermograms and plots of the differential rate loss for all samples measured.

3. Report the decomposition temperature, decomposition rate constant, and order of the reaction.

4. Answer the following questions:

   a. What is the effect of heating rate on the decomposition temperature and rate?

   b. What are the relative merits of isothermal and programmed heating TGA if kinetic data are desired? What differences would you expect between the two methods?

   c. What effect can competitive reactions have on the thermogram?

   d. How could you account for a weight gain in the thermogram?

   e. How do sample size and shape, gas pressure and atmosphere, and ignition of the sample affect the results?

M. Comments

1. Errors and deviations in the TGA curves can result when some of the products are entrapped in the polymer. A further complication making interpretation difficult can occur when a material decomposes in two or more stages with overlapping weight losses.

2. If helium-oxygen mixtures are to be used, the mixtures can easily be charged into the balance using elementary gas-law calculations based on the manometer pressure readings.

3. Many variations of this experiment can be made, especially in the study of the kinetics of polymer degradation: One or more isothermal curves can be measured in addition to the normal weight-loss curve (Anderson 1966); the effect of gaseous atmosphere, especially oxygen, on the decomposition rate or reaction order can be studied (this can be extended to study the effect of an antioxidant, especially for a polymer such as *cis*-polybutadiene); the effect of different thermal stabilizers on the decomposition of poly(vinyl chloride) can be studied; the effluent gases can be analyzed by gas chromatography to provide insight into the degradation mechanism (of particular interest is the study of polymers which decompose to significant amounts of monomer; see the *Textbook*, Table 12-5).

N. General References

Chap. 9E; Wendlandt 1964; Anderson 1966; Doyle 1966; Reich 1971.

O. Bibliography

Anderson 1966. Hugh C. Anderson, "Instrumentation, Techniques and Applications of Thermogravimetry," Chap. 3 in Philip E. Slade, Jr. and Lloyd T. Jenkins, eds., *Techniques and Methods of Polymer Evaluation*, Vol. 1, *Thermal Analysis*, Marcel Dekker, New York, 1966.

Doyle 1966. Charles D. Doyle, "Quantitative Calculations in Thermogravimetric Analysis," Chap. 4 in Philip E. Slade, Jr. and Lloyd T. Jenkins, eds., *Techniques and Methods of Polymer Evaluation*, Vol. 1, *Thermal Analysis*, Marcel Dekker, New York, 1966.

McCaffery 1970. Edward L. McCaffery, *Laboratory Preparation for Macromolecular Chemistry*, McGraw-Hill Book Co., New York, 1970.

Perkin-Elmer. Perkin-Elmer Corporation, Instrument Division, Norwalk, Connecticut 06852.

Reich 1971. Leo Reich and David W. Levi, "Thermogravimetric Analysis," pp. 1-41 in Herman F. Mark, Norman G. Gaylord, and Norbert M. Bikales, eds., *Encyclopedia of Polymer Science and Technology*, Vol. 14, Interscience Division, John Wiley and Sons, New York, 1971.

Wendlandt 1964. Wesley Wm. Wendlandt, *Thermal Methods of Analysis*, Vol. XIX in P. J. Elving and I. M. Kolthoff, eds., *Chemical Analysis*, Interscience Div., John Wiley and Sons, New York, 1964.

D

Experiments on Structure-Property Relations

EXPERIMENT 25. MELT VISCOSITY

A. *Introduction*

In this experiment, melt viscosity is measured as a function of shear rate and temperature for polystyrene or polyethylene. The experiment requires 3 hours.

B. *Principle*

A capillary rheometer of the constant shear-rate type is employed to prepare flow curves at various temperatures. A sample, usually in the form of a powder or granules, is loaded into the thermostatically controlled extrusion reservoir and heated to a specified temperature. After the polymer has reached temperature equilibrium, it is forced out through a capillary die by a plunger fastened to the moving crosshead of a tensile testing machine. The plunger moves at constant speed, which can be pre-selected to cover the shear-rate range of interest. The force on the plunger is a measure of the shear stress. The effect of temperature and rate of shear on the surface characteristics of the extrudate is examined, and the difference in behavior between different materials can be studied. Other useful variations of the experiment are described in step M-1.

C. *Applicability*

This experiment can be carried out with virtually all thermoplastics and elastomers, provided the proper temperature and ranges of shear rate can be selected. The latter depend on the magnitude of the viscosity, and a range of capillary dies of different length and diameter must be available from which a proper selection can be made. Highly crosslinked polymers and thermoset polymers are not suited for this experiment.

D. *Precision and Accuracy*

In the absence of exposure of the polymer to high temperatures or shear rates, viscosity-shear rate curves for polystyrene and polyethylene can be reproduced to within ±5%. This precision depends on achieving temperature control to within ±0.5°C. Attaining temperature equilibrium and keeping the sample free from occluded air are primary considerations in order to achieve good reproducibility.

E. *Safety Considerations*

ALL NORMAL LABORATORY SAFETY PROCEDURES SHOULD BE OBSERVED. SAFETY GLASSES MUST BE WORN AT ALL TIMES IN THE LABORATORY, AND MILL GLOVES MUST BE WORN TO PREVENT BURNS WHEN WORKING WITH EQUIPMENT HEATED TO ELEVATED TEMPERATURES. GAUGE CONTROLS ON THE INSTRUMENT MUST BE SET TO PREVENT THE PLUNGER FROM COMING IN CONTACT WITH THE DIE. FAILURE TO CHECK THIS CAN PERMANENTLY DAMAGE THE LOAD CELL IN ADDITION TO DAMAGING THE DIE AND THE PLUNGER.

F. *Apparatus*

1. Instron Model 3211 Rheometer (Fig. 10-8), Instron Tensile Tester Model TTB or TTC or equivalent with push-button crosshead speed selector and Type MCR Rheometer attachment, or an equivalent rheometer (see step M-2).
2. 0-10,000 lb. load cell.
3. Capillary dies having the following lengths and diameters:

ℓ (in.)	d (in.)
1.000	.05000
2.000	.05000
2.000	.03000
2.000	.06000

4. Micrometer.

G. *Reagents and Materials*

Polystyrene or polyethylene selected from the sources in App. I; or any commercially available material; or polymer prepared in Exp. 7. Approximately 60-100 g is required.

H. *Preparation*

1. To prevent or minimize bubbles in the melt, the polymer samples should be dried at 80°C in a vacuum oven for at least 5 hours and then stored in a sealed container prior to use.
2. The heaters on the rheometer should be turned on at least one hour before the experiment to allow temperature stabilization. Set

the temperature controller to the lowest measurement temperature (see step I-1).

I. Procedure

It is assumed that the rheometer is set up ready for operation, that the load cell has been calibrated according to the manufacturer's instructions, and that the proper die has been selected and inserted in the instrument.

Instrument Check

1. Insert the plunger in the barrel and allow it to equilibrate at the temperature set on the controller. (*Note*: When making runs at more than one temperature, measure at the lowest temperature first. For polystyrene or polyethylene, 160°C is a good starting temperature.)
2. Set the limit of upward crosshead travel by means of the GAGE LENGTH dial. Using the MANUAL POSITIONING knob, move the crosshead up until there is sufficient room for loading the barrel. Set the GAGE LENGTH dial to zero. When run upward automatically, the crosshead will always stop at this selected position.
3. Hold the plunger in place and run the crosshead down. As the plunger approaches the barrel entrance, switch to the MANUAL POSITIONING knob and guide the plunger into the barrel. Use the slowest crosshead speed, 0.02 in./min, at this point. Be sure the PEN MOTOR switch is ON. With the RANGE SELECTOR on the most sensitive scale (500 lb), carefully move the plunger down the barrel until it rests in the capillary entrance and there is a small force (1-2 lb) indicated by the pen. Note the RETURN dial reading and move the crosshead up 0.5 in. Set the RETURN dial to zero at this position of the crosshead. When run down automatically, the crosshead will always stop at this selected position.
4. Weigh out 14.0 g polymer and load it into the rheometer barrel in three or four increments, tamping the material with the special brass rod supplied with the instrument. This operation should take about 1 minute.

Measurement

5. Hold the plunger in place and move the crosshead down automatically until the plunger approaches the barrel. Carefully guide the plunger into the barrel by manual positioning. Set the LOAD SELECTOR to 500 lb full scale, and pack the polymer further by moving the crosshead down until melt is extruded. Care should be taken that the safe load limits of the plunger are not exceeded. (*Note*: Repeated operation of the equipment has established a rule of operation which will prevent plunger damage: *With the plunger less than five inches in the barrel, the load should not be allowed to exceed 500 pounds. With less than eight inches of plunger in the barrel, the load should not be allowed to exceed 1,000 pounds. With the plunger more than eight inches in the barrel, the load should not exceed 2,000 pounds.*)
6. Compress the material until the chart shows a load of about 2000 lb, using the manual crosshead control and a crosshead speed of 10 in./min. The crosshead travel is stopped when 2000 lb is attained,

and 2 min preheat time is allowed for the polymer to reach the set temperature.

7. Select the crosshead speed setting corresponding to the lowest shear rate of interest. Switch ON the CHART MOTOR, move the CLUTCH lever to FORWARD, and depress the DOWN button.

8. Observe the chart load trace. Record the load when it reaches a steady value. Collect a small specimen of the extrudate by cutting it off close to the capillary exit.

9. Select the next higher crosshead speed and continue the run, repeating step 8. Do this for each crosshead speed desired, in order of increasing speed. At the highest crosshead speed, run the plunger down until it is stopped by the downward-travel limit switch.

10. If the same polymer is to be remeasured, proceed as follows:

a. Withdraw the plunger at the fastest possible crosshead speed, being careful not to overload the equipment. Rotate the plunger by hand as it is being withdrawn and select a crosshead speed which allows continued rotation without binding. Clean the plunger with a brass spatula, cloth, and steel wool.

b. Increase the temperature setting by 10°C, allow 15 min for temperature equilibration, and repeat steps 4-9.

c. Continue to repeat step b until the measurements have been made at the highest temperature desired. (*Note*: For polystyrene or polyethylene, 190 or 200°C is a suitable upper temperature limit.) Then proceed to step 11 or step 12, whichever is appropriate.

11. If a different polymer is to be measured, proceed as follows:

a. When the plunger is at its lowest position in the barrel, there still remains a plug of polymer 0.5 in. above the capillary due to the previously set clearance of the plunger above the capillary entrance. Extrude this polymer by moving the crosshead down *slowly* by means of the MANUAL POSITIONING knob. Be careful not to exceed the load capacity of the plunger or load cell. When the plunger tip reaches the capillary entrance a sudden sharp rise in force will be indicated on the chart. (*Note*: *The limit switch which stops downward travel of the crosshead is bypassed with the plunger in this position. Do not use the automatic crosshead travel controls at this point.*) Manually raise the crosshead to zero on the RETURN dial. This will make the downward-travel limit switch operative again and the automatic crosshead travel controls may be used with safety.

b. Repeat step 10a.

c. Repeat steps 4 and 5 with the new material, but run this charge through at the highest crosshead speed to flush out the old polymer in the capillary. Then repeat step a. (*Note*: The capillary may be removed and the barrel cleaned as in step 13, but experience has shown that three to four flush runs of the new material, when the plunger is moved to the capillary entrance in each flush, is sufficient to remove most contamination.)

d. Repeat steps 4-9 for the new polymer material, then proceed to step 10, 11a, or 12, whichever is appropriate.

12. If the last run of the day has been completed, repeat steps 11a and 11b, then proceed to step 13.

Shutdown

13. Repeat steps 11a and 10a. Then remove the capillary and clean

the capillary and barrel with steel wool, then cotton. Shut the MAIN POWER, CHART MOTOR, and PEN MOTOR switches OFF.

J. *Fundamental Equations*

$$\dot{\gamma}_w = (2\ d_p^2/15\ d_c^3)V \tag{10-4}$$

$$s_w = (d_c/\pi\ d_p^2\ \ell)F \tag{10-5}$$

$$s_w = k\ \dot{\gamma}_w^n \tag{10-2}$$

$$\dot{\gamma}_{t,w} = \dot{\gamma}_w(3n + 1)/4n \tag{10-7}$$

K. *Calculations*

1. Calculate the shear rates for the crosshead speed settings and die used in the experiment, using Eq. 10-4.
2. Calculate the shear stress at each shear rate and temperature, using Eq. 10-5.
3. Plot shear stress versus shear rate for the data set at one temperature, and calculate the power law exponent, n, using Eq. 10-2.
4. For the set of data used in step 3, calculate the true wall shear rate using Eq. 10-7.
5. Calculate the apparent viscosity at each shear rate and temperature.
6. Measure the extrudate diameters with a micrometer.

L. *Report*

1. Describe the apparatus and experiment in your own words.
2. Tabulate the shear-rate and shear-stress data for each run.
3. Tabulate the extrudate diameter for each shear rate for each run.
4. Plot log viscosity versus log shear rate for each run.
5. Plot the ratio of the extrudate diameter to the die diameter versus shear rate.
6. Comment on the nature of the extrudate. If melt fracture occurred and if different temperatures were used, comment on the effect of temperature on melt fracture.
7. Comment on the reproducibility of the data.
8. Comment on the magnitude of the Rabinowitsch correction.
9. If different temperatures were used, plot log viscosity versus $1/T$ K and calculate the flow activation energy. Compare the value obtained with those in the literature, 23 kcal/mole for polystyrene and 7 kcal/mole for linear polyethylene (Porter 1965).

10. Include all recorder charts, samples of extrudate, and plots of the data.

11. Answer the following questions:

a. Why does it take longer to reach the steady state at lower crosshead speeds?

b. How can the barrel correction be minimized?

c. Why is it necessary to know the die diameter to 5 significant figures?

M. Comments

1. Variations to this experiment can be made as follows:

a. Dies of different diameter but the same length can be used to measure the end correction.

b. Samples of different molecular weights can be used to establish the effect of molecular weight on melt viscosity and die swell.

c. A comparison can be made of linear and branched polyethylene to establish the effect of branching on melt viscosity and die swell.

2. Other capillary instruments suited for this experiment include the Sieglaff-McKelvey Rheometer (Tinius Olsen), the CIL Rheometer (Mullowney), and the MRC Automatic Rheometer (Monsanto).

N. General References

Chap. 10A; *Textbook*, Chap. 6A; Van Wazer 1963; Lenk 1968; Mendelson 1968; Brydson 1970; Williams 1971, Chap. 12.

O. Bibliography

Brydson 1970. J. A. Brydson, *Flow Properties of Polymer Melts*, Van Nostrand-Reinhold, New York, 1970.

Instron. Instron Corporation, Canton, Massachusetts 02021.

Lenk 1968. R. S. Lenk, *Plastics Rheology*, Interscience Div., John Wiley and Sons, New York, 1968.

Mendelson 1968. Robert A. Mendelson, "Melt Viscosity," pp. 587-620 in Herman F. Mark, Norman G. Gaylord, and Norbert M. Bikales, eds., *Encyclopedia of Polymer Science and Technology*, Vol. 8, Interscience Div., John Wiley and Sons, New York, 1968.

Monsanto. Monsanto Research Corporation, Dayton, Ohio 45407.

Mullowney. F. J. Mullowney Company, Trenton, New Jersey 08600.

Porter 1965. R. S. Porter and J. F. Johnson, "The Entanglement Chain Length and Polymer Composition," pp. 467-477 in E. H. Lee, ed., *Proceedings, 4th International Congress of Rheology (1963)*, Part 2, Interscience Div., John Wiley and Sons, New York, 1965.

Tinius Olsen. Tinius Olsen Testing Machine Company, Willow Grove, Pennsylvania 19090.

Van Wazer 1963. J. R. Van Wazer, J. W. Lyons, K. Y. Kim, and R. E. Colwell, *Viscosity and Flow Measurement, A Laboratory Handbook of*

Rheology, Interscience Div., John Wiley and Sons, New York, 1963.
Williams 1971. David J. Williams, *Polymer Science and Engineering*, Prentice-Hall, Englewood Cliffs, New Jersey, 1971.

EXPERIMENT 26. CONCENTRATED-SOLUTION VISCOSITY

A. *Introduction*

In this experiment the student is introduced to measurements of viscosity and the nature of viscous flow. Solution viscosity is examined as a function of concentration, shear rate and, optionally, temperature. Procedures are given for three different types of viscometer. Each can be carried out in 3 hours.

B. *Principle*

Molecular parameters are generally obtained from measurements on infinitely dilute solutions in order to eliminate any intermolecular interaction. Under these conditions, the viscous contribution is entirely due to the polymer-solvent interaction. As the concentration is increased, a point is reached where the viscous contribution from polymer-polymer interaction becomes appreciable, and with further increase in concentration this becomes the major contribution to the viscous deformation. The critical concentration, above which the viscosity has a much stronger dependence on concentration, provides an insight into molecular interaction and has been used in developing molecular theories of viscosity.

The critical concentration is dependent on the molecular weight, molecular-weight distribution, and other features of the polymer's molecular structure, such as long-chain branching. Further, since molecular dimensions are highly dependent on the nature or "goodness" of the solvent as well as on the temperature, these two parameters must also be considered. In addition, most polymer solutions are non-Newtonian, which requires that the viscosities be examined and compared at the same shear rates. Molecular theories are based on viscosities measured at limiting or zero shear rate, that is, the shear rate at which the viscosity is, for all practical purposes, independent of the rate of shear; in other words, at the shear rate where the material behaves in a Newtonian manner. For a given temperature, solvent, and concentration, changing the molecular weight, molecular-weight distribution, or molecular structure of the polymer will produce a change in the shear rate for the onset of non-Newtonian behavior (shear-rate thinning). As the molecular weight is increased or the molecular-weight distribution broadened, the onset of non-Newtonian flow occurs at progressively lower rates of shear. The critical rate of shear where a material becomes non-Newtonian, that is, where the viscosity starts to decrease with increasing shear rate, also has significance in molecular theories of viscosity, and can be used to measure the molecular weight.

C. Applicability

This experiment can be carried out with any polymer for which a suitable solvent can be found and having molecular weight at least as high as the critical value, M_c, above which entanglement occurs. The main criterion for the solvent is that it have a low vapor pressure at the operating temperature; loss of solvent due to evaporation can change the concentration appreciably and produce erroneous results.

D. Precision and Accuracy

The precision and accuracy will vary with the particular instrument selected for this experiment. With good temperature control (±0.2°C), accuracy of ±1.5 to 3% for absolute viscosity can be attained with the cone-and-plate and couette viscometers, depending on the viscosity and shear-rate range.

E. Safety Considerations

ALL THE NORMAL PRECAUTIONS IN HANDLING VOLATILE AND HAZARDOUS SOLVENTS SHOULD BE EXERCISED. GOOD VENTILATION IS REQUIRED. CONTACT WITH THE SKIN SHOULD BE AVOIDED. SAFETY GLASSES MUST BE WORN AT ALL TIMES IN THE LABORATORY.

F. Apparatus

Procedures are written for the following viscometers, but equivalent instruments can be used.

1. Manual or automatic Ferranti-Shirley cone-and-plate viscometer with the thermostat control (Ferranti) (Fig. 10-10).
2. Rotovisco couette viscometer (Haake), with measuring system NV or MV-1 (Fig. 10-13).
3. Brookfield Synchro-Lectric viscometer (Brookfield), any one of a number of models (Van Wazer 1963) (Fig. 10-15).

G. Reagents and Materials

1. Polymer prepared in Exp. 3, 4, or 7, if enough of the sample is available to prepare the required amount of solution. Polymers selected from those listed in App. I or other commercially available samples are suitable substitutes.
2. Reagent-grade solvent selected for the specific polymer as follows:
   - Polystyrene - Toluene
   - Poly(methyl methacrylate) - Toluene
   - Poly(vinyl chloride) - Cyclohexanone
   - Polyether or polyester polyurethane - Dimethyl formamide
3. Standard oil for checking instrument accuracy (Cannon).

H. Sample Preparation

1. At least 24 hours in advance of this experiment, prepare the minimum amounts of polymer solutions at various concentrations (see step 2), depending upon the molecular weight and instrument used (see step 3). Weigh out polymer in a flask or bottle, and periodically shake until it is dissolved. A laboratory roller or shaker is ideally suited for this. Some polymers may require heating for total solution.

2. a. For polystyrene of $\overline{M}_n$ = 15,000 and $\overline{M}_w$ = 250,000, the following concentrations (by weight) in toluene are suggested: 7, 10, 12, 15, 17, 20, 25, and 30%.

b. For poly(vinyl chloride) of $\overline{M}_n$ = 65,000 and $\overline{M}_w$ = 160,000 (Geon 103), the following concentrations by weight in cyclohexanone are suggested: 3, 5, 7, 10, 12, 13, 14, and 15%.

c. For the polyurethane samples prepared in Exp. 7, the following concentrations by weight in dimethyl formamide are suggested: 5, 8, 12, 13.5, 15, 17, and 20%.

3. The amount of sample required will vary with the particular instrument used:

a. The Ferranti-Shirley cone-and-plate viscometer requires as little as 3 ml of solution at each concentration, but 10 ml is preferred.

b. The minimum size sample required for the Brookfield viscometer varies with the spindle size used. For the lowest viscosity (less than about 10 cp) the largest spindle is used; this requires a minimum of 150 ml. For the smallest spindle, measurements can be made with as little as 25 ml.

c. The sample size requirement for the Rotovisco with the NV measuring system is a minimum of 7 ml at each concentration, but 20 ml is preferred. The MV system requires a minimum of 40-70 ml, but 100 ml is preferred.

4. The constant-temperature thermostat should be turned on and allowed to stabilize for at least 1 hour prior to the experiment.

5. For the Ferranti instrument, the main power to the amplifier should be turned on at least 1 hour in advance.

I. Procedure

Procedure I. Manual Ferranti-Shirley Cone and Plate Viscometer

(*Note*: Full-scale torque and maximum rpm of the instrument should be checked ahead of time by the instructor according to the procedure recommended by the manufacturer and outlined in the manual.)

1. The main power switch and the switch for the motor on the amplifier should be in the ON position and the INDICATE-RECORD switch set in the INDICATE position.

2. The 10-turn potentiometer on the indicating unit, controlling the cone speed, should be set at ZERO and the scale reading should be set in the ×5 position.

3. The gear-change lever on the measuring unit should be in the NEUTRAL position.

4. Check to see if the small motor shaft directly behind the cone is stationary. If it is rotating, *do not proceed* with the experiment,

but call the instructor to make a zero-drift adjustment.

5. Check the gap between cone and plate as follows: With the SCALE READING knob on the indicating unit set on the CONE AND PLATE position, raise the plate *slowly* up to the cone by turning the knurled cylinder counterclockwise. Two turns gives full vertical travel of the plate. At this point, the two red marks on the cylinder and the fixed housing will coincide, and the spring-loaded locating plunger will be felt to click into position. (CAUTION: Care must be exercised when raising the plate for the first time since too small a gap setting can damage the cone apex.) When the cone touches the plate, electrical contact is made and the meter needle will be deflected. Should this occur before the plate is fully raised, the plate can be lowered (gap increased) by turning the micrometer, located underneath the plate, clockwise. Conversely, if the plate is in the full up position and contact is not made, the micrometer should be rotated counterclockwise until the plate just touches the cone.

The micrometer setting should be adjusted until a deflection is just registered on the meter. The micrometer is then backed off (clockwise) 0.0001". Slow rotation of the cone by hand should show no meter deflection.

It is desirable to check the cone-and-plate setting periodically, since resetting may be necessary to compensate for thermal expansion of the metal. The gap setting should also be checked when changing cones.

6. Set the SCALE READING knob on the indicator unit to ZERO CHECK. The meter should read between 2 and 4 scale divisions. If it does not, the instructor should be called to adjust it.

7. The instrument is now ready for checking with a standard oil. Place a standard oil (Cannon) on the plate in sufficient amount to fill the gap between the cone and plate (3-4 ml is sufficient for the medium cone), and raise the plate into position.

8. Allow 2 minutes for the sample to come to temperature equilibrium.

9. Set the GEAR CHANGE lever to HIGH (cone speeds of 0-100 rpm are obtained for this position with "A" gear setting).

10. Set the CONE SPEED at 1 (1 full turn of the speed control 10-turn potentiometer) for a cone speed of 10 rpm, allow 30 sec for equilibration, then record the reading on the meter. If the meter reading is too low, use the 4×, 3×, or 2× scale to increase the sensitivity. Record the scale factor.

11. Change the CONE SPEED settings successively to 2, 3, 4, 5, 6, 7, 8, 9, and 10, allow 30 sec at each setting for equilibration, and record the meter reading. After completing all settings up to 10 (100 rpm), return the CONE SPEED to ZERO.

12. Return the GEAR CHANGE lever to the NEUTRAL position.

13. Turn the knurled cylinder clockwise till the red dot coincides with the green dot on the housing. The plate should break away from the cone at this point. However, the surface forces of the sample may prevent this. A small rotation of the cone by hand will release the plate, after which it can be lowered.

14. Remove the sample with a tissue or cloth and swab with a suitable solvent (toluene).

15. Repeat steps 7 through 14 with each concentration for the test material at each temperature.

16. When all the polymer solutions have been run, clean the cone and plate thoroughly and turn the AMPLIFIER, MOTOR and MAIN POWER switches OFF.

Procedure II. Automatic Ferranti-Shirley Cone-and-Plate Viscometer

1. The MAIN POWER switch and the switch for the MOTOR on the amplifier should be in the ON position and the INDICATE-RECORD switch set in the RECORD position. The switch on the rear side of the control unit (combined with the indicating unit on the manual instrument) should also be set in the RECORD position.

2. The 10-turn potentiometer on the control unit should be set at 10 and the SCALE READING should be set in the ×5 position.

3. The GEAR CHANGE lever on the measuring unit should be in the NEUTRAL position.

4. Check to see if the small motor shaft directly behind the cone is stationary. If it is rotating, *do not proceed* with the experiment, but call the instructor to make a zero-drift adjustment.

5. Adjust the X-Y recorder for shear rate and shear stress as follows, setting the X-Y SWITCH to SPEED CHECK:

a. *Set the Y-R.P.M. axis.* Adjust the SPEED R.P.M. dial to maximum (10 full turns of the speed control potentiometer), place the GEAR CHANGE lever in the NEUTRAL position, and depress the HOLD SPEED switch. Select the 10 sec SWEEP TIME and press the MOTOR START button. Adjust the recorder Y-axis attenuator to give a maximum pen deflection of 10 in. Check that this deflection coincides with a reading of 100 divisions on the indicator instrument when the SPEED CHECK switch is depressed. If this reading is less than 100 divisions, have the instructor make the necessary adjustments.

Intermediate settings of the maximum speed on the SPEED R.P.M. dial should correspond with the readings shown on the instrument, e.g. a setting of 5 full turns on the SPEED R.P.M. dial should give a reading of 50 divisions on the instrument meter.

b. *Set the X-torque axis.* Select the LOW GEAR on the viscometer and adjust the cone speed to zero. Set the SCALE READING switch to ×5 and deflect the cone spindle manually until the pointer reached maximum deflection when the end of the torque potentiometer track is reached. If this maximum deflection is less than 100 divisions, have the instructor make the necessary adjustments.

6. Follow steps 5 through 9 of Procedure I.

7. Select a time scale for accelerating the cone to full speed (60 sec is satisfactory).

8. Lower the pen on the X-Y recorder.

9. Push the START button on the control unit. The indicator light on the control unit will go off. (The cone will now accelerate from 0 to 100 rpm in 60 sec.)

10. When the cone speed returns to zero, the indicator light comes back on indicating the unit is ready for a second measurement.

11. Shift the zero setting on the X-Y recorder axis one division.

12. Repeat step 9 with the same sample still in the viscometer.

13. Remove the sample by following steps 12 through 14 of Procedure I.

14. Repeat steps 6 through 13 for each concentration and temperature.

Procedure III. Rotovisco Viscometer

1. Turn the viscometer on and allow the electrical components to stabilize for 5 min.
2. Turn the thermostat on, actuate the pump to the cup or jacket, and allow 10-15 min for stabilization.
3. Select a rotor and attach it by holding the knurled knob at the bottom of the measuring head and screwing the rotor onto the stem. Adjust the dial pointer to zero as follows: With the GEAR SHIFT in position 3, (cable now turning at 160 rpm) and with the rotor in air (not in the sample), the dial needle is adjusted to zero on the scale by means of the ZERO ADJUSTMENT. This zero should remain constant in air for all speeds and all rotors.
4. Place the test sample in the cup, filling to the ring mark corresponding to the rotor used; e.g., for the largest rotor in a given series, fill to the lowest ring mark, etc.
5. Push the cup up into the frame. Allow 5 minutes for temperature stabilization.
6. Record the dial reading at each speed setting.
7. Repeat steps 4-6 for each sample.

Procedure IV. Brookfield Viscometer

1. Level the viscometer base by adjusting the knurled knobs on the base of the instrument until the bubble is in the center of the level.
2. Select a spindle and attach it to the instrument stem by holding the stem with one hand and turning the spindle counterclockwise until it is finger tight. (*CAUTION*: Use care in attaching the spindle, as the stem is connected to a spring which should not be strained.)
3. Raise the viscometer high enough so that it is above the sample container. Place the sample under the spindle and lower the viscometer until the spindle is immersed in the liquid to the indented mark on the spindle. (*Note*: This must be done carefully to insure that there are no bubbles of air entrapped under the spindle.) If the diameter of the sample container is small, make sure the spindle is in the center of the container.
4. Turn the SPEED knob to the desired speed. (Four speeds are available on the F models, eight on the T models.)
5. Turn the ON-OFF switch to ON, allow the spindle to rotate approximately 10 revolutions or until the needle stops oscillating, and read the position of the needle on the 100 scale. Record the reading, and repeat at other speeds as desired.

(*Note*: At high speeds, the dial cannot be read during rotation, and a clutch is provided on the instrument for use as follows: Wait until the needle reaches equilibrium, depress the CLUTCH lever, and hold it down until the ON-OFF switch is turned OFF, making sure the needle stops in the window.)

(*Note*: If the reading is offscale or at the extreme ends of the scale, the speed or spindle should be changed. The most accurate readings are in the middle of the scale.)

6. Raise the viscometer and remove the spindle by holding the stem in one hand and turning the spindle clockwise, being careful not to put any strain on the spring.

7. Clean the spindle immediately to avoid any change in its surface.

J. Fundamental Equations

$$s = K\,\dot{\gamma}^{n} \tag{10-2}$$

$$\eta_0 = K' c^{n'} \tag{10-3}$$

$$s = 3\,G/2\pi r^3 \tag{10-8}$$

$$\eta = A\,\exp(-\,\Delta E/RT) \tag{10-10}$$

K. Calculations

Procedures I and II. Ferranti Cone-and-Plate Viscometer

1. Calculate the shear stress, shear rate and viscosity constants using Eq. 10-2 and the relations in the instrument manual.
2. Calculate the viscosity of the standard oil at various cone speeds.
3. Calculate the shear stress and shear rate for one set of polymer solutions, and plot log s versus log $\dot{\gamma}$ for each concentration.
4. Calculate the viscosity for each solution at 100 rpm from the data recorded on the x-y recorder or the values read directly on the meter.
5. For those polymer solutions for which the shear stress is *not* directly proportional to the shear rate, calculate the viscosity at all the shear rates measured, and plot the data using a double logarithmic scale, as in Fig. 10-11. Read and record the value of the limiting zero-shear viscosity, η_0. Calculate the power-law exponent using Eq. 10-2.
6. Plot the limiting zero-shear viscosity η_0 versus concentration, using a double logarithmic scale as in Fig. 10-12, and calculate the slopes of the resulting lines using Eq. 10-3.
7. For polystyrene, compare the critical concentration with the results reported by Onogi 1963.

Procedure III. Rotovisco Viscometer

1. Calculate the shear stress, shear rate and viscosity, using the relations in the instrument manual, for each sample at each speed used.
2. Plot log s versus log $\dot{\gamma}$ for those solutions for which the shear stress decreases with increasing shear rate, and calculate the limiting zero-shear viscosity and the power-law exponent using Eq. 10-2.
3. Plot log limiting zero-shear viscosity versus log concentration and calculate the slopes of the resulting lines using Eq. 10-3.
4. If experiments were carried out at different temperatures,

calculate the flow activation energy for each concentration using the limiting zero-shear viscosity.

5. If polystyrene solutions were run, carry out step 7 of Procedures I-II.

Procedure IV. Brookfield Viscometer

1. Calculate the viscosity at the various speeds by multiplying the dial reading by the appropriate factor in the instrument manual.

2. Plot log viscosity versus log speed in rpm, and for those solutions for which the viscosity is a function of speed, estimate the limiting zero-shear viscosity. Calculate the power-law exponent using Eq. 10-2.

3. Plot log limiting zero-shear viscosity versus log concentration and calculate the slopes of the resulting lines using Eq. 10-3.

4. If polystyrene solutions were run, carry out step 7 of Procedures I-II.

L. Report

1. Describe the apparatus and experiment in your own words.

2. Compare the measured viscosity of the standard oil (in the case of the Ferranti-Shirley) with the value reported by Cannon Instruments. Explain any differences.

3. Include all plots and data in tabular form.

4. Comment on the reproducibility of the data.

5. Comment on sources of error in measuring the viscosity with the particular instrument used and answer the following questions:

a. Does the amount of sample under the cone (in the Ferranti) influence the result?

b. How does the size of the container influence the result using the Brookfield?

c. How does the result vary with the height of the sample in the Rotovisco?

6. If temperature dependence was studied, report the activation energy for each concentration and answer the following questions:

a. How does the activation energy change with shear rate?

b. How does the activation energy change with shear stress?

7. Answer the following questions:

a. Does the viscosity change with time at a given shear rate? If so, why? In what direction would you expect the change to be at higher shear rates, and why?

b. How would you check for the effect of a change in concentration during a run?

M. Comments

1. A variation to this experiment can be made by measuring polymers of different molecular weight. In this case, both the concentration and molecular-weight dependence can be studied.

2. Time-dependent effects can be studied with both the Ferranti (most easily with the automatic unit) and the Rotovisco.

N. General References

Chap. 10A; *Textbook*, Chap. 6A; Van Wazer 1963; Fox 1956, 1965; Chinai 1965; Pezzin 1966; Hayahara 1967; Berry 1968; Ferry 1970.

O. Bibliography

Berry 1968. G. C. Berry and T. G Fox, "The Viscosity of Polymers and Their Concentrated Solutions," *Fortschr. Hochpolym. Forsch.* (*Advances in Polymer Sci.*) *5*, 261-357 (1968).

Brookfield. Brookfield Engineering Laboratories, Inc., Stoughton, Massachusetts 02072.

Cannon. Cannon Instrument Company, State College, Pennsylvania 16801.

Chinai 1965. Suresh N. Chinai and William C. Schneider, "Shear Dependence of Viscosity-Molecular Weight Transitions. A Study of Entanglement Effects," *J. Polymer Sci. A3*, 1359-1371 (1965).

Ferranti. Ferranti Electric, Inc., Plainview, New York 11803.

Fox 1956. T. G Fox, Serge Gratch and S. Loshaek, "Viscosity Relationship for Polymers in Bulk and Concentrated Solution," Chap. 12 in Frederick R. Eirich, ed., *Rheology--Theory and Applications*, Vol. 1, Academic Press, New York, 1956.

Fox 1965. T. G Fox, "Polymer Flow in Concentrated Solutions and Melts," *J. Polymer Sci. C9*, 35-41 (1965).

Hayahara 1967. Takuro Hayahara and Seiji Takao, "Relationship Between Polymer Concentration and Molecular Weight in the Viscosity Behavior of Concentrated Solution," *Kolloid Z.--Z. Polymere 225*, 106-111 (1967).

Haake. Haake Instruments, Inc., Saddle Brook, New Jersey 07662.

Onogi 1963. Shigeharu Onogi, Tadashi Kobayashi, Yasuhiro Kojima, and Yosheshige Taniguchi, "Non-Newtonian Flow of Concentrated Solutions of High Polymers," *J. Applied Polymer Sci. 7*, 847-859 (1963).

Pezzin 1966. G. Pezzin and N. Gligo, "Viscosity of Concentrated Polymer Solutions. I. Polyvinyl Chloride in Cyclohexanone," *J. Applied Polymer Sci. 10*, 1-19 (1966).

Van Wazer 1963. J. R. Van Wazer, J. W. Lyons, K. Y. Kim, and R. E. Colwell, *Viscosity and Flow Measurement, A Laboratory Handbook of Rheology*, Interscience Div., John Wiley and Sons, New York, 1963.

EXPERIMENT 27. DYNAMIC MECHANICAL PROPERTIES

A. Introduction

In this experiment the viscoelastic properties of polymers are studied using a direct-reading dynamic viscoelastometer, the Rheovibron (Imass). The storage modulus, loss modulus, and mechanical damping factor are measured as a function of temperature. A minimum of 3 hours is needed for this experiment.

B. *Principle*

All polymers are viscoelastic. When a stress or strain is applied to such a material, part of the energy is stored elastically and part is lost (in the form of heat) in the deformation process. The ratio of the energy lost to that stored is a measure of the damping characteristics of the material. In the Rheovibron, a sinusoidal strain of fixed frequency is applied to one end of a small sample, held in slight tension, and the response or stress is measured at the other end by a transducer. The stress in phase and 90° out of phase with the strain and the phase angle are measured over a wide temperature range.

C. *Applicability*

The dynamic mechanical behavior of most polymers can be measured with the Rheovibron provided the modulus exceeds 10^6 dynes/cm^2 and tan δ is in the range of 0.001 - 1.7.

D. *Precision and Accuracy*

The wide variability in the samples used, and the strong dependence of the modulus and damping behavior on molecular and structural parameters, make it difficult to specify accuracy in this experiment. For a given sample, however, measurements can be reproduced to within 2-4%.

E. *Safety Considerations*

ALL NORMAL LABORATORY SAFETY PRECAUTIONS SHOULD BE OBSERVED. SAFETY GLASSES MUST BE WORN AT ALL TIMES. WHEN USING LIQUID NITROGEN IN EXPERIMENTS AT SUBAMBIENT TEMPERATURES, EXTREME CARE SHOULD BE USED TO AVOID CONTACT WITH THE SKIN: SEVERE FROSTBITE CAN RESULT.

F. *Apparatus*

1. Rheovibron Model DDV II direct-reading dynamic viscoelastometer (Fig. 10-27) (Imass).
2. Oscilloscope (desirable but not essential).
3. Low-temperature accessory (optional).

G. *Reagents and Materials*

Polymer prepared in any of Exp. 3-8, or polymers selected from App. I, or commercial samples, such as polyethylene, poly(ethylene terephthalate), or plasticized poly(vinyl chloride).

H. *Preparation*

1. Prepare the sample, if not already so available, in the form

of a film about 0.1 cm thick or less for crystalline plastics, or up to 0.4 cm thick for rubbers. Preparation methods include casting from solution, milling, or (preferably) compression molding. Residual solvent from casting may affect the results, and milling requires a large sample.

2. Cut a specimen 0.3-0.5 cm wide and about 1.5 cm long (for elevated-temperature experiments) or 2.5 cm long (for subambient-temperature measurements). Use a single stroke with a sharp razor blade, avoiding nicks or rough edges on the sides of the sample.

I. Procedure

Instrument Setup

1. Fill a dewar flask with ice made from distilled or deionized water, and place the reference thermocouples in it.
2. Set the Rheovibron controls as follows: SELECTOR switch OFF, CALIBRATION switch to MEASURE, PHASE/MEASURE switch to MEASURE, AMPLIFICATION FACTOR set to 30, TAN δ RANGE set to 40, FREQUENCY selector as desired (110 Hz is suggested), and HEATER powerstat OFF.
3. Turn the POWER SOURCE switch ON and turn on the STABILIZED ELIMINATOR. Allow 10-15 min warmup time.

Sample Installation

4. Disengage the slider lever (front) and move the slider (front knob) to bring the sample chucks together to check alignment and gap distance. The chucks should just touch when the slider is set at 0.000 cm.
5. Set the slider position to the sample length less 0.5 cm, and lock the front lever. For example, for a sample length of 1.5 cm set at 1.0 cm. This will place the sample length reading at mid-point on the dial gauge (1.000 cm span), allowing accurate length measurement as the sample expands and contracts.
6. Set the sample in the left-hand sample grip (vertically centered), taking care to align the sample with the right-hand grip. Adjust the sample length GAGE KNOB as needed to position the right-hand grip.

Measurement

7. Set the MAIN SELECTOR switch to AMP F and turn the AMPLITUDE ADJUST selector until the Tan δ Meter is at full scale.
8. Set the MAIN SELECTOR switch to DYN F. Turn the DYNAMIC FORCE knob to position the Tan δ Meter reading near the *low* end of the scale.
9. Apply tension to the sample by turning the GAGE knob clockwise. Small movements of the GAGE knob will result in an appreciable increase in the Tan δ Meter reading. Continue applying tension until a small movement of the GAGE knob results in little or no increase in the Tan δ Meter reading. *THIS IS THE PROPER TENSION WHICH IS TO BE APPLIED TO THE SAMPLE PRIOR TO FINAL BALANCE AND RECORDING OF EACH DATA POINT.* (*Note*: If the Tan δ Meter goes off scale, reduce the DYNAMIC FORCE setting to bring the Tan δ Meter reading back on scale.)

10. Increase the DYNAMIC FORCE setting until the Tan δ Meter is at *full scale*.

11. Turn the PHASE ADJUST knob such that the oscilloscope (when used) shows a 45° straight line, or close to this.

12. Turn the HEATER powerstat ON and adjust for a heating rate of about 1°C/min. See Sec. M for further comments on temperature control and the course of the experiment.

13. Record data points using the following procedure:

a. Set the MAIN SELECTOR switch back to AMP F and check that the Tan δ Meter is at full scale. Make any minor adjustments necessary with the AMPLITUDE ADJUST knob.

b. Set the MAIN SELECTOR switch to DYN F and check that the Tan δ Meter is at full scale. Make any minor adjustments necessary with the DYNAMIC FORCE setting.

c. Check and adjust the PHASE ADJUST.

d. Set the MAIN SELECTOR switch to Tan δ, and record tan δ and the temperature. Repeat at 2-4°C intervals through the desired temperature range.

Shutdown

14. At the end of the experiment, turn the HEATER powerstat OFF and set instrument controls as in step 1. When the furnace has cooled, remove the sample and turn the POWER SOURCE and STABILIZED ELIMINATOR switches OFF.

J. *Fundamental Equations*

$$G^* = (2\,\ell/FDA) \times 10^9 \qquad (27\text{-}1)$$

where D is the dynamic force reading, F the amplitude factor, ℓ the length of the sample, and A the sample cross-sectional area.

$$G' = G^* \cos\delta \qquad (\text{from } 10\text{-}17)$$

$$G'' = G^* \sin\delta = G' \tan\delta \qquad (\text{from } 10\text{-}18)$$

K. *Calculations*

1. Calculate G^*, G', and G'' from Eqs. 27-1, 10-17, and 10-18, respectively, at each temperature.
2. Plot G^*, G', G'' and tan δ versus temperature.

L. *Report*

1. Describe the apparatus and experiment in your own words.
2. Tabulate the results and include the plots made in step K-2.
3. Report the value of the glass-transition temperature, if this region was investigated.

4. Comment on the elastic and viscous nature of the material at different temperatures.
5. Comment on the significance of tan δ.
6. Comment on the reproducibility of the results.

M. Comments

1. After each reading, set the MAIN SELECTOR switch to DYN F. At low temperatures where the sample is glassy, (D reads low at F = 1) it is not necessary to release tension between readings. When the sample starts becoming rubbery (D increasing and tan δ increasing) the tension should be lowered slightly and re-applied between readings (step I-9). If the response of the tan δ Meter is not as outlined in step I-9, delay the tension adjustment until a temperature is reached at which tension response is observed.

2. As the temperature increases and the sample becomes rubbery, the DYNAMIC FORCE reading D will increase and eventually reach full span at 1000. When this occurs, the DYNAMIC FORCE amplification can be decreased by simultaneously *raising* the TAN δ RANGE one step and *lowering* the AMPLITUDE FACTOR one step. The value of tan δ is still read on the bottom scale regardless of these range settings. (The TAN δ RANGE, only, can be raised to 50 or 60 to obtain a more accurate reading of tan δ, but it should be returned to its original position for the next reading.)

3. For sub-ambient temperature measurements, adjust the tension at room temperature prior to the initial cooling by steps 4, 5, and 9. As the sample is cooled with liquid nitrogen, and contracts, the tan δ meter will tend to move up-scale. The tension should periodically be lessened slightly and reapplied during the cooling, to maintain the proper tension (as in step 9) until no further change is detected in the tan δ meter reading as the sample continues to cool. This will be the sample-length reading for beginning a run from cold temperature. (The sample shrinkage usually observed will be in the range of 0.05 to 0.08 cm.)

4. To cool for low-temperature runs, liquid nitrogen should be allowed to flow into the furnace cooling-tray at a rate such as to maintain a 1/2" to 1" liquid level in the tray. Generally, a low-temperature run is begun after cooling to about -85°C.

5. Temperatures below 0°C can be read on the temperature-indicating millivoltmeter by changing polarity at the input terminals. The readings can then be converted to the correct temperature using Cu-constantan thermocouple tables.

6. This experiment can be modified by measuring as a function of frequency from 0.01 to 1 Hz and at the fixed frequencies 3.5, 11, 35, and 110 Hz, if the variable-frequency accessory is available.

N. General References

Chap. 10B; *Textbook*, Chap. 6C; Ferry 1958, 1970; Leaderman 1958; Tobolsky 1960; Nielsen 1962; Takayanagi 1962; Meares 1965, Chaps. 9, 11; Rodriguez 1970, Chap. 8; Ward 1971, Chaps. 5, 6; Williams 1971, Chaps. 10, 11.

O. *Bibliography*

Ferry 1958. John D. Ferry, "Experimental Techniques for Rheological Measurements on Viscoelastic Bodies," Chap. 11 in Frederick R. Eirich, ed., *Rheology--Theory and Applications*, Vol. 2, Academic Press, New York, 1958.
Ferry 1970. John D. Ferry, *Viscoelastic Properties of Polymers*, 2nd ed., John Wiley and Sons, New York, 1970.
Imass. Imass, Inc., Accord, Massachusetts 02018.
Leaderman 1958. Herbert Leaderman, "Viscoelasticity Phenomena in Amorphous High Polymeric Systems," Chap. 1 in Frederick R. Eirich, ed., *Rheology--Theory and Applications*, Vol. 2, Academic Press, New York, 1958.
Meares 1965. Patrick Meares, *Polymers: Structure and Bulk Properties*, D. Van Nostrand Co., Princeton, New Jersey, 1965.
Nielsen 1962. Lawrence E. Nielsen, *Mechanical Properties of Polymers*, Reinhold Publishing Corp., New York, 1962.
Rodriguez 1970. Ferdinand Rodriguez, *Principles of Polymer Systems*, McGraw-Hill Book Co., New York, 1970.
Takayanagi 1965. Motowo Takayanagi, "Viscoelastic Behavior of Crystalline Polymers," pp. 161-187 in E. H. Lee and Alfred L. Copley, eds., *Proceedings of the Fourth International Congress on Rheology*, Part 1, Interscience Div., John Wiley and Sons, New York, 1965.
Tobolsky 1960. Arthur V. Tobolsky, *Properties and Structure of Polymers*, John Wiley and Sons, New York, 1960.
Ward 1971. I. M. Ward, *Mechanical Properties of Solid Polymers*, Interscience Div., John Wiley & Sons, New York, 1971.
Williams 1971. David J. Williams, *Polymer Science and Engineering*, Prentice-Hall, Englewood Cliffs, New Jersey, 1971.

EXPERIMENT 28. DIELECTRIC CONSTANT AND LOSS

A. *Introduction*

In this experiment, the dielectric constant and dielectric loss of a polar polymer are measured as a function of frequency through the glass-transition region. The experiment requires 3 hours.

B. *Principle*

The onset of chain-segment mobility at the glass transition can be detected in polar polymers by dielectric measurements, if the temperature of the experiment is selected so that the relaxation times of the motion are comparable to the frequencies used in the experiment. In this case, the dielectric constant ε' decreases with increasing frequency, as the dipoles of the polymer become unable to follow the changes of direction of the electric field. At the same time, the loss factor ε'' goes through a maximum. In the experiment, ε' and ε'' are calculated from the capacitance and dissipation factor of a condenser with the polymer as the dielectric.

Since the frequency range of the equipment is limited, the range 30-50 Hz to 1-3 × 10^5 Hz being most convenient, it is necessary to select a polymer or an operating temperature so that the glass transition falls in the proper range for observation at these high frequencies. Most determinations of T_g (by dilatometry or thermal measurements, for example) are usually so slow that the relaxation times involved are not below 100 sec, in contrast to the range 10^{-2} to 10^{-6} sec for the dielectric method.

Use of the WLF equation (as stated in the *Textbook*, Eq. 6-22) allows calculation of the shift in T_g corresponding to the differences in relaxation times cited above. It is found that a polymer should be selected with T_g as normally measured in the range -5 to -20°C, if the dielectric measurements are to be carried out at room temperature.

C. *Applicability*

The dielectric method is applicable to all polymers, provided that their conductivity is low, that the sample can be prepared in a thickness giving a capacitance within the range of the instrument, and that its dissipation factor also falls in the proper range.

D. *Precision and Accuracy*

Precision and accuracy are determined largely by the care taken in preparing the sample and carrying out the experiment. For the determination of ε', a precision of about ±0.5% can be achieved with a highly uniform sample of accurately known dimensions. For ε'', the dimensions need not be known precisely but sample uniformity is still important; a precision of about ±2% can be reached. The precision of the apparatus itself is of the order of ±0.01%.

E. *Safety Considerations*

GENERAL LABORATORY SAFETY PRECAUTIONS MUST BE OBSERVED, ESPECIALLY IN THE USE OF ORGANIC SOLVENTS FOR SAMPLE PREPARATION, AND IF OPERATION WELL ABOVE OR WELL BELOW ROOM TEMPERATURE IS UTILIZED. SAFETY GLASSES MUST BE WORN IN THE LABORATORY AT ALL TIMES.

F. *Apparatus*

1. Capacitance measuring assembly, Type 1610-B (Fig. 10-30) or 1620-A (Fig. 10-31) (General Radio) (see step M-1).
2. Dielectric Sample Holder, Type 1690-A (Fig. 10-32) (General Radio).
3. Equipment for measurement at elevated or sub-ambient temperature, if desired.
4. Small high-quality calibrated balancing capacitor with capacitance in the range 200-500 pF (not required with 1620-A assembly).
5. Micrometer.

6. Rubber roller.
7. Small bubble level.

G. *Reagents and Materials*

1. Polymer material (*Note*: The experiment is written for poly-(ethylene-*co*-vinyl acetate) with 20-30% ethylene by weight. For other suitable polymers, see step M-2).
2. Acetone, reagent grade.
3. Paraffin oil.
4. Aluminum weighing dishes or radioactivity planchets, at least 6 cm diameter, with perfectly flat bottoms.
5. Aluminum foil.

H. *Preparation*

(*Note*: Sample preparation requires operations at four different times, in two cases with an interval of a few days to a week between operations. Steps 1-4 must be carried out in the indicated time intervals.)

1. Prepare solutions of 10 ± 1% of the polymers to be used, in acetone, in clean flasks or bottles.
2. a. Measure with a micrometer the thicknesses of the bottoms of two aluminum dishes at several points. Calculate and record the average thicknesses for uniform samples. Rinse the dishes with acetone, allow them to dry, and weigh them accurately. Place the dishes in a level dust-free spot (check with bubble level).

   b. Pour into each dish the volume of polymer solution calculated to yield a polymer film about 0.3 mm thick after evaporation of the solvent. Use care to avoid bubbles. Cover the dishes with inverted beakers raised about 1 cm on one side so as to allow free evaporation of solvent but protect them from dust.
3. After a few days, when the solvent has evaporated, place the dishes in a vacuum desiccator or oven and allow the films to dry at room temperature under vacuum for one week.
4. a. Using a template prepared ahead of time and a sharp pencil or scriber, draw a 5.1-cm-diameter circle on the aluminum bottom of the dish, in the area of bubble-free, uniform-thickness film. Cut the dish away with scissors to obtain a 5.1-cm-diameter sample.

   b. Place a drop of paraffin oil 2-3 mm in diameter on the plastic film, and place a piece of aluminum foil on the surface. Carefully press and rub the foil into contact with the film, starting from the center, expelling all air and excess oil. Finish by covering the foil with filter or blotting paper and rolling it firmly in two directions with a rubber roller.

   c. Remove excess foil with scissors or a sharp razor blade. Be sure that no shreds of foil remain to short-circuit the two electrodes.

   d. Measure the total thickness of the sample with foil electrodes, and subtract the thickness of the electrodes (step 2a and thickness of the aluminum foil used in step 4b) to obtain the net sample thickness. Measure in several locations and calculate the average thickness.

I. Procedure

Procedure I. Type 1610-B Assembly

Startup

1. Turn the oscillator power supply POWER switch ON, and turn the null-detector power ON by turning the GAIN knob counterclockwise just enough to activate the switch. Allow 15 min warmup.
2. Set the DECIBELS knob on the oscillator to 0, and the switch below it to 0-45 V. Set the meter switch on the null indicator to LOG, and the METHOD switch on the bridge to DIRECT.
3. Plug the Dielectric Sample Holder (hereafter called the measuring condenser) into the UNKNOWN DIRECT terminals on the bridge, and set the vernier control on its side to zero. (*Note*: This control is not used but must not be adjusted further.)

Sample Insertion

4. Remove the condenser cover plates, raise the upper electrode by rotating the large knob (Fig. 10-32) counterclockwise, and insert the sample. Lower the upper electrode and see that it touches the sample over its entire area. Record the thickness as ℓ_t (total thickness). Replace the cover plates.

Measurement

5. Select the desired frequency by means of the FREQUENCY RANGE selector and FREQUENCY dial on the oscillator. (*Note*: Measure at points approximately equally spaced on a log frequency scale throughout the range 30-30,000 Hz. Suggested sets are 50, 100, 200, 500, ... or 30, 60, 100, 180, 300, ... Hz; but do not measure *exactly* at 60 Hz because of powerline interference).
6. Set the FILTER FREQUENCY selector on the null detector to correspond to the frequency selected, and set the FILTER TUNING knob so that the meter shows maximum deflection. Adjust the GAIN as necessary to obtain this condition.
7. Set the RANGE SELECTOR on the bridge to correspond to the frequency used, e.g., between 30 and 300 Hz use the 100c setting. (*Note*: Use the 1 kc-1 setting (M = 1) for frequencies between 300 and 3000 Hz.) Set the DISSIPATION FACTOR switch to 0 or to the approximate dissipation factor of the sample, if known.
8. Balance the bridge by adjusting the CAPACITANCE and DISSIPATION FACTOR dials alternately to obtain minimum deflection on the null detector. Set the METER switch on the detector to LINEAR, then increase the GAIN on the detector as the adjustments are made, so that the final null is obtained at maximum gain. If the DISSIPATION FACTOR dial goes offscale, increase its selector switch setting by one step. Finally, reduce the null detector GAIN setting to zero to avoid overloading the detector should the balance be disturbed.
9. Read the capacitance to the nearest 0.2 pF by adding the drum

and dial readings; e.g., if the drum is between 250 and 300 and the dial reads 13.6, the capacitance is 263.6 pF.

10. Read the dissipation factor selector and dial, add these readings, and multiply the sum by the ratio of the frequency used to the RANGE SELECTOR frequency. Express the result (which is in percent) as a fraction and round to 3 significant figures to obtain tan δ. Example: When measuring at 180 Hz with the RANGE SELECTOR at 100c, for a selector reading of 10 and a dial reading of 2.2, the dissipation factor is (10 + 2.2)(180)/(100) = 21.96%, and tan δ = 0.2196, but report tan δ = 0.220.

11. Repeat steps 5-10 for each frequency selected.

Condenser Calibration

12. Remove the sample from the condenser, set the spacing of the electrodes to the *net* (not total) sample thickness, and replace the cover plates. Measure its capacitance at 1000 Hz by repeating steps 5-10. If the capacitance is above 100 pF, proceed to step 16; otherwise (as is likely if the sample thickness is greater than 0.008 in.), do steps 13-15.

13. Remove the measuring condenser and replace it with the calibrated balancing capacitor. Determine the capacitance and dissipation factor of this capacitor at 1000 Hz, following steps 5-10. *Do not* disconnect it from the bridge.

14. Set the METHOD switch to SUBST (for substitution). Turn the measuring condenser upside down, and connect its ground plug only (now on top) to the UNKNOWN SUBST ground terminal (bottom) on the bridge. Balance the bridge by steps 5-11 and record the values as C_2 and tan δ_2.

15. Remove the measuring condenser, turn it right side up, and plug it into the UNKNOWN SUBST terminals so that both sides are connected. Balance the bridge, following steps 5-10. Record the results as C_1 and tan δ_1.

16. Repeat steps 3-15 for subsequent samples.

Shutdown

17. Turn the POWER switch on the power supply OFF, and turn the detector GAIN knob fully counterclockwise.

Procedure II. Type 1620-A Assembly

Startup

1. Turn the oscillator power supply POWER switch ON, and turn the null-detector power ON by turning the GAIN knob counterclockwise just enough to activate the switch. Allow 15 min warmup.

2. Set the MAXIMUM OUTPUT switch of the oscillator to 1 V and the OUTPUT LEVEL control fully clockwise. Set the METER switch on the null indicator to LOG.

3. Set the TERMINAL SELECTOR switch on the bridge to 2 TERMINAL, and set the MULTIPLY EXTERNAL STANDARD control to 0. Plug the Dielectric Sample Holder (hereafter called the measuring condenser) into the UNKNOWN H and L terminals. (*Note*: In 2-terminal operation, the L

terminal is grounded.

Sample Insertion

4. Remove the condenser cover plates, raise the upper electrode by rotating the large knob (Fig. 10-32) counterclockwise, and insert the sample. Lower the upper electrode and see that it touches the sample over its entire area. Record the thickness as ℓ_t (total thickness). Replace the cover plates.

Measurement

5. Select the desired frequency by means of the FREQUENCY SELECTOR on the audio oscillator.

6. Set the Filter Frequency selector on the null detector to correspond to the frequency selected, and set the FILTER TUNING knob so that the meter shows maximum deflection. Adjust the GAIN as necessary to obtain this condition.

7. Set the C MAX range selector to 1000 pF and the D MAX range selector to 0.1. Set all the C and D control levers to zero. Then set the left-hand C control lever at the position giving minimum deflection on the meter of the null indicator. Then set the remaining C controls, proceeding from left to right, to obtain the minimum deflection. Turn the null-detector METER switch to LINEAR - MAX SENS, and increase its GAIN as required.

8. Adjust the right-hand D control lever as in step 7, and proceed from right to left to obtain minimum deflection. Final setting should be made at full GAIN setting. Immediately reduce the GAIN setting to zero after the balance is made, to protect the null detector from possible overload.

9. Read the capacitance in pF directly from the C dials. Read the D dials and multiply their value by the frequency of the measurement in kHz to obtain the dissipation factor.

10. Repeat steps 5-9 for the remaining frequencies. At higher frequencies, increase the MAXIMUM OUTPUT setting to obtain higher sensitivity, but do not exceed 30 V per kHz.

Condenser Calibration

11. Remove the sample from the condenser, set the spacing of the electrodes to the *net* (not total) sample thickness, and replace the cover plates. Measure its capacitance at 1000 Hz by repeating steps 5-9.

Shutdown

12. Turn the POWER switch on the power supply OFF, and turn the detector GAIN knob fully counterclockwise.

J. Fundamental Equations

$$C = C_2 - C_1 \qquad (28\text{-}1)$$

$$\tan\delta = (C_2/C_1)(\tan\delta_2 - \tan\delta_1) \qquad (28\text{-}2)$$

$$C_{air} = 1/(0.00141\ \ell) \qquad (28\text{-}3)$$

$$C_1 = C - C_{air} \qquad (28\text{-}4)$$

$$\tan\delta_1 = (C/C_1)\tan\delta \qquad (28\text{-}5)$$

$$C_s = C - C_1 \qquad (28\text{-}6)$$

$$\tan\delta_s = (C/C_s)\tan\delta \qquad (28\text{-}7)$$

$$\varepsilon' = 0.00141\ C_s\ \ell \qquad (\text{from } 10\text{-}32)$$

$$\varepsilon'' = 0.00141\ C\ \ell\ \tan\delta \qquad (\text{from } 10\text{-}33)$$

$$\tau = 1/\omega_{max} = 1/2\pi f_{max} \qquad (28\text{-}8)$$

where capacitances are in pF and thickness ℓ is in mils.

K. Calculations

1. If the substitution method was used (steps I-13 to I-15, Procedure I), calculate values of C and $\tan\delta$ for the empty condenser using Eqs. 28-1 and 28-3.
2. From C and $\tan\delta$ from step 1 or from direct measurement (step I-12, Procedure I, or step I-11, Procedure II), calculate the condenser constants C_1 and $\tan\delta_1$, using Eqs. 28-2 to 28-5.
3. Calculate C_s and $\tan\delta_s$ for the sample using Eqs. 28-6 and 28-7, and ε' and ε'' using Eqs. 10-32 and 10-33.
4. Plot ε', ε'', and $\tan\delta$ as a function of $\log\omega$ for all the samples measured.
5. Determine relaxation time τ using Eq. 28-8.
6. Construct Cole-Cole plots for all samples.
7. Plot $\varepsilon''_{max}/\varepsilon'$ versus $\log\omega$ for all samples.

L. Report

1. Describe the apparatus and experiment in your own words.
2. Tabulate δ, C, $\tan\delta$, C_s, $\tan\delta_s$, ε', and ε'' for each sample.
3. Include all graphs and plots.
4. Discuss the differences among the samples measured.

M. Comments

1. The General Radio 1610-B Assembly (Fig. 10-30) consists (top to bottom) of a 1210-C oscillator with 1203-B power supply, a 1232-A null detector, a 716-C capacitance bridge. Since the 1690-A sample holder is a two-terminal device, the guard circuit is bypassed by connecting the cable from the null detector directly to the DETECTOR terminals on the bridge. The cables between the bridge and the guard circuits should be disconnected. The 1610-B2 assembly has no guard circuit, and is identical in configuration to the 1610-B as modified.

The 1620-A assembly consists of a 1311-A audio oscillator, 1232-A null detector and tuned amplifier, and 1615-A capacitance bridge. The major difference between the two assemblies is the more sensitive direct-reading bridge in the 1620-A. As supplied, the 1311-A audio oscillator supplies only 11 fixed frequencies from 50 Hz to 10 kHz, but a 1310-A oscillator, as in the 1610-B assemblies, can be substituted to provide more flexibility in selection of frequencies.

2. Among the polymers having glass transitions in the frequency range available at room temperature are poly(vinylidene chloride), poly(vinyl fluoride), and lower poly(vinyl ethers), the lower polyacrylates, and some elastomers of the balata or gutta-percha type.

N. General References

Chap. 10C; Tucker 1965; Mathes 1966; McCrum 1967; Hill 1969.

O. Bibliography

General Radio. General Radio Company, Concord, Massachusetts 01742.

Hill 1969. Nora E. Hill, Worth E. Vaughan, A. H. Price, and Mansel Davies, *Dielectric Properties and Molecular Behavior*, Van Nostrand-Reinhold, New York, 1969.

Mathes 1969. K. N. Mathes, "Electrical Properties," pp. 528-628 in Herman F. Mark, Norman G. Gaylord, and Norbert M. Bikales, eds., *Encyclopedia of Polymer Science and Technology*, Interscience Div., John Wiley and Sons, New York, 1966.

McCrum 1967. N. G. McCrum, B. E. Read, and G. Williams, *Anelastic and Dielectric Effects in Polymeric Solids*, John Wiley and Sons, New York, 1967.

Tucker 1965. Robert W. Tucker, "Dielectric Constant and Loss Measurements," Chap 7 in John V. Schmitz, ed., *Testing of Polymers*, Vol. I, Interscience Div., John Wiley and Sons, New York, 1965.

EXPERIMENT 29. STRESS-STRAIN PROPERTIES

A. *Introduction*

The stress-strain properties of polyethylene are measured at different strain rates in this experiment. A standard tensile testing machine is used. The time for this experiment is 3 hours.

B. *Principle*

A sample of known length and cross-sectional area is held between two jaws and extended at a constant rate until failure occurs. The stress is measured as a function of the elongation. Analysis of measurements made at different strain rates provides an insight into the strength, toughness, and ultimate properties of the material. A standard method for this experiment is given in ASTM D 638.

C. *Applicability*

This experiment can be carried out with any polymer provided suitable specimen holders are available or devised to handle the samples.

D. *Precision and Accuracy*

For homogeneous and uniform specimens, reproducibility to better than ±5% can be achieved. However, normal variability in the specimen preparation and the operation raises this figure somewhat.

E. *Safety Considerations*

NORMAL LABORATORY SAFETY PROCEDURES SHOULD BE OBSERVED. SAFETY GLASSES MUST BE WORN IN THE LABORATORY AT ALL TIMES. CAUTION SHOULD BE TAKEN TO KEEP HANDS AWAY FROM MOVING ELEMENTS OF THE APPARATUS.

F. *Apparatus*

1. Instron Tensile Tester with proper jaws, load indicator, extension indicator, and area compensator (optional); or equivalent testing machine (see step M-1).
2. Equipment for sample preparation, as required (see Sec. H).
3. Micrometer.

G. *Reagents and Materials*

Linear and branched polyethylene, crosslinked polymers prepared in Exp. 8, other polymers selected from App. I, or commercial material.

H. *Preparation*

Prepare ahead of time standard tensile test specimens (see ASTM D 638) by molding or cutting from sheet or slabs of material. Prepare 5 specimens for each strain rate to be used.

I. *Procedure*

(*Note*: It is assumed that the student will familiarize himself with the operation of the tensile tester, with the assistance of the instructor and reference to the instruction manual. The procedure is therefore presented only in broad outline.)

1. Measure the width and thickness of each specimen with a micrometer, and calculate and record the cross-sectional areas at the point of minimum cross-section.
2. Adjust the crosshead so that the grips are 4.5 in. apart, and place a specimen in the grips.
3. Set the crosshead speed at 2 in./min.
4. Start the crosshead and record the load as a function of the extension. (This is done automatically with an extension indicator.)
5. Return the crosshead to its starting position manually, if automatic return is not available on the tester.
6. Repeat steps 2-5 for the remaining 4 samples.
7. Change crosshead speeds first to 5, then to 10 and 20 in./min and repeat steps 2-6 at each crosshead speed.

J. *Fundamental Equations*

$$\gamma = (L - L_0)/L_0 = kt/L_0 \tag{10-34}$$

$$A = A_0 L_0/L \tag{10-35}$$

$$E = s/\gamma = FL_0/A(L - L_0) \tag{10-36}$$

K. *Calculations*

1. Calculate the strain and stress as a function of time, for elongations below about 10%, using Eqs. 10-34 and 10-35.
2. Calculate the modulus using Eq. 10-36.
3. Calculate the stress and strain at break.
4. Repeat steps 1-3 for each specimen tested.

L. *Report*

1. Describe the apparatus and experiment in your own words.
2. Tabulate the results and include all graphs in your report.
3. Explain the reason for running 5 duplicate samples.
4. Answer the following questions:
   a. What is the effect of strain rate on the ultimate

properties?

b. What observation can you make on differences in behavior between the linear and branched polyethylene, if these materials were measured?

M. Comments

1. Other instruments which are suitable for this experiment include the Model CRE Scott Tester (Precision) and the Tinius Olsen Testers.

N. General References

Chap. 10D; *Textbook,* Chap. 4G; Nielsen 1962, Chap. 5; Thorkildsen 1964; Rodriguez 1970, Chap. 9; Bikales 1971; Ward 1971; Williams 1971, Chap. 11.

O. Bibliography

ASTM D 638. *Standard Method of Test for Tensile Properties of Plastics*, ASTM Designation: D 638. American Society for Testing and Materials, Philadelphia, Pennsylvania 19103.

Bikales 1971. Norbert M. Bikales, ed., *Mechanical Properties of Polymers (Encyclopedia Reprints)*, Wiley-Interscience, New York, 1971.

Instron. Instron Corporation, Canton, Massachusetts 02021.

Nielsen 1962. Lawrence E. Nielsen, *Mechanical Properties of Polymers,* Reinhold Publishing Corp., New York, 1962.

Precision. Precision Scientific Company, Chicago, Illinois 60647.

Rodriguez 1970. Ferdinand Rodriguez, *Principles of Polymer Systems,* McGraw-Hill Book Co., New York, 1970.

Thorkildsen 1964. R. L. Thorkildsen, "Mechanical Behavior," Chap. 5 in Eric Baer, ed., *Engineering Design for Plastics*, Reinhold Publishing Corp., New York, 1964.

Tinius Olsen. Tinius Olsen Testing Machine Company, Willow Grove, Pennsylvania 19090.

Ward 1971. I. M. Ward, *Mechanical Properties of Solid Polymers*, Wiley-Interscience, New York, 1971.

Williams 1971. David J. Williams, *Polymer Science and Engineering,* Prentice-Hall, Englewood Cliffs, New Jersey, 1971.

EXPERIMENT 30. SWELLING OF NETWORK POLYMERS

A. Introduction

This experiment is designed to determine the number of effective network chains (crosslinks) per unit volume of polymer. Two useful pieces of information are obtained: The resistance of the polymer to the solvent, and the crosslink density of the polymer. The time for this experiment will vary depending upon the crosslinking level. It can be carried out in two hours total time but may require an elapsed time of 24 hours.

B. Principle

Swelling is the first step in the solubilization of a polymer. The degree of swelling depends upon the polymer-solvent interaction parameter χ_1 and the molecular weight of the polymer. In the case of a crosslinked polymer, solubilization cannot take place, and an equilibrium degree of swelling is attained. The swelling experiment can be run gravimetrically or volumetrically; for this experiment, the gravimetric technique is employed. Knowledge of the value for the polymer-solvent interaction parameter χ_1 allows the crosslink density per unit volume of polymer or the molecular weight between crosslinks to be calculated. Pinner (1961) describes a similar experiment, his Exp. C.3.3.

C. Applicability

This technique is applicable to most crosslinked polymers provided a suitable swelling solvent can be found. The method works best for low crosslink densities, since the fewer the crosslinks, the greater the swelling.

D. Precision and Accuracy

The precision of this technique can be such as to detect small amounts of swelling (0.2%) with a reproducibility of 1%. Accuracy has little meaning in this experiment. See steps M-1 to M-3.

E. Safety Considerations

CARE MUST ALWAYS BE TAKEN TO USE ORGANIC SOLVENTS IN WELL-VENTILATED AREAS AWAY FROM SOURCES OF FLAME OR SPARKS. AVOID BREATHING FUMES OR EXCESSIVE CONTACT OF SOLVENT WITH SKIN. SAFETY GLASSES MUST BE WORN AT ALL TIMES IN THE LABORATORY.

F. *Apparatus*

1. Analytical balance.
2. Stoppered flasks, wide mouth, any convenient size.
3. Filter paper.

G. *Reagents and Materials*

1. Crosslinked polymers prepared in Exp. 8, or suitable substitute such as vulcanized rubber.
2. Toluene, reagent grade.

H. *Preparation*

The samples should be in the form of sheets which can be cut with scissors or razor blade.

I. *Procedure*

1. Cut three specimens weighing approximately 2 g each, accurately weigh each specimen, and place it in a stoppered flask containing enough toluene to cover the specimen.
2. Periodically remove the specimen, blot it dry with filter paper, and rapidly weigh it. (*Note*: Work rapidly to avoid loss of solvent by deswelling and evaporation, causing the sample weight to drift downward.) Repeat until the sample reaches its equilibrium degree of swelling. Use the final constant weight in the calculations to follow.

J. *Fundamental Equations*

$$V_s = w_0/\rho_2 + (w_s - w_0)/\rho_1 \qquad (10\text{-}37)$$

$$c = V_s w_0/\rho_2 \qquad (10\text{-}39)$$

$$M_x = -\, v_1\rho_2(c^{1/3} - c/2) \;/\; [\ln(1-c) + c + \chi_1 c^2] \qquad (\text{from } 10\text{-}38)$$

K. *Calculations*

1. For each sample, calculate the equilibrium volume after swelling using Eq. 10-37, the equilibrium concentration of polymer in the swollen gel using Eq. 10-39, and the average molecular weight between crosslinks, M_x, using Eq. 10-38, and taking $\chi_1 = 0.44$ for polystyrene in toluene (Huggins 1943).

L. Report

1. Describe the experiment in your own words.
2. Report the data in tabular form.
3. Explain the cause of variability in the results.
4. Answer the following questions:
 a. What effect does solubility of the polymer have on the result?
 b. How can the interaction parameter, χ_1, be determined (see Chap. 7A)?

M. Comments

1. The presence of any uncrosslinked material in the polymer network, which may dissolve and change the concentration, can be a source of error. A sample of the solution should be tested for dissolved polymer.

2. It should be recognized that inherent in the analysis are the experimental and theoretical difficulties encountered with the Flory-Huggins thermodynamic theory, upon which the development of Eq. 10-38 is based.

3. A variation of this experiment is to determine the cohesive energy density, as described by Bristow (1958*a*, *b*).

N. General References

Chap. 10E; *Textbook*, Chap. 2A; Flory 1953; Treloar 1958; Ellis 1963*a*, *b*; Miller 1966, Chap. 7; Smith 1968.

O. Bibliography

Bristow 1958a. G. M. Bristow and W. F. Watson, "Viscosity-Equilibrium Swelling Correlations for Natural Rubber," *Trans. Faraday Soc. 54*, 1567-1573 (1958).

Bristow 1958b. G. M. Bristow and W. F. Watson, "Cohesive Energy Densities of Polymers. I. Cohesive Energy Densities of Rubbers by Swelling Measurements," *Trans. Faraday Soc.* 54, 1731-1741 (1958).

Ellis 1963a. Bryan Ellis and G. N. Welding, "Estimation from Swelling of the Structural Contributions of Chemical Reactions in the Vulcanization of Natural Rubber, Part I, General Methods," pp. 35-45 in *Techniques of Polymer Science*, S.C.I. Monograph No. 17, Gordon and Breach Science Publishers, New York, 1963.

Ellis 1963b. Bryan Ellis and G. N. Welding, "Estimation from Swelling of the Structural Contributions of Chemical Reactions in the Vulcanization of Natural Rubber. Part II, Estimation of Equilibrium Degree of Swelling," pp. 46-53 in *Techniques of Polymer Science*, S.C.I. Monograph No. 17, Gordon and Breach Science Publishers, New York, 1963.

Flory 1953. Paul J. Flory, *Principles of Polymer Chemistry*, Cornell University Press, Ithaca, New York, 1953.

Huggins 1943. Maurice L. Huggins, "Thermodynamic Properties of Solutions of High Polymers. The Empirical Constant in the Activity Equation," *Ann. N. Y. Acad. Sci. 44*, 431-443 (1943).
Miller 1966. M. L. Miller, *The Structure of Polymers*, Reinhold Publishing Corp., New York, 1966.
Pinner 1961. S. H. Pinner, *A Practical Course in Polymer Chemistry*, Pergamon Press, New York, 1961.
Smith 1968. Derek A. Smith, "Characterization of Structure by Measurement of Bulk Properties," Chap. 10 in Derek A. Smith, ed., *Addition Polymers: Formation and Characterization*, Plenum Press, New York, 1968.
Treloar 1958. L. R. G. Treloar, *The Physics of Rubber Elasticity*, 2nd ed., Clarendon Press, Oxford, England, 1958.

EXPERIMENT 31. PERMEABILITY OF POLYMERS TO GASES

A. *Introduction*

In this experiment, the rate of permeability of carbon dioxide through a polymer film is determined at different pressures. This experiment can be carried out in 3 hours.

B. *Principle*

The rate of gas permeation through a film is measured by determining the volume of gas permeating through the film under a pressure differential. A high gas pressure is applied on one side of the membrane under study, and the permeated gas is allowed to expand on the opposite side against atmospheric pressure. The volume of the permeating gas is measured as a function of time by following the displacement of a short column of mercury in a glass capillary. A standard method for this experiment is described in ASTM D 1434.

C. *Applicability*

This method is applicable to most polymers which can be obtained as suitable films.

D. *Precision and Accuracy*

The precision of this experiment varies with the permeability, since it depends on a volume measurement. In practice, it is found that for high permeabilities, a precision of ±5% can be attained, while for low permeabilities the precision is closer to ±15%. Absolute accuracy has no meaning in this experiment.

E. *Safety Considerations*

NORMAL LABORATORY SAFETY PRECAUTIONS SHOULD BE OBSERVED. CARE SHOULD BE TAKEN WHEN HANDLING BOTTLED GASES. THE GLASS CAPILLARY ATTACHED TO THE PERMEABILITY CELL IS EASILY BROKEN. CARE MUST BE TAKEN TO AVOID BREAKAGE AND POSSIBLE CUTS DURING HANDLING OF THE CELL. EXTRA CARE MUST BE TAKEN TO AVOID SPILLAGE OF MERCURY. SAFETY GLASSES MUST BE WORN IN THE LABORATORY AT ALL TIMES.

F. *Apparatus*

1. Permeability Cell Model CS-135 (Custom) (Fig. 10-37) (Stern 1964) (see step M-1).
2. CO_2 gas cylinder with pressure regulator.
3. Constant-temperature water bath.
4. Stop watch.
5. Circular die 92 mm in diameter (optional).
6. Micrometer.

G. *Reagents and Materials*

1. Polymer in film form, preferably polyethylene or plasticized poly(vinyl chloride), with thickness in the range 0.02-0.5 mm, prepared from materials selected from App. I, or obtained from commercial material.
2. Mercury, reagent grade, clean (see step M-2).

H. *Preparation*

1. Dry the polymer film samples prior to use.
2. Prepare film specimens and sheets of filter paper cut to 92 mm diameter.
3. Determine the film thickness in at least 10 different areas, measuring to ±2 μm, and record the average film thickness.

I. *Procedure*

1. Remove the top half of cell by removing the six wing nuts. Leave the upper and lower vents OPEN.
2. Place the flat rubber gasket in place on the face of the cell bottom. Place the polymer film specimen evenly on the gasket. Place two sheets of filter paper on top of the film. (*Note*: Make sure the filter paper fits evenly to the inside diameter of the rubber gasket.)
3. *Carefully* guide the top half of the cell over the six studs on the bottom half. Make sure the cell seats together evenly all around. being careful not to disturb the position of the sample.
4. Replace the wing nuts and tighten them evenly by alternately tightening opposite nuts. Normally, finger tight will suffice. Overtightening may damage the cell or cause leakage. (See step M-3.)
5. Connect the gas cylinder to the main inlet line (see Fig. 10-37*a*).

6. Place the cell in the thermostated water bath.

7. Connect the inlet lines to inlet side of the cell. Both top and bottom vents on the cell should be OPEN for purging.

8. Open the regulator valve.

9. *SLOWLY* open the cylinder valve until gas is noted coming out of the vents of the cell.

10. Purge for 2-3 min, then *slowly* close the HIGH PRESSURE (bottom) vent valve.

11. Adjust the cylinder regulator until the pressure gauge reads the desired test pressure.

12. Allow the sample to condition to the test gas for at least 30 min. (*Note*: Thick samples may require longer conditioning times.)

13. Start the electric vibrator. Then close the LOW PRESSURE (top) vent valve, allowing the permeated gas to begin moving the mercury plug in the capillary.

14. With a stop watch, measure the time required for the mercury plug to reach arbitrarily selected heights up to 25 cm. (*Note*: This part of the experiment should take no more than 10 min. If longer times are found, the pressure should be increased or the specimen thickness decreased, or both.)

15. Repeat runs are performed rapidly by opening the *upper* vent line (causing the mercury plug to drop to the bottom of the capillary), and then closing the vent. Make three runs at each pressure setting.

16. Upon completing the experiment, close the main gas inlet valve. Close the upper vent valve on the cell and *SLOWLY* release gas from the bottom of the cell by opening the lower vent valve. (*Note*: Pressure should be relieved from both sides of the cell as evenly as possible. *Failure to do this will cause the membrane to drop and draw the mercury plug into the cell.*)

17. After depressurizing the cell, remove the gas inlet lines and remove the film from the cell by reversing the procedure of steps 3 and 4.

J. *Fundamental Equations*

$$P = (\Delta Q/\Delta t)\ \ell/A(p_1 - p_2) \qquad (10\text{-}49)$$

K. *Calculations*

1. Plot, for each sample and run, the height of the gas column versus time (see Fig. 10-38). Draw the best straight line through the points.

2. Calculate the slope of the line drawn in step 1.

3. Convert the slope $\Delta(\text{height})/\Delta t$ to $\Delta Q/\Delta t$ as follows:

   a. Convert height to observed gas volume, using the diameter of the capillary on the apparatus.

   b. Convert observed volume to volume at standard temperature and pressure by means of simple gas-law calculations.

4. Calculate permeability by Eq. 10-49.

L. *Report*

1. Describe the apparatus and experiment in your own words.
2. Tabulate the data and include all tables and graphs in the report.
3. Answer the following questions:
   a. What effect does the permeating gas pressure have on the permeability?
   b. Why is the conditioning in step I-12 necessary?

M. *Comments*

1. A simple apparatus can easily be fabricated (see, for example, Peterson 1968).
2. Fluids other than mercury can be used to follow the increase in height with time. For polyethylene, methyl isobutyl ketone containing a small amount of dye is satisfactory. Its use eliminates the need for the electric vibrator necessary to prevent mercury sticking in the capillary.
3. Leaks in the low-pressure section can be detected both visually (since the cell is immersed in a water bath) and from a sharp decrease in the rate of movement of the mercury plug in the capillary. If this occurs, remove the cell from the water bath. A pinhole in the film specimen has the opposite effect, causing anomalously high rates of travel of the mercury plug.
4. The rate of permeation can be varied by appropriate choice of pressure gradient and capillary diameter.
5. A variation of this experiment is to measure the permeability at three different temperatures.

N. *General References*

Chap. 10F; Geddes 1960; Rogers 1964, 1965; Stern 1964; Stannett 1965; Lebovits 1966; Crank 1968.

O. *Bibliography*

ASTM D 1434. *Standard Methods of Test for Gas Transmission Rate of Plastic Film and Sheeting*, ASTM Designation: D 1434. American Society for Testing and Materials, Philadelphia, Pennsylvania 19103.

Crank 1968. J. Crank and G. S. Park, eds., *Diffusion in Polymers*, Academic Press, New York, 1968.

Custom. Custom Scientific Instruments, Inc., Whippany, New Jersey 07981.

Geddes 1960. A. L. Geddes and R. B. Pontius, "Determination of Diffusivity," Chap. XVI in Arnold Weissberger, ed., *Physical Methods of Organic Chemistry*, Vol. I of A. Weissberger, ed., *Technique of Organic Chemistry*, 3rd ed., Part II, Interscience Div., John Wiley and Sons, New York, 1960.

Lebovits 1966. Alexander Lebovits, "Permeability of Polymers to Gases, Vapors and Liquids," *Modern Plastics 43* (7), 139-211 (11 pages) (March, 1966).
Peterson 1968. Clarke M. Peterson, "A Simplified Gas Permeability Apparatus," *J. Applied Polymer Sci. 12*, 2669-2671 (1968).
Rogers 1964. C. E. Rogers, "Permeability and Chemical Resistance of Polymers," Chap. 9 in Eric Baer, ed., *Engineering Design for Plastics*, Reinhold Publishing Corp., New York, 1964.
Rogers 1965. C. E. Rogers, "Solubility and diffusivity," Chap. 6 in David Fox, Mortimer M. Labes, and Arnold Weissberger, eds., *Physics and Chemistry of the Organic Solid State*, Vol. II, Interscience Div., John Wiley and Sons, New York, 1965.
Stannett 1965. V. Stannett and H. Yasuda, "The Measurement of Gas and Vapor Permeation and Diffusion in Polymers," Chap. 13 in John V. Schmitz, ed., *Testing of Polymers*, Vol. I, Interscience Div., John Wiley and Sons, New York, 1965.
Stern 1964. S. A. Stern, T. F. Sinclair and P. J. Gareis, "An improved permeability apparatus of the variable-volume type," *Modern Plastics 42* (2) 154-206 (4 pages) (October, 1964).

EXPERIMENT 32. INFRARED SPECTROSCOPY

A. *Introduction*

Although infrared spectroscopy can be widely used for the qualitative identification of polymers (see McCaffery 1970, Exp. 27), this experiment is limited to the quantitative determination of the composition of the styrene-methyl methacrylate copolymers prepared in Exp. 5. This determimation requires approximately 2 hours. Since infrared spectroscopy is in many respects a standard analytical technique, operating instructions for a specific spectrometer are not supplied; it is assumed that the student will be familiar with these instruments.

B. *Principle*

The quantitative analysis of any spectroscopic information is based on the combined Beer-Lambert law, which states that, at a given wavelength, the absorbance (negative logarithm to the base 10 of the transmittance of the sample) is the product of the molar (or weight) absorptivity, the molar (or weight) concentration, and the path length of the absorption cell or the sample thickness. In this experiment, the last-named variable is held constant, the molar absorptivity of the methyl methacrylate repeat unit is determined at a wavelength of approximately 5.7 μm from measurement of pure poly(methyl methacrylate) (see also Fig. 5-3), and the compositions of styrene-methyl methacrylate copolymers are calculated from their absorbances at the same wavelength.

C. Applicability

Qualitative examination by infrared spectroscopy is applicable to all polymers from which appropriate samples can be prepared (Hummel 1966; Henniker 1967). The specific procedure of this experiment is limited to copolymers of styrene and methyl methacrylate only.

D. Precision and Accuracy

Although absorbances can be determined with a precision of 0.5 - 1.0% with most infrared spectrometers, the accuracy of the determination of the copolymer composition is probably no better than ±5%.

E. Safety Considerations

CHLOROFORM IS A HIGHLY TOXIC SOLVENT. ALL OPERATIONS INVOLVING ITS USE MUST BE CARRIED OUT IN A LOCATION WITH ADEQUATE VENTILATION. SAFETY GLASSES MUST BE WORN IN THE LABORATORY AT ALL TIMES.

F. Apparatus

1. Infrared spectrometer, any of many models having good resolution in the 5.7 μm region.
2. Liquid cell, path length 0.2 mm, equipped for filling and emptying by syringe.
3. Two 2-ml syringes and one needle.
4. 10-ml volumetric flasks, stoppered.

G. Reagents and Materials

1. Poly(methyl methacrylate) and styrene-methyl methacrylate copolymers prepared in Exp. 5.
2. Chloroform, reagent grade.

H. Preparation

Prepare ahead of time chloroform solutions of 0.5 g/10 ml of each of the polymers described in step G-1. Follow the procedure of Exp. 11-13, sec. H.

I. Procedure

1. Remove the liquid infrared cell from the desiccator (see step M-1), remove the stoppers, and place it on the bench with the outlet end slightly higher than the inlet end.
2. Place protective collars of filter paper around the inlet and outlet syringe connectors, to avoid accidentally spilled solution from flowing onto the window, where it would leave a polymer film on evapo-

ration.

3. Draw about 0.5 ml polymer solution into a syringe, remove the needle, express air bubbles, and attach the syringe to the lower inlet to the cell.

4. Fill the cell slowly while watching the liquid level. When the cell is full but before it overflows, stopper the upper outlet, lay the cell flat, remove the syringe, and stopper the inlet. (*Note*: If any solution is spilled on the cell, quickly place it upright and wash the window with a stream of solvent from the second syringe and needle. Repeat the washing at least twice.)

5. Put the cell in the spectrometer and proceed to step 7 without delay.

6. While step 7 is in progress, wash the syringe carefully with solvent.

7. Carry out the infrared scan over at least the wavelength range 4-8 μm, following instructions for the instrument used.

8. Remove the cell, place it flat on the bench, and remove both stoppers. Place an empty syringe on one end, and a syringe filled with solvent on the other. Flow all of the solvent slowly through the cell. Discard the effluent, and repeat at least twice. Wash the stoppers to the cell thoroughly with solvent.

9. Draw out all the solvent, empty and dry both syringes, attach both to the cell, and dry the cell by passing air from one syringe to the other.

10. Repeat steps 2-9 for each of the remaining polymer solutions.

11. Finally replace the stoppers and place the cell in a desiccator (see step M-1).

J. *Fundamental Equations*

$$A = \varepsilon bc \tag{32-1}$$

where A is absorbance, ε molar absorptivity, b the cell path length (0.2 mm), and c the molar concentration of the absorbing substance (see step M-2).

K. *Calculations*

1. Calculate the molar concentration c of methyl methacrylate repeat units in the homopolymer from the weight concentration of its solution.

2. Determine the effective value of ε for the methyl methacrylate unit from the absorbance A of the homopolymer, using c from step 1 and Eq. 32-1.

3. From the absorbances of the copolymers, calculate the molar concentration of methyl methacrylate units in each.

4. From the data of step 3, the molecular weights of methyl methacrylate and styrene, and the weight concentration of each solution, calculate the molar composition of each copolymer.

L. *Report*

1. Describe the apparatus and experiment in your own words. Include a description of the infrared spectrometer and its mode of operation.
2. Include all spectra obtained.
3. Tabulate total absorbance; baseline absorbance (see step M-2); their difference, *A*; the relative values of *A* for the homopolymer and each copolymer; and the copolymer compositions.

M. *Comments*

1. The cell windows are damaged irreversibly by moisture. Remove the cell from the desiccator only for the minimum time needed for the measurements. *Never* touch the windows with your fingers!
2. If the instrument is balanced to zero absorbance with no cell in place, some absorbance will still be registered even if there is no methyl methacrylate in the solution, as a result of surface reflections at the cell windows, etc. It is customary to draw a base line between portions of the spectrum adjacent to the 5.7 μm band exhibiting lowest absorbance. The value of *A* in Eq. 32-1 is taken as the difference in absorbance between the absorbance of this base line at 5.7 μm and the maximum absorbance of the band.

N. *General References*

Chap. 1F; *Textbook*, Chap. 4B; Stanley 1959; Sixma 1964, pp. 203-221; Zbinden 1964; Hummel 1966, 1971; Henniker 1967.

O. *Bibliography*

Henniker 1967. J. C. Henniker, *Infrared Spectrometry of Industrial Polymers*, Academic Press, New York, 1967.

Hummel 1966. Dieter O. Hummel, *Infrared Spectra of Polymers in the Medium and Long Wavelength Regions*, Interscience Div., John Wiley and Sons, New York, 1966.

Hummel 1971. Dieter O. Hummel, *Infrared Analysis of Polymers, Resins and Additives. An Atlas*, Wiley-Interscience, New York, 1971.

McCaffery 1970. Edward M. McCaffery, *Laboratory Preparation for Macromolecular Chemistry*, McGraw-Hill Book Co., New York, 1970.

Sixma 1964. F. L. J. Sixma and Hans Wynberg, *A Manual of Physical Methods of Organic Chemistry*, John Wiley and Sons, New York, 1964.

Stanley 1959. Edward L. Stanley, "Acrylic Plastics," pp. 1-16 in Gordon M. Kline, ed., *Analytical Chemistry of Polymers*, Part I, Interscience Publishers, New York, 1959.

Zbinden 1964. Rudolph Zbinden, *Infrared Spectroscopy of High Polymers*, Academic Press, New York, 1964.

Appendix I

Standard Polymer Samples

TABLE AI-1. Polymer standards available from ArRo

Cat. No.	Polymer Type	Nominal M	$\overline{M}_n$	$\overline{M}_w$	$\overline{M}_w/\overline{M}_n$
500-2	Polystyrene[a]	600			<1.1
500-6	Polystyrene	2,100			<1.1
500-8	Polystyrene	3,600			<1.1
500-10	Polystyrene	10,300			<1.06
500-12	Polystyrene	20,400			<1.06
500-14A	Polystyrene	37,000			<1.06
500-16A	Polystyrene	111,000			<1.06
500-18	Polystyrene	200,000			<1.06
500-20B	Polystyrene	390,000			<1.1
500-20A	Polystyrene	498,000			<1.2
500-22	Polystyrene	670,000			<1.1
500-26	Polystyrene	1,800,000			<1.2
500-28	Polystyrene	2,000,000			<1.3
700-1	Polyester	670			<1.3
700-3	Polyester	3,550			<1.4
700-4	Polyester	4,850			<1.3
400-2	Poly(vinyl chloride)		25,500	68,600	
400-3	Poly(vinyl chloride)		41,000	118,200	

[a] See note to Table AI-5.

TABLE AI-2. Polymer standard reference materials available from NBS

Sample No.	Polymer Type	$\overline{M}_w$ Light Scattering	$\overline{M}_n$ Osmometry	[η] dl/g
705	Polystyrene	179,000	171,000	0.74[a]
706	Polystyrene[b]	258,000	136,000	0.94[a]
1475	Linear polyethylene	52,000	18,300[c]	0.89[d]
1476	Branched polyethylene	--	--	0.81[d]
1483	Narrow-distribution polyethylene[e]			

[a]Benzene, 25°C.

[b]Sedimentation equilibrium $\overline{M}_z$ = 288,000 (NBS value); other experimental values are $\overline{M}_n$ = 136,000 quoted above and the following GPC values: $\overline{M}_n$ = 86,000; M_w = 287,000; $\overline{M}_z$ = 510,000.

[c]GPC value.

[d]α-Chloronaphthalene, 130°C.

[e]Tentative sample.

References: Wagner 1971, Hoeve 1972.

TABLE AI-3. Polymer reference standards available from Phillips

Description	Polymer type	$\overline{M}_w$ Light Scattering	$\overline{M}_n$ Osmometry	Isomer Content, Percent *Cis*	*Trans*	*Vinyl*
High *cis*	Polybutadiene	--	--	95.3	1.4	3.3
High *trans*	Polybutadiene	--	--	8.9	88.5	2.6
High *vinyl*	Polybutadiene	--	--	7.0	1.5	91.5
17 M	Polybutadiene	17,000	16,100	43.5	49.1	7.4
170 M	Polybutadiene	170,000	135,000	47.1	44.5	8.4
272 M	Polybutadiene	272,000	206,000	49.8	43.5	6.7
332 M	Polybutadiene	332,000	226,000	48.7	43.6	7.7
423 M	Polybutadiene	423,000	286,000	51.7	41.7	6.6
16 MH	Hydrogenated polybutadiene	16,400	14,100	--	--	--
108 MH	Hydrogenated polybutadiene	108,000	82,000	--	--	--
194 MH	Hydrogenated polybutadiene	194,000	126,000	--	--	--
420 MH	Hydrogenated polybutadiene	420,000	158,000	--	--	--

TABLE AI-4. *Samples available from the Polymer Bank*

Sample No.	Description	Data
6003	Cotton linters	99.5% α-cellulose
6014	Cellulose wood pulp	97.4% α-cellulose
6001	Branched polyethylene	$\overline{M}_w$ = 130,000
6012	Branched polyethylene	$\overline{M}_w$ = 480,000
6018	Branched polyethylene	$\overline{M}_n$ = 13,500
6007	Branched polyethylene	$\overline{M}_n$ = 20,500
6026	Poly(ethylene-*co*-ethyl acrylate)	80:20 ethylene : ethyl acrylate
6032	Poly(ethylene-*co*-vinyl acetate)	70:30 ethylene : vinyl acetate
6037	Poly(chlorotrifluoroethylene)	
6039	Poly(chlorotrifluoroethylene-*co*-vinylidene fluoride)	3% vinylidene fluoride
6042, 6045	Poly(chlorotrifluoroethylene-*co*-vinylidene fluoride)	70% vinylidene fluoride
6036	Poly(methyl methacrylate)	$\overline{M}_n$ = 48,600; $[\eta]$ = 0.250
6038	Poly(methyl methacrylate)	$\overline{M}_n$ = 19,400; $[\eta]$ = 0.144
6041	Poly(methyl methacrylate)	$\overline{M}_n$ = 160,000; $[\eta]$ = 0.436
6005	Nylon molding powder	Rel. visc. = 69.8
6004	Polypropylene	$[\eta]$ = 3.5
6016	Polypropylene	$[\eta]$ = 2.3
6023	Polypropylene	$[\eta]$ = 1.8
6043	Polypropylene	$[\eta]$ = 2.95

Sample No.	Description	Data
6022	Polystyrene	$[\eta] = 0.75$; higher impact
6029	Polystyrene	$[\eta] = 0.75$; lower impact
6013	Poly(vinyl chloride)	$[\eta] = 0.89$
6024	Poly(vinyl chloride)	$[\eta] = 0.97$
6031	Poly(vinyl chloride)	$[\eta] = 1.07$
6002	Poly(vinyl chloride-*co*-vinyl acetate)	$[\eta] = 0.44$; 85-88% vinyl chloride
6047	Poly(vinyl chloride-*co*-vinyl acetate)	$[\eta] = 0.420$; 87.4% vinyl chloride
6049	Poly(vinyl chloride-*co*-vinyl acetate)	$[\eta] = 0.483$; 87.0% vinyl chloride
6052	Poly(vinyl chloride-*co*-vinyl acetate)	$[\eta] = 0.528$; 85.5% vinyl chloride
6053	Polypropylene	$[\eta] = 3.08$
6054	Polypropylene	$[\eta] = 2.47$
6057	Polypropylene	$[\eta] = 1.87$
6056	Polypropylene	$[\eta] = 4.14$
6006	Poly(vinyl chloride)	$[\eta] = 0.975$
6017	Poly(vinyl chloride)	$[\eta] = 1.07$
6025	Poly(vinyl chloride-*co*-vinyl acetate)	$[\eta] = 0.475$
10322	Nylon 6, molding grade	viscosity = 3.20

Note: Viscosity specifications, as quoted, are generally incomplete.

TABLE AI-5. Polymer samples available from Pressure

Sample No.	Polymer Type	Nominal *M*	$\overline{M}_w/\overline{M}_n$
16a	Polystyrene[a]	600	1.10
12b	Polystyrene[a]	2,100	1.10
11b	Polystyrene[a]	4,000	1.10
8b	Polystyrene[a]	10,000	1.06
2b	Polystyrene[a]	20,400	1.06
7b	Polystyrene[a]	37,000	1.06
4b	Polystyrene[a]	110,000	1.06
1c	Polystyrene[a]	200,000	1.06
3b	Polystyrene[a]	390,000	1.10
5a	Polystyrene[a]	498,000	1.20
13a	Polystyrene[a]	670,000	1.15
14b	Polystyrene[a]	2,000,000	1.30
3ac	Polyethylene	7,100	1.7
4af	Polyethylene	17,900	1.95
5af	Polyethylene	27,700	2.0
6af	Polyethylene	35,700	4.2
7af	Polyethylene	32,000	6.0
8ac	Polyethylene	41,700	3.7

[a]*Note*: These anionic narrow-distribution polystyrene standards, produced by Pressure, are further characterized and resold by ArRo and Waters.

Reference: Altares 1964.

TABLE AI-6. Polymers available from RAPRA

Sample No.	Polymer Type	Description
PS 1	Polystyrene	Broad distribution
PS 2	Polystyrene	Narrow distribution
PS 3	Polystyrene	Heat resistant
PVC 1	Poly(vinyl chloride)	Paste polymer
PVC 2	Poly(vinyl chloride)	Suspension polymer
LDPE 1	Branched polyethylene	MFI[a] 0.12
LDPE 2	Branched polyethylene	MFI 2
LDPE 3	Branched polyethylene	MFI 2.5
LDPE 4	Branched polyethylene	MFI 20
LDPE 5	Branched polyethylene	MFI 20
LDPE 6	Branched polyethylene	MFI 50
HDPE 1	Linear polyethylene	MFI 2.5
HDPE 2	Linear polyethylene	MFI 0.9
HDPE 3	Linear polyethylene	MFI 5.0
HDPE 4	Linear polyethylene	Unstabilized HDPE 2
HDPE 5	Linear polyethylene	Unstabilized HDPE 3
A 1	Polyoxymethylene	Copolymer
A 2	Polyoxymethylene	Homopolymer
N 1	66 Nylon	Medium mol. wt.
N 2	66 Nylon	Lower than medium mol. wt.
PC 1	Polycarbonate	Medium mol. wt.
PMMA 1	Poly(methyl methacrylate)	Medium mol. wt.

[a]Melt flow index.

TABLE AI-7. *Polymer fractions and standards available from Waters*

Catalog No.	Polymer Type	Nominal M	GPC Size,[a] A	$\overline{M}_w/\overline{M}_n$
26971	Polystyrene[b]	2,000	51	Approx. 1.05
25169	Polystyrene[b]	5,000	117	Approx. 1.05
25171	Polystyrene[b]	10,300	244	Approx. 1.05
25168	Polystyrene[b]	19,850	480	Approx. 1.05
25170	Polystyrene[b]	51,000	1,220	Approx. 1.05
41995	Polystyrene[b]	96,200	2,360	Approx. 1.05
41984	Polystyrene[b]	173,000	4,160	Approx. 1.05
25166	Polystyrene[b]	411,000	9,800	Approx. 1.05
25167	Polystyrene[b]	867,000	19,600	Approx. 1.05
61970	Polystyrene[b]	2,145,000	48,000	Approx. 1.05
41993	Poly(propylene glycol)	790	51	Narrow
41994	Poly(propylene glycol)	1,220	78	Narrow
41985	Poly(propylene glycol)	2,020	126	Narrow
41983	Poly(propylene glycol)	3,900	241	Narrow
27229	Polystyrene	860,000	--	<1.009
27230	Polystyrene	411,000	--	<1.009
27231	Polystyrene	160,000	--	<1.009
27232	Polystyrene	97,000	--	<1.009
27233	Polystyrene	51,000	--	<1.009
27234	Polystyrene	20,000	--	<1.009
27235	Polystyrene	10,000	--	<1.009
27236	Polystyrene	4,000	--	<1.009
27237	Polystyrene	2,000	--	<1.009
26101	Poly(dimethyl siloxane)	--	10,000	1.26
26102	Poly(dimethyl siloxane	--	3,800	1.35
26103	Poly(dimethyl siloxane)	--	1,000	1.21

Catalog No.	Polymer Type	Nominal M	GPC Size,[a] A	$\overline{M}_w/\overline{M}_n$
26104	Poly(dimethyl siloxane)	--	600	1.20
26105	Poly(dimethyl siloxane)	--	320	1.21
26106	Poly(dimethyl siloxane)	--	135	1.20
26107	Poly(dimethyl siloxane)	--	95	1.20
26108	Poly(dimethyl siloxane)	--	50	1.31
26109	Poly(dimethyl siloxane)	--	420	2.65
26110	Poly(dimethyl siloxane)	--	2,300	3.72
26120	Poly(dimethyl siloxane)	--	6,400	1.08
26121	Poly(dimethyl siloxane)	--	2,100	1.15
26122	Poly(dimethyl siloxane)	--	690	1.06
26123	Poly(dimethyl siloxane)	--	170	1.04
26124	Poly(dimethyl siloxane)	--	82	1.04

[a]See definition of this size parameter in Sec. 7C.
[b]See note to Table AI-5.

BIBLIOGRAPHY

Altares 1964. T. Altares, Jr., D. P. Wyman, and V. R. Allen, "Synthesis of Low Molecular Weight Polystyrene by Anionic Techniques and Intrinsic Viscosity--Molecular Weight Relations Over A Broad Range In Molecular Weight," *J. Polymer Sci. A 2*, 4533-4544 (1964).

ArRo. ArRo Laboratories, Inc., 1107 West Jefferson, Joliet, Illinois 60435.

Hoeve 1972. C. A. J. Hoeve *et al*., "The Characterization of Linear Polyethylene SRM 1475," Parts I-X, J. Res. Natl. Bur. Stds. *76A*, 137-170 (1972).

NBS. Office of Standard Reference Materials, Room B314, Chemistry Building, National Bureau of Standards, Washington, D.C. 20234.

Phillips. Special Products Division, Chemical Department, Phillips Petroleum Company, Bartlesville, Oklahoma 74003.

Polymer Bank. Polymer Bank, c/o Dr. H.-G. Elias, Director, Midland Macromolecular Institute, 1910 West St. Andrews Drive, Midland, Michigan 48640.

Pressure. Pressure Chemical Company, 3419 Smallman St., Pittsburgh, Pennsylvania 15201.

RAPRA. Polymer Supply and Characterization Centre, Rubber and Plastics Research Association of Great Britain, Shawbury, Shrewsbury, SY4 4NR, England.

Wagner 1971. Herman L. Wagner, "Standard Reference Materials for Polymer Characterization at the National Bureau of Standards," *Polymer Preprints, ACS Div. Polymer Chem. 12* (2), 770-774 (1971).

Waters. Waters Associates, Inc., 61 Fountain Street, Framingham, Massachusetts 01701.

Appendix II

Properties of Selected Polymers

Appendix III

Properties of Monomers and Solvents

TABLE II. Properties of selected polymers

Property	ASTM Method		ABS Resin	66 Nylon	Poly-carbonate	Polyester	Branched Poly-ethylene
Specific gravity, g/ml	D 792		1.01-1.04	1.13-1.15	1.2	1.37-1.38	0.910-0.925
Tensile strength, psi	D 1708		3500-6200	11,200	8000-9500	10,400	600-2300
Elongation, percent	D 1708		5-60	300	100-130	100-300	90-800
Tensile modulus, 10^5 psi	D 412		2.0-3.5	1.75	3.0-3.5	--	0.14-0.38
Impact strength, ft-lb/in.	D 256, 1/2", notched		6-8	2-2.5	12-18	0.8	No Break
Hardness, Rockwell units (M, R) or Shore units (A, D)	D 785		R75-105	R120	M73-78 R115-125	M98	R10 D41-46
T_g, °C	--		100-110	50	150	Wide Range	-80
T_m, °C	--		--	265	220	Wide Range	115
Heat deflection temp., °C	D 648	66 psi stress	102-118	105	130-140	85	32-41
		264 psi stress	113-122	244	132-143	116	38-49
Thermal conductivity, 10^{-4} cal/sec cm °C	C 177		4.5-8	5.8	4.6	6.9	8.0
Specific heat, 23°C, cal/g °C			0.3-0.4	0.11	0.3	0.3	0.55
Thermal expansion coefficient, 10^{-5} in./in. °C	D 696		9.5-13	8.0	6.6	6.0	10-20
Dielectric constant	D 150	60 Hz	2.4-5	--	2.97-3.17	--	2.25-2.35
		1 kHz	2.4-4.5	5.2	3.02	3.46	2.25-2.35
		1 MHz	2.4-3.8	4.1	2.92-2.93	3.28	2.25-2.35
Dissipation factor	D 150	60 Hz	0.003-0.008	0.04	0.0009	--	<0.0005
		1 kHz	0.004-0.007	0.04	0.0021	0.0005	<0.0005
		1 MHz	0.007-0.015	0.05	0.010	0.0208	<0.0005
Electrical resistivity ohm-cm	D 257		$1\text{-}5\times10^{16}$	10^{13}	2×10^{16}	3×10^{16}	$>10^{16}$
Water absorption, percent/24 hr	D 570, 1/8" thick bar		0.20-0.45	1.1-1.5	0.15-0.18	0.02	<0.015
Flammability, in./min	D 635		Slow	Self-ext.	Self-ext.	Slow	1.04

REFERENCE

Sidney Gross, ed., *Modern Plastics Encyclopedia 1971-1972* (McGraw-Hill Book Co., New York), *48* (*10A*), October, 1971.

Linear Poly-ethylene	Poly-(methyl methacrylate)	Polyoxy-methylene	Poly-styrene	Polytetra-fluoro-ethylene	Polyurethane	Poly-(vinyl chloride) (Rigid)	Urea-formaldehyde Resin
0.941-0.965	1.17-1.20	1.42	1.04-1.09	2.14-2.02	1.05-1.25	1.30-1.45	1.47-1.52
3100-5500	8000-11,000	10,000	5000-12,000	2000-5000	4500-8400	5000-9000	5500-13,000
20-1000	2-7	25-75	1.1-2.5	200-400	10-650	2-40	0.5-1
0.6-1.8	3.5-4.5	5.2	4.0-6.0	0.58	0.1-3.5	3.5-6.0	10.0-15.0
0.5-20	0.3-0.4	1.4-2.3	0.25-0.4	3	No Break	0.4-2.0	0.25-0.40
D60-70	M80-100	M94 R120	M 65-80	D50-56	A30-70 M28, R60	D65-85	M110-120
-80	105	-68	100	125(?)	Wide Range	83	--
135	--	180	--	327	235	240	--
43-54 60-88	72-102 74-113	124 170	105 max 82-110	-- 121	Wide Range	54-79 57-82	127-143 --
11.0-12.4	4.0-6.0	5.5	2.4-3.3	6.0	1.7-7.4	3.0-7.0	7.0-10.0
0.55	0.35	0.35	0.32	0.25	0.4-0.45	0.2-0.3	0.4
11-13	5-9	8.1	6-8	10	10-20	5-18.5	2.2-4
2.30-2.35	3.5-4.5	--	2.45-3.1	<2.1	5.4-7.6	3.2-3.6	7.0-9.5
2.30-2.35	3.0-3.5	3.7	2.40-2.65	<2.1	5.6-7.6	3.0-3.3	7.0-7.5
2.30-2.35	2.1-3.2	3.7	2.40-2.7	<2.1	4.2-5.1	2.8-3.1	6.8
<0.0005	0.05-0.06	--	0.0001-0.0006	<0.0002	0.0015-0.048	0.007-0.020	0.035-0.043
<0.0005	0.04-0.06	0.0048	0.0001-0.0003	<0.0002	0.043-0.060	0.009-0.017	0.025-0.035
<0.0005	0.02-0.03	0.0048	0.0001-0.0004	<0.0002	0.050-0.075	0.006-0.019	0.25-0.35
$>10^{16}$	$>10^{15}$	1×10^{15}	$>10^{16}$	10^{18}	2×10^{11}-1×10^{13}	$>10^{16}$	10^{12}-10^{13}
<0.01	0.2-0.4	0.25	0.03-0.10	0.00	0.7-0.9	0.007-0.04	0.4-0.8
1.0-1.04	0.9-1.3	1.1	<1.5	None	Self-ext.	Self-ext.	Self-ext.

TABLE AIII-1. Properties of solvents used in the experiments

Solvents	Density 20°C d^{20}	Refractive Index n_d^{20}	T_m °C	T_b °C	Vapor Pressure 25°C, torr	Distillation Conditions[a]
Acetone	0.791	1.3588	-95	56	226	a, f
Benzene	0.879	1.5011	5	80	100	f
2-Butanone	0.805	1.3791	-87	80	91	f
Carbon tetrachloride	1.594	1.4601	-23	77	115	f
Chloroform	1.489	1.4453	-64	61	195	f
m-Cresol	1.034	1.5392	12	202	0.14	f
Ethyl acetate	0.901	1.3701	-84	77	92	f
Formic acid	1.220	1.3714	8	101	43	f
Methanol	0.792	1.3306	-94	65	125	f
Phenol	1.072	1.5509	41	182	0.35	f, s, v
Tetrachloroethylene	1.623	1.5053	-22	121	19	f, v
Tetrahydrofuran	0.889	1.4050	-109	66	197	f, v
Toluene	0.867	1.4961	-95	111	29	f, v
Water	0.997	1.3330	0	100	24	--

[a] a, azeotropic; f, fractional; s, steam; v, vacuum.

[b] d, distillation; r, reflux; s, storage.

[c] NO ALKALIS!

[d] NO $CaCl_2$, P_2O_5!

[e] NO "DRIERITE"!

Drying Agents and Conditions[b]	Health Hazard		Flammability	
	Permissible Conc. in Air, Max, ppm	Principal Hazard if Inhaled	Conc. Limits for Burning, Vol. Percent	Min. Ignition Temp. in Air °C
$Ca(NO_3)_2$, $CaCl_2$, P_2O_5, d	1000	Narcosis	2.5-12.8	561
$CaCl_2$, P_2O_5, d; Na-Pb alloy, s	25	Toxic	1.4-7.1	562
Na_2SO_4, d, s; K_2CO_3, d	200	Irritant	2.1-11.0	505
$CaCl_2$, d, s	25	Toxic	Nonflammable	
$CaCl_2$, Na_2SO_4, K_2CO_3, d, s; P_2O_5, s[c]	50	Toxic	Nonflammable	
Na_2SO_4, d, s	5	Toxic	--	--
K_2CO_3, P_2O_5, d, s	400	Irritant	2.2-11.5	524
B_2O_3, $CuSO_4$, s, d[d]	5	Irritant	--	--
Na, Mg, d, s; CaH_2, s[e]	200	Toxic, narcosis	6.7-36.5	867
any, in desiccator, s	5	Toxic	--	715
Na_2CO_3, $CaCl_2$, d, s	100	Narcosis	Nonflammable	
NaOH, Na_2SO_4, $CaCl_2$, d, s; Na, Na-K alloy, r, d	200	Toxic, irritant	1.8-11.8	321
$CaCl_2$, P_2O_5, d; Na-Pb alloy, s	200	Toxic, irritant	1.2-7.1	536
--	--	--	--	--

TABLE AIII-2. Properties of monomers and additives used in the experiments

Monomer or Additive	Density 20°C d_{20}	Refractive Index n_d^{20}	T_m °C	T_b °C	Vapor Pressure 25°C, torr	Health Hazard: Permissible Conc. in Air, Max, ppm	Health Hazard: Principal Hazard if Inhaled
ω-Aminoundecanoic acid	--	--	189-191	--	--	Not fully investigated	
1,4-Butanediol	1.017	1.4465	20	235	Low	--	--
n-Butyl mercaptan	0.834	1.4440	-117	99	--	10	Toxic
4,4'-Diphenylmethane diisocyanate	1.18	--	38	--	Low	--	Irritant
Ethyl acrylate	0.923	1.4068	-71	100	38	25	Toxic, irritant
Hexamethylene diamine	--	--	39	204	--	--	Irritant
Methyl methacrylate	0.943	1.4146	-48	100	38	100	Irritant
Polyether-glycol (Polymeg 1000)	0.98	--	37	--	Low	--	--
Sebacoyl chloride	1.12	1.4678	-6	220	--	--	Irritant
Styrene	0.909	1.5458	-31	145	6.3	100	Irritant, narcosis
Trioxane	1.17	--	64	115	--	--	Irritant

BIBLIOGRAPHY

Castille 1971. Y. P. Castille, "Physical Properties of Monomers," pp. VIII-1 to VIII-27 in J. Brandrup and E. H. Immergut, eds., with the collaboration of H.-G. Elias, *Polymer Handbook*, Interscience Div., John Wiley and Sons, New York, 1966.

Marsden 1963. C. Marsden, ed., with the collaboration of Seymour Mann, *Solvents Guide*, Interscience Div., John Wiley and Sons, New York, 1963.

Riddick 1970. John A. Riddick and William S. Bunger, *Organic Solvents, Physical Properties and Methods of Purification*, Vol. II in A. Weissberger, ed., *Techniques of Chemistry*, 3rd ed., Interscience Div., John Wiley and Sons, New York, 1970.

Steere 1967. Norman V. Steere, ed., *CRC Handbook of Laboratory Safety*, Chemical Rubber Company, Cleveland, Ohio, 1967.

Weast 1971. Robert C. Weast, ed., *CRC Handbook of Chemistry and Physics*, Chemical Rubber Company, Cleveland, Ohio, 1971.

Appendix IV

Errors and the Statistical Treatment of Data

A. *Errors*

It is inherent in any measurement process that small deviations from the true result of the measurement will be observed. Inevitably, one must inquire about the sources and magnitude of these errors and the reliability of the experimental results.

Sources of error

As a general rule, significant errors can be introduced at every step in an experiment or measurement process. In Exp. 10-32, for example, the first possibilities of error lie in the preparation of the sample. What is its purity? Are there any foreign materials present which will interfere with the determination? Is it homogeneous, and does the portion weighed out reflect the average properties of the entire sample? These and other parameters, such as the concentrations of any solutions prepared, must be known and kept within limits commensurate with the accuracy desired in the final results.

The same attention must be paid to experimental conditions, such as temperature and pressure. Needless to say, the condition and state of calibration of the apparatus are also of crucial importance. In this manual, responsibility for these matters is normally left to the instructor, but it is always true that no result can be more reliable than the calibration information in the instrument.

Random versus systematic errors

One of the difficulties in assessing the reliability of an experimental result is to distinguish between random and systematic errors. There is little problem in defining the former, but there is no truly satisfactory definition of a systematic error. Perhaps it can be said that if an error occurs with a certain regularity or can be detected in an entire series of determinations, it is systematic. An example might be an incorrect zero reading which adds a constant increment to all determinations, or a temperature which changes slowly during the experiment.

The elimination of systematic errors is a part of experimental skill which is built up only with practice and exposure. Some standard things to check on are the real constancy of supposedly constant parameters and the identity of zero readings in all determinations.

Often the experiment is designed to eliminate many quantities which are uncertain by the use of a standard sample for comparison; an example is the elimination of the need to know scan speed and instrument sensitivity accurately in the determination of specific heats by DSC in Exp. 22. In general, the measurement of a standard sample just prior to determination of the unknown is of great value in all experiments.

Need for multiple determinations

When only a single determination of an experimental quantity is made, one can scarcely estimate its reliability except by reference to previous experience. That is, the reliability of previous measurements made under the same conditions is assigned to this determination. The conditions selected must be standardized and known to lead to results with an acceptable precision. An example is the determination of a melting point by DTA (Exp. 21) at a standard heating rate.

The soundest method of assessing experimental precision is the statistical evaluation of the results of a series of identical (as nearly as possible, or in some cases systematically varied) experiments. This is the subject of the remainder of this appendix.

Practically, it is important to know how many experiments are needed to establish an average result with a desired degree of confidence. Or, from another point of view, it is necessary to know the confidence limits corresponding to a given number of determinations. If the limits selected are too broad, confidence in the result is unnecessarily undermined, whereas if they are too narrow, emphasis on precision is overexaggerated.

While the formal application of statistics yields information on the *precision* of a result, in terms of the confidence that another determination will yield the same result, it gives no assurance whatsoever that the result is *accurate*, sound, or reliable. Questions of accuracy relate to the validity of values assigned to calibration samples, which usually do not vary from one determination to another, and to assumptions concerning the nature of the sample and the theory underlying the experiment. Examples of such assumptions are that the polymers whose molecular weight is being measured in Exp. 10-13 do not associate, that microgel is absent in the light-scattering experiment, and that unwanted small molecules are not present in the sample selected for vapor-phase osmometry. Common sense must be used in judging the validity of every step in an experimental procedure and the subsequent treatment of the data.

B. *The Statistical Treatment of Data*

The statistical methods covered in this section are based on least-squares analysis, which is well explained in a number of textbooks and manuals, including those cited as general references. Here we give only the essential equations, and examples based on the experiments to illustrate their use. The equations are written in a form (Fisher 1969) which reduces the number of numerical operations required for their solution. It should be noted that all the quantities discussed, since they are calculated from a finite number of observations, are only best estimates of the truth, and can be relied upon only with a certain (though often high) probability.

It is assumed throughout that a number n of observations is made, and the subscript n is omitted. Unless otherwise specified, summations Σ are carried out for all subscripts $i = 1, 2, \ldots, n$, and these subscripts are omitted in the summation symbol and the variables following it.

Determination of drift

It is sometimes necessary to determine whether a parameter is fluctuating about a constant value, or whether this average is changing with time. For small fluctuations and long times, visual examination of the data, as from a graph or recorder chart, is adequate. For shorter times and larger fluctuations, the following criterion (Youden 1954) can be used: The true value of a variable x can be suspected of drifting if the following inequality holds.

$$\Sigma(x_i - x_{i+1})^2 \,/\, \Sigma(x_i - \bar{x})^2 < 2 \qquad \text{(AIV-1)}$$

where x_i and x_{i+1} are successive readings and $\bar{x}$ is the average of n readings. Bennett (1954) has tabulated limits on the inequality corresponding to various confidence intervals.

Example 1 The galvanometer zero readings of an instrument are suspected of drifting during its 1-hr warmup period. Twelve readings at 5-min intervals were +1, +1, -1, -3, +1, -2, +2, -1, +1, +1, +1, and -2 scale divisions. The expression on the left of Eq. AIV-1 is calculated to be 71/29. Since this is greater than 2, there is no evidence for drift, within the accuracy of the measurements.

Determination of a constant parameter

Average The best result from the n determinations of a parameter x is the average, $\bar{x}$ (the mean, the best estimate of the true value of x):

$$\bar{x} = (\Sigma x) \,/\, n \qquad \text{(AIV-2)}$$

Standard deviation An estimate of the precision of a single determination is given by the adjusted root-mean-square *standard deviation*:

$$s = [\Sigma(\bar{x} - x_i)^2 \,/\, (n - 1)]^{1/2} \qquad \text{(AIV-3)}$$

For large n, about 67% of the *experimental values* of x fall between $\bar{x} - s$ and $\bar{x} + s$, about 95% fall between $\bar{x} - 2s$ and $\bar{x} + 2s$, and about 99% fall between $\bar{x} - 3s$ and $\bar{x} + 3s$.

Confidence limits such as these can be used to provide a criterion for the rejection of improbable values. For example, a value of x falling outside the range $\bar{x} - 3s$ to $\bar{x} + 3s$ can be assumed to be dubious, temporarily eliminated in the calculation of $\bar{x}$, and investigated further to obtain final justification for its rejection. No value should ever be rejected on purely statistical grounds. More complete tables of confidence limits (Bowker 1959, Ehrenfeld 1964) should be used in considerations of rejection.

Standard error The adjusted error, best estimate of the standard error, or briefly the *standard error*, of the mean is given by

$$S = [\Sigma(\bar{x} - x_i)^2 \;/\; n(n - 1)]^{1/2} \qquad \text{(AIV-4)}$$

In contrast to the standard deviation s, the standard error S determines the confidence limits for the *true value* of x: For large n, there is about a 67% chance that the true value of x falls within the limits $\bar{x} - S$ and $\bar{x} + S$; similar statements to those above can be made about the 95% and 99% confidence limits. Estimates of the precision of the experiments, as stated in Sec. D. Precision and Accuracy, are based approximately on the 95% confidence limits $\bar{x} - 2S$ to $\bar{x} + 2S$.

Example 2 Four weighings m of a sample are made, for example for the DSC determination in Exp. 22. The values of m are 10.1, 10.1, 10.1, 9.7 mg. The mean is $\bar{m}$ = 10.0 mg, and the standard deviation is s = 0.2 mg. Thus the value of 9.7 mg *cannot* reasonably be rejected. The standard error S = 0.1 mg, and the weight of the sample can be written as $\bar{m} \pm 2S$ or 10.0 ± 0.1 mg. Note, however, that for this small value of n, it is not proper to assign 95% confidence to these limits; tables of confidence limits must be used to assign these values for n less than, say, 20 or so.

Propagation of errors

When a calculated quantity y is a function of several experimental values $x_1 \pm S_1$, $x_2 \pm S_2$, etc., the standard error of y must be estimated as

$$S(y) = [\Sigma(\partial y/\partial x_i)^2 \; S_i^2]^{1/2} \qquad \text{(AIV-5)}$$

Example 3 Consider the calculation of specific heat from an equation analogous to Eq. 9-8*b*,

$$c_p = 60k/qm \qquad \text{(from 9-8}b\text{)}$$

where k is the rate of heat delivery to the sample in mcal/sec, q the heating rate in deg/min, and m the sample weight in mg. As the results of repeated measurements, these quantities were determined to be k =

0.50 ± 0.02 (±4%) mcal/sec, q = 10.0 ± 0.2 (±2%) °C/min, and m = 10.0 ± 0.1 (±1%) mg. Then Eq. 9-8*b* yields c_p = 0.30 cal/g and, using Eq. AIV-5,

$$S(c_p) = [(\partial c_p/\partial k)^2\, S(k)^2 + (\partial c_p/\partial q)^2\, S(q)^2 + (\partial c_p/\partial m)^2\, S(m)^2]^{1/2}$$

$$= [(0.6)^2\,(0.02)^2 + (-0.03)^2\,(0.2)^2 + (-0.03)^2\,(0.1)^2]^{1/2}$$

$$= 0.014.$$

We can thus write c_p = 0.30 ± 0.03 (±10%) cal/g.

Linear regression

The use of linear regression usually implies that more experimental points have been measured than the minimum two necessary to determine a straight line. The additional points provide an important check on the accuracy of the underlying theory which calls for a linear relationship as well as increase the precision of the determination.

Straight line known to pass through the origin The slope a of the line $y = ax$ can be calculated from observations y_i and corresponding values of the independent variable x_i as

$$a = \Sigma\, xy \,/\, \Sigma\, x^2 \qquad \text{(AIV-6)}$$

The standard error of a is given by

$$S(a) = [S_0/(n-1)\, \Sigma\, x^2]^{1/2} \qquad \text{(AIV-7)}$$

where

$$S_0 = \Sigma\, y^2 - [(\Sigma\, xy)^2 \,/\, \Sigma\, x^2] \qquad \text{(AIV-8)}$$

Example 4 The reactivity constant k for the polycondensation of Exp. 1 is determined (at constant temperature) from the reaction time t and degree of polymerization $\bar{x}_n$ using the equation

$$\bar{x}_n - 1 = kc_0t \qquad \text{(from 1-4)}$$

Typical data are given in Table AIV-1 and treated by Eqs. AIV-6 to AIV-8:

TABLE AIV-1. Typical data from Exp. 1 arranged for linear regression calculations

	t, min	t^2	$\bar{x}_n-1$	$(\bar{x}_n-1)^2$	$(\bar{x}_n-1)t$
	20	400	19	361	380
	30	900	42	1,764	1,260
	45	2,025	69	4,761	3,105
	60	3,600	82	6,724	4,920
	90	8,100	109	11,881	9,810
Sum	--	15,025	--	25,491	19,475

$$kc_0 = \Sigma(\bar{x}_n-1)t \;/\; \Sigma\, t^2 = 1.296 \text{ min}^{-1}$$

$$S_0 = \Sigma(x_n-1)^2 - [\Sigma(x_n-1)t]^2 \;/\; \Sigma\, t^2 = 251$$

$$S(kc_0) = [S_0 \;/\; 4\, \Sigma\, t^2]^{1/2} = 0.064 \text{ min}^{-1}$$

We can write with assurance $kc_0 = 1.30 \pm 0.13$ min^{-1}. After the initial monomer concentration c_0 has been determined from the monomer density, k can be calculated and $S(k)$ determined with the aid of Eq. AIV-5.

Note an important general rule: Even though the precision of the final result is not expected to be high, it is essential to carry all significant figures in intermediate results, and double this number in calculating squares. For example, in calculating S_0 above, the small difference between two 5-digit numbers must be taken. Obviously, the slide rule is not an appropriate instrument to use in carrying out these calculations.

General straight line The best estimates of the coefficients a and b of the general straight line relation $y = a + bx$ are calculated from observations y_i and corresponding values of the independent variable x_i as follows:

$$a = [n\, \Sigma\, xy - (\Sigma x)(\Sigma y)] \;/\; D \qquad \text{(AIV-9)}$$

where

$$D = n\, \Sigma\, x^2 - (\Sigma x)^2 \qquad \text{(AIV-10)}$$

and

$$b = \overline{y} - a\overline{x} \tag{AIV-11}$$

The standard errors are

$$S(a) = [nS_0'/(n-2)D]^{1/2} \tag{AIV-12}$$

and

$$S(b) = [S_0' \ \Sigma \ x^2/(n-2)D]^{1/2} \tag{AIV-13}$$

where

$$S_0' = \Sigma \ y^2 - [(\Sigma y)^2/n] - a[\Sigma \ xy - (\Sigma x \Sigma y)/n] \tag{AIV-14}$$

Example 5 The number-average molecular weight $\overline{M}_n$ and the second virial coefficient A_2 are to be determined by membrane osmometry (Exp. 11) using the typical data in Table 7-3. Equation 7-6 can be rewritten as

$$\pi/c = RTA_2c + RT/\overline{M}_n \qquad \text{(from 7-6)}$$

from which we may identify $y = \pi/c$, $x = c$, $a = RTA_2$, and $b = RT/\overline{M}_n$. Equations AIV-9 through AIV-14 can then be applied, for $n = 7$. The quantity RT is taken as 31.015 ℓ at./mole, and the factor 0.785/1033 = 0.0007599 converting cm xylene to at. must be included from the beginning in order to simplify the estimates of error. The pertinent data are given in Table AIV-2. The following quantities are calculated:

$$D = 7 \ \Sigma \ c^2 - (\Sigma c)^2 = 167.44$$

$$RTA_2 = [7 \ \Sigma \ \pi - (\Sigma c)(\Sigma \pi)]/D = 4.766 \times 10^{-5}$$

$$\overline{c} = 4.38 \qquad \overline{\pi/c} = 9.54 \times 10^{-4}$$

$$RT/\overline{M}_n = \overline{\pi/c} - RTA_2\overline{c} = 7.45 \times 10^{-4}$$

$$S_0' = \Sigma(\pi/c)^2 - [(\Sigma\pi/c)^2/n] - RTA_2[\Sigma\pi \ - (\Sigma c)(\Sigma\pi/c)/n]$$

$$= 1.36 \times 10^{-9}$$

TABLE AIV-2. Typical data from Table 7-3 arranged for linear regression calculations

	c, g/ℓ	c^2	π, cm xylene	π, at.	$\pi/c \times 10^3$	$(\pi/c)^2 \times 10^6$
	1.81	3.276	2.00	0.00152	0.840	0.70560
	2.53	6.401	2.88	0.00219	0.866	0.74996
	3.00	9.000	3.49	0.00265	0.883	0.77969
	4.42	19.536	5.54	0.00421	0.952	0.90630
	5.64	31.810	7.60	0.00578	1.025	1.05063
	6.26	39.188	8.33	0.00633	1.011	1.02212
	7.00	49.000	10.15	0.00771	1.101	1.21220
Sum	30.66	158.211	--	0.03039	6.678	6.42650

$$S(RTA_2) = [nS_0'/(n-2)D]^{1/2} = 3.37 \times 10^{-6}$$

$$S(RT/\overline{M}_n) = [S_0' \, \Sigma \, c^2/(n-2)D]^{1/2} = 1.60 \times 10^{-5}$$

Since T is known accurately, any contribution to the above errors from an uncertainty in T will be small. We can simplify the propagation of error calculations by assuming T constant. Then

$$A_2 = RTA_2/RT = 1.54 \times 10^{-6} \text{ ℓ g}^{-2}$$

$$S(A_2) = S(RTA_2)/RT = 0.109 \times 10^{-6} \text{ ℓ g}^{-2}$$

$$\overline{M}_n = RT/(RT/\overline{M}_n) = 4.16 \times 10^4 \text{ g/mole}$$

and from Eq. AIV-5, the standard error of $\overline{M}_n$ is calculated from

$$S(\overline{M}_n) = \{[\partial \overline{M}_n/\partial(RT/\overline{M}_n)]^2 \, S^2(RT/\overline{M}_n)\}^{1/2}$$

$$= [(\overline{M}_n^2/RT)^2 \, S^2(RT/\overline{M}_n)]^{1/2} = 890 \text{ g/mole}$$

We can therefore write

$$A_2 = (1.54 \pm 0.11) \times 10^{-6} \ \ell \ g^{-2}$$

$$\overline{M}_n = (4.16 \pm 0.1) \times 10^4 \ \text{g/mole}$$

The limits of $\pm 2S$ round up to $\pm 5\%$ for $\overline{M}_n$ and $\pm 15\%$ for A_2. With errors of this size, it is unjustifiable to keep the last significant figures above, and one should write $\overline{M}_n = 4.2\ (\pm 5\%) \times 10^4$ and $A_2 = 1.5\ (\pm 15\%) \times 10^{-6}$.

Linear regression is relatively simple compared to error analysis for more complicated functions, and transformations of other functions into linear form should be made whenever possible to simplify the analysis. In fact, the situation in Example 5 results from just such a transformation, if one considers as the fundamental equation

$$\pi = RT(A_1 c + A_2 c^2 + \cdots) \tag{7-5}$$

Another transformation useful in membrane osmometry in a good solvent is suggested by Eq. 3-7 in the *Textbook*, where $(\pi/c)^{1/2}$ is plotted against c to obtain a linear relation. Still another example is in the treatment of light-scattering data, where $\sin^2(\theta/2)$ rather than θ is selected as the angle variable in order to obtain a linear plot.

Multiple linear regression

More complicated regression techniques can be applied to data in any number of dimensions, but we consider only the case of two sets of independent variables, x and y, describing together a third quantity, z. Such a situation can be represented by a plane whose equation is $z = a + a_x x + a_y y$. The quantity z can be thought of as the height above the x, y plane for any specified pair of values of x and y. The analysis follows Deming (1938) and Fisher (1969). We make the following definitions:

$$A_{xx} = \Sigma\, x^2 - [(\Sigma x)^2/n] \tag{AIV-15}$$

$$A_{xy} = \Sigma\, xy - [\Sigma x\ \Sigma y/n] \tag{AIV-16}$$

$$A_{yy} = \Sigma\, y^2 - [(\Sigma y)^2/n] \tag{AIV-17}$$

$$\Delta = A_{xx}A_{yy} - A_{xy}^2 \tag{AIV-18}$$

$$C_x = \Sigma\, xz - [\Sigma x\ \Sigma z/n] \tag{AIV-19}$$

$$C_y = \Sigma\, yz - [\Sigma y\ \Sigma z/n] \tag{AIV-20}$$

The best estimates of the coefficients a, a_x and a_y are given by

$$a_x = (A_{yy}C_x - A_{xy}C_y)/\Delta \qquad \text{(AIV-21)}$$

$$a_y = (A_{xx}C_y - A_{xy}C_x)/\Delta \qquad \text{(AIV-22)}$$

$$a = \overline{z} - a_x\overline{x} - a_y\overline{y} \qquad \text{(AIV-23)}$$

where $\overline{x}$, $\overline{y}$, and $\overline{z}$ are averages as defined in Eq. AIV-2.

The standard errors are given as follows:

$$S_0'' = \Sigma\ z^2 - [(\Sigma z)^2/n] - a_xC_x - a_yC_y \qquad \text{(AIV-24)}$$

$$S(a) = \{S_0''[\Sigma\ x^2\ \Sigma\ y^2 - (\Sigma xy)^2]\ /\ n(n-3)\Delta\}^{1/2} \qquad \text{(AIV-25)}$$

$$S(a_x) = [S_0''A_{yy}\ /\ (n-3)\Delta]^{1/2} \qquad \text{(AIV-26)}$$

$$S(a_y) = [S_0''A_{xx}\ /\ (n-3)\Delta]^{1/2} \qquad \text{(AIV-27)}$$

Example 6 The data from a light-scattering experiment are usually plotted as a Zimm plot (Fig. 7-14). Instead, consider them plotted in three dimensions with axes c, $\sin^2(\theta/2)$, and $kc/\Delta R_\theta$. We can identify by comparison with Eqs. 7-15 and 7-17 that $z = kc/\Delta R_\theta$, $a = 1/\overline{M}_w$, $a_x = 2\,A_2$, $x = c$, $a_y = 16\pi^2 \overline{s_z^2}/3\lambda_s^2\overline{M}_w$, and $y = \sin^2(\theta/2)$. The data from Table 7-5F have been transcribed for the multiple regression calculation in Table AIV-3. The constant $K = 0.925 \times 10^{-7}$ mole cm^{2}g^{-2} has been introduced at this point to effect a later simplification.

It should be noted that in the table, one line corresponds to each pair of concentration and angle. Of course, the procedure does not require that these pairs fall on rows with c or θ constant, though construction of the Zimm plot does.

We calculate:

$$A_{xx} = \Sigma\ c^2 - (\Sigma c)^2\ /\ n = 5.4781 \times 10^{-6}$$

$$A_{yy} = \Sigma\ \sin^4(\theta/2) - [\Sigma\ \sin^2(\theta/2)]^2\ /\ n = 1.328$$

$$A_{xy} = \Sigma\ c\ \sin^2(\theta/2) - [\Sigma\ c\ \Sigma\ \sin^2(\theta/2)]\ /\ n = 0$$

(exactly, since x and y are independent variables).

TABLE AIV-3. Typical data from Table 7-5 arranged for multiple regression calculations

θ	c, g/ml	$c^2 \times 10^6$	$\sin^2(\theta/2)$	$\sin^4(\theta/2)$	$kc/\Delta R_\theta \times 10^7$	$(kc/\Delta R_\theta)^2 \times 10^{12}$	$c \sin^2(\theta/2) \times 10^5$	$kc^2/\Delta R_\theta \times 10^{10}$	$kc \sin^2(\theta/2)/\Delta R_\theta \times 10^7$
30	0.60	0.3600	0.067	0.004489	6.244	0.38987	4.020	3.74640	0.41835
30	1.08	1.1664	0.067	0.004489	7.705	0.59367	7.236	8.32140	0.51624
30	1.47	2.1609	0.067	0.004489	8.760	0.76738	9.849	12.87720	0.58692
30	1.80	3.2400	0.067	0.004489	9.990	0.99800	12.060	17.98200	0.66933
30	2.08	4.3264	0.067	0.004489	10.869	1.18135	13.936	22.60752	0.72822
60	0.60	0.3600	0.250	0.062500	7.983	0.63728	15.000	4.78980	1.99575
60	1.08	1.1664	0.250	0.062500	9.602	0.92198	27.000	10.37016	2.40050
60	1.47	2.1609	0.250	0.062500	10.453	1.09265	36.750	15.36591	2.61325
60	1.80	3.2400	0.250	0.062500	11.655	1.35839	45.000	20.97900	2.91375
60	2.08	4.3264	0.250	0.062500	12.626	1.59416	52.000	26.26208	3.15650
90	0.60	0.3600	0.500	0.250000	10.332	1.06750	30.000	6.19920	5.16600
90	1.08	1.1664	0.500	0.250000	11.812	1.39523	54.000	12.75696	5.90600
90	1.47	2.1609	0.500	0.250000	12.765	1.62945	73.500	18.76455	6.38250
90	1.80	3.2400	0.500	0.250000	14.153	2.00307	90.000	25.47540	7.07650

θ	c,g/ml	$c^2 \times 10^6$	$\sin^2 (\theta/2)$	$\sin^4 (\theta/2)$	$kc/\Delta R_\theta \times 10^7$	$(kc/\Delta R_\theta)^2 \times 10^{12}$	$c \sin^2(\theta/2) \times 10^5$	$kc^2/\Delta R_\theta \times 10^{10}$	$kc \sin^2(\theta/2)/\Delta R_\theta \times 10^7$
90	2.08	4.3264	0.500	0.250000	14.846	2.20404	104.000	30.87968	7.42300
120	0.60	0.3600	0.750	0.562500	12.534	1.57101	45.000	7.52040	9.40050
120	1.08	1.1609	0.750	0.562500	14.051	1.97431	81.000	15.17508	10.53825
120	1.47	2.1609	0.750	0.562500	14.911	2.22338	110.250	21.91917	11.18325
120	1.80	3.2400	0.750	0.562500	16.169	2.61437	135.000	29.10420	12.12675
120	2.08	4.3264	0.750	0.562500	16.909	2.85914	156.000	35.17072	12.68175
Sum	28.12	45.0148	7.835	4.397451	234.369	29.07623	1101.601	346.267	103.883

$$\Delta = A_{xx}A_{yy} - A_{xy}^{2} = 7.275 \times 10^{-6}$$

$$C_x = \Sigma\, kc^2/\Delta R_\theta - (\Sigma c\ \Sigma kc/\Delta R_\theta)\ /\ n = 1.674 \times 10^{-9}$$

$$C_y = \Sigma\, kc\ \sin^2(\theta/2)/\Delta R_\theta - [\Sigma\ \sin^2(\theta/2)\ \Sigma\ kc/\Delta R_\theta]\ /\ n$$

$$= 1.207 \times 10^{-6}$$

$$a_x = 2\ A_2 = 3.056 \times 10^{-4}$$

$$a_y = 16\ \pi^2\ \overline{s_z^2}\ /\ 3\lambda_s^2\ \overline{M}_w = 9.089 \times 10^{-7}$$

$$\overline{x} = \overline{c} = 1.406 \times 10^{-3}$$

$$\overline{y} = \overline{\sin^2(\theta/2)} = 0.3918$$

$$\overline{z} = \overline{kc/\Delta R_\theta} = 1.172 \times 10^{-6}$$

$$a = 1/\overline{M}_w = 3.86 \times 10^{-7}$$

whence

$$\overline{M}_w = 2.591 \times 10^{6}$$

The standard errors of the coefficients are determined from Eqs. AIV-24 to AIV-27. Again, the need for high precision in the calculations is emphasized: the 2-digit value for S_0'' below is in fact the small difference between two 6-digit figures.

$$S_0'' = 3.2 \times 10^{-15}$$

$$S(a) = 1.0 \times 10^{-8}$$

$$S(a_x) = 5.9 \times 10^{-6}$$

$$S(a_y) = 1.2 \times 10^{-8}$$

Determination of the standard errors of $\overline{M}_w$, A_2, and $\overline{s_z^2}$ requires consideration of the propagation of errors through Eq. AIV-5, as follows:

$$S(\overline{M}_w) = [\partial\overline{M}_w / \partial(1/\overline{M}_w)]^2 \, S^2(a)]^{1/2}$$

$$= [\overline{M}_w^4 \, S^2(a)]^{1/2} = 6.7 \times 10^4$$

$$S(A_2) = S(a_x)/2 = 3.0 \times 10^{-6}$$

Inserting $\lambda_s = 5.46 \times 10^{-5}/1.3856 = 3.94 \times 10^{-5}$ cm,

$$\overline{s_z^2} = 3\, a_y \, \lambda_s^2 \, \overline{M}_w \, / \, 16\, \pi^2 = 6.94 \times 10^{-11} \text{ cm}^2$$

$$S(\overline{s_z^2}) = \{[\partial(s_z^2) \, / \, \partial a_y]^2 \, S^2(a_y) + [\partial(\overline{s_z^2}) \, / \, \partial\overline{M}_w]^2 \, S^2(\overline{M}_w)\}^{1/2}$$

$$= [(3\, \lambda_s^2 \, \overline{M}_w/16\pi^2)^2 \, S^2(a_y) + (3\, \lambda_s^2 \, a_y/16\, \pi^2) \, S^2(\overline{M}_w)]^{1/2}$$

$$= 2.0 \times 10^{-12} \text{ cm}^2$$

We can write the final results using $2S$ limits and rounded-up percentage errors as follows:

$$\overline{M}_w = (2.6 \pm 0.15) \times 10^6 \text{ g/mole or } (\pm 6\%)$$

$$A_2 = (1.53 \pm 0.06) \times 10^{-4} \text{ ml mole g}^{-2} \text{ or } (\pm 4\%)$$

$$\overline{s_z^2} = (6.9 \pm 0.4) \times 10^{-11} \text{ cm}^2 \text{ or } (\pm 6\%)$$

A useful rule of thumb for the number of significant figures to be kept is that for $2S$ below about 5%, keep 3 significant figures, while for $2S$ above 5-6%, keep only 2 figures.

GENERAL REFERENCES

Deming 1938; Bennett 1954; Bowker 1959; Freund 1960; Ehrenfeld 1964; Barford 1967; Fisher 1969.

BIBLIOGRAPHY

Barford 1967. N. C. Barford, *Experimental Measurements: Precision, Error, and Truth*, Addison-Wesley, New York, 1967.

Bennett 1954. C. A. Bennett and N. L. Franklin, *Statistical Analysis in Chemistry and Chemical Industry*, John Wiley and Sons, New York, 1954.

Bowker 1959. Albert H. Bowker and Gerald J. Lieberman, *Engineering Statistics*, Prentice-Hall, Englewood Cliffs, New Jersey, 1959.

Deming 1938. W. Edwards Deming, *Statistical Adjustments of Data*, John Wiley and Sons, New York, 1938.

Ehrenfeld 1964. Sylvain Ehrenfeld and Sebastian B. Littauer, *Introduction to Statistical Methods*, McGraw-Hill Book Co., New York, 1964.

Fisher 1969. Otto Fisher, "Method of Least Squares. Fitting Curves to Empirical Data. Elements of the Calculus of Observations," pp. 1285-1321 in Karel Rektorys, ed., *Survey of Applicable Mathematics*, M.I.T. Press, Cambridge, Massachusetts, 1969.

Freund 1960. John E. Freund, Paul E. Livermoore, and Irwin Miller, *Manual of Experimental Statistics*, Prentice-Hall, Englewood Cliffs, New Jersey, 1960.

Youden 1954. W. J. Youden, "Instrumental Drift," *Science 120*, 627 (1954).

Index